T0192710

SHOCK WAVE–BOUNDARY-LAYER INTERACTIONS

Shock wave–boundary-layer interaction (SBLI) is a fundamental phenomenon in gas dynamics that is observed in many practical situations, ranging from transonic aircraft wings to hypersonic vehicles and engines. SBLIs have the potential to pose serious problems in a flowfield; hence they often prove to be critical – or even design-limiting – issues for many aerospace applications.

This is the first book devoted solely to a comprehensive state-of-the-art explanation of this phenomenon with coverage of all flow regimes where it occurs. The book includes a description of the basic fluid mechanics of SBLIs plus contributions from leading international experts who share their insight into their physics and the impact they have in practical flow situations. This book is for practitioners and graduate students in aerodynamics who wish to familiarize themselves with all aspects of SBLI flows. It is a valuable resource for specialists because it compiles experimental, computational, and theoretical knowledge in one place.

Holger Babinsky is Professor of Aerodynamics at the University of Cambridge and a Fellow of Magdalene College. He received his Diplom-Ingenieur (German equivalent of an MS degree) with distinction from the University of Stuttgart and his PhD from Cranfield University with an experimental study of roughness effects on hypersonic SBLIs. From 1994 to 1995, he was a Research Associate at the Shock Wave Research Centre of Tohoku University, Japan, where he worked on experimental and numerical investigations of shock-wave dynamics. He joined the Engineering Department at Cambridge University in 1995 to supervise research in its high-speed flow facilities. Professor Babinsky has twenty years of experience in the research of SBLIs, particularly in the development of flow-control techniques to mitigate the detrimental impact of such interactions. He has authored and coauthored many experimental and theoretical articles on high-speed flows, SBLIs, and flow control, as well as various low-speed aerodynamics subjects. Professor Babinsky is a Fellow of the Royal Aeronautical Society, an Associate Fellow of the American Institute of Aeronautics and Astronautics (AIAA), and a Member of the International Shock Wave Institute. He serves on a number of national and international advisory bodies. Recently, in collaboration with the U.S. Air Force Research Laboratories, he organized the first AIAA workshop on shock wave–boundary-layer prediction. He has developed undergraduate- and graduate-level courses in Fluid Mechanics and received several awards for his teaching.

John K. Harvey is a Professor in Gas Dynamics at Imperial College and is a visiting professor in the Department of Engineering at the University of Cambridge. He obtained his PhD in 1960 at Imperial College for research into the roll stability of slender delta wings, which was an integral part of the Concorde development program. In the early 1960s, he became involved in experimental research into rarefied hypersonic flows, initially with Professor Bogdonoff at Princeton University and subsequently back at Imperial College in London. He has published widely on the use of the direct-simulation Monte Carlo (DSMC) computational method to predict low-density flows, and he has specialized in the development of suitable molecular collision models used in these computations to represent reacting, ionized, and thermally radiating gases. He has also been active in the experimental validation of this method. Through his association with CUBRC, Inc., in the United States, he has been involved in the design and construction of three major national shock tunnel facilities and in the hypersonic aerodynamic research programs associated with them. Professor Harvey has also maintained a strong interest in low-speed experimental aerodynamics and is a recognized expert on the aerodynamics of F1 racing cars. Professor Harvey is a Fellow of the Royal Aeronautical Society and an Associate Fellow of the American Institute of Aeronautics and Astronautics.

Cambridge Aerospace Series

Editors:
Wei Shyy and Michael J. Rycroft

Shock Wave–Boundary-Layer Interactions

Edited by

Holger Babinsky
University of Cambridge

John K. Harvey
Imperial College

CAMBRIDGE
UNIVERSITY PRESS

CAMBRIDGE
UNIVERSITY PRESS

32 Avenue of the Americas, New York NY 10013-2473, USA

Cambridge University Press is part of the University of Cambridge.

It furthers the University's mission by disseminating knowledge in the pursuit of education, learning and research at the highest international levels of excellence.

www.cambridge.org
Information on this title: www.cambridge.org/9781107646537

© Cambridge University Press 2011

First published 2011
First paperback edition 2014

A catalogue record for this publication is available from the British Library

Library of Congress Cataloguing in Publication data

Shock wave–boundary-layer interactions / [edited by] Holger Babinsky, John Harvey.
   p.   cm. – (Cambridge aerospace series)
Includes bibliographical references and index.
ISBN 978-0-521-84852-7 (hardback)
1. Shock waves.   2. Boundary layer.   I. Babinsky, Holger.   II. Harvey,
John (John K.)   III. Title.   IV. Series.
TL574.S4S575   2011
629.132'37–dc22   2011001978

ISBN 978-0-521-84852-7 Hardback
ISBN 978-1-107-64653-7 Paperback

To my late mother – Holger Babinsky

Brief Contents

Contents

Contributors

Holger Babinsky (Editor and Chapter 3) Department of Engineering, University of Cambridge, Cambridge CB2 1PZ, UK hb@eng.cam.ac.uk

John K. Harvey (Editor and Chapter 8) Department of Aeronautics, Imperial College, London SW7 2AZ, UK; Department of Engineering, University of Cambridge, Cambridge CB2 1PZ, UK jkh28@cam.ac.uk

Graham V. Candler (Chapter 7) Department of Aerospace Engineering & Mechanics, University of Minnesota, Minneapolis, MN 55455-0153, USA candler@aem.umn.edu

J. F. Debiève (Chapter 9) Institut Universitaire des Systèmes Thermiques Industriels, Université d'Aix-Marseille, UMR CNRS 6595, Marseille, France

Jean Délery (Chapters 2 and 3) ONERA, 29 Avenue Division Le Clerc 92320 Chatillon, France jean.delery@free.fr

P. Dupont (Chapter 9) Institut Universitaire des Systèmes Thermiques Industriels, Université d'Aix-Marseille, UMR CNRS 6595, Marseille, France

J. P. Dussauge (Chapter 9) Institut Universitaire des Systèmes Thermiques Industriels, Université d'Aix-Marseille, UMR CNRS 6595, Marseille, France Jean-Paul.Dussauge@polytech.univ-mrs.fr

Michael S. Holden (Chapter 6) CUBRC, 4455 Genesee Street, Buffalo, NY 14225, USA holden@cubrc.org

George V. Inger (Chapter 10) Formerly at Department of Aerospace and Ocean Engineering, Virginia Polytechnic Institute and State University, Blacksburg, VA 24060, USA

Doyle D. Knight (Chapters 4 and 5) Department of Mechanical and Aerospace Engineering, Rutgers – The State University of New Jersey, Piscataway, NJ 08854-8058, USA knight@soemail.rutgers.edu

Alexander A. Zheltovodov (Chapters 4 and 5) Khristianovich Institute of Theoretical and Applied Mechanics, Siberian Branch of Russian Academy of Science, Russia Novosibirsk 630090, Russia zhelt@itam.nsc.ru

1 Introduction

Holger Babinsky and John K. Harvey

Shock wave–boundary-layer interactions (SBLIs) occur when a shock wave and a boundary layer converge and, since both can be found in almost every supersonic flow, these interactions are commonplace. The most obvious way for them to arise is for an externally generated shock wave to impinge onto a surface on which there is a boundary layer. However, these interactions also can be produced if the slope of the body surface changes in such a way as to produce a sharp compression of the flow near the surface – as occurs, for example, at the beginning of a ramp or a flare, or in front of an isolated object attached to a surface such as a vertical fin. If the flow is supersonic, a compression of this sort usually produces a shock wave that has its origin within the boundary layer. This has the same affect on the viscous flow as an impinging wave coming from an external source. In the transonic regime, shock waves are formed at the downstream edge of an embedded supersonic region; where these shocks come close to the surface, an SBLI is produced.

In any SBLI, the shock imposes an intense adverse pressure gradient on the boundary layer, which causes it to thicken and possibly also to separate. In either case, this increases the viscous dissipation within the flow. Frequently, SBLIs are also the cause of flow unsteadiness. Thus, the consequences of their occurrence almost invariably are detrimental in some respect. On transonic wings, they increase the drag and they have the potential to cause flow unsteadiness and buffet. They increase blade losses in gas-turbine engines, and complicated boundary-layer control systems must be installed in supersonic intakes to minimize the losses that they cause either directly by reducing the intake efficiency or indirectly because of the disruption they cause to the flow entering the compressor. These systems add weight to an aircraft and absorb energy. In hypersonic flight, SBLIs can be disastrous because at high Mach numbers, they have the potential to cause intense localized heating that can be severe enough to destroy a vehicle. In the design of scramjet engines, the SBLIs that occur in the intake and in the internal flows pose such critical issues that they significantly can limit the range over which vehicles using this form of propulsion can be deployed successfully. This list of examples is by no means exhaustive.

Our aim in writing this book is to establish a general understanding of the aerodynamic processes that occur in and around SBLIs, concentrating as much as possible on the physics of these flows. We seek to explain which factors determine their structure under a variety of circumstances and also show how they impact on other parts of their flowfield, influencing parameters such as the drag, the surface-flux distributions, and the overall body flow. Our intention is to develop an understanding of which circumstances lead to their formation, how to estimate their effect, and how to manage them if they do occur. We demonstrate how the present state of our understanding has resulted through contributions from experiments, computational fluid dynamics (CFD), and analytical methods. Because of their significance for many practical applications, SBLIs are the focus of numerous studies spanning several decades. Hence, there is a considerable body of literature on the subject. We do not attempt to review all of it in this book but we aim to distill from it the information necessary to fulfill our aims.

1.1 Structure of the Book

The first chapter of the book explains the fundamental aerodynamic concepts relevant to all SBLIs. Subsequent chapters examine in more detail the interactions in specific Mach-number regimes, beginning with transonic flows, followed by supersonic flows, and finally hypersonic and rarefied flows. Throughout the chapter, examples are cited that demonstrate how the nature of the interaction varies with these changes. Because of the wide range of knowledge and disciplines involved, we do not attempt to do this entirely alone; we have enlisted several prominent internationally recognized experts in the field who very generously contributed to the preparation of the book. They were asked specifically to give their perspective on critical experimental, computational, and analytical issues associated with SBLIs in their particular area. In all, six chapters are contributions from other authors and we gratefully acknowledge their assistance. Although we edited the material provided, we do not attempt to unify the writing style but instead seek to retain the flavor of individual contributions as much as possible.

Chapter 2 explains the fundamental aerodynamic concepts relevant to SBLIs throughout the Mach-number range. This chapter was written by Professor Jean Délery, the former Head of Aerodynamics at ONERA in France. Although it is not our intention to produce a conventional textbook, this chapter comes close in that it provides a wide-ranging overview of the background aerodynamics relevant to SBLIs. In writing this chapter, Professor Délery emphasized the explanation of the underlying physics of the flows, and his contribution is an invaluable platform for subsequent chapters.

Chapter 3 addresses *transonic SBLIs*. This topic is of particular relevance to the super-critical wings that are used widely on many current aircraft and to gas-turbine-blade design. We wrote this chapter in conjunction with Professor Délery and it includes new results on SBLIs in this range. Again, the emphasis is on establishing an understanding of the physical processes taking place within these interactions because this is considered a necessary prerequisite for devising effective control strategies to minimize detrimental effects.

Chapters 4 and 5 are devoted to *supersonic interactions and their numerical modeling*. Chapter 4 concentrates on two-dimensional interactions and is followed by a discussion of three-dimensional SBLIs in Chapter 5. Both chapters were written by Professor Doyle D. Knight from Rutgers University in New Jersey, USA, and Professor Alexander A. Zheltovodov from the Khristianovich Institute of Theoretical and Applied Mechanics in Novosibirsk, Russia. They chose to describe in detail a number of fundamental flowfields to explain how the SBLI structure changes across the parameter range. More complex flowfields can be understood as combinations of one or more of these fundamental elements. In particular, three-dimensional interactions are explained from a basis of comparison with equivalent two-dimensional flow cases described in Chapter 4. The capabilities of CFD to predict these complex supersonic flows also are assessed in this part of the book.

The next three chapters comprise a section devoted to *hypersonic SBLIs*. Predicting when and how they develop in this speed range is especially important because of the impact on vehicle design. Chapter 6 is written by Dr. Michael Holden from CUBRC in Buffalo, New York, USA. For several decades, he has been acknowledged as the leading experimentalist in this area. He presents a wide range of results from which he develops a detailed insight into the impact that SBLIs make on vehicle aerodynamics at high Mach numbers and to what extent the outcome can be predicted.

Chapters 7 and 8 focus on numerical simulation, including the influence of real-gas effects, rarefaction, and chemical reactions on the interactions. Professor Graham V. Candler of the University of Minnesota, Minneapolis, MN, USA is the author of Chapter 7. He shows that despite the success of current advanced computational methods in predicting hypersonic flows, accurate simulation of SBLIs remains a major challenge. He discusses the physics of hypersonic SBLI flows and emphasizes how understanding this bears on the effective numerical simulation of them. Chapter 8 addresses the way in which the very low ambient density that occurs in the upper atmosphere impacts vehicle flows involving SBLIs. Under these circumstances, the conventional Navier-Stokes methods fail and particle-simulation methods, specifically Direct Simulation Monte Carlo (DSMC), must be used as an effective alternative predictive tool. Chapter 8 cites results for hypersonic flows from which the influence that rarefaction and chemical reactions have on SBLIs can be assessed.

The book concludes with two chapters that address the specialised topics of *flow unsteadiness* associated with SBLIs and the use of *analytical treatments*. Chapter 9 was written by Dr. Jean-Paul Dussauge in collaboration with Drs. P. Dupont and J. F. Debiève from the Institut Universitaire des Systèmes Thermiques Industriels, Université d'Aix-Marseille in France. In their chapter, they consider turbulent interactions in the transonic and lower supersonic range and explore how flow structures in the SBLI and external stimuli (e.g., upstream turbulence and acoustic disturbances) lead to flow unsteadiness and downstream disturbances.

1.1.1 George Inger

Very sadly, Professor George Inger, who produced the final chapter, died on November 6, 2010 before this book was published.[1] His contribution is written from

---

[1] When George Inger died he was serving as a Visiting Professor at Virginia Polytechnic Institute and State University in Blacksburg, Virginia, where he had previously taught in the 1970s. Before that he occupied the Glenn Murphy Chair of Engineering at Iowa State University where he had been a researcher, teacher, and consultant in the field of aero-thermodynamics for more than 30 years.

a very personal perspective and describes those areas where he considered analytical methods could be applied effectively to the SBLI problem. We consider it a fitting tribute to him that we are able to include his material in its entirety as, in some measure, it summarizes a significant portion of his life's work, that of applying asymptotic expansion methods to fluid mechanics. Building on the legacy of Lighthill, Stewartson, and Neiland who developed the so-called triple deck method in the late 1960s, George extended their concepts well beyond their early achievements to, for example, SBLIs including turbulent interactions; a section devoted to this is included in his chapter as hitherto unpublished work. He was a great enthusiast for analytical methods and he firmly believed that they complemented and were an essential adjunct to experiments and numerical methods, being an effective means of enhancing our insight into the physics of a flow. While acknowledging that for complex flows such as SBLIs these methods have proved to have had significant limitations, they nevertheless continue to provide us with valuable interpretations of observed phenomena and predict flow behavior over a wide range of conditions. For this reason we were delighted to be able to have his contribution to this book.

1.2 Intended Audience

This book is targeted to technologists, research workers, and advanced-level students working in industry, research establishments, and universities. It is our intention to provide a single source that presents an informed overview of all aspects of SBLIs. In preparing the book, we endeavored to explain clearly the relevant fluid mechanics and ensure that the material is accessible to as wide an audience as possible. However, we assumed that readers have a good working knowledge of basic fluid mechanics and compressible flow. This is the first book solely devoted to the subject and it incorporates the latest developments, including material not previously publicized.

Before entering academia, he had worked in the aerospace industry with McDonnell-Douglas, Bell Aircraft, and the GM Research Laboratories. Over his career, he published extensively and became a pioneer in the basic theory of high temperature chemically reacting gas flows and propulsion in space.

2 Physical Introduction

Jean Délery

2.1 Shock Wave–Boundary-Layer Interactions: Why They Are Important

The repercussions of a shock wave–boundary layer interaction (SBLI) occurring within a flow are numerous and frequently can be a critical factor in determining the performance of a vehicle or a propulsion system. SBLIs occur on external or internal surfaces, and their structure is inevitably complex. On the one hand, the boundary layer is subjected to an intense adverse pressure gradient that is imposed by the shock. On the other hand, the shock must propagate through a multilayered viscous and inviscid flow structure. If the flow is not laminar, the production of turbulence is enhanced, which amplifies the viscous dissipation and leads to a substantial rise in the drag of wings or – if it occurs in an engine – a drop in efficiency due to degrading the performance of the blades and increasing the internal flow losses. The adverse pressure gradient distorts the boundary-layer velocity profile, causing it to become less full (i.e., the shape parameter increases). This produces an increase in the displacement effect that influences the neighbouring inviscid flow. The interaction, experienced through a viscous-inviscid coupling, can greatly affect the flow past a transonic airfoil or inside an air-intake. These consequences are exacerbated when the shock is strong enough to separate the boundary layer, which can lead to dramatic changes in the entire flowfield structure with the formation of intense vortices or complex shock patterns that replace a relatively simple, predominantly inviscid, unseparated flow structure. In addition, shock-induced separation may trigger large-scale unsteadiness, leading to buffeting on wings, buzz for air-intakes, or unsteady side loads in nozzles. All of these conditions are likely to limit a vehicle's performance and, if they are strong enough, can cause structural damage.

In one respect, shock-induced separation can be viewed as a compressible manifestation of the ubiquitous flow-separation phenomenon: The shock is simply an associated secondary artefact. From the perspective of viscous-flow, the behaviour of the separating boundary layer is basically the same as in incompressible flow, and the overall topology is identical. Nevertheless, the most distinctive and salient features of shock-separated flows are linked to the accompanying shock patterns formed in the contiguous inviscid outer flow. The existence of these shocks may have major consequences for the entire flowfield; in practice, it is difficult to completely separate SBLIs from the phenomena that arise due to the intersection of shock waves – usually referred to by the generic term *shock-shock interference*. SBLIs can occur at any Mach number ranging from transonic to hypersonic, but it is in the latter category that the shocks have particularly dramatic consequences due to their greater intensity.

It is not inevitable that SBLIs or, more generally, shock wave/shear layer interactions have entirely negative consequences. The increase in the fluctuation level they cause can be used to enhance fuel-air mixing in scramjet combustion chambers or to accelerate the disorganisation of hazardous flows, such as wing-trailing vortices. Also, because interactions in which separation occurs can lead to smearing or splitting of the shock system, the phenomenon can be used to decrease the wave drag associated with the shock. This last point illustrates a subtle physical aspect of the behaviour of SBLIs. Shock waves also form in unsteady compressible flows by focusing compression waves, as seen in the nonlinear acoustic effects in rocket combustion chambers or the compression caused by a high-speed train entering a tunnel. Extreme cases are associated with explosions or detonations in which interactions occur in the boundary layer that develops on the ground or the surface behind the propagating blast wave.

SBLIs are a consequence of the close coupling between the boundary layer – which is subjected to a sudden retardation at the shock-impact point – and the outer, mostly inviscid, supersonic flow. The flow can be influenced strongly by the thickening of the boundary layer due to this retardation. Although in many instances these flows can be computed effectively with modern computational fluid dynamics (CFD), the methods are certainly not infallible, especially if the flow is separated. For this reason, it is necessary to clearly understand the physical processes that control these phenomena. With this understanding, good designs for aerodynamic devices can be produced while avoiding the unwanted consequences of these interactions or, more challengingly, exploiting the possible benefits. An effective analysis of both the inviscid flow and boundary layer must be obtained to achieve this understanding. Therefore, the next section summarizes the basic results from shock-wave theory and describes the relevant properties of boundary-layer flows and shock-shock interference phenomena.

2.2 Discontinuities in Supersonic Flows

2.2.1 Shock Waves

The discontinuities that can occur within supersonic flow take several forms, including shear layers and slip lines as well as shock waves. The governing equations are presented in Appendix A of this chapter and include the Rankine-Hugoniot equations that govern shock waves and other discontinuities. From these equations, we can establish the following results, which have direct relevance when considering aerodynamic applications involving SBLIs. When the flow crosses a shock wave, it entails the following:

1. A discontinuity in flow velocity, which suddenly decreases.
2. An abrupt increase in pressure, which has several major practical consequences including:

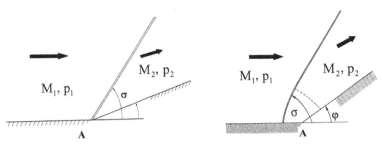

a- attached shock wave (φ < limit deflection)   b- detached shock wave (φ > limit deflection)

Figure 2.1. Attached and detached planar shock wave.

    i. A boundary layer at a surface that is hit by a shock suffers a strong adverse pressure gradient and therefore will thicken and may separate.

    ii. The structure of the vehicle is submitted to high local loads, which can fluctuate if the shock oscillates.

3. A rise in flow temperature, which is considerable at high Mach numbers, so that:

    i. The vehicle surface is exposed to localised high-heat transfer.

    ii. At hypersonic speeds, this heating is so intense that the fluid can dissociate, become chemically reactive, and possibly ionise downstream of the shock.

4. A rise in entropy or, equivalently, a decrease in the stagnation pressure. This is a significant source of drag and causes a drop in efficiency (i.e., the maximum recovery pressure diminishes).

The Rankine-Hugoniot conservation equations provide an inviscid description of a discontinuity, whether a shock wave or a slip line. This analysis conceals the fact that such phenomena, in reality, are dominated by viscosity at work either along the slip line or inside the shock wave. This is a region of rapid variation of flow properties but of finite thickness, roughly 10 to 20 times the incident flow molecular mean free path. This fact explains why there is an entropy rise through a shock wave: In an adiabatic and nonreacting flow, the only source of entropy is viscosity.

2.2.2 The Shock-Polar Representation

Valuable physical insight about how the shock patterns associated with SBLIs develop can be gained by considering the so-called shock polar [1], which provides a graphical representation of the solution to the Rankine-Hugoniot equations for oblique shocks. Consider a uniform supersonic flow with Mach number M_1 and pressure p_1 flowing along a rectilinear wall with direction φ_1 (by convention, we assume $\varphi_1 = 0$), as shown in Fig. 2.1a. At **A**, the wall exhibits a change of direction, $\Delta\varphi = \varphi - \varphi_1 = \varphi$. As long as this deflection is not too large, **A** will be the origin of a plane-oblique shock wave that separates upstream flow (1) from downstream state (2), with states (2) and (1) connected by the Rankine-Hugoniot equations. Shock polar (Γ) is the locus of the states connected to upstream state (1) by a shock wave; the shape of (Γ) depends on the upstream Mach number M_1 and the value of the specific heat ratio γ. There are several such representations; the most convenient form is a plot of the shock-pressure rise (or the pressure ratio p_2/p_1) versus the velocity deflection φ through the shock. Shock polars defined this way are

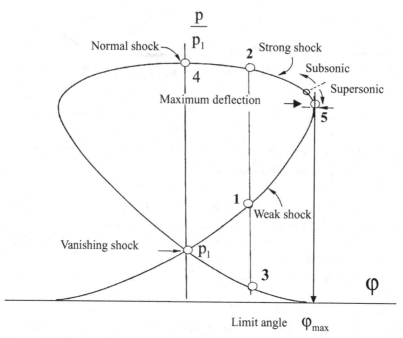

Figure 2.2. The shock-polar representation in term of flow deflection and pressure jump.

closed curves that are symmetrical with respect to the axis $\varphi = 0$ (if φ_1 is assumed to be equal to zero), as shown in the example plotted in Fig. 2.2. At the origin, the polar has a double point corresponding to a vanishing shock (i.e., Mach wave). For a given value of the deflection angle φ, there are two admissible solutions, (1) and (2). (A third solution for which the pressure through the shock decreases is rejected because it fails to satisfy the Second Law of Thermodynamics.) Solution (1), which leads to the smaller pressure jump, is called the *weak solution*; the second is the *strong solution*. For $\varphi = 0$, the strong solution is the normal shock – that is, point (4) on the shock polar.

There is a maximum deflection φ_{max} beyond which an attached shock at **A** is no longer possible. If the deflection φ imparted by the ramp is greater than φ_{max}, then a detached shock is formed starting from the wall upstream of **A** (see Fig. 2.1b). In this case, the flow downstream of the shock does not have a unique image point on the polar but instead follows an arc extending from the normal shock image (i.e., for the shock foot at the wall) to the image corresponding to the shock away from the wall. Another particular point about the polar is the image of the shock for which the downstream flow is sonic. This point is slightly below the maximum deflection location and it separates shocks with supersonic downstream conditions from those with subsonic downstream conditions. A shock polar exists for every upstream Mach number; the shape of the curves for several examples is illustrated in Fig. 2.3. The slope of shock polar $(dp/d\varphi)_0$ at the origin passes through a minimum $(dp/d\varphi)_{min}$ for an upstream Mach number $M_1 = \sqrt{2} = 1.414$, with $(dp/d\varphi)_0 > (dp/d\varphi)_{min}$ for $M_1 > \sqrt{2}$ and $(dp/d\varphi)_0 > (dp/d\varphi)_{min}$ for $M_1 < \sqrt{2}$. Thus, when $M_1 > \sqrt{2}$, each successive polar is above the previous one as the Mach number increases. The order

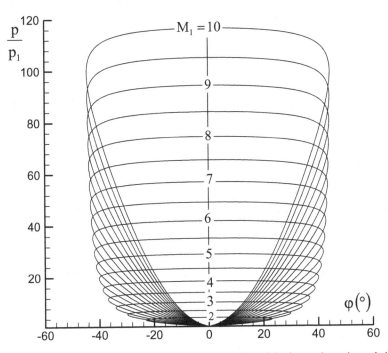

Figure 2.3. Shock polars for varying upstream flow Mach numbers ($\gamma = 1.4$).

reverses for $M_1 < \sqrt{2}$, with the polar for a lower Mach number now above the previous polar (Fig. 2.4). This fact is significant when considering the shock penetration of a boundary layer (see Section 2.5).

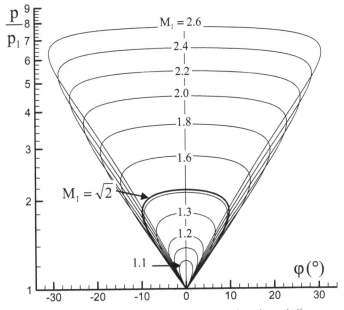

Figure 2.4. Relative positions of the shock polars ($\gamma = 1.4$).

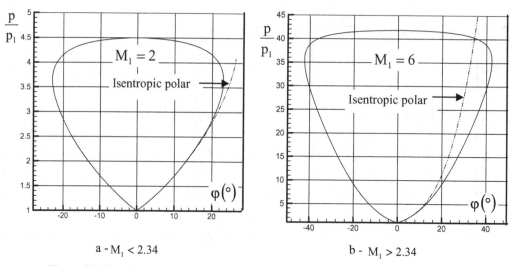

a - $M_1 < 2.34$ b - $M_1 > 2.34$

Figure 2.5. Relative location of the shock polar and the isentropic polar ($\gamma = 1.4$).

Any flow of gas in equilibrium undergoing isentropic changes from known stagnation conditions is completely defined by two independent variables: pressure p and direction φ. This flow has a unique image point in the plane $[\varphi, p]$, which is considered a *hodographic plane*. Passing through a shock wave entails a change in entropy of the fluid, resulting in a jump of its image in the $[\varphi, p]$ plane. The new point must lie on the shock polar attached to the upstream state. An interesting property of the hodographic representation $[\varphi, p]$ is that two contiguous flows separated by a slip line have coincident images because – according to the Rankine-Hugoniot equations – the condition for the flows to be compatible is that they have the same pressure and direction. Similarly, a simple isentropic expansion or compression can be represented by an isentropic polar in the $[\varphi, p]$ plane. For a planar two-dimensional flow of a calorically perfect gas, such a curve is defined by the following characteristic equation:

$$\omega(M, \gamma) \pm \varphi = \text{constant},$$

where $\omega(M, \gamma)$ is the Prandtl-Mayer function:

$$\omega(M, \gamma) = \sqrt{\frac{\gamma + 1}{\gamma - 1}} \tan^{-1} \sqrt{\frac{\gamma - 1}{\gamma + 1}(M^2 - 1)} - \tan^{-1} \sqrt{M^2 - 1}$$

The polar representing an isentropic compression from the same Mach number is plotted with the shock polar in Fig. 2.5. It can be demonstrated that at the origin, the two curves have a third-order contact, so they remain very close until relatively high deflection angles. At moderate Mach numbers, weak-type shock solutions can be considered as almost isentropic. The isentropic polar is below the shock polar for upstream Mach numbers that are less than about 2.34 and it passes above for higher Mach numbers.

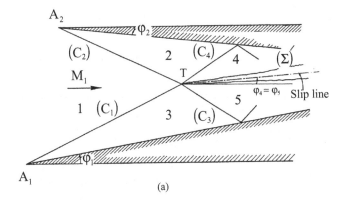

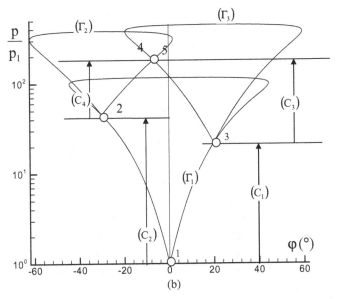

Figure 2.6 (a) Type I shock-shock interference. Physical plane. (b) Type I shock-shock interference. Plane of polars ($M_1 = 10$, $\varphi_1 = 20°$, $\varphi_2 = -30°$).

2.2.3 Shock Intersections and the Edney Classification of Shock-Shock Interferences

Distinctive features of shock-induced separation are the shock patterns that occur in the contiguous inviscid flow as a consequence of the behaviour of the boundary layer during the interaction process. The patterns produced when two shocks intersect, or interfere, were classified by Edney [2] into what are now commonly acknowledged as six types, although variants may exist in some circumstances. These types can be interpreted by referring to the discontinuity theory and by considering their shock-polar representation.

Type I interference occurs when two oblique shocks from opposite families (or opposite directions) cross at point **T**, as shown in Fig. 2.6a. Shock (C_1) provokes a pressure jump from p_1 to p_3 and an upward deflection $\Delta\varphi = \varphi_1$, whereas shock (C_2) causes the pressure to increase from p_1 to p_2 with a downward deflection $\Delta\varphi = \varphi_2$. In general, flows (1) and (2) downstream of (C_1) and (C_2) are not compatible because their pressures and directions are different. Thus, flows (1) and (2) must be deflected such that they adopt a common direction, $\varphi_3 = \varphi_4$, which is achieved across the two transmitted shocks, (C_3) and (C_4), emanating from the point of intersection $\mathbf{T}$. At the same time, the pressures increase from p_3 to p_5 and from p_2 to p_4. For flows (4) and (5) to be compatible, $p_4 = p_5$, which determines shocks (C_3) and (C_4). The situation at point $\mathbf{T}$ in the plane of the shock polars is represented in Fig. 2.6b. The images of flows (2) and (3) are points (2) and (3) situated on polar (Γ_1) attached to upstream flow (1). Shocks (C_3) and (C_4) are represented by polars (Γ_2) and (Γ_3) attached to flows (2) and (3), respectively. To equalise the pressures and flow directions, the images of states (5) and (4) must coincide with the intersection of (Γ_2) and (Γ_3). In the general case, shocks (C_1) and (C_2) do not have the same intensity; therefore, the increases in entropy across $(C_1) + (C_3)$ and $(C_2) + (C_4)$ are different. Accordingly, flows (4) and (5) have different stagnation pressures and are separated by slip line (Σ) emanating from $\mathbf{T}$, across which the fluid properties are discontinuous (see Appendix A).

Type II interference occurs in the following conditions: If the intensity of shock (C_2) (or shock (C_2)) increases, image point (2) moves to the left of polar (Γ_1); as the Mach number M_2 of flow (2) decreases, the size of polar (Γ_2) contracts. Accordingly, (Γ_1) and (Γ_2) are first tangential and then they no longer intersect. A Type I solution is no longer possible and the entire flow is reconfigured such that compatibility conditions are satisfied again. Because compatible states (4) and (5) can no longer exist, intermediate states must be introduced between states (4) and (5), with the flow adopting the pattern as represented in Fig. 2.7a. This structure, called Type II interference, can be interpreted by considering the situation in the hodographic plane shown in Fig. 2.7b. Polars (Γ_2) and (Γ_3) intersect the strong-shock branch of (Γ_1) and a near-normal shock (C_5) forms in upstream flow (1) joining triple points $\mathbf{T}_1$ and $\mathbf{T}_2$, with the image being the arc of (Γ_1) included between states (4) and (5). Shock (C_5) is a strong-oblique shock of variable intensity between $\mathbf{T}_1$ and $\mathbf{T}_2$.

Type III interference occurs when a weak-oblique shock crosses a strong near-normal shock. The situations in the physical and hodographic planes are represented in Figs. 2.8a and 2.8b. In this case, polar (Γ_2) intersects the strong-shock branch of (Γ_1). A Type I solution is impossible because of downstream conditions that force the strong-shock solution for (C_2). Shock (C_3) causes a jump from state (2) to state (4) such that the image of state (4) is at the intersection of polars (Γ_1) and (Γ_2). Downstream of the two weak-oblique shocks (C_1) and (C_3), flow (4) is still supersonic, whereas in flow (3), it is subsonic. Therefore, a strong discontinuity in velocity exists on either side of slip line (Σ_1) separating flows (3) and (4) and stemming from triple point $\mathbf{T}_1$. The situation shown in Fig. 2.8a is a more complex case in which slip line (Σ_1) impinges a nearby surface. Then, the region of impact is the seat of large pressure and heat-transfer peaks in flows with high Mach numbers.

The characteristic feature of Type IV interference is the existence of a supersonic jet embedded between two subsonic regions (Figs. 2.9a and 2.9b). Up to region (4), the structure of the field is similar to that for Type III interference with the formation of a shear layer. In this case, shock wave (C_4) terminates at region (4); the

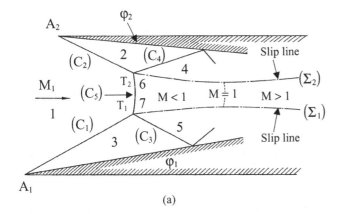

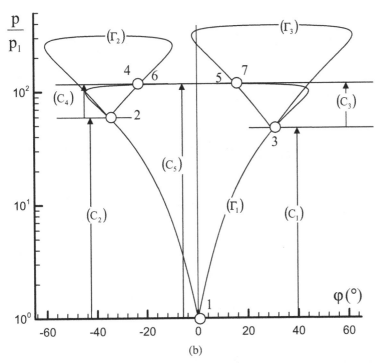

Figure 2.7 (a) Type II shock-shock interference. Physical plane. (b) Type II shock-shock interference. Plane of polars ($M_1 = 10$, $\varphi_1 = 30°$, $\varphi_2 = -35°$).

flow is still supersonic downstream of (C_4); and a supersonic jet bounded by two slip lines, or jet boundaries, (f_1) and (f_2) are formed. The jet is surrounded by subsonic flows in which the pressure is virtually constant. As in the previous case, the flow contains two triple points, T_1 and T_2. To maintain continuity in pressure when shock (C_4) impacts boundary (f_2), a centered expansion must form to offset the pressure jump across (C_4). This expansion is reflected by the opposite boundary (f_1) as a compression wave, which, in turn, is returned by (f_2) as an expansion wave, and so on.

Type V interference occurs when incident shock (C_1) crosses shock (C_2) in a region where (C_2) is a strong-oblique shock; the interference involves two oblique

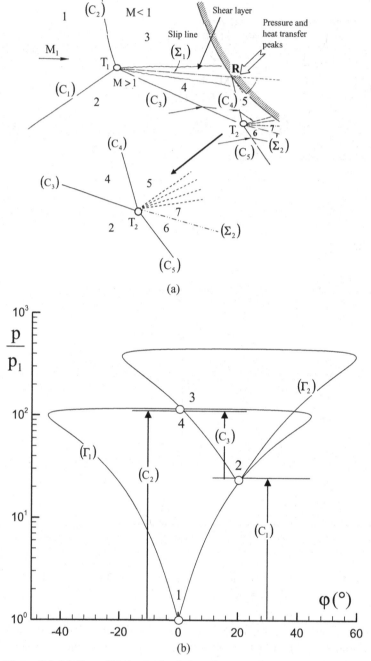

Figure 2.8 (a) Type III shock-shock interference. Physical plane. (b) Type III shock-shock interference. Plane of polars ($M_1 = 10$, $\varphi_1 = 20°$).

shocks from the same family. As shown in Fig. 2.10, the resulting field adopts a complex structure with two multiple points $\mathbf{T}_1$ and $\mathbf{T}_2$ similar to those associated with Type II interference; however, a supersonic jet leaves from $\mathbf{T}_2$ instead of a simple slip line. This complex structure also can be interpreted by considering a

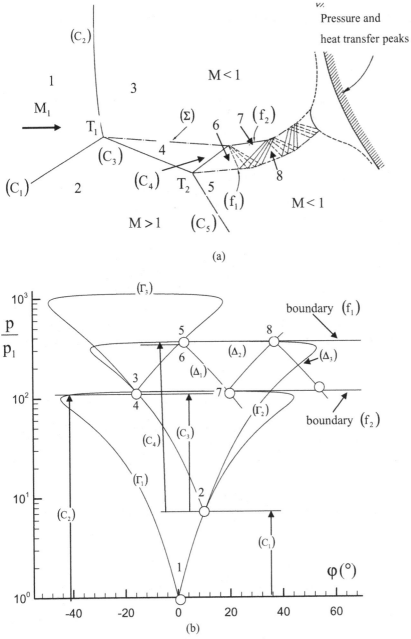

Figure 2.9 (a) Type IV shock-shock interference. Physical plane. (b) Type IV shock-shock interference. Plane of polars ($M_1 = 10$, $\varphi_1 = 10°$).

shock-polar diagram. Because Type V interference is rarely encountered, this rather lengthy exercise is not undertaken here.

Type VI interference occurs when shocks (C_1) and (C_2) cross in a region where they are both weak-oblique shocks from the same family. The corresponding pattern is represented in Fig. 2.11a. The flow organisation is simpler than in the previous cases. The two shocks, (C_1) and (C_2), meet at triple point **T** from which shock (C_3) leaves, causing a jump from state (1) to state (3) with conditions (p_3, φ_3). State (4),

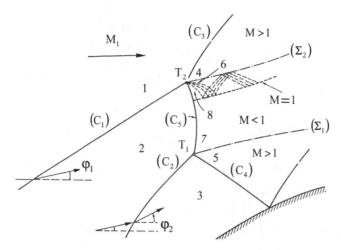

Figure 2.10. Type V shock-shock interference. Physical plane.

which exists downstream of (C_2), and state (3) are incompatible and intermediate state (6) must be introduced, which is connected to state (4) by an expansion wave, as shown in Figs. 2.11a and 2.11b. At moderately supersonic Mach numbers (i.e., maximum close to 2), the transition between states (4) and (6) may occur across a shock wave, which is usually of very low intensity (Fig. 2.11c).

2.2.4 Shock Waves, Drag, and Efficiency: The Oswatitsch Relationship

An equation devised by Oswatitsch [3] establishes a link between the drag of a vehicle and the entropy and stagnation enthalpy it introduces into the flow. If V_∞ designates the uniform upstream velocity of incident flow, T_∞ the upstream flow temperature, s the specific entropy of the fluid, h_{st} the specific total enthalpy, and $dq_m(=\rho\vec{V}\vec{n}ds)$ is the elementary mass flow, then the following relationship exists between the generalized force $\mathbf{F}$ on the vehicle in the drag direction and the flux of entropy and stagnation enthalpy through a surface surrounding it at large distance:

$$F = \frac{1}{V_\infty} \iint_{(s)} (T_\infty \delta s - \delta h_{st}) dq_m$$

The quantities δs and δh_{st} represent variations of specific entropy and total enthalpy, respectively, on the surface relative to the uniform upstream state. The aim of a propulsion system is to 'inject' total enthalpy into the flow to produce thrust in order to compensate for the entropy term, thereby allowing the integral to be equal to zero or to change its sign so that the thrust exceeds the drag. If we concentrate on drag alone, then:

$$D = \frac{1}{V_\infty} \iint_{(s)} T_\infty \delta s \, dq_m,$$

and it is immediately evident that drag and entropy production are closely linked. Conversely, if the flow is adiabatic and the fluid is in an equilibrium state, then the

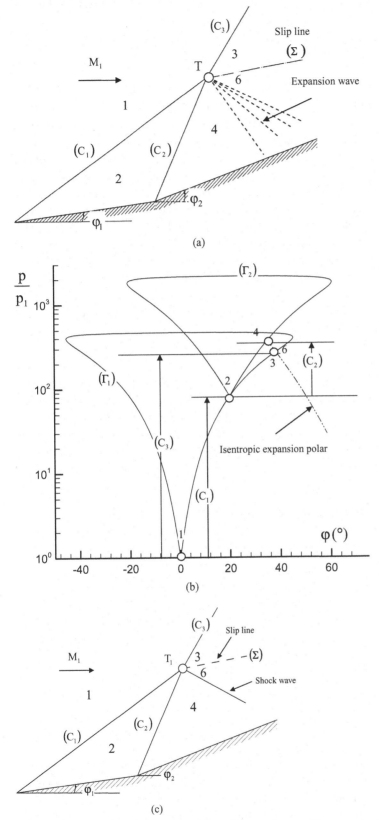

Figure 2.11 (a) Type VI shock-shock interference (case with expansion). Physical plane. (b) Type VI shock-shock interference. Plane of polars ($M_1 = 10$, $\varphi_1 = 20°$, $\varphi_2 = 35°$). (c) Type VI shock-shock interference (case with shock). Physical plane.

energy equation written with the specific entropy as variable is as follows:

$$\rho T \frac{ds}{dt} = \tau_{ij} \frac{\partial u_i}{\partial x_i},$$

where τ_{ij} is the stress tensor and d/dt is the total (particular) derivative. This equation shows that the only source of entropy is viscosity through the shear stresses. These relationships establish the link among drag, entropy, and viscosity. Because viscosity is active in regions of rapid velocity variation, the origin of drag is found in either the discontinuities such as shock waves and slip lines or the regions close to body surfaces – namely, the boundary layers – that constitute special slip lines. It is typical to distinguish between the drag generated by entropy produced in the boundary layers, which is called the *friction drag*, and the drag resulting from entropy production in shock waves, which is called *wave drag*. Because drag is directly connected to entropy production, shock waves comprise the major source in high-speed flows. First, the wave drag can represent a substantial proportion of the total drag in high Mach number flows (i.e., almost 20 percent for a supersonic transport aircraft flying at Mach 2 and significantly more for hypersonic vehicles). Second, the interaction of the shocks with the boundary layer can enhance its entropy production, thereby increasing the friction drag. An aim of controlling SBLIs is to act on both of these terms to reduce the drag of a vehicle.

In internal aerodynamics, the concern about improving the flow is expressed in terms of efficiency or pressure loss. The main aim of supersonic air-intake design is to minimize the drop in stagnation pressure occurring in the shock waves and boundary layers so that the pressure recovered in the engine-entrance section (2) is at maximum. This fact is expressed through the following efficiency coefficient:

$$\eta = \frac{p_{st_2}}{p_{st_\infty}}$$

defined as the ratio between the (mean) stagnation pressure at engine level p_{st_2} and the upstream flow stagnation pressure p_{st_∞}, which is the maximum recoverable pressure. For a calorically perfect gas, the specific entropy is expressed by:

$$s = C_p \ln T_{st} - r \ln p_{st},$$

where C_p is the coefficient of specific heat and r is the gas constant. Because the stagnation temperature is conserved in the flow, the following relationship exists between the entropy rise and the stagnation pressure ratio through a shock wave:

$$\Delta s = s_2 - s_1 = r \ln \left(\frac{p_{st_2}}{p_{st_1}} \right).$$

Hence:

$$\frac{p_{st_2}}{p_{st_1}} = \exp \left(-\frac{s_2 - s_1}{r} \right).$$

This demonstrates that efficiency losses also are directly connected to entropy production (hence, viscosity) and that they are the internal aerodynamics equivalent to the drag in external flows. Pressure recovery is not the only concern in air-intake design; the quality of the flow entering the engine also is of utmost importance. Thus, a good uniformity, or small distortion, is required, which implies the absence of

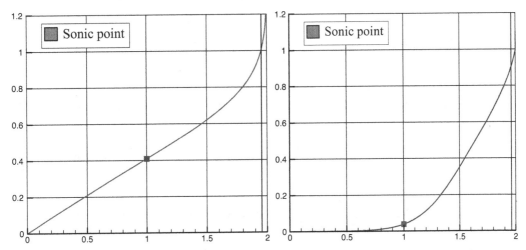

Figure 2.12. Mach-number distribution in a flat-plate boundary layer (outer Mach number 2).

large turbulent eddies, vortices, or discontinuities resulting particularly from strong SBLIs. Such interactions also may be the source of or contribute to flow unsteadiness such as buzz, which can be harmful for the engine by provoking combustion extinction or even physical destruction in extreme conditions.

2.3 On the Structure of a Boundary-Layer Flow

2.3.1 Velocity Distribution through a Boundary Layer

Velocity distribution through a laminar flat-plate boundary is represented by the classical Blasius solution. In principle, this is valid only for incompressible flows, but it continues to provide a good representation of the distribution even at high Mach numbers. (Compressible solutions to this equation exist but considering them here is not useful for our purpose) For most of its range, the Blasius profile is nearly linear and, as shown in Fig. 2.12, has distinct differences from the turbulent distribution, especially in the lower velocity part of the boundary layer. By assuming a Crocco-type law for the temperature distribution through the boundary layer, the Mach number distribution can be determined and, hence, the position of the sonic point on the profiles. These heights are indicated in Fig. 2.12 and are compared with a turbulent profile (for an adiabatic flow).

The turbulent boundary layer has a more complex structure consisting of an excessively thin viscous or laminar layer in contact with the wall, a logarithmic region above it, and a wake-like velocity distribution in the outer part of the layer. The relative importance of the various regions depends of several factors – mainly, the Reynolds number and the externally imposed pressure gradient. This structure is illustrated for a 'well-behaved' boundary layer formed on a flat plate at a high Reynolds number in Fig. 2.13, in which a blending region was added to ensure a continuous variation between the logarithmic region and the laminar layer.

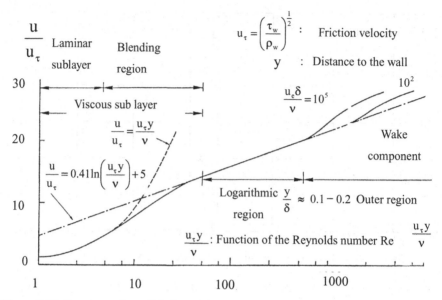

Figure 2.13. Structure of a well-behaved flat-plate turbulent boundary layer.

It is often convenient to use the following analytical formula proposed by Coles [4] to represent a turbulent-boundary-layer velocity profile. It combines a logarithmic law and a wake law but is not valid very close to the wall:

$$\frac{\bar{u}}{\bar{u}_e} = 1 + \left(\frac{1}{k}\right)\left(\frac{u_\tau}{\bar{u}_e}\right)\ln\left(\frac{y}{\delta}\right) - \left(\frac{\delta^*}{\delta} - \frac{u_\tau}{k\bar{u}_e}\right)\left[2 - w\left(\frac{y}{\delta}\right)\right]$$

Here, δ^* is the boundary-layer displacement thickness and u_τ is the friction velocity such that:

$$\frac{u_\tau}{u_e} = \sqrt{\frac{C_f}{2}}$$

$k = 0.41$ is the von Kàrmàn constant, C_f is the skin-friction coefficient, and $w\,(y/\delta)$ is the so-called wake component of the form:

$$w\left(\frac{y}{\delta}\right) = 1 - \cos\left(\pi\frac{y}{\delta}\right)$$

This relationship can be expressed more conveniently as:

$$\frac{\bar{u}}{\bar{u}_e} = 1 + \left(\frac{1}{0.41}\right)\sqrt{\frac{C_f}{2}}\ln\left(\frac{y}{\delta}\right) - \left(\frac{\delta^*}{\delta} - \frac{1}{0.14}\sqrt{\frac{C_f}{2}}\right)\left[2 - w\left(\frac{y}{\delta}\right)\right]$$

Coles's relationship can be used for compressible flows by replacing C_f with the 'incompressible' skin friction C_{f_i}, which is related to C_f by:

$$C_{f_i} = C_f\left(1 + \frac{\gamma - 1}{2}M_e^2\right)^{\frac{1}{2}}$$

where M_e is the Mach number at the boundary-layer outer edge.

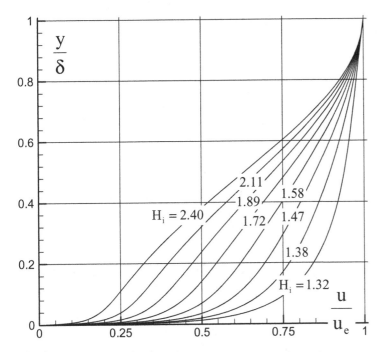

Figure 2.14. Cole's turbulent-velocity distribution.

Because the response of a boundary layer to the action of a shock depends to a considerable degree on velocity distribution, it is convenient to introduce the incompressible velocity-profile shape factor, defined as follows:

$$H_i = \frac{\delta_i^*}{\theta_i}$$

where δ_i^* and θ_i are the so-called incompressible displacement and momentum thicknesses defined by:

$$\delta_i^* = \int_0^\delta \left(1 - \frac{u}{u_e}\right) dy, \quad \theta_i = \int_0^\delta \frac{u}{u_e} \left(1 - \frac{u}{u_e}\right) dy$$

The parameter H_i characterises the velocity shape and should be used in preference to the alternative shape factor H, which is computed with the compressible or true integral thicknesses. The compressible shape factor is sometimes used but it is a strong function of the Mach number, which makes it less practical as a universal parameter. Examples of Coles's turbulent profiles are represented in Fig. 2.14 for increasing values of the incompressible shape parameter H_i.

There is a progressive distortion of the profiles with increasing H_i and a general decrease in the velocity levels – particularly in the lower part of the boundary layer. This evolution is representative of the behaviour of a layer submitted to an adverse pressure gradient. As shown in Fig. 2.15, the incompressible shape parameter for a turbulent boundary layer depends weakly on the outer Mach number and slowly decreases when the Reynolds number increases. A computation similar to the one for the laminar boundary layer shows that the sonic point in turbulent boundary layers is much closer to the wall, even for moderately supersonic outer Mach numbers

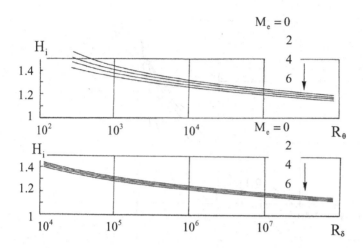

Figure 2.15. Flat-plate turbulent boundary-layer incompressible-shape parameter (Cousteix [6]).

(see Fig. 2.12). The distance from the wall of the sonic line in a turbulent velocity profile as a function of the shape parameter is shown in Fig. 2.16 for two values of the Mach numbers and different ratios of the wall temperature to the outer-flow stagnation temperature. More precisely, the thermal condition at the wall should be characterised by the ratio of the wall temperature T_w to the recovery temperature T_r corresponding to adiabatic conditions at the wall; that is, the temperature acquired by the fluid at the wall for zero heat transfer (the recovery temperature is inferior to the outer-flow stagnation temperature but not very different from it). For flat-plate conditions ($H_i \approx 1.3$) and even for a moderate outer Mach number, the sonic point is close to the surface. For cold-wall ($T_w/T_r < 1$) conditions, this point is even closer because of the lower sound speed. However, the location of the sonic point in the boundary layer rapidly moves away from the wall if it is heated ($T_w/T_r > 1$).

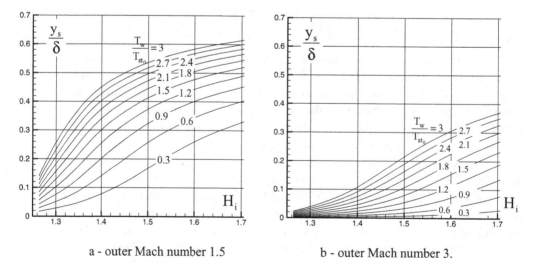

a - outer Mach number 1.5   b - outer Mach number 3.

Figure 2.16. Height of the sonic line in a flat-plate turbulent boundary layer (outer stagnation temperature 1,000K).

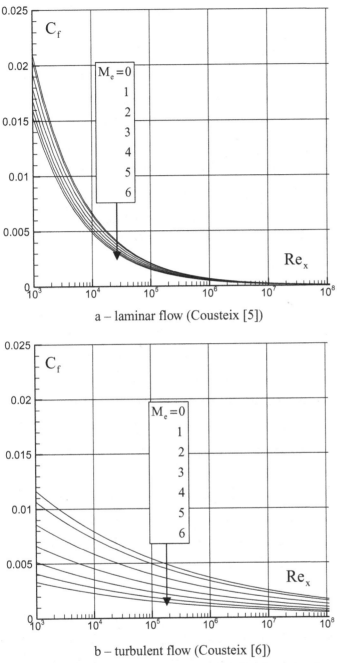

a – laminar flow (Cousteix [5])

b – turbulent flow (Cousteix [6])

Figure 2.17. Skin-friction coefficient for a flat-plate boundary layer (outer stagnation temperature 3,000K, adiabatic case).

The response of the boundary layer to an adverse pressure gradient is also dependent on the viscous forces, which are usually characterised by the wall-shear stress or the skin-friction coefficient. As shown in Fig. 2.17, the skin-friction coefficient for a flat-plate turbulent boundary layer is a strong function of the Mach number and decreases rapidly when the Reynolds number is increased. Because the

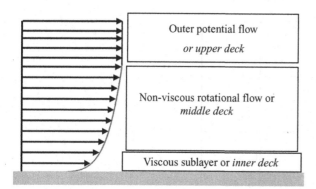

Figure 2.18. The interacting flow multi-layer structure or triple deck (Lighthill [7], Stewartson-Williams [8]).

shear forces tend to oppose the retardation effect of the shock, we can infer that the resistance of a laminar boundary layer to an imposed pressure gradient will diminish with an increasing Reynolds number. For the turbulent boundary layer, the situation is more subtle, as discussed subsequently.

2.3.2 The Multilayer Structure

The flow along a solid surface can be viewed as a structure composed of three layers, (Fig. 2.18), as follows:

1. An outer inviscid layer that usually is *irrotational* (i.e., isentropic) and there-fore obeys the Euler equations or alternatives such as the potential equation. However, there are exceptions in which this part of the flow is rotational – for example, downstream of the curved shock formed ahead of a blunted leading edge, where what is referred to as an *entropy layer* is formed. A similar rota-tional layer can occur behind the near-normal but curved shock that forms on a transonic aerofoil.
2. Closer to the surface and deeper within the boundary layer, first is an outer por-tion where, over a streamwise distance of several boundary-layer thicknesses, the flow can be considered as inviscid but rotational. In this part of the flow, vis-cosity contributes to create entropy and, consequently, vorticity, in agreement with Crocco's equation connecting the gradient of entropy s with the rotational vector in a nonviscous flow, as follows:

$$T\overrightarrow{grads} = -\vec{V} \times \overrightarrow{rot}\vec{V}$$

Simply stated, this layer is a region of variable stagnation pressure and stagna-tion temperature. Because the flow is considered inviscid, the stagnation con-ditions are constant along streamlines because entropy is a transported quan-tity. The static pressure is constant across the boundary layer; hence, the layer behaves like an inviscid flow through which the velocity – and, hence, the Mach number – decreases steadily from the outer value M_e at the boundary-layer edge ($y = \delta$) towards zero at the wall.
3. The third layer is in contact with the wall and, to ensure the transition between the previous region and the surface, viscosity again must play a role. This vis-cous layer must be introduced to avoid inconsistencies because it is not possible for a nonviscous flow to decrease its velocity without an increase in the static

pressure; at the wall, the stagnation pressure is equal to the static pressure (i.e., the velocity is equal to zero because of the no-slip condition).

The structure described here was first suggested by Sir James Lighthill [7] in a famous article published in 1953. A more formal justification was proposed in 1969 by Stewartson and Williams [8] for the case of a laminar boundary layer using an asymptotic expansion approach. They introduced the *triple-deck* terminology to designate such a structure: The outer deck is the outer irrotational flow, the middle deck is the inviscid rotational layer, and the inner deck is the viscous layer in contact with the wall. This representation is valid only if the viscous forces have not contributed to modify the entropy level of the boundary-layer streamlines (except in the inner deck). This implies that the time scale of any phenomenon considered with this approach is short compared to the time scale over which the viscous terms take effect. This is the case for SBLIs, in which the shock imparts a sudden retardation to the flow. Such a model is also valid for a rapid acceleration, as in the centred expansion wave that can occur at the base of a vehicle. In a turbulent boundary layer, the middle deck represents the greatest part of the boundary layer, even at a moderate Mach number, so that the behaviour of an interaction can be described (for the most part) by considering a perfect-fluid model. However, for reasons cited previously, such an inviscid model becomes inadequate close to the wall, where viscosity must be considered.

2.3.3 The Boundary-Layer Response to a Rapid Pressure Variation

During the first part of an SBLI, most of the flow – including a greater part of the boundary layer – behaves as an inviscid flow for which the pressure and inertia terms of the Navier-Stokes equations are predominant compared to the viscous terms. Thus, many aspects of the boundary-layer response can be interpreted with perfect fluid arguments and by considering the boundary-layer mean properties defined previously. This description of the boundary-layer behaviour calls on the concept of rapid interaction and is justified by the fact that in an SBLI, important changes occur over a short streamwise distance, the extent of the interaction being on the order of 10 times the boundary-layer thickness δ for the laminar flow and less in the turbulent case. This fact has several major consequences, including the following:

- Streamwise derivatives are comparable to derivatives in the direction normal to the wall, whereas in a classical boundary layer, they are considered to be of a lower order. This fact also influences the mechanism for turbulence production because the normal components of the Reynolds stress tensor now may play a role comparable to that of the turbulent shear stress – which, in general, is the only quantity considered.
- The turbulent Reynolds stresses do not react instantly to changes in the mean flow imparted via the pressure gradient. In the first phases of the interaction, there is a lag in the response of the turbulence; hence, a disconnection occurs between the mean velocity and turbulent fields. Reciprocally, the velocity field is weakly affected by the shear stress; the action of viscosity is confined to a thin layer in contact with the wall. Thereafter, the turbulence level increases and can reach high levels if separation occurs. This explains the difficulty in devising

adequate turbulence models for SBLIs, especially for the inception part of the process.

An inviscid-fluid analysis provides an explanation for basic features of an interaction; however, it is not entirely correct in that viscous forces cannot be neglected over the entire extent of the interaction. Viscous terms must be retained in the region in contact with the wall; otherwise, we are confronted with inconsistency. Nevertheless, the neglect of viscosity is justifiable in describing the penetration of the shock into the boundary layer. However, as discussed later, there are a number of situations in which the shock does not penetrate the boundary layer, as in transonic or shock-induced interactions (except at a very high Mach number). In these cases, viscosity may have sufficient time to influence the flow behaviour even outside the near-wall region. This is also true for interactions with large separated regions, where the flow depends on its viscous properties to determine the longitudinal extent of the interaction.

2.4 Shock Waves and Boundary Layers: The Confrontation

2.4.1 The Basic SBLI in Two-Dimensional Flows

Five basic interactions can occur between a shock wave and a boundary layer in two-dimensional flows. These occur when there is:

- an impinging oblique-shock reflection
- a ramp flow
- a normal shock
- an imposed pressure jump
- an oblique shock induced by a forward-facing step

We consider each interaction in more detail, as follows:

1. In an oblique-shock reflection at a flat surface (Fig. 2.19a), the approaching supersonic flow of Mach number M_1 undergoes a deflection $\Delta\varphi_1$ through incident shock (C_1). For the downstream flow to remain parallel to the wall (i.e., a Euler type or slip-boundary condition for a nonviscous fluid), the formation of a reflected shock (C_2) is required. The deflection $\Delta\varphi_2$ across this is such that $\Delta\varphi_2 = -\Delta\varphi_1$. Shock patterns like this occur inside a supersonic air-intake of the mixed-compression type or at the impact of the shock generated by any obstacle on a nearby surface.
2. In the ramp flow (Fig. 2.19b), a discontinuous change in the wall inclination is the origin of a shock through which the incoming flow undergoes a deflection $\Delta\varphi_1$ equal to the wedge angle α. Such a shock occurs at a supersonic air-intake compression ramp, at a control surface, or at any sharp change in the direction of a surface.
3. A normal shock wave can be produced in a supersonic flow by a back pressure forcing the flow to become subsonic. In channel flow with a two-throat system (e.g., in a supersonic wind tunnel), a normal shock is formed when choking downstream necessitates a stagnation pressure loss to satisfy mass conservation.

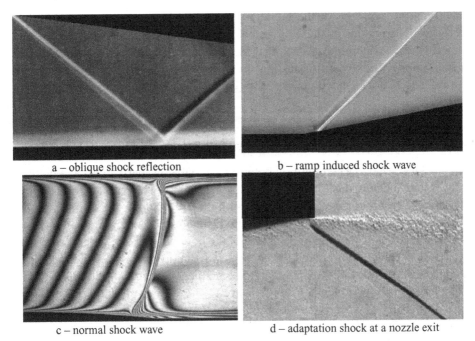

a – oblique shock reflection

b – ramp induced shock wave

c – normal shock wave

d – adaptation shock at a nozzle exit

Figure 2.19. Basic SBLIs (ONERA documents).

The distinctive feature of a normal shock is that it decelerates the flow without imparting deflection to the velocity vector – the Mach number behind the shock is subsonic. However, in most practical cases, the shock is not perfectly normal (i.e., the interferogram in Fig. 2.19c); rather, it is a strong oblique-shock solution to the Rankine-Hugoniot equations – even if the shock intensity is very weak as in transonic flows! In these situations, the velocity deflection through the shock is so small that the shock is said to be normal. Normal or near-normal shocks are found in channel flows (e.g., turbomachine cascades, air intakes, and supersonic diffusers), in shock tubes, and over transonic profiles in which a near-normal shock terminates the supersonic pocket. Interactions in which the down-stream flow is totally (or partially) subsonic lead to specific problems. These are of special interest because of the possibility that downstream disturbances can influence the shock and initiate an interactive process, which then can result in large-scale unsteadiness involving the entire flow (e.g., in transonic buffeting or air-intake buzz).

4. An oblique shock is produced if a supersonic flow encounters a change in pressure (e.g., at the exit of an overexpanded nozzle). In this case, the pressure discontinuity induces a flow deflection; whereas in cases 1 and 2, the pressure discontinuity is induced by the deflection (Fig. 2.19d). This is the mirror problem of the duality [deflection–pressure jump].

5. When a flow encounters an obstacle such as a forward-facing step, it separates upstream of the step. The extent of the separation region is a direct function of the step height. When the Mach number is supersonic, a shock wave forms at the separation location.

Concerning the response of the boundary layer to the shock, there are no basic differences between all of these situations – except perhaps case 4 in which the interacting flow communicates with an atmosphere. Therefore, we do not distinguish among cases when discussing the viscous-flow behaviour in subsequent sections. The major distinctions are between interactions with and without separation.

2.4.2 The Boundary-Layer–Shock-Pressure-Jump Competition

SBLI can be viewed as a competition between a variable property flow – the boundary layer, in which viscous forces are (or have been) at work – and an abrupt pressure rise. The result of this conflict depends on the pressure-rise amplitude and the boundary-layer characteristics. It is typical to distinguish between laminar and turbulent interactions according to the nature of the boundary layer meeting the shock wave. If we consider an averaged turbulent flow – that is, of the Reynolds or Favre averaging, which filters out the fluctuating components – there are no basic differences between the two types of flow relative to the overall physics and topology of the flow. Thus, any description of the interaction of one nature can be applied, *mutatis mutandis*, to the interaction of the other type. Therefore, in subsequent sections, laminar and turbulent interactions are examined globally with the choice of the more commonly occurring turbulent interaction used to illustrate the descriptions. However, if quantitative properties are involved, then dramatic differences between laminar and turbulent flows render the nature of the incoming boundary layer an essential parameter. Basically, the streamwise-length scales involved in a laminar interaction are considerably longer than in a turbulent interaction, for reasons that are explained in the following discussion.

The boundary-layer equation for the streamwise momentum for a steady flow is as follows:

$$\rho u \frac{\partial u}{\partial x} + \rho v \frac{\partial u}{\partial y} = \frac{\partial}{\partial x}\left(\rho u^2\right) + \frac{\partial}{\partial y}(\rho uv) = -\frac{dp}{dx} + \frac{\partial \tau}{\partial y},$$

where ρ is the density, u and v are the x-wise and y-wise velocity components (y is normal to the wall), p is the pressure, and τ is the shear stress. The central part of this equation expresses the streamwise derivative of the flow momentum. We see that an adverse pressure gradient tends to make the momentum along the boundary-layer streamlines decrease. In a flat-plate boundary layer, the shear stress is nearly constant close to the wall and then steadily decreases towards the boundary-layer edge. Thus, in the first part of an interaction process, the boundary-layer flow is retarded under the combined action of the shock-induced pressure gradient and the shear stress; the retardation is more important in the near-wall region because of lower momentum. The result is a velocity profile presenting an inflection point with an inner part in which the derivative $\partial \tau / \partial y$ is positive and an outer part in which $\partial \tau / \partial y$ is negative. In this situation, illustrated in Fig. 2.20, the shear stress in the boundary-layer inner part counteracts the action of the pressure gradient by transferring momentum from the outer high-velocity to the inner low-velocity regions.

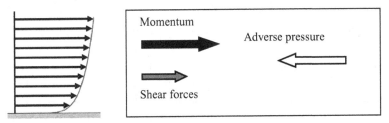

Figure 2.20. Forces at work in a SBLI.

A more simplistic analysis can be proposed by neglecting the contribution of the normal velocity component – that is, by considering the boundary layer as a parallel flow. Then:

$$\rho u \frac{\partial u}{\partial x} = \frac{\partial}{\partial x}\left(\rho u^2\right) = -\frac{dp}{dx} + \frac{\partial \tau}{\partial y}$$

Hence, by integrating between a lower boundary δ_i close to the wall and the boundary-layer outer edge, we obtain the following relationship:

$$\frac{d}{dx}\int_{\delta_i}^{\delta}\rho u^2 dy = -\frac{dp}{dx}(\delta - \delta_i) + (\tau_\delta - \tau_{\delta_i}) \approx -\frac{dp}{dx}(\delta - \delta_i) \approx -\frac{dp}{dx}\delta$$

In this equation, the shear stress is neglected and the inner boundary is assumed to be very close to the wall ($\delta << \delta_i$). As shown herein, such a situation corresponds to a turbulent boundary layer in which the shear stress has a minor role across most of the boundary layer. This simplified relationship highlights the competition between the flow momentum and the pressure gradient. Of course, this analysis is wrong if the normal component v – or, more precisely, its derivative $\partial v/\partial y$ – and the shear stress are no longer negligible. This is the case in the inner part of a turbulent boundary layer and over almost the entire thickness of a laminar boundary layer. Also, the situation changes when the shock induces separation of the boundary layer, as discussed in the next section. Then, the viscous contribution is essential.

A major cause of the differences observed between laminar and turbulent flows is in the velocity distribution in the boundary layer. During an interaction, the flow behaviour is dictated by the resistance of the boundary layer to the pressure jump imparted by the shock; therefore, it is clear that a turbulent boundary layer having a 'full' velocity distribution – and thus carrying more momentum – will react less than a laminar boundary layer the profile of which is far less full. This is illustrated by the velocity profiles plotted in Fig. 2.12, which are representative of flat-plate laminar and turbulent boundary layers. The incompressible-shape parameter H_i (see Section 2.3.1) characterises how full the boundary-layer profile is; a low value indicates a fuller profile. For a flat-plate boundary layer, H_i is close to 2.5 in laminar flow and ranges from 1.3 to 1.4 for turbulent flows. When a boundary layer is submitted to an adverse pressure gradient, the retardation causes the shape parameter to increase; this effect is more pronounced in the parts of the layer where the velocity is low. In addition, the interaction mechanism depends on the thickness of the subsonic part of the boundary layer. Thus, a laminar boundary layer the subsonic layer of which is proportionally thicker will respond differently to a turbulent layer with a much thinner subsonic layer for the same outer Mach number.

The boundary-layer velocity distribution is perhaps the most important factor influencing an interaction, but it is not the only one. The shear forces also play a role because they counteract the retardation imparted by the shock. Their role is predominant in the case of a laminar boundary layer, which is termed a *viscous-dominated flow*. As the Reynolds number increases, the relative magnitude of the viscous forces decreases. Thus, a laminar boundary layer is less resistant to the influence of the shock at high Reynolds numbers than at lower values (see Section 2.7.1). In turbulent flows, the influence of the shear forces is less obvious and the interaction depends weakly, in general, on the Reynolds number. This means that for a well-established turbulent regime, there is practically no influence of the Reynolds number in contrast to laminar interactions. In turbulent flow, the influence of the Reynolds number is experienced mainly through the effect on the incompressible shape parameter; it determines the value of H_i at the interaction onset. Therefore, the Reynolds number is a 'history' parameter, and its influence is experienced through the development of the boundary layer before it enters the interaction region. Other factors can have an historical role, such as a previous interaction (which could have distorted the boundary-layer profile), an adverse pressure gradient (which has 'emptied' its velocity distribution), wall heating, or any boundary-layer manipulation (see Section 2.12).

The wall temperature within or upstream of the interaction region also is an important factor that influences the flow through several mechanisms. This is especially true in the hypersonic regime, in which large differences exist between the temperature of the wall and the outer-flow stagnation temperature. The wall-temperature level (or, more precisely, the ratio of the wall temperature T_w to the recovery temperature T_r) is an essential parameter. Most experiments show that the velocity profile in the boundary layer is almost independent of the wall temperature. Consequently, the Mach-number distribution is affected principally through the wall-temperature influence on the distribution of the sound velocity across the boundary layer. On a cooled wall, the speed of sound is lower; hence, the Mach number is higher in the region close to the wall. This is illustrated in Fig. 2.21, which shows the variation of the sonic-point location with the wall temperature for a turbulent boundary layer at Mach 10 and an outer-flow stagnation temperature of 3,000 K. We see that there is a rapid drop of the sonic-point location when the wall temperature becomes less than the recovery temperature. At the same time, the density is greater; hence, there is increased momentum and the wall shear stress is reduced because of the lower viscosity. The overall effect of wall cooling is to make the boundary layer more resistant, whereas wall heating makes it more fragile and provokes a dilatation of the interaction domain. Conversely, it is clear that turbulence has a central role in determining the interaction because the turbulent eddies operate a transfer of momentum from the outer high-speed flow to the inner low-speed part of the boundary layer. Hence, there is greater resistance to the shock and the separated region, if it forms, is less extensive. This aspect also explains the behaviour of transitional interactions in which the laminar-to-turbulent transition occurs in the interaction domain.

Although the boundary-layer response is determined by the intensity of the pressure jump (or, more precisely, the pressure gradient) – whatever its origin or cause – the overall flowfield structure, and not only the boundary-layer region,

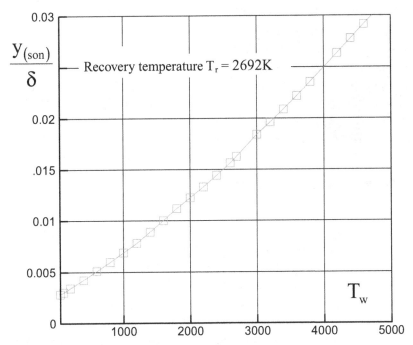

Figure 2.21. The variation of the sonic-point location with the wall temperature for a turbulent boundary layer at Mach 10, stagnation temperature of 3,000K.

greatly depends on the way the shock is generated and the Mach number. If it increases to hypersonic levels, distinctive phenomena occur due to the intensity of the shocks and their small inclination relative to the velocity vector, as well as due to the high enthalpy level of the outer flow. This is discussed briefly later in this chapter and in greater depth in Chapter 6.

2.5 Interactions without Separation: Weakly Interacting Flows

2.5.1 The Incident-Reflecting Shock

Overall Flow Organisation
The interaction resulting from the reflection of an oblique shock wave from a turbulent boundary layer is illustrated by the schlieren visualisation in Fig. 2.22. A similar structure would be seen for a laminar boundary layer, but the streamwise extent of the interaction domain would be greater. (The apparent thickening of the incident shock is due to its interaction with the boundary layers on the test-section side windows; its true location is indicated by the sharp deflection in the superimposed streamline.) The flowfield organisation is illustrated in Fig. 2.23. Incident shock (C_1) can be seen penetrating the rotational inviscid part of the boundary layer, where it progressively bends because of the local Mach number decrease. Correspondingly, the intensity weakens and vanishes altogether when it reaches the boundary-layer sonic line. At the same time, the pressure rise through (C_1) is experienced upstream of where the incident shock would have impacted the wall in the absence of a boundary layer. This upstream-influence phenomenon is predominantly an inviscid mechanism; the pressure rise caused by the shock is transmitted upstream through the

Figure 2.22. Schlieren photograph of a shock reflection at Mach 1.95 (ONERA document).

subsonic part of the boundary layer. This leads to a spreading of the wall-pressure distribution over a distance on the order of the boundary-layer thickness, compared with the purely inviscid-flow solution. As shown in Fig. 2.24, the pressure starts to rise upstream of the inviscid pressure jump, after which it steadily increases and tends towards the downstream inviscid level. In this case, the viscous (or real) solution does not depart far from the purely inviscid solution. Accounting for the viscous effect would be a mere correction to a solution that is already close to reality. Such behaviour is said to be a *weak interaction process* in the sense that the flow is weakly affected by viscous effects. The dilatation of the boundary-layer subsonic region is experienced by the outer supersonic flow, which constitutes the major part of the boundary layer if the flow is turbulent. It acts like a ramp inducing compression waves (η) that coalesce to form the reflected shock (C_2). The thickness of the subsonic layer depends on the velocity distribution; hence, a fuller profile – which has a thinner subsonic channel – also has a shorter upstream-influence length. In addition, a boundary-layer profile with a small velocity deficit has a higher momentum and, therefore, greater resistance to the retardation imparted by an adverse pressure gradient.

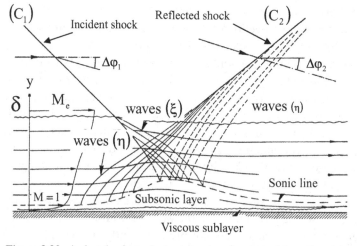

Figure 2.23. A sketch of a turbulent shock reflection without boundary-layer separation.

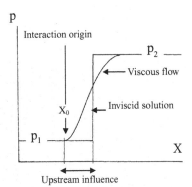

Figure 2.24. The corresponding pressure distribution.

Shock Penetration in a Rotational Layer

For the case of an incident-reflecting shock, the mechanism for the penetration of a shock wave into a boundary layer was studied in great detail by Henderson [9], who used perfect-fluid arguments. His analysis is more pertinent to turbulent flow because the boundary-layer velocity profile is very full and the sonic line is close to the wall even at a moderate outer Mach number, as discussed previously. Most of the boundary layer behaves like an inviscid rotational flow, its entropy being constant along each streamline. For simplicity, Henderson split the boundary-layer flow into N layers made of uniform and parallel constant-pressure streams of different Mach numbers. The shock propagation was analysed by considering a shock-polar representation (Fig. 2.25). Penetration of the incident shock into the rotational layer results in a succession of transmitted shocks (i.e., the incident-shock system) and expansion waves. As the upstream Mach number of the layers decreases on approach to the wall, the angle of the transmitted shock increases so that the incident shock bends and becomes steeper until it vanishes on approach to the sonic line.

The propagation of a shock wave in a turbulent boundary layer is illustrated by perfect-fluid calculations using the rotational method of characteristic. This provides

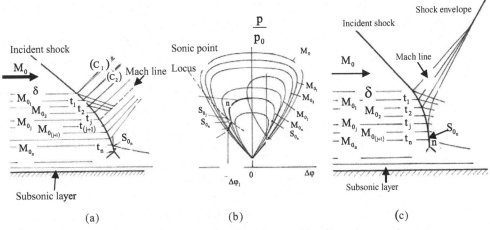

Figure 2.25. Shock-wave refraction into a rotational layer with reflected expansion waves: a, wave pattern; b, corresponding shock-polar diagram; c, wave pattern for a fuller boundary-layer profile (Henderson [9]).

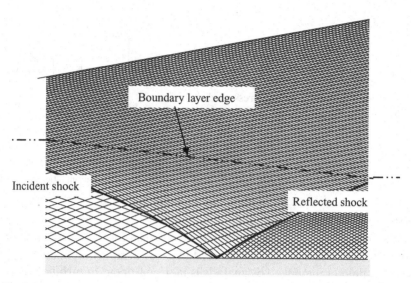

Fig. 2.26. Method of characteristic calculation of a shock reflection in a rotational layer. Wave system and shocks. Turbulent boundary-layer profile (upstream Mach number 4, primary deflection –6°)

both high accuracy (i.e., the shock is fitted) and a picture of the wave's propagation in the supersonic flows. Calculations were made for a turbulent velocity distribution represented by Coles's analytical expression (see Section 2.3.1), with the outer Mach number equal to 4. The part of the boundary layer of which the Mach number is less than 1.8 was removed (i.e., this cut-off distance from the wall was chosen to avoid singular shock reflection). The behaviour of the viscous sublayer is neglected, which is justified for moderate shock strengths at high Mach numbers. The calculation corresponds to the reflection on a rectilinear wall of a shock producing a downward deflection of –6 degrees in the outer irrotational stream. The characteristic mesh represented in Fig. 2.26 shows the bending of the shock through the rotational layer and the waves coming from the wall downstream of the reflection. The wall-pressure distribution plotted in Fig. 2.27b shows that the pressure first jumps at the impact point to an intermediate value and then progressively reaches the constant level that corresponds to shock reflection in a Mach 4 uniform flow. This behaviour, which is observed in high-Mach-number flows, thus can be interpreted by inviscid arguments. At a lower Mach number, below 2.5, an overshoot is observed in the wall-pressure distributions, which cannot be explained simply by rotational effects. In these circumstances, the influence of the subsonic layer close to the wall as well as the viscous inner layer can no longer be neglected, and a purely inviscid analysis captures only part of the solution. The contours in Fig. 2.28 confirm that behind the shock, there is a static-pressure decrease from the outer flow down to the wall. The analysis proposed by Henderson and the method of characteristic calculations are instructive because they describe the complex wave pattern that is generated when a shock traverses a boundary layer considered as a rotational inviscid stream. However, this scenario does not consider the upstream transmission through the

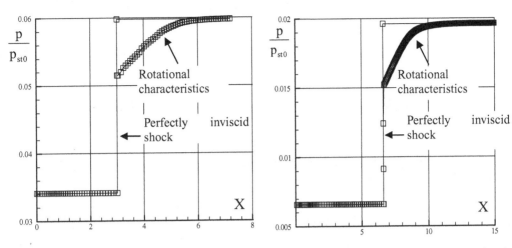

a – ramp induced shock (upstream Mach number 2.85 - wedge angle deflection 6°)

b – shock reflection (upstream Mach number 4-primary deflection -6°)

Figure 2.27. Method of characteristic calculation of a shock reflection in a rotational layer. Entropy gradient effect on the wall-pressure distribution.

subsonic part of the boundary layer with the subsequent generation of compression waves, which coalesce to produce the reflected shock.

2.5.2 Ramp-Induced Shock

As in the previous case, the pressure rise for an SBLI induced by a ramp (Fig. 2.29) associated with shock (C_1) is transmitted upstream of point **A** (i.e., the origin of the ramp) through the boundary-layer subsonic channel. This causes a dilatation of the subsonic channel, which then induces compression waves (η_1) in the contiguous outer supersonic part. In fact, shock (C_1) results from the coalescence of these waves. Thus, the intensity of (C_1) increases steadily with distance from the surface until it reaches the value corresponding to the entirely inviscid solution. However, at high Mach numbers, the subsonic channel has a minor effect on the overall interaction structure, and most of the physics can be interpreted in terms of an inviscid process, as in the incident-shock induced interaction. This point is illustrated in Fig. 2.30 for a ramp in a Mach 2.85 flow with a turbulent incoming boundary layer. Concerning the shock reflection, the rotational method of characteristics calculation predicts the bending of the shock through the rotational layer. This pattern

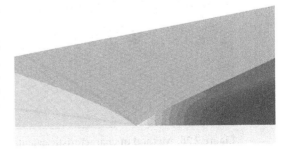

Figure 2.28. Method of characteristic calculation of a shock reflection in a rotational layer. Static pressure contours. Turbulent boundary-layer profile (upstream Mach number 4, primary deflection –6°).

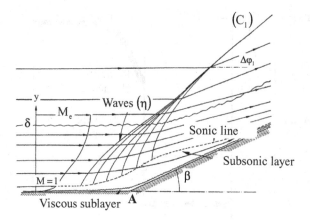

Figure 2.29. Ramp-induced shock without boundary-layer separation.

has remarkable agreement with the shadowgraph of the flow (Fig. 2.31). The wall-pressure distribution in Fig. 2.27a and the static-pressure contours in Fig. 2.32 exhibit the same features as in the shock reflection.

2.5.3 Normal Shock and Transonic Interactions

In transonic flow, the shock most often belongs to the strong-oblique shock solution of the Rankine-Hugoniot equations (although the shock intensity can be minor; e.g., on a transonic aerofoil). As shown in the interferogram in Fig. 2.33 and the sketch in Fig. 2.34, the same upstream transmission mechanism is at work within the boundary layer as in the previous cases. The compression waves induced by the thickening of the subsonic layer coalesce into near-normal shock (C_1). Because the compression in the lower part of the interaction is almost isentropic (except in the viscous layer in contact with the wall), in this region the entropy on each streamline is less than the entropy behind the shock at some distance from the wall.

Consequently, the Mach number is higher near the wall (before penetrating deeper into the boundary layer), where a more-or-less extended pocket of supersonic flow subsists. Starting from the centre of the channel, at some distance behind the interaction, the stagnation-pressure level first corresponds to the drop behind a

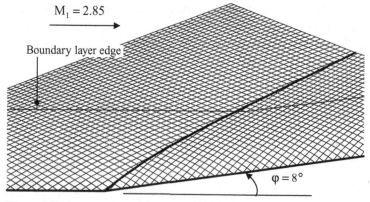

Figure 2.30. Method of characteristic calculation of a ramp-induced shock. Wave system and shock. Turbulent boundary-layer profile (upstream Mach number 2.85, ramp deflection 8°).

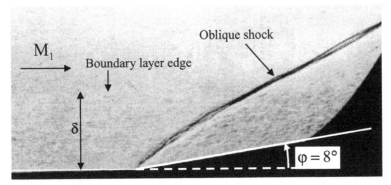

Figure 2.31. Shadowgraph visualisation of ramp-induced shock in a Mach 2.85 flow (Settles [10]).

Figure 2.32. Method of characteristic calculation of a ramp-induced shock. Static pressure contours. Turbulent boundary-layer profile (upstream Mach number 2.85, ramp deflection 8°).

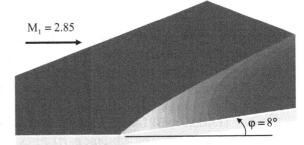

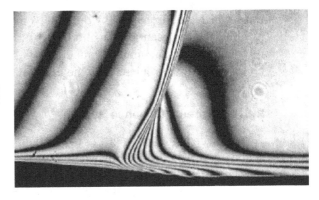

Figure 2.33. Normal shock interaction without separation. Interferogram of flowfield (ONERA document).

Figure 2.34. Normal shock interaction without separation. Sketch of flowfield.

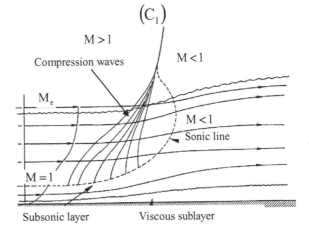

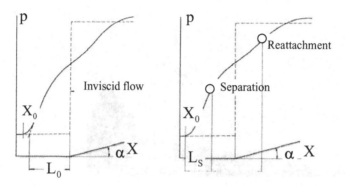

a – upstream interaction length   b – separation length

Figure 2.35. Characteristic lengths of a supersonic interaction.

normal shock. It then rises for the streamlines that have travelled through the lower part of the interaction, where the compression is essentially isentropic; finally, it rapidly decreases close to the surface on penetrating the boundary-layer inner part. This typical stagnation-pressure (or pressure-loss) distribution has great importance when considering SBLI control.

Transonic SBLIs have a specific character because of the partly subsonic nature of the outer flow. The subsonic part causes the phenomenon to be dependent on the downstream conditions; however, this is not the case for purely supersonic interactions, which are 'protected' by the supersonic nature of the outer inviscid flow that prevents the upstream propagation of disturbances. Transmission, of course, exists through the subsonic inner channel of the boundary-layer flow close to the wall; however, for the turbulent case, the upstream transmission length is very short.

2.5.4 Upstream Influence Scaling

An upstream influence length L_0 can be defined as the distance separating the interaction onset (i.e., the location where the wall pressure starts to rise) from the shock foot in the inviscid-flow model (Fig. 2.35). It is a measure of the spreading of the interaction caused by the boundary layer. The main parameters likely to influence L_0 are as follows (Green [11]):

- upstream Mach number M_0
- Reynolds number Re_{δ_0}
- shock intensity, expressed (for example) as the flow deflection $\Delta\varphi_1$ for oblique shocks
- thickness of the incoming boundary layer δ_0
- incoming boundary-layer incompressible-shape parameter H_i, which is probably more significant than the Reynolds number

If we presuppose that any typical streamwise length L_0 will scale with the incoming boundary-layer thickness or its displacement thickness, there remain the three farther parameters – M_0, Re_{δ_0}, and $\Delta\varphi_1$ – that can influence the interaction. If we focus our attention on the dimensionless length L/δ_0, then for a fixed value of Re_{δ_0}:

- L/δ_0 increases with $\Delta\varphi_1$ for a fixed Mach number M_0
- L/δ_0 decreases when M_0 increases for a fixed flow deflection $\Delta\varphi_1$
- for a laminar boundary layer, L/δ_0 increases when Re_{δ_0} increases

For a turbulent boundary layer, the influence of the Reynolds number on L/δ_0 is less clear. However, it is agreed that L/δ_0 increases when Re_{δ_0} increases at a moderate Reynolds number and decreases with Re_{δ_0} at a higher Reynolds number, with the changeover between the two tendencies occurring at around $Re_{\delta_0} \approx 10^5$. In fact, the Reynolds number can be a misleading parameter in that it characterises two competing effects. Considering a flat-plate turbulent boundary layer, a high Reynolds number means reduced importance of the viscous forces. Hence, the boundary layer has reduced resistance to an adverse pressure gradient; thus, the interaction length tends to increase. Conversely, as illustrated in Figure 2.15, a high Reynolds number means a fuller velocity profile (i.e., lower shape-parameter H_i) and, hence, higher momentum of the boundary-layer flow as well as a decrease in the subsonic-channel thickness. Thus, the interaction length tends to contract (Settles [10]). In the first case, the Reynolds number determines the character of the local flow; in the second case, it characterises a history effect.

2.6 Interaction Producing Boundary-Layer Separation: Strongly Interacting Flows

2.6.1 Separation Caused by an Incident Shock

Overall Flow Organisation

A boundary layer is a flow within which the stagnation pressure decreases when approaching the wall and where – at least, for short distances – it can be considered constant along each streamline. Neglecting compressibility (which is, of course, an oversimplification), we can write the Bernoulli equation for each streamline as follows:

$$p_{st} = p + \frac{\rho}{2}V^2$$

Thus, any rise in p provokes greater retardation in regions where the stagnation pressure p_{st} is lowest – that is, in the boundary-layer inner part. By imposing an adverse pressure gradient, a situation can be reached in which the flow adjacent to the wall is stagnated or reversed so that a separated region forms. An incident-shock wave can readily induce separation this way – for example, in the Mach 2 flow for which a schlieren picture is presented in Fig. 2.36. (The apparent thickness of the shock waves is due to the interactions occurring on the test-section side windows.) The structure of this flow is illustrated in Fig. 2.37. Downstream of separation point **S** is a recirculating 'bubble' flow bounded by a dividing streamline (S), which separates the recirculating flow from the flow streaming from upstream to downstream 'infinity'. The streamline (S) originates at separation point **S** and ends at reattachment point **R**. Due to the action of the strong mixing taking place in the detached shear layer emanating from **S**, a mechanical-energy transfer occurs from the outer high-speed flow towards the separated region. As a consequence, the velocity U_s on the dividing streamline (S) steadily increases until the deceleration associated with the reattachment process starts.

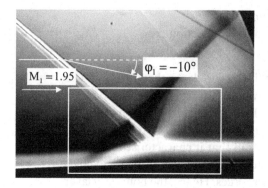

Figure 2.36. Schlieren visualisation of an incident-reflecting shock at Mach 1.95 (ONERA document).

Transmitted shock (C_4) penetrates the separated viscous flow, where it is reflected as an expansion wave because there is a near-constant pressure level in the bubble. This causes a deflection of the shear layer towards the wall, where it eventually reattaches at **R**. At this point, the separation bubble vanishes and the flow on (S) is decelerated until it stagnates at **R**. This process is accompanied by a sequence of compression waves that coalesce into a reattachment shock in the outer stream. This shock pattern is described in more detail in a subsequent section. As shown in Fig. 2.38, the wall-pressure distribution initially exhibits a steep rise, associated with separation, followed by a plateau typical of separated flows. A second, more progressive pressure rise occurs during reattachment. In this situation, the flowfield structure is markedly different from what it would be for the purely inviscid case, and the shock reflection is said to be a *strong viscous-inviscid interaction*. This means that the viscous effects must be fully considered when predicting the flow. They no longer comprise a simple adjustment to an already near-correct inviscid solution, but they have a central role in establishing the solution. It is evident that there has been a hierarchy reversal.

The Outer Inviscid-Flow Structure

The separated configuration described previously can be associated with an equivalent perfect-fluid pattern in which the viscous part of the flow is replaced by an

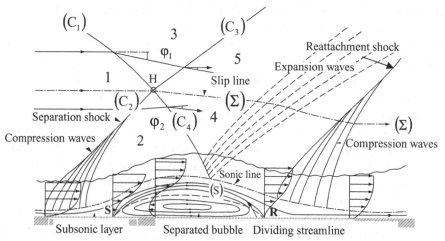

Figure 2.37. Sketch of the flow induced by a shock reflection with separation.

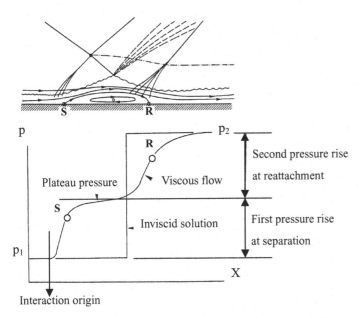

Figure 2.38. Wall-pressure distribution in a shock-separated flow.

isobaric region at pressure p_2. This is isolated from the outer supersonic stream by a slip line, which is now an isobaric (i.e., constant-pressure) boundary (Fig. 2.39). This representation can be useful in improving our understanding of the strong coupling between the separated boundary layer and the outer flow that leads to the formation of specific shock patterns. If the separated region is sufficiently large, the pressure in it is equivalent to the plateau pressure of the free-interaction theory (see Section 2.7.1). Under these conditions, the pressure rise at separation and the initial part of the interaction do not depend on the downstream conditions, even for the transonic case. The effect of the downstream constraints is to fix the location of separation, not its nature.

Because plateau pressure p_2 is higher than upstream pressure p_1, the isobaric frontier (f) (i.e., the slip line) limiting the isobaric region starts at the 'inviscid' separation point with an angle, inducing what is called *separation shock* (C_2). This shock intersects incident shock (C_1) at point **H**, where (C_1) undergoes a deflection (i.e., refraction) to become shock (C_4); separation shock (C_2) similarly becomes shock (C_3). Shock (C_4) meets the isobaric frontier at point **I**. There, to ensure continuity

Figure 2.39. The inviscid flow pattern associated with shock reflection and separation.

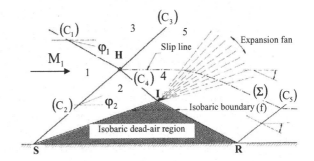

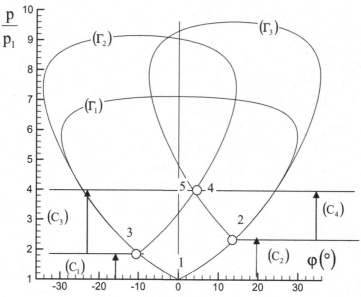

Figure 2.40. Shock-pattern interpretation in the shock-polar diagram. upstream Mach number 2.5. Separation shock deflection 14°; incident shock deflection –10°.

of pressure, the pressure rise produced by (C_4) must be compensated for by a centred expansion emanating from **I**. This expansion provokes a deflection of isobaric frontier (f), which is turned towards the wall such that the impact is at the 'inviscid' reattachment point **R**. There, a new deflection occurs with the formation of reattachment shock (C_5). In addition, a slip line emanates from intersection point **H**. For this case, the two-shock system of the perfect-fluid oblique-shock reflection – which comprises simply an incident plus reflected shock – is replaced by a pattern involving five shock waves.

The pattern, made by shocks (C_1), (C_2), (C_3), and (C_4), is a Type I shock-shock interference according to Edney's classification (see Section 2.2.3), which can be understood best by considering the shock-polar representation shown in Fig. 2.40. The figure corresponds to an incoming uniform flow of Mach number 2.5. The separation-shock deflection is given by a turbulent-separation criterion, which fixes the Mach number behind the separation shock (see Section 2.7.2). This angle, which is approximately 14 degrees, does not depend on the intensity of the shock that caused the separation. Polar (Γ_1) is associated with upstream uniform state 1 and represents any shock forming in state 1, specifically incident shock (C_1). The image of downstream flow 3 is point 3 on (Γ_1), and the deflection imparted by (C_1) is negative (i.e., the velocity is deflected towards the wall). Separation shock (C_2) is also represented by (Γ_1) because the upstream state is 1. The image of downstream flow 2 is at point 2 on (Γ_1), and the deflection $\Delta\varphi_2$ is upward. The situation downstream of **H** is at the intersection of polars (Γ_3) and (Γ_2) attached to states 3 and 2, respectively. Their intersection is the image of two states 4 and 5 with the same pressure $(p_4 = p_5)$ and the same direction $(\varphi_4 = \varphi_5)$; hence, they are compatible with the Rankine-Hugoniot equations. The set of successive shocks $(C_1) + (C_3)$ is different from the set $(C_2) + (C_4)$; therefore, the flows that traversed each set have undergone

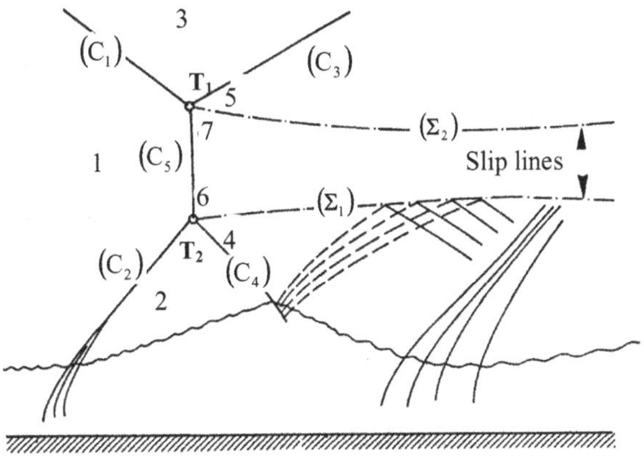

Figure 2.41. Shock reflection with singular shock intersection or Mach phenomenon: Schematic view of the situation in the physical plane.

different entropy increases. Thus, slip line (Σ) is formed separating flows 4 and 5, which have different velocities, densities, temperatures, and Mach numbers (but identical pressures). In a real flow, a shear layer develops along (Σ) thereby ensuring a continuous variation of the flow properties between states 4 and 5. The fluid that flows along a streamline passing under point **H** and belonging to the inviscid part of the field crosses three shock waves: (C_2) and (C_4) plus reattachment shock (C_5). Thus, the final entropy level is lower than for the entirely inviscid case, in which the fluid would have traversed only the incident plus reflected shocks. This is also the case for an interaction without separation, which is close to the inviscid model at some distance from the wall. The conclusion is that entropy production through the shock system is smaller in a shock-induced, separated interaction than in an interaction without separation or in the limiting case of the inviscid model. This result is exploited by control techniques that aim to reduce wing drag or efficiency losses in internal flows.

If for a fixed upstream Mach number the strength of the incident shock is increased, a situation is reached in which the two polars (Γ_2) and (Γ_3) do not intersect. Then, an Edney Type II interference occurs at the crossing of shocks (C_1) and (C_2) and a near-normal shock, or a Mach stem, is formed between the two triple points T_1 and T_2, as shown in Fig. 2.41. The singular shock interaction in Fig. 2.42 is for an upstream Mach number of 2.5; the separation shock deflection of 14 degrees results from the separation criterion. The Mach reflection is obtained by increasing the incident-shock deflection. Downstream states 4 and 6 located at the intersection of polars (Γ_1) and (Γ_2) are separated in the physical plane by slip line (Σ_1), whereas downstream states 5 and 7 at the intersection of (Γ_1) and (Γ_3) are separated by slip line (Σ_2). The subsonic channel downstream of Mach stem (C_5) is accelerated under the influence of the contiguous supersonic flows such that a sonic throat appears after which the flow is supersonic (Fig. 2.41). In this case, the interaction produces a completely different outer-flow structure with the formation of a complex shock pattern replacing the simple, purely inviscid-flow solution. The occurrence of a Mach phenomenon can be detrimental in hypersonic air-intakes because the stagnation-pressure loss behind the normal shock is much greater than behind the oblique shocks.

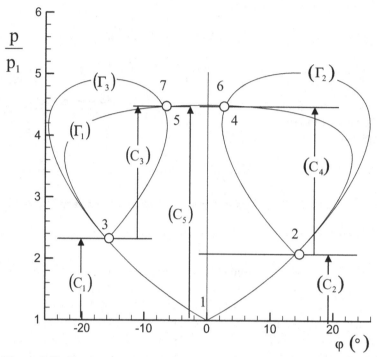

Figure 2.42. Shock reflection with singular shock intersection or Mach phenomenon: Situation in the shock-polar plane ($M_1 = 2$, $\varphi_1 = -16°$, $\varphi_2 = 14°$).

2.6.2 Ramp-Induced Separation

The case of separation induced by a ramp is illustrated by the flow visualisation in Fig. 2.43 and the sketch in Fig. 2.44. If deflection α imparted by the ramp induces a shock the strength of which exceeds the capacity of the boundary layer to withstand the compression, separation occurs at a point **S** located upstream of the ramp apex. As in the shock reflection, separation shock (C_1) is formed due to the focussing of the compression waves induced by the separation process. Downstream of **S**, the fluid in the boundary layer near the wall recirculates, and the bubble topology is identical to that of the previous case. Reattachment at **R** on the ramp gives rise to reattachment shock (C_2), which is less inclined than the separation shock because of the change in flow direction and because the Mach number downstream of the separation shock is lower.

The equivalent inviscid representation of this case is presented in Fig. 2.45. A simpler two-shock system forms with a separation shock emanating from the inviscid

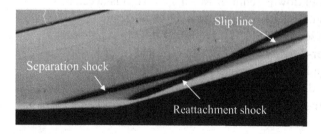

Figure 2.43. Schlieren visualisation of a ramp flow with laminar separation at Mach 5 (ramp angle 15°, Reynolds number $Re_L = 1.5 \times 10^5$ (ONERA document).

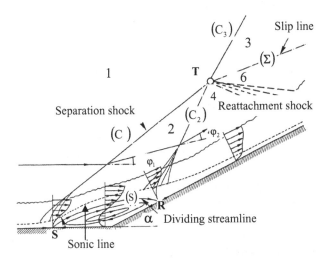

Figure 2.44. Sketch of the flow induced by a ramp with separation.

separation point and a reattachment shock from the inviscid reattachment point. If the Mach number is high enough, the two converging shocks intersect at a small distance from the wall, giving a Type VI interference, as illustrated in Fig. 2.45. The situation at point **T** where the two shocks (C_1) and (C_2) meet is represented in the shock-polar diagrams in Fig. 2.46. The solution is made of shock (C_3), which is seen as the shock induced by the ramp at a great distance from the wall and by states 3 and 6, at the same pressure, and separated in the physical plane by slip line (Σ). Polars (Γ_1) and (Γ_2) are distinct; therefore, an intermediate state 4 must be introduced between states 2 and 6. In the case shown in Fig. 2.44, polar (Γ_2) is above polar (Γ_1), and downstream-compatible states 6 and 3 are found at the intersection of (Γ_1) and polar (Δ_1), representing an isentropic expansion from state 4. In this case, a centred expansion emanates from triple point **T** and propagates in the direction of the wall, on which it is reflected as a new expansion wave. This situation is the most common; however, at low supersonic Mach numbers (the limit is around 2; see Section 2.2.3), the relative position of polars (Γ_1) and (Γ_2) change, the first one now above the other in the region of interest. In this case, compatibility is achieved through a fourth shock (C_4), which is very weak. It emanates from **T** and propagates towards the wall.

Figure 2.45. The inviscid flow pattern associated with a ramp flow with separation.

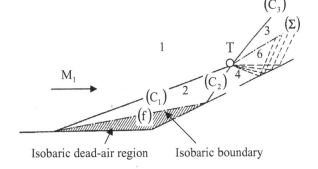

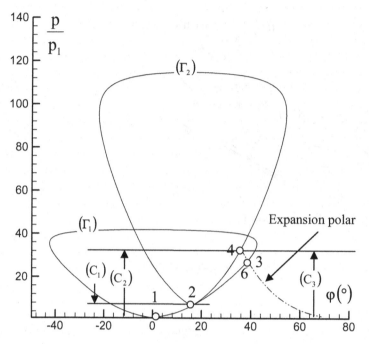

a – shock interference with expansion starting from the triple point **T**.

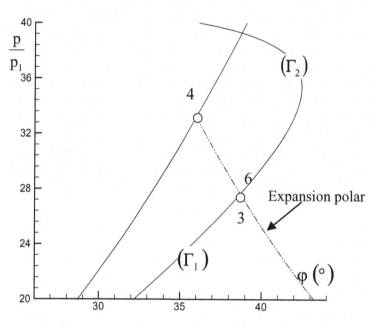

b – close up of the polars in the triple point region

Figure 2.46. The shock-polar diagrams corresponding to the flow shown in Figure 2.45 ($M_1 = 6$, $\varphi_1 = 16°$, $\varphi_2 = 36°$).

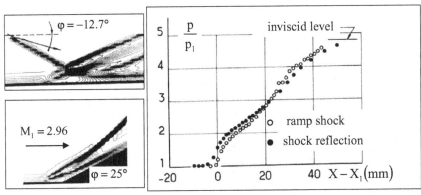

Figure 2.47. Shock- and ramp-induced separation of a turbulent boundary layer. Navier-Stokes calculations (Shang et al. [12]).

Concerning the boundary-layer behaviour, there are no basic differences among the separations induced by an incident shock, a ramp, and a normal shock. In all instances, the boundary layer responds in the same way to a given pressure rise regardless of the cause. The wall-pressure distribution for a separated ramp flow is nearly coincident with that produced by a shock reflection if they are both subjected to the same overall pressure rise. This behaviour is illustrated by the laminar-flow calculations for an identical upstream Mach number shown in Fig. 2.47.

The wall-pressure distributions are similar for a ramp flow and a shock reflection, provided that the total deflection undergone by the flow is the same in the two cases. In this example, it is 25 degrees for the ramp and 2×12.27 degrees (i.e., 24.54 degrees) for the shock reflection. A similar observation can be made for transonic interactions for which the wall-pressure distributions and the boundary-layer properties behave similarly. From the perspective of the boundary layer, what is experienced is a strong adverse pressure gradient that induces separation. Whether reattachment subsequently occurs depends on the flow circumstances downstream of the interaction. For example, in the case of an overexpanded nozzle, reattachment most often does not occur. Thus, any specific properties must be looked at in light of the separation process and not with regard to the complete separated-flow structure, which will depend on a coupling with the complete outer field. In some respects, shock-induced separation can be viewed as the compressible facet of the wider category of flows involving boundary-layer separation but occurring at supersonic or transonic Mach numbers, with the shock wave simply being an associated secondary artefact. However, what makes the phenomenon specific and different among configurations is the shock pattern associated with the interaction; hence, this merits special attention.

2.6.3 Normal Shock-Induced Separation or Transonic Separation

The schlieren picture in Fig. 2.48 shows a transonic interaction with shock-induced separation taking place in a channel. In addition to the near-normal shock, the main flowfield features are the oblique shock induced by separation, the shear layer issuing from the separation point, and the development of turbulent eddies that survive

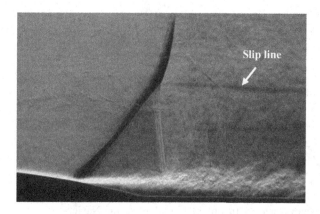

Figure 2.48. Schlieren visualisation of a transonic interaction with separation (ONERA document).

into the far-downstream part of the interaction. The sketch of this flow in Fig. 2.49 shows separation shock (C_1), which is followed by a supersonic region terminated by near-normal shock (C_2). In contrast to the previous examples, it is not possible to identify a shock linked to reattachment. The two shocks that do exist meet at a point from which slip line (Σ) starts. For the outer supersonic stream, the upstream part of the separated region behaves like a viscous wedge producing the oblique shock after which the flow is still supersonic. Shock (C_1) meets the normal shock (C_3) – which causes the separation – at point **T** and results in a pattern similar to the Edney Type VI interference. The situation at point **T** in the shock-polar diagram is shown in Fig. 2.50. Compatibility conditions downstream of **T** entail the formation of 'trailing' shock (C_2) represented on shock polar (Γ_2) attached to state 2. The two compatible states 3 and 4 are separated by slip line (Σ). At transonic velocity, where the upstream Mach number $M_0 \approx 1.4$–1.5, the flow downstream of (C_1) is weakly supersonic $(M_2 \approx 1.20$–$1.10)$. Shock (C_2) satisfies the strong solution to the oblique-shock equations but, in fact, the intensity is very weak. Downstream of (C_2), the flow may be subsonic or still supersonic with a Mach number close to unity. The remainder of the compression is nearly isentropic with a continuous transition to subsonic

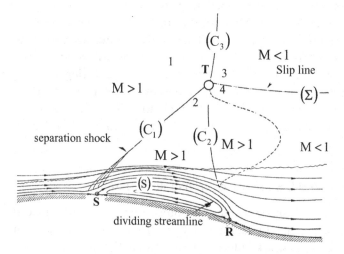

Figure 2.49. Sketch of the flow induced by a normal shock interaction.

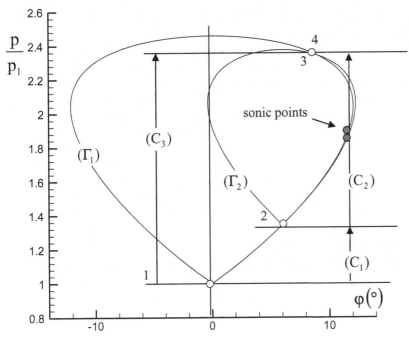

a – shock intersection at the triple point with subsonic downstream flow

$(M_1 = 1.5 \; \varphi_1 = 6°)$

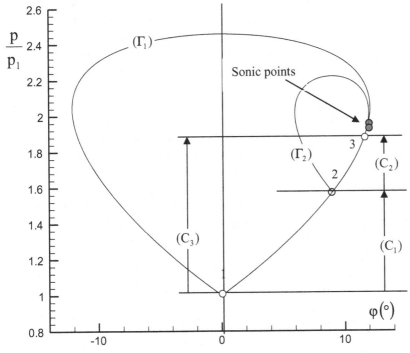

b – shock intersection at the triple point with supersonic downstream flow

$(M_1 = 1.5 \; \varphi_1 = 9°)$

Figure 2.50. The shock-polar diagram of shock-shock interference induced by separation in transonic flow: the lambda shock pattern.

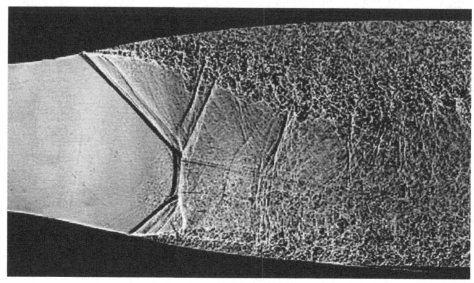

Figure 2.51. Nonsymmetrical shock-induced separation in a planar supersonic nozzle. Short exposure time Schlieren photograph (ONERA document).

velocities, although shocklets often are observed in this region, which is sometimes called the *supersonic tongue*. The extent of the supersonic domain behind (C_2) depends on local and downstream boundary conditions. This structure, typical of shock-induced separation in transonic flow, is termed a *lambda shock pattern*. The shape of the downstream part of the isobaric-separated region is more conjectural because the reattachment process depends on the coupling with an outer flow, which primarily is subsonic.

Strong-normal shock interactions are encountered in internal flows such as air-intakes and missile/space-launcher nozzles operating in overexpending conditions. Such a flow is shown in Fig. 2.51 by a short-exposure-time shadowgraph of the flow in a planar supersonic nozzle. For reasons not fully understood, the separated flow adopts an asymmetric structure; in this case, the separation on the upper wall takes place earlier than on the lower wall, although the geometry is symmetric (this point is not discussed here). The separations taking place on each wall are supersonic in nature, with the Mach number at the interaction origins close to 1.6. In the present situation, the intersection of the two separation oblique shocks is singular, leading to a Mach reflection. The large separation on the upper wall results in a lambda shock pattern, with the flow behind the trailing shock (C_3) still supersonic (Fig. 2.52). Separation on the lower wall causes a smaller lambda pattern associated with a reduced separated region. Here, the intersection of separation shocks (C_1) and (C_2) is Type II with the existence of a Mach reflection made of near-normal shock (C_5) with the two triple points $\mathbf{T}_1$ and $\mathbf{T}_2$. Slip lines (Σ_1) and (Σ_2) emanating from $\mathbf{T}_1$ and $\mathbf{T}_2$ form a fluidic subsonic channel between two supersonic streams. Due to conditions imposed by the adjacent supersonic flows, the flow in this channel accelerates until it reaches the sonic speed at a throat (minimum of area). Thereafter, the cross section of the fluidic channel increases and the expansion continues as a supersonic flow. The penetration of (C_3) into the separated shear layer generates a reflected

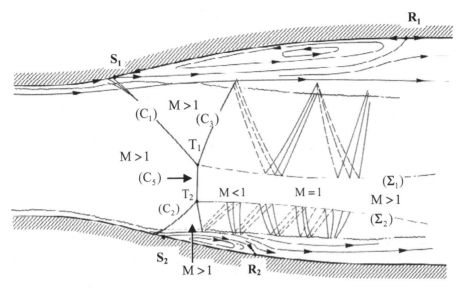

Figure 2.52. Nonsymmetrical shock-induced separation in a planar supersonic nozzle. Schematic representation of flowfield.

expansion wave, which is reflected by slip line (Σ_1) into a wave made of converging compression waves. In turn, these waves are reflected by the nearly isobaric separated region as expansion waves and the pattern is repeated over a certain distance several times. A similar pattern can be observed near the lower wall. The present configuration is an example that demonstrates the coupling between the separated regions and the inviscid part of the flow that generates complex structures because of the confinement of the flow. Channel flows of this type can occur in supersonic diffusers, compressor cascades, and propulsive nozzles.

2.7 Separation in Supersonic-Flow and Free-Interaction Processes

2.7.1 The Free-Interaction Theory

Supersonic separation is a local self-induced free-interaction process between the boundary layer and the outer inviscid stream. Stewartson and Williams [8] published an elaborate theory for this in 1968 that called on an asymptotic expansion technique. In the 1950s, Chapman [13] published a simplified analysis describing the interaction, which is worth a reminder because of its major importance for the physical understanding of separation in supersonic flows. The first equation used is the boundary-layer momentum equation, as follows:

$$\rho u \frac{\partial u}{\partial x} + \rho v \frac{\partial u}{\partial y} = -\frac{dp}{dx} + \frac{\partial \tau}{\partial y},$$

where u and v are the velocity components along x and y, respectively; x and y are the streamwise and normal coordinates with respect to the surface; p is the pressure; and τ is the shear stress (including the turbulent contribution). The previous equation written at the wall ($y = u = v = 0$) gives the following exact relationship between the streamwise pressure gradient and the normal shear-stress gradient at the wall:

$$\frac{dp}{dx} = \left(\frac{\partial \tau}{\partial y}\right)_w$$

An x-wise integration of the previous equation from the interaction origin x_0 gives:

$$p(x) - p(x_0) = \int_{x_0}^{x} \left(\frac{\partial \tau}{\partial y}\right)_w dx$$

The physical variables are rendered dimensionless by introducing the following appropriate scales:

- the dynamic pressure: q_o for the pressure
- the wall shear stress at the interaction onset: τ_{w_0} for the shear stress
- the boundary-layer displacement thickness at $x_0 : \delta_0^*$ for the ordinate
- a length L typical of the interaction streamwise extent for the abscissa

This makes the change of variables:

$$\bar{\tau} = \frac{\tau}{\tau_{w_0}}, \quad \bar{y} = \frac{y}{\delta_0^*}, \quad \bar{x} = \frac{x - x_0}{L}$$

and introduces the skin-friction coefficient at $x_0 : C_{f_0} = \frac{\tau_{w_0}}{q_0}$ and defines the dimensionless function:

$$f_1(\bar{x}) = \int_{\bar{x}_0}^{\bar{x}} \left(\frac{\partial \bar{\tau}}{\partial \bar{y}}\right)_w d\bar{x}$$

so we arrive at a first equation representing the boundary-layer response to the pressure rise:

$$\frac{p(\bar{x}) - p(\bar{x}_0)}{q_0} = C_{f_0} \frac{L}{\delta_0^*} f_1(\bar{x}) \tag{2.1}$$

The second equation, which links the boundary-layer thickening and the pressure variation in the contiguous inviscid flow, is obtained from the relationship between pressure and flow direction in a supersonic simple wave flow, as follows:

$$\frac{\sqrt{M^2 - 1}}{\gamma M^2} \frac{dp}{p} - d\varphi = 0$$

which, in its linearised form, is:

$$\frac{\sqrt{M_0^2 - 1}}{\gamma M_0^2} \frac{\Delta p}{p_0} = \Delta\varphi$$

The flow deflection is determined through the displacement concept according to which the outer inviscid flow streams along an effective surface, which is the body shape augmented by the boundary-layer displacement thickness. This gives:

$$\Delta\varphi = \varphi = \tan^{-1}\left(\frac{d\delta^*}{dx}\right) \simeq \frac{d\delta^*}{dx}$$

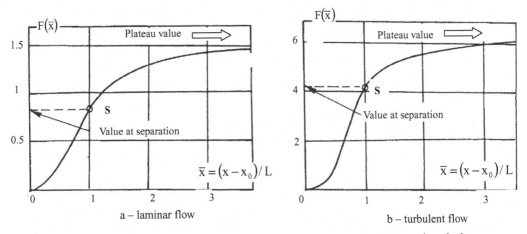

Figure 2.53. Free-interaction theory. Correlation functions for the pressure rise during separation.

By introducing scaled quantities and the dimensionless function:

$$f_2(\bar{x}) = \frac{d\bar{\delta}^*}{d\bar{x}},$$

a second independent equation is obtained after the following simple calculations:

$$\frac{p(\bar{x}) - p(\bar{x}_0)}{q_0} = \frac{2}{\sqrt{M_0^2 - 1}} \frac{\delta_0^*}{L} f_2(\bar{x}) \qquad (2.2)$$

From the product of equations (2.1) and (2.2) and introducing the correlation function:

$$F(\bar{x}) = \sqrt{f_1(\bar{x})\, f_2(\bar{x})},$$

we obtain the following expression for the pressure rise during the interaction:

$$\frac{p - p_0}{q_0} = F(\bar{x})\sqrt{\frac{2C_{f_0}}{(M_0^2 - 1)^{1/2}}}$$

The dimensionless function $F(\bar{x})$ exhibits an initial steep rise from the interaction onset to the separation-point location (Fig. 2.53), followed by a more progressive rise to reach an asymptotic-plateau value corresponding to the isobaric-separated region.

The free-interaction theory is confirmed by experiment; the pressure rise at separation for various conditions is correlated by two functions, for the laminar and turbulent cases (see Fig. 2.53). The following table gives the values of $F(\bar{x})$ corresponding to the separation point and the plateau value reached in an extended separation:

	F (separation point)	F (plateau value)
Laminar flow	0.8	1.5
Turbulent flow	4.2	6

As shown, the values of $F(\bar{x})$ are much higher in turbulent than in laminar flow, which indicates the important susceptibility of laminar flows to separate. This analysis establishes that the pressure rise undergone by the boundary layer is of the following form:

$$\frac{p - p_0}{q_0} \propto (C_{f_0})^{1/2} \left(M_0^2 - 1\right)^{-1/4}$$

In hypersonic flows $(M_0^2 - 1)^{-1/4} \approx M_0^{-1/2}$ and because $q_0 = \frac{\gamma}{2} p_0 M_0^2$, we have:

$$\frac{p - p_0}{p_0} \propto (C_{f_0})^{1/2} M_0^{3/2}$$

It is usual to take as the streamwise scale the distance $L = x_s - x_0$ between the interaction origin and the separation-point location. Dividing the equations [i.e., (2.1)/(2.2)] and considering the separation-point location where $\bar{x} = 1$, then:

$$1 = \frac{C_{f_0}\sqrt{M_0^2 - 1}}{2} \frac{L^2}{(\delta_0^*)^2} \frac{f_1(1)}{f_2(1)}$$

From this, we obtain a relationship for the interaction extent L of the following form:

$$L \propto \delta_0^* (C_{f_0})^{-1/2} (M_0^2 - 1)^{-1/4}$$

This equation suggests that the pressure rise Δp_S at separation and the extent of the first part of the interaction depend only on the flow properties at the interaction onset and not on the downstream conditions, particularly the shock intensity. During the first part of the interaction, the flow is a consequence of the reciprocal and mutual influence, or coupling, between the local boundary layer and the inviscid contiguous stream, not the further development of the interaction – hence, the portrayal of this phenomenon is as a *free-interaction* or *free-separation process*. This important result, well verified by experiment, explains many features of interactions with shock-induced separation.

A major consequence of the interaction is to split the pressure jump Δp_T imparted by the shock into an initial compression Δp_S at separation, with associated shock (C_1), and a second compression Δp_R at reattachment, with the overall pressure rise such that $\Delta p_S + \Delta p_R = \Delta p_T$ (Fig. 2.54). The extent of the separated region is dictated by the ability of the shear layer that has its origin at separation point **S** to overcome the pressure rise at reattachment. This ability is a function of the momentum available at the start of the reattachment process. In this analysis, the pressure rise up to separation does not depend on downstream conditions. Thus, an increase in the overall pressure rise imparted to the boundary layer or a rise in the incident-shock strength requires a greater pressure rise at reattachment. This can be achieved only by an increase of the maximum velocity $(U_s)_{max}$ attained on the dividing streamline; hence, an increase in the shear-layer length is needed to achieve a greater transfer of momentum from the outer flow. The length of the separated region therefore grows in proportion to the pressure rise at reattachment and the separation point moves in the upstream direction.

Because the relative importance of the viscous forces decreases with increasing Reynolds numbers, the free-interaction theory predicts an increase in the

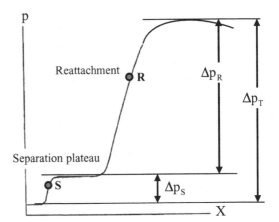

Figure 2.54. The splitting of the compression in an interaction with separation and reattachment.

interaction extent and a decrease in the overall pressure rise as the Reynolds number increases. A consequence for both laminar and turbulent regimes is that a stronger shock is required to separate the boundary layer at lower Reynolds numbers than at higher values. This behaviour is confirmed by experiment for laminar and turbulent flows as long as the local Reynolds number Re_δ is less than about 10^5. Above this value, the tendency is reversed and the interaction domain contracts when the Reynolds number is increased and the turbulent boundary layer becomes more resistant to separation. This conflict is resolved when we recall that during the interaction process, the boundary-layer behaviour is a consequence of the action of both inertia and viscous forces, as explained in Section 2.4.2. The free-interaction theory favours the viscous forces because it involves only the skin-friction coefficient. The predominance of viscosity is apparent in laminar flows or low-Reynolds-number turbulent flows; however, at a high Reynolds number, the momentum transported by the boundary layer becomes the dominant factor in the interaction with the shock. Because the boundary-layer profile becomes fuller when the Reynolds number increases, the resistance to the retarding influence of the shock increases.

2.7.2 Incipient Shock-Induced Separation in Turbulent Flow

Determining the onset of shock-induced separation is a major concern because for practical applications, it is important to know the maximum shock intensity that a boundary layer can withstand without separating. This limit most often is defined in the plane of two variables: the ramp angle (or equivalent angle leading to the same total pressure rise for shock reflection) and the Reynolds number. A different curve corresponding to each value of the upstream Mach number M_0 is required because the shock intensity that results in separation increases with Mach number for a fixed Reynolds number (Délery and Marvin [14]). As a consequence, for a given upstream Mach number, the shock strength required to separate a turbulent boundary layer first decreases as the Reynolds number increases; however, above $Re_\delta \approx 10^5$, this trend is reversed. The limit of shock strength increases with the Reynolds number but the dependence becomes very weak and nearly nonexistent.

Several criteria predict incipient shock-induced separation in supersonic flow; the most widely used are as follows:

1. A criterion deduced from the free-interaction theory, which states that the pressure p_1 behind separation is such that:

$$\frac{p_1}{p_0} = 1 + 6\frac{\gamma}{2}M_0^2\sqrt{\frac{2C_{f_0}}{\left(M_0^2 - 1\right)^{1/2}}}$$

This criterion considers the influence of the Reynolds number through the skin-friction coefficient.

2. The criterion proposed by Zhukoski [15], which is simply:

$$\frac{p_1}{p_0} = 1 + 0.5M_0$$

where there is no Reynolds number influence.

3. For separation in rocket-engine nozzles, the Schmucker [16] criterion is used frequently:

$$\frac{p_1}{p_0} = (1.88M_0 - 1)^{0.64}$$

For reasons already discussed, special attention is focussed on transonic interaction. The limit at which normal shock-induced separation can occur is of great importance in airfoil design and in the mechanism of unsteady phenomena such as buffeting and air-intake buzz (see Section 2.11). Incipient separation in turbulent flow occurs when the Mach number upstream of the shock reaches a value close to 1.3. This is nearly independent of the boundary-layer incompressible shape parameter H_i, which appears to contradict the previous discussion. In reality, because a higher H_i entails a spreading of the interaction domain and, hence, a weakening of the adverse pressure gradient, separation is postponed compared with low H_i situations in which the pressure gradient is more intense. The two opposite tendencies compensate such that the limit for shock-induced separation essentially depends only on the upstream Mach number.

2.8 Transitional SBLIs

For a hypersonic vehicle flying at high altitude and therefore at very low ambient density, the combination of a high Mach number and a low Reynolds number can produce entirely laminar SBLIs. However, with the decrease in altitude during reentry, the Reynolds number will rise so that transition – which first occurs well downstream – encroaches the interaction region. There are other circumstances of practical interest in which the Reynolds number is such that the SBLI is transitional in the sense that transition occurs somewhere during the interaction (e.g., on compressor blades or laminar airfoils).

Transition and SBLI is a complex double-faceted problem: On the one hand, the shock acts as a perturbation triggering a premature boundary-layer transition that otherwise would take place farther downstream of the shock origin or impingement point. On the other hand, when transition occurs in the interaction domain,

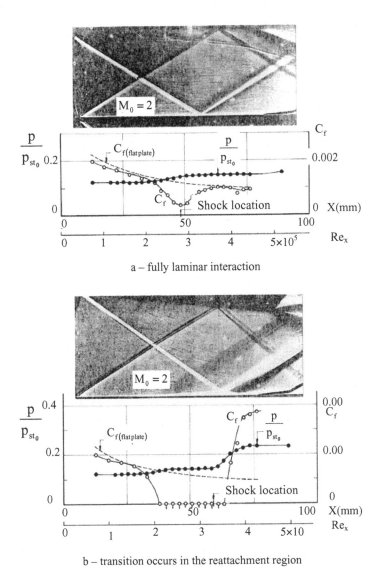

a – fully laminar interaction

b – transition occurs in the reattachment region

Figure 2.55. Transitional interaction caused by a reflecting shock (Hakkinen et al. [17]).

it profoundly affects the phenomenon by promoting momentum exchanges through the development of instabilities degenerating into turbulence. Despite its practical significance, the interaction between shock waves and transition is a delicate question that is far from fully elucidated. Most of our understanding of the effect is based on experimental evidence.

The progressive development of the transitional regime with increasing Reynolds number is illustrated by first considering a shock reflection that is fully laminar, with boundary-layer transition occurring well downstream of the interaction region (Fig. 2.55a). With an increase in the Reynolds number or shock strength, this moves upstream into the interaction domain – more precisely, the vicinity of reattachment (Fig. 2.55b). Meanwhile, the extent of the separated region has increased as well because the incoming boundary layer, still being laminar, is

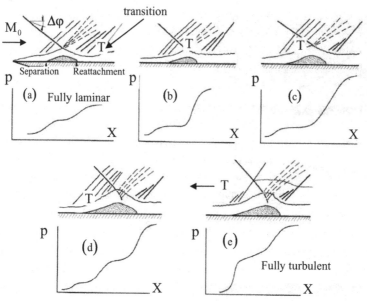

Figure 2.56. Schematic representation of transition motion on a shock-separated flow (Gadd et al. [18]).

less resistant to separation at the higher Reynolds number (see Section 2.7.1). As transition moves farther forward and takes place approximately at the shock-impact point, the flow can accommodate a steeper pressure rise at reattachment. The consequences of this forward movement are twofold: The streamwise scale of the reattachment domain shrinks and the associated pressure rise, along with the accompanying pressure gradient, is amplified. Conversely, the overall extent of the separated flow increases.

The situations encountered as this process takes place are illustrated in Fig. 2.56, in which the effect of a Reynolds-number increase for a reflecting shock of constant intensity is shown. Starting from a fully laminar interaction (a), a rise in the Reynolds number Re_L provokes a displacement of the transition in the upstream direction until it reaches reattachment region (b). The peak pressure and heat transfer then become much higher than in the fully laminar interaction. At the same time, a reversal in the Reynolds-number dependence occurs and the separation extent decreases with increasing Re_L. Transition first stays in the reattachment region until a limit value of Re_L is reached, beyond which it suddenly moves to separation region (d). With further increase, transition occurs upstream of the interaction, profoundly affecting flow structure (e). For instance, the separated zone may disappear because the shock strength is no longer sufficient to separate the boundary layer. During the transitional phase of the interaction, the peak heat transfer at reattachment in a hypersonic flow can be higher than in the fully turbulent case (c). Such an overshoot, also observed during boundary-layer transition over a flat plate, is due to the existence of large and well-organised structures, denoting a pre-turbulence state and enhancing transfer mechanisms. Such flow structures then are broken into smaller eddies when the turbulent regime is established. Most of the observed so-called laminar hypersonic interactions are, in fact, transitional because maintaining a laminar regime throughout the interaction domain is difficult due to the extreme sensitivity of the separated shear layer to disturbances. This transition produces a mixed inter-action in which separation has the feature of a laminar flow (i.e., decrease of the heat transfer), whereas reattachment exhibits turbulent behaviour (i.e., higher pressure and heat-transfer peaks).

2.9 Specific Features of Hypersonic Interactions

2.9.1 Shock Pattern and Flowfield Organisation

Although Chapter 6 is devoted to a detailed discussion of hypersonic SBLIs, we briefly consider them here in the context of the general review of the physical char-acteristics of the interactions. The high specific-enthalpy level, typical of hypersonic conditions, has three direct consequences for SBLIs, as follows:

1. When the wall temperature is well below the outer-stream stagnation temper-ature, a cold-wall situation arises that may significantly affect the interaction properties.
2. Heat-transfer processes take on dramatic importance, especially in separated flows where the impact to the surface of reattaching shear layers leads to very high heat fluxes.
3. Real-gas effects that result from the intense heating produced by the shocks modify the thermodynamic and transport properties of the fluid (most often, air) in a way that may influence the interaction.

We saw previously that if the approaching boundary layer is turbulent, the sonic point is close to the wall. If, in addition, the incident Mach number is increased, this point moves even closer; therefore, for hypersonic flow, the boundary-layer subsonic channel is exceedingly thin (see Fig. 2.16). The consequence is that the upstream propagation distance for any disturbance is very short. Thus, the com-pression waves formed by the flow deflection in the vicinity of the separation and reattachment points coalesce very rapidly to produce shock waves from so deep within the boundary layer that they seem to originate almost from the wall. Also, because these waves propagate at a small angle with respect to the streamlines, it is possible for much of the shock pattern to be embedded within the boundary layer, as shown in Fig. 2.57. The intersection of the separation and reattachment shocks usu-ally produces a Type VI shock-shock interference pattern with a centred expansion emanating from triple point **T** (see Section 2.6.2). The signature of this expansion on the nearby wall is denoted by a sharp pressure decrease following the rise at reat-tachment. This is shown in Fig. 2.58, in which the pressure distributions on the wall are plotted for a ramp-induced interaction for increasing value of the ramp angle. The Type VI interference also produces a jet of high-velocity fluid that moves in the direction of the surface and enhances the heat flux downstream of reattachment.

The wall-pressure distributions plotted in Fig. 2.58 illustrate the behaviour explained in Section 2.7.1. When the wedge angle is large enough to induce sepa-ration, the pressure rise at separation is considered independent of the value of this angle. When the intensity of the shock provoked by the wedge increases, the effect is a displacement of the separation point in the upstream direction. At the same time,

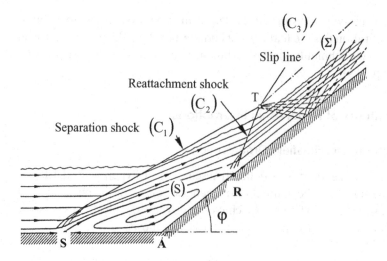

Figure 2.57. A sketch of a ramp-induced separation in a high Mach number flow.

the pressure rise at reattachment increases to reach the downstream level resulting from the wedge deflection (after an overshoot, as explained previously).

2.9.2 Wall-Temperature Effect

Experiments performed at a high Mach number on a laminar ramp-induced interaction show that wall cooling ($T_w/T_r < 1$) provokes a contraction of the interaction domain compared to the adiabatic case (Fig. 2.59) (Lewis et al. [20]). The same tendency was observed in turbulent interactions for which wall cooling reduces the separation distance or the upstream-interaction length (Spaid and Frishett [21]). Results for SBLIs on a heated wall ($T_w/T_r > 1$) are scarce because this situation is encountered less frequently. (It can occur at low altitude on a hypersonic

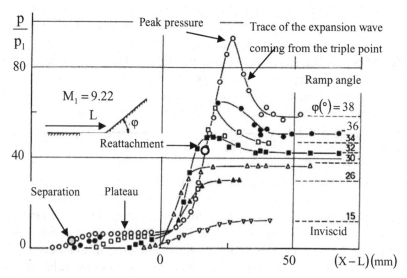

Figure 2.58. Wall-pressure distribution in a hypersonic ramp flow for increasing ramp angle (Elfstrom [19]).

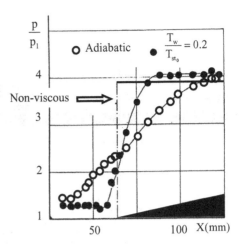

Figure 2.59. Wall-temperature effect on a laminar high Mach number interaction (Lewis et al. [20]).

vehicle releasing the heat stored during the earlier high-speed part of the reentry trajectory.) The results show that wall heating results in a lengthening of the interaction domain, which confirms *a contrario* observations made on a cooled wall (Délery [22]). There is no unique explanation of the decrease in L_s/δ_0 with the wall temperature (or its increase when T_w is raised). This tendency, however, is in agreement with the free-interaction theory (see Section 2.7.1) because a lowering of the wall temperature provokes an increase in the skin-friction coefficient and a decrease in the boundary-layer displacement thickness (i.e., the density level in the boundary layer is raised): hence, a contraction of L_s. However, the observed dependence on wall temperature is beyond the scope of what can be predicted by the free-interaction theory. The contraction of the interaction domain is also a consequence of the thinning of the boundary-layer subsonic part (see Fig. 2.16) because the Mach number in the inner part of the boundary layer is greater due to lower sound speed. The same argument holds in reverse for a heated wall because then the sound velocity is increased and the Mach number is reduced.

The wall temperature also has a more subtle influence through its effect on the laminar or turbulent state of the incoming boundary layer. Surface-pressure distributions taken from ramp-flow experiments show that compared to the adiabatic case, the separated region is more extensive when the model is cooled, with the pressure plateau forming well ahead of the ramp origin (Délery and Coët [23]). At the same time, the compression on the ramp is more spread out. At first sight, these tendencies appear to contradict the previous conclusions. The present behaviour must be attributed to the fact that in these flows, laminar-to-turbulent transition has occurred within the interaction domain. Because wall cooling tends to delay transition, the boundary layer developing on the cooled model is 'more laminar' than on an adiabatic model. For this reason, in the transitional flow, the separated zone is more extensive when the boundary layer is cooled.

2.9.3 Wall-Heat Transfer in Hypersonic Interactions

The salient feature of hypersonic interactions is the existence of high heat-transfer rates in the interaction region, especially when there is separation [24]. This

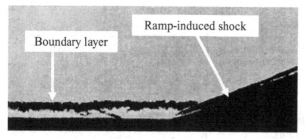

a – interaction without separation

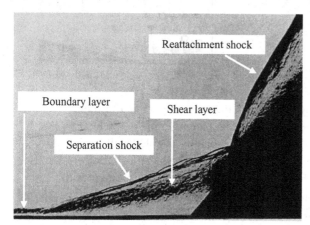

Figure 2.60. Shadowgraphs of a turbulent ramp flow at Mach 9.22 (Elfstrom [19]).

b – interaction with separation

problem, which is crucial for the sizing of the thermal protection in parts of a vehicle where such interactions are likely to occur, has been studied by many investigators for both laminar and turbulent flows.

Shadowgraphs taken for two turbulent hypersonic interactions on a compression ramp at Mach 9.22 are shown in Fig. 2.60. In the lower illustration, the flow is separated, and it is clear how the shear layer that develops on the upper edge of the separated region impacts the ramp at a steep angle. Heat transfer is particularly intense in the vicinity of reattachment, and a sharp rise is associated with the stagnation at **R** of the shear layer developing from the separation point (see Fig. 2.57). In some respects, the situation at **R** is similar to a nose-stagnation point, the difference being that the flow that impinges on the ramp (or control surface) has been compressed through a succession of oblique shocks at separation and reattachment instead of a single normal shock. Consequently, the (average) stagnation pressure is significantly higher, whereas the (average) stagnation temperature is comparable to that of the outer flow so that the transfer processes are more efficient, causing higher levels of heat transfer.

The surface-heat transfer most often is represented in a nondimensional form by the Stanton number defined as:

$$S_t = \frac{q_w}{\rho_\infty U_\infty \left(h_{st_\infty} - h_w \right)},$$

where q_w is the wall-heat transfer (in W/m^2), ρ_∞, U_∞ is the density and velocity of the upstream flow, h_{st_∞} is the upstream-flow stagnation enthalpy, and h_w is the

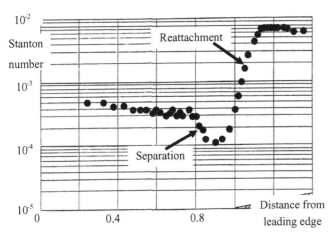

a – laminar incoming boundary layer ($M_\infty = 10$, ramp angle 15°)

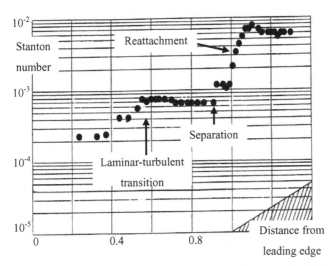

b – turbulent incoming boundary layer ($M_\infty = 5$, ramp angle 35°)

Figure 2.61. Wall-heat-transfer distribution in a ramp-induced interaction (Délery and Coët [23]).

flow enthalpy at the wall. The laminar heat-transfer distribution shown in Fig. 2.61a was measured at $M_\infty = 10$ in a two-dimensional ramp-induced separation (i.e., the Reynolds number computed with the distance L from the model's sharp leading edge to wedge apex $Re_L = 2.3 \times 10^6$). A semi-logarithmic plot is used to emphasise the phenomena in the first part of the interaction. The heat transfer decreases slowly on the upstream part of the cylinder, in agreement with the hypersonic strong/weak viscous-interaction theory (Hayes and Probstein [25]). A more rapid decrease starts at a location coincident with the separation onset. This decrease is typical of shock-induced separation in laminar flows. The heat transfer experiences a minimum in the separated region, then rises during reattachment; the peak value is achieved downstream of the reattachment point.

Results for a turbulent ramp-induced interaction at Mach 5 for $Re_L = 10^7$ are presented in Fig. 2.61b. The initial rise in heat transfer, followed by a slow decay,

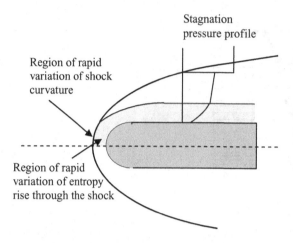

Stagnation
pressure profile

Region of rapid
variation of shock
curvature

Region of rapid
variation of entropy
rise through the shock

Figure 2.62. Entropy layer in a high-Mach-number flow.

is due to the transition from laminar to turbulent flow in the approaching boundary layer ahead of the interaction. A second sharp rise occurs at the separation location; this behaviour is opposite to that observed in laminar flow and is typical of turbulent shock-induced separation. This can be explained by the amplification of turbulence in the vicinity of the separation point and farther downstream. After this point, the flow then 'leaves' the surface and large eddies develop, which promote exchanges between the wall region and the outer high-enthalpy flow, leading to a rise in heat transfer. Farther downstream, the heat transfer sharply increases during reattachment to reach a peak value downstream of the reattachment point.

2.9.4 Entropy-Layer Effect

There are situations in which the outer inviscid flow is rotational and within which an entropy gradient has been created by the flow crossing a curved shock wave generated upstream of the interaction – for example, by the body having a blunt leading edge (Fig. 2.62). The existence of a region of entropy variation makes it difficult to differentiate the boundary with the outer nonviscous flow because (as discussed previously) in a rapid interaction process, the major part of a turbulent boundary layer behaves like a rotational inviscid flow. In this case, the triple-deck decomposition becomes questionable, with only two decks effectively present.

Most hypersonic vehicles have a blunt nose or wings or control surfaces with rounded leading edges to reduce the heat-transfer rate in attachment regions. As a consequence, a detached shock forms in front of these obstacles and produces a region of increased entropy that then envelopes the vehicle. This so-called *entropy layer* is an inviscid-fluid feature of particular significance in high-Mach-number flows because of the rapid variation of the shock angle in the detachment region and because of the large entropy production through strong shock waves. This impact has significant repercussions for SBLIs occurring downstream, which is illustrated by the wall-pressure coefficient and Stanton-number distributions plotted in Fig. 2.63, measured at Mach 10 in a ramp-induced interaction. The stagnation conditions were such that the boundary layer was laminar at the interaction onset. The wall-pressure distribution is affected dramatically by the blunting of the leading edge of the plate

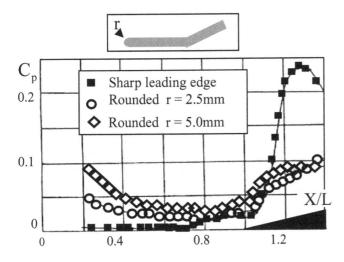

a – wall pressure distribution

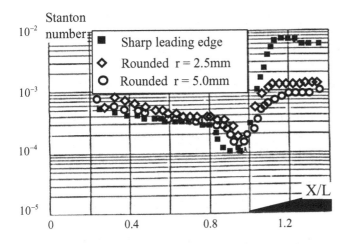

b – wall heat transfer distribution

Figure 2.63. Entropy-layer effect on a ramp-induced separated flow: Mach number 10, ramp angle 15, (Coët et al. [26]).

that supports the ramp. The pressure on the plate upstream part is increased as a consequence of the Mach number decrease, whereas the pressure level on the ramp is reduced. With the blunt leading edge, there is a reduction by a factor of 10 of the peak heat transfer at reattachment. There also is a contraction of the separated zone, the origin of which – denoted by a decrease in heat transfer – moves in the downstream direction when the leading edge is rounded. This entropy-layer effect is linked to the stagnation-pressure loss through the detached shock ahead of the leading edge. The changes in stagnation pressure due to the rounding of the leading edge lead to a decrease in the local Reynolds number and, consequently, greater resistance of the flow to separation in agreement with free-interaction theory (see Section 2.7.1). This compensates the opposite effect of the local-Mach-number reduction. In addition, lowering the Reynolds number contributes to maintaining a laminar regime throughout the interaction domain, whereas the interaction is transitional when the leading edge is sharp.

2.9.5 Real-Gas Effects on SBLI

At hypersonic speeds, the flow over a vehicle exhibits real-gas effects due to dissociation, chemical reactions, and ionisation provoked by the passage of the air through the intense shock waves. These processes involve absorbing or releasing large amounts of heat that modify the thermodynamic equilibrium of the flow. Such effects also are present after the very strong shock waves produced by explosions or in shock tubes. Hence, any SBLIs occur in a gas whose composition and physical properties differ from an equilibrium nondissociated gas. In this situation, real-gas effects are coupled with complex viscous/inviscid interactions. If a calorically perfect gas (i.e., a gas with a constant ratio of specific heats γ throughout the flowfield) is taken as the benchmark, the real-gas effects are experienced in the following two ways:

1. Because the thermodynamic properties are not the same, the structure of the inviscid part of the flow will be modified compared to the constant γ case.
2. Dissociation and chemical phenomena affect the transport properties (i.e., viscosity, heat conduction, and diffusion coefficients), which have repercussions on the viscous part of the flow.

Thus, at high enthalpy, nonequilibrium vibrational excitation, chemical reactions, and ionisation affect the scaling of a separated region through changes in the shock angle and the thickness and profile of the incoming boundary layer.

Compared with the perfect-gas case, there are few experimental results showing the impact of real-gas effects on SBLIs. Basic experiments are difficult to perform because they require high-enthalpy facilities, of which there are few in the world and that are costly to operate. In addition, it is difficult to make parametric investigations in these facilities because operating at different enthalpy levels entails changing other flow parameters (e.g., upstream composition and thermodynamic properties and Mach and Reynolds numbers). As a consequence, it has been difficult to establish a clear picture of the influence of real-gas effects alone on the interactions. To a great extent, information had to be obtained from computations. However, as explained in Chapter 7, these effects are difficult to incorporate into effective CFD solutions and it is possible with any degree of certainty only for fully laminar interactions. Nevertheless, it was established that for a ramp-type flow under the assumption of chemical equilibrium for dissociated air, a smaller separated region forms because of the weaker shock waves. Moreover, the heat-transfer rates are lower because of the reduced temperatures [27]. In the case of an impinging-reflecting shock, results obtained for air using nonequilibrium chemistry suggest that real-gas effects only weakly affect the interaction [28] at low Reynolds numbers. Thus, under these circumstances, the flow can be computed with a degree of accuracy by assuming a constant local value of γ. This assumption may be invalid if the reflection becomes singular (i.e., the occurrence of a Mach reflection phenomenon). Then, an accurate calculation of the adjacent inviscid flow is necessary.

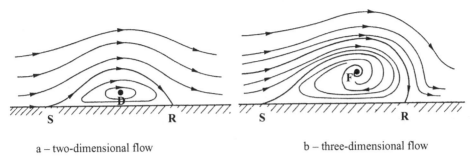

a – two-dimensional flow   b – three-dimensional flow

Figure 2.64. Simple conceptions of separation and separated flows.

Conversely, at a high Reynolds number, chemistry effects lead to substantial differences in the wall-pressure and heat-transfer distributions, with an increase of the heat-transfer levels. Major differences between the noncatalytic and catalytic wall conditions are anticipated because for the latter, the high-energy release that occurs in the separated region strongly affects the interaction. This provokes a dilatation of the separation bubble (similar to the dilatation of the separated region on a heated wall) and a spectacular increase in the wall-heat transfer.

2.10 A Brief Consideration of Three-Dimensional Interacting Flows

2.10.1 Separation in Three-Dimensional Flow

In two dimensions, separated flow is defined by the existence of a bubble containing closed streamlines circling around a common point **D** and bounded by a separation streamline (*S*). This starts at separation point **S** and ends at reattachment point **R**. This description is inadequate for three-dimensional flows, and the closed configuration shown in Fig. 2.64a must be replaced by the open bubble shown in Fig. 2.64b, in which the streamlines are no longer closed curves but rather they spiral around a common point or focus **F** into which they disappear, the flow escaping laterally from **F**. Mass conservation, or topological consistency, then requires that streamline (*S*) issuing from the separation point be distinct from streamline (*A*) stagnating at reattachment point **R**. In three-dimensional flows, the boundary layer can develop crosswise velocity profiles as shown in Fig. 2.65, with the velocity vector turning in the boundary layer from the outside direction to a direction at the wall where it is tangent to what is called the *limit streamline*. Thus, a flow that initially was two-dimensional (e.g., the boundary layer developing on a flat plate) now has the capability of escaping in the spanwise direction when it is confronted by an adverse pressure gradient. The skin friction is now a vector, with the set of skin-friction vectors constituting a field the trajectories of which are the *skin-friction lines*. (It can be demonstrated that the skin-friction line coincides with the limit streamline on the surface.)

For these reasons, it is necessary to reconsider the definition of separation on a three-dimensional obstacle, which can be accomplished by using the *critical point theory* (Legendre [29]). This theory focuses on the skin-friction lines on an object, which are the lines tangent to the local skin friction. More precisely, we examine the

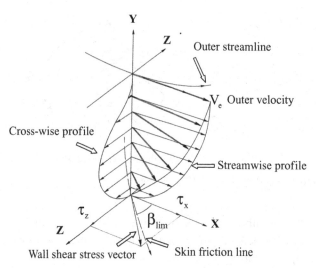

Figure 2.65. Three-dimensional boundary-layer velocity distribution.

behaviour of the skin-friction lines in the vicinity of a point where the skin friction vanishes. This point, termed a *critical point*, can be of the *node*, *saddle-point*, or *focus* type according to the behaviour of the skin-friction lines in its vicinity. These points have the following properties:

- All skin-friction lines pass through a *node* (Fig. 2.66a) that can be of the *separation* or *attachment* type according to flow direction. An attachment node occurs at the origin of the skin-friction lines on an object and a separation node is found where they end. At a node, all of the skin-friction lines except one have a common tangent.
- An *isotropic node*, at which the skin-friction lines have distinct tangents (Fig. 2.66b), corresponds to an axisymmetric attachment/separation point.
- Only two skin-friction lines run through *saddle point* **S**; all of the other lines avoid **S** by taking a hyperbolic shape (Fig. 2.66c). These special skin-friction lines, called *separators*, are of the separation or attachment type according to the flow behaviour in their vicinity.
- All of the skin-friction lines spiral around the *focus* where they eventually terminate (Fig. 2.66d). A focus is the surface trace of a tornado-like vortex and is a key feature of separated flow in three dimensions.
- If the flow is two-dimensional or axisymmetric, the focus degenerates into a *centre* (Fig. 2.66e).

Critical-point theory, which also can be applied to the velocity field, is a powerful tool for rationally describing the organisation, or topology, of three-dimensional flows. Within this framework, a flow is said to be separated if the skin-friction-line pattern contains at least one saddle point through which a separation line passes. The flow in the vicinity of a separation saddle point is illustrated in Fig. 2.67a. The skin-friction lines run towards the saddle point, where they separate into two families flowing along what is called the *separation line*. On approaching the separation line, the streamlines close to the surface tend to lift off, flowing in the innermost

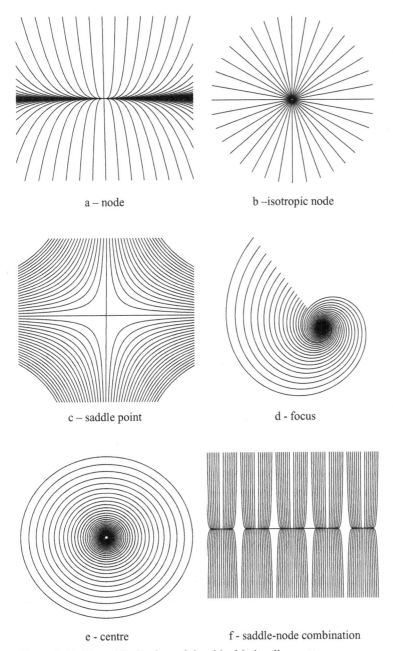

a – node

b –isotropic node

c – saddle point

d - focus

e - centre

f - saddle-node combination

Figure 2.66. The critical points of the skin-friction-line pattern.

part of the boundary layer and moving up into the outer stream. The separation line is the trace on the surface of a stream surface (i.e., the separation surface), which rolls up to form a vortical structure typical of three-dimensional separated flows. In the attachment process, the direction of the skin-friction lines is reversed; they first follow what is now called an attachment line and then flow away from the saddle point (Fig. 2.67b). At the same time, the outer flow dives towards the surface. The attachment line is the trace of an attachment surface.

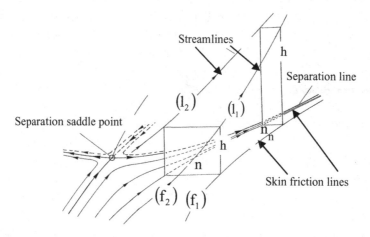

a – flow in the vicinity of a separation saddle point

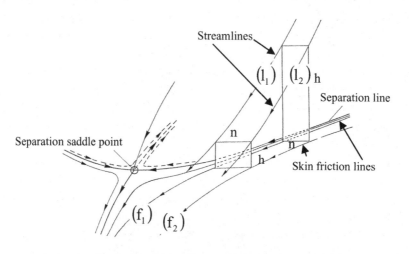

b – flow in the vicinity of an attachment saddle point

Figure 2.67. Flow behaviour in the vicinity of separation and reattachment (Délery [30]).

2.10.2 Topology of a Three-Dimensional Interaction

Examples of basic three-dimensional SBLIs that occur in the flows associated with a swept wedge, a sharp fin at incidence, a blunt fin, and a slanted blunt fin are shown in Fig. 2.68.

For an example, we consider the flow produced by a blunt fin at zero angle of attack placed normal to a flat plate. This flow contains most of the physics of three-dimensional shock-induced separation. Here, the upstream Mach number is equal to 1.97 and the flat-plate boundary layer is turbulent (Barberis and Molton [31]). The schlieren picture and sketch in Fig. 2.69 show the structure of the flow in a vertical plane containing the fin plane of symmetry. The blunt leading edge provokes separation of the boundary layer well ahead of the fin and a three-dimensional lambda shock pattern forms in the inviscid part of the flowfield. A projection can be made in the plane of symmetry of the pattern of separation shock (C_1), trailing shock (C_2), and shock (C_3) (Fig. 2.69b). These shocks meet at triple point $\mathbf{T}$, from which a shear

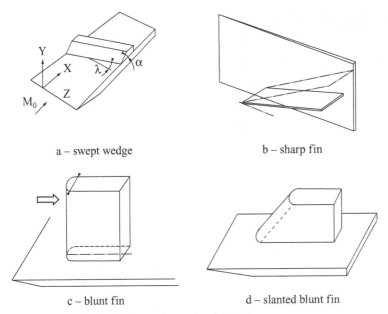

a – swept wedge   b – sharp fin

c – blunt fin   d – slanted blunt fin

Figure 2.68. The basic three-dimensional SBLIs.

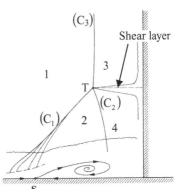

Figure 2.69. Blunt-fin–induced separation in a Mach 1.97 flow (Barberis and Molton [31]).

a – Schlieren photograph of flow field

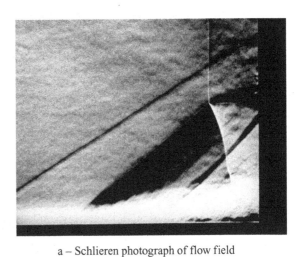

b – sketch of flow organisation in the symmetry plane

a – surface flow visualisation (ONERA document)

Figure 2.70. Surface-flow topology of blunt-fin–induced separation in a Mach 1.97 flow.

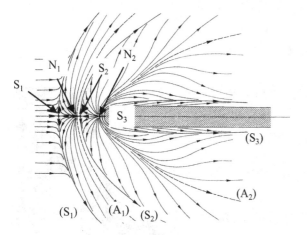
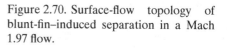

b – surface flow pattern topology

layer emanates before striking the fin leading edge. Information on the interaction topology is provided by the surface-flow visualisation showing the skin-friction-line pattern (Figs. 2.70a and 2.70b). Separation saddle point S_1 occurs in front of the obstacle, through which the primary separation line (S_1) passes. This separates the skin-friction lines originating upstream from those originating at attachment node N_1 situated behind S_1. A second saddle point S_2 exists downstream of S_1 with a secondary separation line (S_2). For topological reasons, attachment line (A_1) originates at node N_1 and separates the skin-friction lines flowing towards (S_1) from those flowing towards (S_2). A third separation line (S_3) is present close to the fin. The flow topology in the plane of symmetry is illustrated in Fig. 2.71a (for clarity, the vertical scale is greatly increased). The main outer-flow separation line (S_1) springs from separation point S_1 and spirals around focus F_1. Attachment line (A_1) ends at the half-saddle point coincident with node N_1 in the surface. The streamlines flowing between (S_1) and (A_1) disappear into focus F_1. Two other similar structures exist associated with separation lines (S_2) and (S_3). The three foci F_1, F_2, and F_3 are the

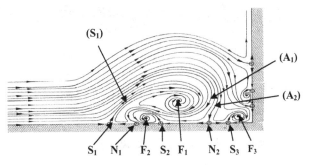

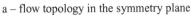

a – flow topology in the symmetry plane

Figure 2.71. Outer flow topology of blunt-fin–induced separation in a Mach 1.97 flow.

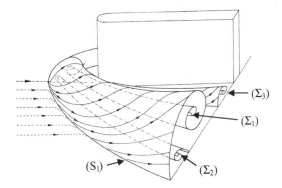

b – separation surfaces and vortex formation

traces in the symmetry plane of three horseshoe vortices surrounding the obstacle, as illustrated in Fig. 2.71b. These vortices can be identified by the spiralling separation surfaces (Σ_1), (Σ_2), and (Σ_3), which are associated with the three separation lines that lie along the surface. The topology of the flow in the planes almost normal to (S_1) is similar, the difference being that the lines in question are projected and not actual streamlines.

2.10.3 Reconsideration of Two-Dimensional Interaction

Within the framework of this general definition, two-dimensional separation can be viewed as a particular case in which the separation line is made of an infinite number of identical saddle point–node combinations, as shown in Fig. 2.66f. The separation point is at the crossing of this line with the plane containing the two-dimensional flow. This situation is extremely unlikely to occur in a three-dimensional world, even with two-dimensional geometries and a uniform upstream boundary. Separated flows invariably appear to adopt a three-dimensional organisation, at a scale either macroscopic or microscopic. In planar two-dimensional channels, the skin-friction-line pattern most often has a macroscopic organisation, as shown in Figs. 2.72a and 2.72b. Saddle point S_1 exists in the test-section symmetry plane through which separation line (S) passes before spiralling into foci F_1 and F_2. Reattachment takes place along attachment line (A) going through reattachment saddle point S_2. If the ratio of the test-section width to the incoming boundary-layer thickness is insufficient, the surface-flow pattern can be highly three-dimensional, as shown in

a – surface flow visualisation (IMP-Gdansk document)

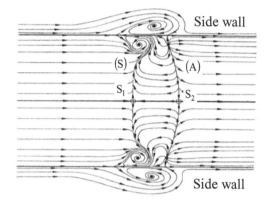

b – topology of the skin friction line pattern

Figure 2.72. The three-dimensional macro-structure of a nominally two-dimensional flow.

the figure. For axisymmetric configurations, the flows are less prone to show three-dimensional effects. If they do, the periodicity condition imposes an organisation in which a finite number of critical points of the node and saddle types is distributed in succession on the reattachment line (Figs. 2.73a and 2.73b). This pattern can be interpreted as the trace on the surface of Görtler-type vortices, the origin of which is unclear. These are intensified by the concave curvature effect resulting from the reattachment of the shear layer. It is probable that these 'microstructures' – which are scaled by the incoming boundary-layer thickness – have a weak influence on the overall flow organisation. Therefore, in reality, the flows that exhibit these structures are not far from the axisymmetric idealisation; this is generally not the case for planar, supposedly two-dimensional configurations.

2.11 Unsteady Aspects of Strong Interactions

The unsteady aspects of SBLI are a subject of major concern because of their potentially dangerous effects on a vehicle's behaviour and structural integrity. A distinction should to be made between large-scale unsteadiness affecting the entire flowfield and small-scale fluctuations influencing only the interaction region and the nearby flow. These unsteady phenomena have a bearing on acquiring a physical understanding of turbulent interactions because there is a close correlation between them and the fluctuating nature of a turbulent boundary layer. The flow unsteadiness subjects the interacting shock to a variable incident flow, which reacts

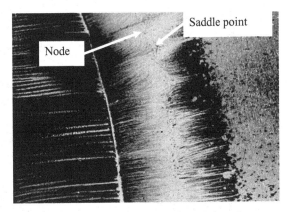

a – surface flow visualisation of reattachment at a flare cylinder junction (ONERA document)

Figure 2.73. The three-dimensional microstructure of an axisymmetric flow.

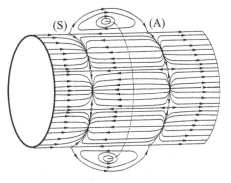

b – topology of the skin friction line pattern of a separating-reattaching flow

accordingly. This provokes several questions concerning the interaction: What is the nature of the relationship between the turbulence and the induced-shock oscillations? Is it possible for such a mechanism to affect a transfer of energy from the outer flow to the turbulent field, thereby enhancing the turbulence level? If this is true, what is the validity of the classical turbulence models, which do not incorporate such a mechanism? Answering these questions is a demanding task and requires sophisticated experimental techniques and advanced theoretical approaches.

The fluctuation levels in the interaction region can be detected by unsteady pressure measurements and/or high-speed cinematography; it has been found that these increase when the flow separates. If we consider the example of a shock reflection with induced separation at Mach number 2.3, as shown in Fig. 2.74, we can distinguish three regions. The farthest upstream part of the interaction appears to be where high-frequency fluctuations originate within the turbulence of the incoming boundary layer. As the separated bubble develops, the fluctuating field is progressively dominated by lower frequencies that correlate with the large eddies shed by the shear layer emanating from the separation point. The size of these eddies increases in proportion to the extent of the separated region. When the reattachment process begins, the dominant fluctuation frequency increases as a new boundary-layer-type structure is recovered progressively. The existence of a

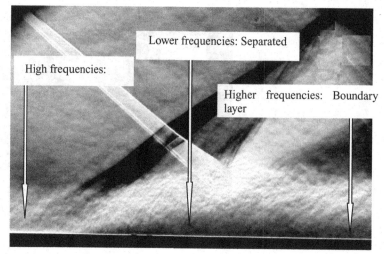

High frequencies:

Lower frequencies: Separated

Higher frequencies: Boundary layer

Figure 2.74. Instant shadowgraph of a separated flow induced by shock reflection. Incoming Mach number 2.3, incident-shock deflection 8° (Dupont et al. [32]).

separated bubble, which is probably unstable, leads to amplification of the flow unsteadiness, which is transmitted upstream through the mechanism explained in Section 2.4. This feedback process entails a general rise in the fluctuation level. This behaviour is observed in shock-induced separation, with the amplitude of the shock motion on the order of the incoming boundary-layer thickness. In addition, large eddies are formed at the shock foot that propagate downstream and emit pressure waves, which – in the case of a transonic interaction – can propagate upstream and influence the shock on its downstream face. These waves are visible in the spark-schlieren photograph shown in Fig. 2.75. This excitation leads to a shock motion, which contributes to the formation of the large structures and also determines their emission frequency. The feedback mechanism that results from this coupling is typical of transonic flows in which downstream information has a greater possibility of travelling upstream than at higher Mach numbers.

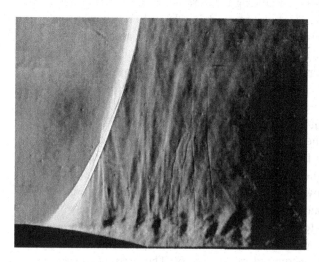

Figure 2.75. Instantaneous shadowgraph of a transonic interaction showing the pressure waves emitted by the turbulent eddies (ONERA document).

In some circumstances, the mechanism described herein can be amplified dramatically; in such cases, the entire flowfield is affected by large-scale fluctuations that can be periodic or not, depending on the conditions. This phenomenon is at work in transonic airfoil buffeting when the normal shock moves over a significant portion of the chord length, and the separated bubble disappears and reappears periodically. A scenario for such a periodic unsteadiness is proposed in Fig. 2.76, which is obtained from Navier-Stokes calculations. In the first frame, the shock occupies a downstream position but is moving upstream. Thus, the flow velocity relative to the shock is increased because the shock velocity adds to the upstream flow velocity. If the relative Mach number is larger than 1.3, separation occurs (see Section 2.7.2) and a large separation bubble forms due to the compression on the profile towards the rear of the airfoil. This separation has the effect of reducing the effective aerodynamic incidence so that the shock weakens as it continues to move upstream until it stops. Simultaneously, the separation bubble starts to shrink after reaching its greatest size (frame 3); this leads to an increase in the aerodynamic incidence, with the shock now moving downstream. The relative shock strength is then less and separation at the shock foot is suppressed (frame 4). When the shock slows down, separation occurs and is amplified when the shock starts to move upstream again. A similar scenario is at work in the buzz of supersonic air-intakes. In this case, the unsteadiness is a periodic large amplitude motion of the shock system in which shock-induced separation is the triggering factor. In reality, there are two possible origins of air-intake buzz: (1) shock-induced separation (the so-called Dailey's scenario); and (2) the result of swallowing of the slip line due to Type IV interference between the cowl shock and a shock formed by a compression ramp (the so-called Ferri's scenario). Overexpanded propulsive nozzles are affected by unsteady and asymmetric shock-induced separation, which can be the source of high side loads during the start-up transient.

Large-amplitude oscillations also occur within rotating machines, such as compressors, turbines, and helicopter rotors. The shock oscillation is forced by the device itself, which raises a question concerning the interplay between this motion of the body and the fluctuation frequencies of the SBLI, including those due to turbulence. A similar coupling mechanism is found in the aeroelastic response of a structure such as compressor blades. In this case, the deformation of the structure induces shock displacements and a subsequent change in the pressure load, which can lead to a divergent process or flutter.

2.12 SBLI Control

2.12.1 Mechanisms for Control Action

Because it is often difficult to avoid detrimental SBLIs occurring within a flow, the idea soon arose during the early history of the development of compressible-fluid mechanics of controlling the phenomenon by an appropriate 'manipulation' of the flow, either before or during the interaction process (Regenscheit [34]; Fage and Sargent [35]). The target of the control techniques was mainly to either prevent shock-induced separation or stabilise the shock when it occurred in naturally

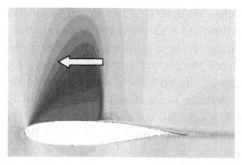

a – the shock moves upstream and induces separation

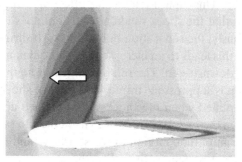

b – the shock reaches its most upstream location

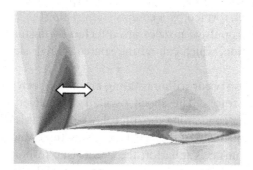

c – the shock stops and starts to move downstream

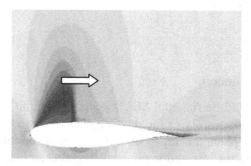

d – the shock moves downstream and does not induce separation

Figure 2.76. A scenario for transonic buffeting over a profile. Navier-Stokes calculations (Furlano [33]).

unsteady configurations (see Section 2.11). The upstream influence of the shock and the resistance of a turbulent boundary layer to separation depend mainly on the momentum (see Section 2.4). Thus, one way to limit the shock's effect is to increase the boundary-layer momentum prior to interaction with the shock, which can be done by appropriate boundary-layer manipulation techniques, such as the following:

- mass injection (or boundary-layer blowing) through one or several slots located upstream of the shock origin or impact point
- distributed suction applied over a specified boundary-layer length upstream of the interaction, which lowers the shape parameter, thereby producing a fuller velocity profile
- removal of the low-speed part of the boundary layer by applying strong suction through a slot located within or slightly upstream of the interaction region
- use of vortex generators upstream of the shock, which transfer momentum from the outer high-speed flow thereby reenergising the boundary-layer low-velocity parts; this transfer enhances resistance to an adverse pressure gradient
- localised suction also can be applied within or in the immediate vicinity of the interaction

A key factor in controlling the interaction is to determine the velocity that is achieved on the separating streamline (S). Any action changing its magnitude influences the interaction and modifies the shear-layer reattachment (see Section 2.6). If some fluid is sucked through the wall, topological considerations lead to a flow structure like that illustrated in Fig. 2.77b. The streamline (S_2), which stagnates at reattachment point **R**, originates at a greater distance from the wall, and the velocity $(U_S)_{max}$ is greater than in the basic case shown in Fig. 2.77a. This enhances the ability of the flow to withstand more significant compression and causes a subsequent contraction of the interaction domain. For the case of fluid injection at low velocity, illustrated in Fig. 2.77c, the velocity on (S_2) is reduced because (S_2) reaches a lower altitude on the velocity profile. The effect is to lengthen the separated bubble. However, if the injected-mass flow is increased, there is a reversal of this effect. The velocity on the lower part of the profile (in particular, $(U_S)_{max}$) is increased if the mass flow fed into the separated region exceeds a certain threshold.

We also can contract the interaction domain by cooling the wall on which the interaction occurs because the boundary layer is more resistant to separation on a cold wall (see Section 2.3). This control technique would be practical for vehicles using cryogenic fuels (e.g., hypersonic planes and space launchers).

2.12.2 Examination of Control Techniques

When considering SBLI control, the objectives must be stated clearly. The control can be used to prevent separation and/or stabilise the shock in a duct or a nozzle. Boundary-layer blowing, suction, or wall cooling can be effective for this purpose. If the aim is to decrease the drag of a profile in transonic flow, the situation is more subtle because the drag that is a consequence of the entropy production (see Section 2.2.4) originates in both the shock (i.e., the wave drag) and the boundary layer (i.e., the friction drag). The problem is similar for internal flows such as air-intakes

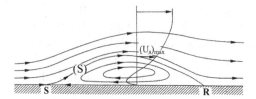

a – separation bubble in the basic case

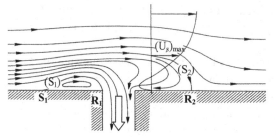

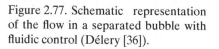

Figure 2.77. Schematic representation of the flow in a separated bubble with fluidic control (Délery [36]).

b – separation bubble with fluid suction

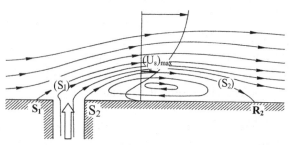

c – separation bubble with fluid injection

a – reference case without control

c – active control by suction through a slot

b – passive control

d – contoured wall or bump

Figure 2.78. SBLI control in a transonic flow (Stanewsky et al. [37]).

because, here again, the efficiency loss is the result of entropy production through the compression shocks and boundary layer.

Active Control

By removing the low-energy part of the boundary layer, the upstream propagation mechanism is inhibited and the thickening of the boundary layer is reduced. The flow behaviour for transonic flows then tends towards the perfect-fluid solution with a near-normal shock extending down close to the surface. Thus, any action that energises the boundary layer also tends to strengthen the shock because any spreading caused by the interaction is reduced. This effect is illustrated in Figs. 2.78a and 2.78b, which show a comparison between a transonic interaction first without and then with control through a suction slot. In the latter case, the entropy production through the shock is increased; thus, the wave drag is higher. Conversely, because the downstream boundary-layer profile is fuller, the momentum loss in the boundary layer is reduced.

Passive Control

As described in Section 2.6, when separation occurs, the smearing of the shock system and the splitting up of the compression process that is achieved by the interaction reduce the wave drag and efficiency loss due to the shock. However, the momentum loss in the separated boundary layer is far greater than for an attached boundary layer. Thus, the separation can result in an increase in drag or an overall efficiency loss. However, because separation has a favourable effect on the wave drag, we can envisage replacing a strong but unseparated interaction with a separated-flow organisation (or one that mimics the separated flow) to gain the advantages of the low-wave drag. Passive-control schemes have been devised to exploit these concepts, which combine the two effects by spreading the shock system while minimising (in theory) the boundary-layer thickening. The most common form of passive control involves replacing a part of the surface with a perforated plate installed over a closed cavity. The plate is positioned so that it is in the shock region of the interaction. A natural circulation occurs – via the cavity – from the downstream high-pressure part of the interaction to the upstream low-pressure part. The upstream transpiration provokes growth of the boundary-layer displacement thickness, which causes an oblique shock (referred to previously as (C_1)). The situation is similar to the case of the natural shock-induced separation considered in Section 2.6.3, with the strong-normal shock being replaced by an oblique two-shock system in the vicinity of the surface; hence, the wave drag is reduced. The detrimental effect caused by thickening the boundary layer is limited by the suction operated in the downstream part of the perforated plate. The effect of passive control is illustrated in Fig. 2.78c, in which the smearing of the transonic shock and the thickening of the boundary layer are visible. Passive control can be effective in stabilising a shock, although the advantage in terms of drag reduction is questionable. This concept can be improved by adding a suction slot downstream of the passive-control cavity. The device combines the advantages of passive control, which reduces the wave drag and the effectiveness of suction to lower the friction losses.

Wall Contouring

Because friction-drag production in passive control is generally unacceptable and because fluid suction requires an energy supply that can compromise the economical benefit of control, mimicking the separated flow structure with a local deformation of the surface can be considered. For example, a bump with a double-wedge shape could reproduce the flow characteristics of the viscous separated fluid induced by a shock reflection (see Section 2.6.1). In transonic flow, bumps with a more progressive upstream concave contour that achieve a nearly isentropic compression are effective because they weaken the normal shock forming at their location (Fig. 2.78d). This only slightly affects the boundary layer while substantially reducing the wave drag.

2.13 Concluding Remarks

In high-speed flows, the occurrence of SBLIs is an almost inevitable outcome of the presence of shocks. The structure of these interactions is predominantly a consequence of the response of the boundary layer to the sudden local compression imparted by the shock; it reacts as a nonuniform flow in which viscous and inertial terms combine intricately. The most significant result is the spreading of the pressure discontinuity caused by the shock so that the influence is experienced well upstream of where it would have been located in an inviscid-fluid model. When a shock is strong enough to separate the boundary layer, the interaction has dramatic consequences for the development of the boundary layer and for the contiguous inviscid flowfield. Complex shock patterns are formed that involve shock-shock interferences, the nature of which depends on the Mach number and how the primary shock is produced (i.e., shock reflection, ramp, or normal shock). In these circumstances, the most salient feature of shock-induced separation is probably not the behaviour of the boundary layer but rather the pattern of shocks produced – even though this is a secondary phenomenon associated with the process. The boundary layer behaves more or less as it would for any other ordinary separation and essentially the same as in subsonic flows. It obeys the specific laws dictated mainly by the intensity of the overall pressure rise imparted by the shock, regardless of the way in which it is generated. A striking feature of these interactions is the overwhelming repercussion of the shock on the contiguous inviscid supersonic stream, which can be spectacular for internal flows. Although the basic flow topology is the same, laminar and turbulent interactions have distinctly different properties that stem from the greater resistance of a turbulent boundary layer to flow retardation and, hence, separation. Interactions in hypersonic flows have specific features coming from the high enthalpy level of the outer flow.

A detrimental consequence of SBLI is the occurrence of flow unsteadiness, which can be of high intensity when the shock is strong enough to induce separation. Such unsteadiness can occur at high frequencies when associated with turbulent fluctuations and/or separated-bubble instabilities. In other circumstances, the unsteadiness occurs at low frequency when the fluctuating motions involve the entire aerodynamic field (e.g., in transonic buffeting or air-intake buzz). Such large-scale unsteadiness seems to be a special feature of transonic interactions when the downstream subsonic flow allows a forward transmission of perturbations that excite the shock wave. In fully supersonic interactions, the higher Mach number of the outer flowfield tends to isolate the interaction domain from downstream perturbations; such interactions are mostly free of large-scale instabilities.

The physics of perfect-gas SBLI in two-dimensional flows is considered to be well understood. However, there is no room for complacency because nearly all of the practical situations are three-dimensional. The obstacle is then the difficulty in comprehending the structure of three-dimensional flows so as to arrive at a consistent topological description of the flowfield organisation. At the same time, the definition of separation in three-dimensional flow is far more subtle than in two-dimensional flows, where it is associated with the skin-friction coefficient being reduced to zero. The description and study of three-dimensional SBLIs first must consider this basic question, which necessitates calling on the critical-point theory.

Predicting turbulent interactions is still limited because theoreticians are confronted with the frustrating problem of turbulence modelling. The usual two-transport equation models perform poorly as soon as a noticeable separated region forms. Turbulence in interacting flows involves many aspects: compressibility terms in the time-averaged equations, shock/turbulence interaction, history effects, flow unsteadiness, strong anisotropy, and transfer processes, to name only the more important – to which turbulence/chemistry coupling should be added for hypersonic flows. Substantial improvements result from using Reynolds Stress Equation models transporting the full Reynolds stress tensor or models using nonlinear explicit expansion of the Reynolds stress tensor in terms of the strain and vorticity tensors (the Boussinesq law is the first term of this expansion). Also, modelling based on the capture of the unsteady travelling big eddies, such as Large Eddy Simulation (LES) and Detached Eddy Simulation (DES), results in substantial improvement by providing a more faithful representation of the turbulent behaviour in the shock-separated region. However, we must be cautious when drawing overly pessimistic conclusions from a poor agreement between theory and experiment. Many of the existing results are instructive for the physical understanding of interactions; although they appear unsatisfactory from code-validation assessments, the discrepancies are frequently attributable to poorly identified flow conditions, undesirable side effects, or unwanted transition occurring in the course of the interaction.

Appendix A: Discontinuities in Supersonic Flow and the Rankine-Hugoniot Equations

The origin of shock waves can be explained within the framework of gas dynamics and compression-wave propagation theory, a subject not considered here. In inviscid-fluid theory, where the flow is assumed to obey the Euler equations, shock waves are postulated as surfaces of discontinuity across which the flow properties undergo sudden changes. To satisfy the conservation equations, the flow properties downstream of the shock are linked to their corresponding upstream values by the Rankine-Hugoniot equations. These generally take the following form in which brackets [] denote a jump in the quantities across a surface (Σ), $\vec{n}$ is the unit vector normal to (Σ), $\vec{V}$ is the velocity vector, ρ is the density, and h_{st} is the specific stagnation enthalpy (enthalpy per unit of mass).

Mass conservation:

$$[\rho \vec{V}.\vec{n}] = 0$$

Momentum equation:

$$[\vec{V}]\rho \vec{V}.\vec{n} + [p]\vec{n} = 0$$

Energy equation:

$$[h_{st}]\rho \vec{V}.\vec{n} = 0$$

The above equations are satisfied by several types of discontinuity and not exclusively shock waves. We consider the various possibilities:

1. If, $\vec{V} \cdot \vec{n} = 0$ then the velocity component normal to (Σ) is zero. The component of the momentum equation in a plane tangent to (Σ) gives:

$$[\vec{V}]\rho \vec{V} \cdot \vec{n} = [V_t] \times 0 = 0$$

This shows that there is no condition on the velocity component tangent to the surface, which can be discontinuous. The normal projection of the momentum equation gives:

$$0 + [p] = 0$$

which means that the pressure must be the same on each side of (Σ). Similarly, the energy equation shows that:

$$0 \times [h_{st}] = 0$$

Thus $[h_{st}]$ is arbitrary, as are also the temperature and the density (hence entropy). Such a surface is a **vortex sheet** or, in two-dimensional flow, a **slip line**.

2. If $V_t = 0$ one obtains the special case a **contact surface** as in one-dimensional unsteady flow.

3. If $\vec{V}.\vec{n} \neq 0$, the normal velocity component is non-zero and there is a mass flux through the surface which is of the **shock wave** type of discontinuity. In this case, one has the following relations:

$[V_t] = 0$: the tangential velocity component does not change through the shock.
$[V_n]\rho V_n + [p] = 0$ for the normal component of the momentum equation.

The energy equations now gives $[h_{st}] = 0$, which establishes that the stagnation enthalpy remains constant when crossing a shock wave.

The Rankine-Hugoniot equations must be supplemented by the following condition on the entropy jump $(s_2 - s_1)$ through the shock (application of the Second Law of Thermodynamics):

$$s_2 - s_1 \geq 0$$

Such a global condition avoids invoking the Navier-Stokes equations because the dissipative terms, including viscosity and thermal conductivity, impose the correct sense to the variation in entropy. Shock-structure analysis (applied to a shock normal to the direction x) leads to the following expression for the change in the entropy flux through the shock region, which is a zone of rapid but continuous variation of the flow properties (Délery [38]):

$$m\,(s_2 - s_1) = \int_{-\infty}^{+\infty} \frac{1}{T^2} \left[\frac{4}{3}\mu T \left(\frac{du}{dx} \right)^2 + \lambda \left(\frac{dT}{dx} \right)^2 \right] dx,$$

where s is the specific entropy, $m = \rho u$ is the constant mass-flow rate (per unit section), μ is the molecular viscosity, λ is the thermal conductivity, x is the streamwise coordinate, u is the x-wise velocity component, and T is the temperature. Because μ and λ are positive (i.e., the condition imposed to satisfy the Second Law of Thermodynamics), the integrated quantities are strictly positive. Thus, the Navier-Stokes equations predict that an entropy rises through the shock consistent with the law.

REFERENCES

[1] A. H. Shapiro. *The Dynamics and Thermodynamics of Compressible Fluid Flow*, Vols. 1 and 2 (New York: The Ronald Press Company, 1953).

[2] B. Edney. Anomalous heat transfer and pressure distributions on blunt bodies at hypersonic speeds in the presence of an impinging shock. *Aeronautical Research Institute of Sweden*, FFA Report 115, Stockholm, 1968.

[3] K. Oswatitsch. *Der Luftwiderstand als Integral des Entropiestromes* (Drag Expressed as the Integral of Entropy Flux). Presented by Ludwieg Prandtl; Kaiser-Wilhelm Institute for Research, 1945.

[4] D. E. Coles. The law of the wake in the turbulent boundary layer. *J. Fluid Mech.*, 2 (1956), 191–226.

[5] J. Cousteix. *Couche limite laminaire* (Toulouse: Cépaduès-Editions, 1988).

[6] J. Cousteix. *Turbulence et couche limite* (Toulouse: Cépaduès-Editions, 1989).

[7] M. J. Lighthill. On boundary layer upstream influence. Part II: Supersonic flows without separation. *Proc. Roy. Soc. A*, 217 (1953), 478–507.

[8] K. Stewartson and P. G. Williams. Self-induced separation. *Proc. Roy. Soc., A*, 312 (1969), 181–206.

[9] L. F. Henderson. The reflection of a shock wave at a rigid will in the presence of a boundary layer. *J. Fluid Mech.*, 30 (1967), 4, 699–722.

[10] G. S. Settles. An experimental study of compressible boundary-layer separation at high Reynolds number. Ph.D. thesis, Princeton University, Princeton, NJ (1975).

[11] J. E. Green. Interaction between shock waves and turbulent boundary layers. *Progress in Aerospace Science*, 11 (1970), 235–340.

[12] J. S. Shang, W. L. Hankey Jr., and C. H. Law. Numerical simulation of shock wave/turbulent boundary-layer interaction. *AIAA J.*, 14 (1976), 10, 1451–7.

[13] D. R. Chapman, D. M. Kuhen, and H. K. Larson. Investigation of separated flows in supersonic and subsonic streams with emphasis on the effect of transition. *NACA TN-3869* (1957).

[14] J. Délery and J. G. Marvin. Shock Wave/Boundary Layer Interactions. *AGARDograph* 280 (1986).

[15] E. E. Zhukoski. Turbulent boundary-layer separation in front of a forward-facing step. *AIAA J.*, 5 (1967), 10, 1746–53.

[16] R. H. Schmucker. Side loads and their reduction in liquid rocket engines. TUM-LRT TB-14; 24th International Astronautical Congress, Baku, USSR, October 7–13, 1973.

[17] R. J. Hakkinen, I. Greber, L. Trilling, and S. S. Abarbanel. The interaction of an oblique shock wave with a laminar boundary layer. *NASA Memo* 2–18-59W (1959).

[18] G. E. Gadd, D. W. Holder, and J. D. Regan. An experimental investigation of the interaction between shock waves and boundary layers. *Proc. Roy. Soc. A*, 226 (1954), 226–53.

[19] G. M. Elfstrom. Turbulent hypersonic flow at a wedge compression corner. *J. Fluid Mech.*, 53 (1972), 1, 113–29.

[20] J. E. Lewis, T. Kubota, and L. Lees. Experimental investigation of supersonic laminar two-dimensional boundary layer separation in a compression corner with and without cooling. *AIAA Paper* 67–0191. Also *AIAA J.*, 6 (1967), 1, 7–14.

[21] F. W. Spaid and J. C. Frishett. Incipient separation of a supersonic, turbulent boundary layer, including effects of heat transfer. *AIAA J.*, 10 (1972), 7, 915–22.

[22] J. Délery. *Etude expérimentale de la réflexion d'une onde de choc sur une paroi chauffée en présence d'une couche limite turbulente* (Experimental investigation of the reflection of a shock wave on a heated surface in presence of a turbulent boundary layer). *La recherche Aérospatiale*, 1992–1 (1992), pp. 1–23 (French and English editions).

[23] J. Délery and M.-C. Coët. Experiments on shock wave/boundary layer interactions produced by two-dimensional ramps and three-dimensional obstacles. *Workshop on Hypersonic Flows for Reentry Problems*, Antibes, France (1990).

[24] M. Holden. Shock wave/turbulent boundary layer interaction in hypersonic flow. *AIAA Paper* 77–0045 (1977).

[25] W. D. Hayes and R. F. Probstein. *Hypersonic flow theory,* Vol. 1: Inviscid Flows (New York: Academic Press, 1966).

[26] M.-C. Coët, J. Délery, and B. Chanetz. Experimental study of shock wave/boundary layer interaction at high Mach number with entropy layer effect. *IUTAM Symposium on Aerothermochemistry of Spacecraft and Associated Hypersonic Flows*, Marseille, France, 1992.

[27] F. Grasso and G. Leone. Chemistry effects in shock wave/boundary layer interaction problems. *IUTAM Symposium on Aerothermochemistry of Spacecraft and Associated Hypersonic Flows*, Marseille, France, 1992.

[28] S. G. Mallinson, S. L. Gai, and N. R. Mudford. High enthalpy, hypersonic compression corner flow. *AIAA J.*, 34 (1996), 6, 1130–7.

[29] R. Legendre. *Lignes de courant en écoulement permanent: Décollement et séparation* (Streamlines in permanent flows: Detachment and separation). *La Recherche Aérospatiale*, No. 1977–6 (Novembre–Décembre 1977).

[30] J. Délery. Robert Legendre and Henri Werlé: Toward the elucidation of three-dimensional separation. *Ann. Rev Fluid Mech.*, 33 (2001), 129–54.

[31] D. Barberis and P. Molton. Shock wave/turbulent boundary layer interaction in a three-dimensional flow. *AIAA Paper* 95–0227 (1995).

[32] P. Dupont, C. Haddad, and J.-F. Debiève. Space and time organization in a shock-induced boundary layer. *J. Fluid Mech.*, 559 (2006), 255–77.

[33] F. Furlano. *Comportement de modèles de turbulence pour les écoulements décollés en entrée de tremblement* (Behaviour of turbulence models for buffet onset in separated flows). Ph.D. Thesis, Ecole Nationale Supérieure de l'Aéronautique et de l'Espace (2001).

[34] B. Regenscheit. Versuche zur Widerstandsverringerung eines Flügels bei hoher Machscher – Zahl durch Absaugung der hinter dem Gebiet unstetiger Verdichtung abgelösten Grenzschicht. *ZWB, Forschungsbericht* #1424 (1941), English translation. *NACA TM* No. 1168.

[35] A. Fage and R. F. Sargent. Effect on aerofoil drag of boundary-layer suction behind a shock wave, *ARC R&M* 1913 (1943).

[36] J. Délery. Shock-wave/turbulent boundary-layer interaction and its control. *Progress in Aerospace Sciences*, 22 (1985), 209–80.

[37] E. Stanewsky, J. Délery, J. Fulker, and W. Geissler. EUROSHOCK: Drag Reduction by Passive Shock Control. *Notes on Numerical Fluid Mechanics*, Vol. 56 (Wiesbaden: Vieweg, 1997).

[38] J. Délery. *Handbook of Compressible Aerodynamics*. ISTE – WILEY & Sons, 2010.

3 Transonic Shock Wave–Boundary-Layer Interactions

Holger Babinsky and Jean Délery

3.1 Introduction to Transonic Interactions

By definition, transonic shock wave–boundary layer interactions (SBLIs) feature extensive regions of supersonic and subsonic flows. Typically, such interactions are characterized by supersonic flow ahead of the shock wave and subsonic flow downstream of it. This mixed nature of the flow has important consequences that make transonic interactions somewhat different from supersonic or hypersonic interactions.

The key difference between transonic interactions and other SBLIs is the presence of subsonic flow behind the shock wave. Steady subsonic flow does not support waves (e.g., shock waves or expansion fans), and any changes of flow conditions are gradual in comparison to supersonic flow. This imposes constraints on the shock structure in the interaction region because the downstream flow conditions can feed forward and affect the strength, shape, and location of the shock wave causing the interaction. The flow surrounding a transonic SBLI must satisfy the supersonic as well as subsonic constraints imposed by the governing equations. The interaction also is sensitive to downstream disturbances propagating upstream in the subsonic regions. In contrast, supersonic interactions are "shielded" from such events by the supersonic outer flow.

Because transonic interactions are defined as having a subsonic postshock flow, this discussion is limited to normal or near-normal shock waves with low sweep. Larger shock-sweep angles invariably lead to the postshock flow being supersonic and these interactions are better understood in the framework of three-dimensional supersonic interactions (see Chapter 4).

Typical transonic SBLIs, therefore, are exemplified by the interaction of a normal shock wave interacting with a boundary layer. Whereas normal shocks exist at all Mach numbers, normal SBLIs are more common at moderate supersonic speeds; in practice, transonic interactions are observed mainly at Mach numbers below 2.

3.2 Applications of Transonic SBLIs and Associated Performance Losses

The most common example of a transonic SBLI is found on transonic-aircraft wings where the wing shock (Fig. 3.1) interacts with the wing-boundary layer. A closely

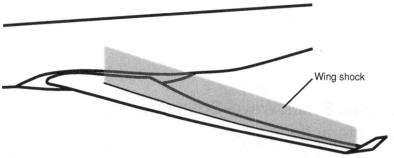

Figure 3.1. Shock wave on transonic aircraft wing.

related example is the interaction inside transonic turbine- and compressor-blade cascades, where local regions of supersonic flow are terminated by shock waves that interact with the blade-boundary layers. However, transonic SBLIs also can be found in the engine intakes of supersonic aircraft, in which the oncoming supersonic flow is compressed and decelerated to subsonic velocities. Transonic SBLIs also are encountered in industrial flows such as inside ducts and diffusers.

3.2.1 Transonic Airfoils and Cascades

An airfoil shape immersed in a free stream approaching the speed of sound is likely to feature flow regions in which the local velocity is supersonic. On wings and turbomachinery blades (Figs. 3.2 and 3.3), these supersonic regions typically are found on the suction side, where low pressures and high velocities prevail. Inside such regions, the local flow is strongly dependent on the surface curvature of the airfoil. Because the curvature is likely to be convex, expansion waves are generated that act to further decrease the local pressure and increase the local Mach number. However, on reaching the edge of the supersonic region (i.e., the sonic line), the expansion waves reflect as waves of equal strength and opposite sign (i.e., compression waves). This is because the sonic line effectively acts as a constant-pressure boundary. The following discussion explains why it is easy to see that this is the case.

Outside the supersonic region and away from the boundary layers and the viscous wake, the flow is *isentropic* (i.e., there are no shock waves upstream of the

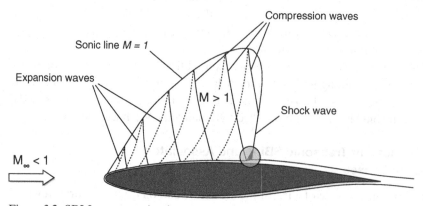

Figure 3.2. SBLI on transonic wing.

Figure 3.3. Transonic SBLI in a turbine cascade.

sonic line). In isentropic flow, pressure is only a function of stagnation pressure p_0 and Mach number M according to the following:

$$\frac{p}{p_0} = \left(1 + \frac{\gamma - 1}{2} M^2\right)^{\frac{-\gamma}{\gamma - 1}} \tag{3.1}$$

Because the stagnation pressure is uniform everywhere in the oncoming flow and the Mach number along the sonic line is equal to unity by definition, the pressure on the sonic line is fixed.

An alternative, more physical explanation for the wave reflection is that at the boundary of the supersonic and the subsonic regions, there is a potential pressure "mismatch." In the outer subsonic region, the pressure is determined by the overall flowfield and influenced from all directions. In the supersonic region, pressure information can travel only downstream (along waves), which can cause a "mismatch" in pressures at the boundary that is resolved by the generation of waves – which are reflections of the incoming waves.

Compression waves returning towards the airfoil surface reflect as compression waves, unless they are cancelled out by further expansion waves originating from convex surface curvature. It is impossible for compression waves to reach the sonic line, a fact beautifully argued by Shapiro [1] in his classic textbook. In practice, compression waves within the supersonic region catch up with one another to form a (normal or near-normal) shock wave that terminates the supersonic flow. This shock wave interacts with the boundary layer on the wing surface in a typical transonic SBLI.

The size of the supersonic region around transonic airfoils or turbine blades determines the location of the shock wave and, indirectly, its strength. In general, the larger the supersonic region, the higher the flow Mach number immediately

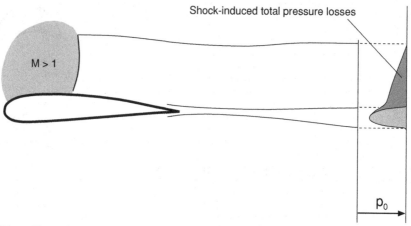

Figure 3.4. The link between stagnation-pressure losses and wave drag.

ahead of the shock wave. For this reason, the severity of the SBLI on a typical airfoil depends on the airfoil shape, the free-stream Mach number, and the angle of attack. Thicker and more highly curved airfoils, higher free-stream Mach numbers, and greater angles of attack all cause stronger shock waves. On aircraft wings, the strength of the shock is kept relatively low by design, in the range of M = 1.1–1.2. More severe shocks can be encountered under off-design conditions.

Shock Losses on Transonic Wings

The presence of shock waves on a transonic airfoil causes reductions of total pressure that are directly linked to the generation of drag. As a result, a contribution to the wake-pressure profile of a transonic airfoil is from the viscous drag on the airfoil and the pressure losses caused by the shock, as shown schematically in Fig. 3.4. The contribution to the overall drag due to the shock is referred to generally as *wave drag*.

Wave drag often is considered an inviscid-flow phenomenon. However, any stagnation-pressure loss in an adiabatic flow is related to an entropy rise. Here, the entropy increase is achieved across the shock wave by viscous actions on a microscopic scale. Thus, even wave drag is caused by viscous actions (i.e., d'Alembert's statement about no drag in a truly inviscid flow remains valid) and, consequently, the classification of wave drag as an inviscid-drag contribution is incorrect. One reason why this terminology remains in widespread use is that wave drag can be predicted with numerical methods solving the (inviscid) Euler equations (because they correctly model shock waves) or other analytical techniques that incorporate the Rankine-Hugoniot shock equations (supplemented by the Second Law of Thermodynamics to impose an entropy rise); however, viscous terms are explicitly excluded.

The magnitude of wave drag can be estimated by integrating the stagnation-pressure losses across the shock. These losses are not uniform along the shock because the upstream Mach number (and shock strength) reduces with distance from the surface. Typically, the greatest losses are observed close to the airfoil surface. With increasing distance from the airfoil, the shock strength decays until it turns into an isentropic sound wave as it reaches the sonic line. Here, the stagnation-pressure loss vanishes. If we consider the stagnation-pressure loss in terms of the local-entropy increase across the shock wave, this is a good illustration of the Oswatitsch relationship between drag and entropy rise discussed in Chapter 2.

To compute the wave drag, it is necessary to know the Mach-number distribution ahead of the shock. In practice, this information is not often available. Lock [2] derived an approximate equation for the wave drag by estimating the Mach-number variation between the surface and the sonic line from knowledge of the local airfoil-surface radius of curvature at the shock location, the free-stream Mach number, and the Mach number M_s at the foot of the shock. The resulting equation for the drag contribution of a shock wave on an airfoil is as follows:

$$c_{DW} = \frac{0.243}{c\kappa_W} \left(\frac{1 + 0.2 M_\infty^2}{M_\infty} \right)^3 \frac{(M_s - 1)^4 (2 - M_s)}{M_s (1 + 0.2 M_s^2)} \tag{3.2}$$

where κ_W is the surface curvature at the shock location.

This equation assumes a constant radius of surface curvature and concentric streamlines ahead of the shock. In reality, the curvature along the airfoil varies, which changes the Mach-number distribution from the idealized assumption incorporated in Lock's equation. However, because most of the pressure loss is generated close to the surface where the shock is strongest, the error caused by this assumption is generally small. Lock developed strategies to improve on this estimate by considering more of the airfoil shape; readers are referred to his paper for more information [2].

The SBLI on airfoils is the cause of significant adverse pressure gradients for the boundary layer on the surface. Once shocks are sufficiently strong, this can cause either local separations in which the boundary layer manages to reattach downstream or a complete flow breakdown in which the flow is separated all the way to the trailing edge. The latter case is often referred to as "shock stall" because of the associated sudden loss of lift and increase in drag.

However, this is not the only mechanism by which an SBLI can cause flow separation. On most airfoil shapes, there is a significant adverse-pressure-gradient region in the subsonic part of the flow between the location of the shock wave and the trailing edge. The presence of a shock indirectly can affect this flow because it thickens the boundary layer, which makes it more sensitive to the adverse-pressure gradient. This can be observed by considering the following nondimensional scaling parameter for adverse-pressure gradients:

$$\frac{\delta^* \, dp}{q \, dx} \tag{3.3}$$

where q is the dynamic pressure. A greater displacement thickness δ^* effectively enhances the adverse-pressure-gradient effect and the boundary layer is thus more likely to separate. It is therefore possible that a shock indirectly may cause a separation near the trailing edge even before separation at the shock foot is reached.

Pearcey [3] categorized transonic-airfoil separation behavior into two types depending on the location of separation onset, as shown schematically in Fig. 3.5. In "Model A," the flow first separates underneath the shock but then reattaches shortly afterwards to form a small separation bubble. With increasing shock strength

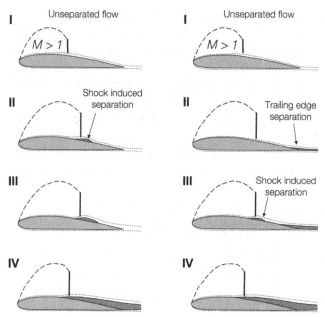

Figure 3.5. Pearcey's models of transonic airfoil separation. *Left:* Model A, where separation first originates at the shock. *Right:* Model B, where separation originates at the trailing edge.

(caused by either an increase in the free-stream Mach number or the angle of attack), this separation bubble grows and eventually bursts to cause separation all the way from the shock foot to the trailing edge. In "Model B," the combined effects of the shock and the subsequent adverse pressure gradient to the trailing edge cause the flow to separate first at the rear of the upper surface. With increasing shock strength, the trailing-edge separation grows, eventually leading to significant separation spanning from the shock foot to the trailing edge. Depending on the presence of a separation bubble under the shock, Pearcey suggested further subdividing Model B into three categories, which are described in reference [3].

Generally, a small separation bubble underneath the shock wave does not incur major performance penalties, whereas trailing-edge separations cause additional drag and loss of lift (which, in turn, affects the shock position and strength). Once the flow is completely separated between the shock location and the trailing edge, flow breakdown has occurred with severe effects on lift and drag. This, in turn, causes the shock wave to move forward and the supersonic region to shrink. Under these conditions, when the overall flow is considerably affected by the presence of separation, it is possible that shock-induced buffet occurs.

Depending on the strength of the shock wave on a transonic airfoil, it can increase drag via two mechanisms: (1) the shock wave introduces a stagnation-pressure drop that leads to wave drag; and (2) the presence of separations causes additional viscous drag.

Although airfoil drag is often (somewhat erroneously) decomposed into viscous and wave-drag components, it may be better to think of the various drag contributions as skin friction and pressure drag. Wave drag is transmitted to the airfoil via pressure drag because the loss of stagnation pressure across the shock reduces

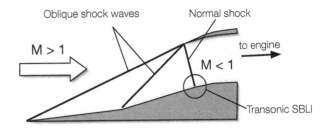

Figure 3.6. External compression inlet.

surface pressures over the rear of the airfoil. Similarly, separations (whether shock-induced or otherwise) also change the pressure distribution over an airfoil, thereby contributing to pressure drag. Skin-friction drag, however, is affected by the state of the boundary layer, which can be reduced by maintaining laminar flow over more of the airfoil surface.

Modern supercritical airfoils are designed carefully to limit the shock strength to avoid the detrimental effects of SBLIs as well as wave drag. The problems associated with both of these effects increase rapidly with Mach number; for this reason, shock-related problems are the major constraint on the cruise-Mach-number or cruise-lift coefficient. For civil-transonic aircraft, it is also necessary to retain sufficient margin before any severe adverse effects occur (e.g., shock stall or buffet); typical design shock Mach numbers for transonic cruise range from $M = 1.1$–1.2. A better understanding of SBLIs and control of their adverse effects is the key to improving transonic-airfoil performance.

One method by which a reduction in shock strength can be achieved is to utilize the compression waves that reflect from the sonic line. These waves decrease the Mach number in the supersonic region, thereby reducing the shock Mach number. This process often is referred to as isentropic compression to distinguish it from shock compression. In principle, it is possible to achieve a shock-free isentropic airfoil [4] in which the compression at the end of the supersonic region is achieved entirely by compression waves. In practice, however, this is difficult because the compression waves are prone to merging into a shock wave.

3.2.2 Supersonic Engine Intakes

For jet aircraft operating at supersonic speeds, it is necessary to decelerate and compress the oncoming air to subsonic velocities before entering the engine. This is achieved via the intake. The simplest form of compression is via a normal shock ahead of a Pitot inlet, but this incurs significant stagnation-pressure losses, which render this form of intake impractical for Mach numbers greater than 2. A better approach is to generate a series of oblique shock waves that can increase the pressure and reduce the Mach number before eventually changing the flow state to subsonic through a terminating near-normal shock. For a given incoming flow Mach number, a series of multiple shock waves incurs a smaller entropy production (and, thus, lower losses) than a single normal shock wave. Depending on whether the oblique shock waves are generated outside the intake or within the inlet duct, such designs are referred to as *external* or *internal compression inlets* (Figs. 3.6 and 3.7). In practice, the terminating shock is generally an oblique shock wave of

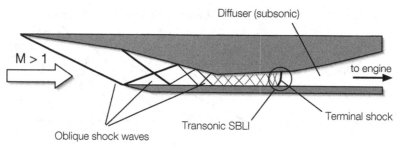

Figure 3.7. Internal compression inlet.

the strong-shock solution; however, for simplicity, we refer to it as a near-normal shock wave.

In either case, shock waves interact with the boundary layer growing along the aircraft or inlet surface. Most of the interactions feature oblique shock waves with supersonic flow on both sides of the interaction (see Chapter 4). However, in each inlet design, there also is a final terminating, near-normal shock wave that switches the flow from supersonic to subsonic speeds in a transonic SBLI. Typically, these terminating transonic shock waves are stronger than those observed on transonic wings or in turbomachinery with upstream Mach numbers ranging from 1.3 to 2. Although it is preferable to design inlets with weaker terminating shock waves, constraints on overall system size generally limit the compression that can be achieved through oblique shock waves.

Such strong interactions pose considerable problems for inlet efficiency. First, strong normal or near-normal shock waves incur considerable entropy increase and stagnation-pressure loss, which is a direct performance loss for the system. Second, the boundary layer already has experienced a number of adverse-pressure-gradient regions in previous oblique SBLIs. This makes the boundary layer more vulnerable to flow separation when encountering the final shock. Flow separation has an obvious detrimental impact on inlet performance. In addition to the introduction of additional stagnation-pressure losses, it introduces considerable nonuniformity in the flow entering the subsonic diffuser or the engine. This is shown schematically in Fig. 3.8. Because the flow separation occurs just at the entrance to the subsonic diffuser, which constitutes a further adverse-pressure gradient, the reversed flow region typically reaches far downstream and thus generates a highly nonuniform flow at the engine face. This nonuniformity, or flow distortion, is often the prime concern for inlet designers because it not only causes significant performance degradation; it also can prove extremely harmful to the engine.

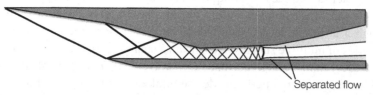

Figure 3.8. Separation originating from the terminal shock wave in an internal compression inlet.

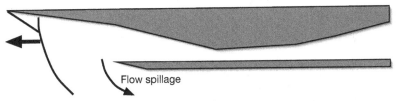

Figure 3.9. Inlet unstart.

Finally, any flow separations are also likely to introduce considerable unsteadiness into the flow, which can lead to unacceptable dynamic loads on the engine. If the terminal (near-normal) shock oscillation is so extreme that it reaches the converging part of the inlet geometry, it becomes unstable. At this point, it moves rapidly upstream, making more of the flow inside the inlet subsonic, until it is eventually expelled from the intake causing *unstart* (or buzz, if this phenomenon is periodic), as illustrated in Fig. 3.9. This is comparable to shock stall or shock-induced buffet on transonic wings; such a violent event is extremely damaging to the engine.

To avoid the problems associated with strong transonic SBLIs in inlets, designers make use of flow control to enable the boundary layer to stay attached even when the shock waves have considerable strength. The most popular control method is boundary-layer suction, or bleed. Flow control for transonic SBLIs is discussed in more detail later in this chapter, but we emphasize here that modern inlets cannot function successfully without some form of SBLI control.

3.2.3 Internal Flows

In internal transonic flows, SBLIs are experienced when a supersonic flow is decelerated to subsonic speeds such as in diffuser flows (Fig. 3.10) or supersonic-nozzle flows for strongly overexpanded conditions – that is, when the pressure at the nozzle exit is much lower than the ambient pressure (Fig. 3.11). In certain industrial applications, it is also possible to encounter supersonic flow in ducts.

A particular problem is blockage effects caused by the confinement of the flow in a restricted area. In cases in which the boundary-layer thickness is relatively large compared to the duct size, it is possible to encounter multiple shock waves in close succession – the so-called shock trains (see Section 3.3.4).

3.3 Normal SBLIs in Detail

In most practical situations, the surface curvature in the vicinity of transonic SBLIs is relatively small and the shock wave is almost normal to the oncoming flow. Even

Figure 3.10. Transonic SBLI in a diffuser flow.

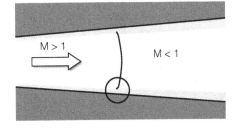

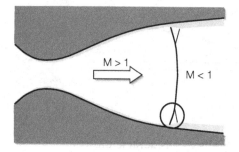

Figure 3.11. Transonic SBLI in a nozzle.

on cambered airfoils or turbine blades, strong surface curvature in the shock region generally is avoided by design. For this reason, it is convenient to concentrate on the idealized case of a normal shock wave interacting with a boundary layer on a flat surface. The effects of surface curvature and three-dimensionality or sweep are discussed in Section 3.3.4.

In most practical situations, transonic SBLIs tend to occur where the boundary layer is turbulent. In principle, the flow structure is similar for laminar and turbulent transonic interactions. However, because turbulent boundary-layer profiles are much fuller, they also exhibit a smaller upstream influence and interaction size compared to laminar interactions. Conversely, laminar boundary layers cannot withstand adverse-pressure gradients very well; therefore, it is rare to see attached, transonic interactions for laminar flow.

3.3.1 Attached-Flow Interaction

The basic situation considered herein is that of a boundary layer on a flat plate intersected by a normal shock wave at a streamwise position, as sketched schematically in Fig. 3.12. By definition, this means that the flow ahead of the shock wave is supersonic, whereas it is subsonic behind the shock wave. If we treat the shock wave as an inviscid event, then we can approximately distinguish between an inviscid outer-flow region and the viscous region close to the surface.

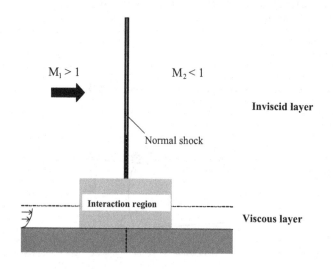

Figure 3.12. Normal SBLI.

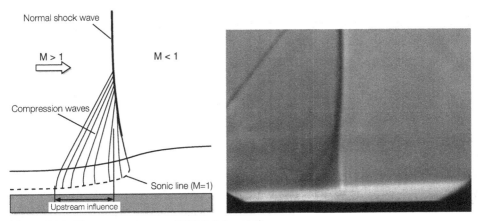

Figure 3.13. Unseparated normal SBLI ($M_s = 1.3$).

In the free stream, the shock wave constitutes a sudden pressure jump (ignoring the small thickness of the shock, which is on the order of a few molecular-mean-free paths); thus, the adverse-pressure gradient is infinitely strong. Because boundary-layer separation depends primarily on the gradient of adverse pressure rather than the magnitude of the pressure rise, any boundary layer would separate if the shock extended unchanged into the boundary layer. In practice, however, this is not the case due to the way the boundary layer and the shock interact.

Inside the boundary layer, there is a "sonic line" that separates the supersonic from the subsonic portion of the boundary layer. Below the sonic line, pressure information can travel upstream; this spreads the pressure increase ahead of the shock location. In response, the boundary-layer profile changes shape according to the adverse-pressure gradient experienced. This results in an increase in shape factor and thus an increase in displacement thickness. Consequently, the streamlines in the (supersonic) flow above the sonic line are deflected away from the surface, generating compression waves. The compression waves act to increase the local boundary-layer-edge pressure, which in turn affects the boundary-layer development. Conversely, the compression waves also decrease the Mach number ahead of the shock, thereby weakening its strength. As a result of this interplay, the flow settles into an equilibrium position, shown schematically in Fig. 3.13. The sketch can be compared to a schlieren image of a normal shock–turbulent boundary-layer interaction at $M = 1.3$.

The compression waves form in the supersonic portion of the incoming boundary layer from where they propagate into the outer flow. In this region of the boundary layer, the flow behaves almost like an inviscid supersonic flow with a nonuniform (and rotational) inflow. Viscosity has little bearing on the flow development because the interaction process is taking place over a short distance (comparable to a boundary-layer thickness), which allows viscous forces little time to act. Only when very close to the wall do viscous forces remain significant – and, of course, they cause the nonuniformity of the inflow in the first place.

The fact that the shock wave is "smeared" into a series of compression waves near the boundary-layer edge has important consequences. First, the shock-induced pressure rise along the surface is spread over a streamwise distance. This reduces

Figure 3.14. Surface-pressure distribution underneath a normal SBLI at $M = 1.3$.

the pressure gradient experienced by the boundary layer. Second, the flow passing through the smeared shock foot undergoes near-isentropic compression with little if any shock compression. In transonic interactions, it is commonly observed that the main shock deteriorates to almost a sonic wave at the boundary-layer edge. Third, the streamlines ahead of the main shock are deflected away from the surface (in line with the increase in displacement thickness). The schlieren image in Fig. 3.13 reflects this in the shape of the main shock, which is tilted slightly backwards as the boundary-layer edge is approached (remaining approximately normal to the local flow direction).

As more of the shock wave is replaced by compression waves in the shock foot, the pressure jump through the remaining shock reduces as the wall is approached. At the boundary-layer edge, the shock is almost completely reduced to a Mach wave and the surface pressure is close to the critical pressure at sonic conditions ($p/p_0 = 0.528$). The remainder of the overall shock-induced pressure rise imposed by the outer flow occurs in the subsonic region following the shock. Here, flow changes are gentler due to the nature of subsonic flow and the absence of waves. As a result, the initial surface-pressure increase is more rapid in the supersonic portion of the interaction than in the later part; this is reflected in a break or kink in the surface-pressure distribution, as shown in Fig. 3.14. Particularly in confined flows (i.e., channel flows), it can take a long time before the inviscid pressure jump is reached at the wall.

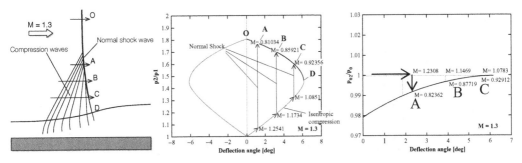

a) flowfield

b) pressure change across isentropic compression followed by normal shock wave

c) stagnation pressure change (isentropic compression followed by normal shock)

Figure 3.15. Static- and stagnation-pressure changes through the smeared shock foot.

The smearing of the shock foot causes the pressure immediately behind the shock to vary with wall distance. This is illustrated in Fig. 3.15, which shows the pressure-deflection diagram for an isentropic compression followed by a normal shock. Here, we consider a Mach number of 1.3, which is at the high end for an unseparated turbulent interaction.

Far from the wall, the flow experiences a normal shock wave and the associated pressure jump (marked "O"). In position "A," there are only a few compression waves before the main shock, which has reduced minimally in strength. With decreasing wall distance (moving to "B" and "C"), more of the shock is replaced by compression waves; this has considerable impact on the overall pressure jump as well as the local flow direction (which becomes more inclined away from the surface). A compression by isentropic waves alone is close to the actual situation at the boundary-layer edge (marked "D"). The pressure is only slightly more than half of the inviscid pressure jump across the main shock ("O"). This variation on postshock pressures generates a strong pressure gradient in wall-normal direction, which must be supported by convex streamline curvature in this region and is shown in Fig. 3.16.

Behind the smeared shock foot, other flow parameters such as Mach number and stagnation pressure also vary, as indicated in Fig. 3.15. The effect is seen clearly in measurements such as those shown in Fig. 3.17, where the stagnation pressure behind the smeared-shock system is greater than the free-stream postshock value. At the boundary-layer edge, the stagnation pressure is almost identical to the inflow value, which confirms that the compression is achieved almost isentropically.

Although the effect is felt in the inviscid flow, the reduction in stagnation-pressure losses through the smeared-shock foot is a consequence of the *viscous* interaction. This highlights the fact that the viscous forces experienced inside the boundary layer are only part of the viscous effects in SBLIs. Inviscid-prediction methods that cannot correctly account for the SBLI therefore are likely to overpredict shock losses (or wave drag on airfoils).

The near-sonic velocity at the boundary-layer edge makes the development of secondary supersonic regions behind the shock foot probable. As discussed previously, this is a region in which convex streamline curvature is expected to support the vertical pressure gradient as well as because the flow direction has

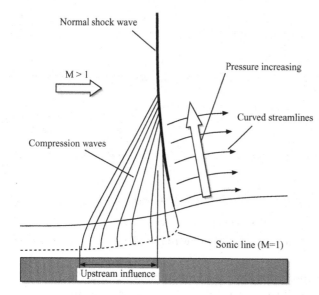

Figure 3.16. Pressure gradient and streamline curvature behind the shock foot.

been deflected away from the wall by the interaction and needs to return to the horizontal. Similar to a transonic airfoil, the combination of convex curvature and near-sonic velocities can lead to the development of supersonic regions (i.e., "supersonic tongues") in which expansion waves are generated at the surface that reflect as compression waves from the sonic line. Similar to supersonic regions on transonic airfoils, supersonic tongues often are terminated by weak secondary shocks. Their size and magnitude depend on the degree of flow curvature after the shock (which may be enhanced by confinement effects), the curvature of the boundary-layer displacement surface, and the extent to which the main shock is replaced by compression waves (which relates to the growth rate of the boundary layer ahead of the shock). If the supersonic region is relatively short, then the terminating secondary shock is likely to be in a region where there is continued streamline curvature and other (but smaller) supersonic regions with additional (and even weaker) shocks can be observed. This gives the appearance of a string of ever-weaker "shocklets" just downstream of the smeared shock foot. An example is shown in Fig. 3.18, which includes a contrast-enhanced schlieren image to emphasize these features.

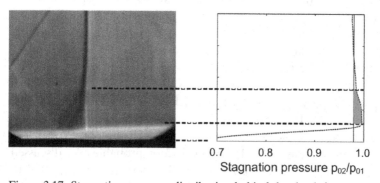

Figure 3.17. Stagnation-pressure distribution behind the shock foot.

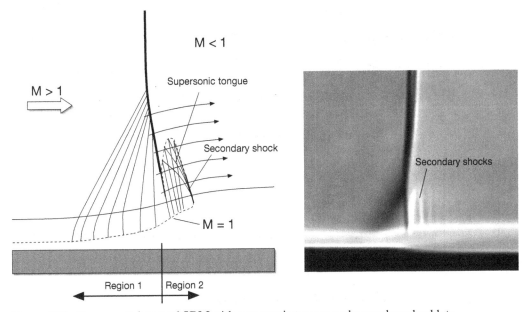

Figure 3.18. Unseparated normal SBLI with supersonic tongue and secondary shocklets.

Secondary shocks aside, the main effect of the supersonic tongue on the boundary layer is to delay the pressure increase downstream of the shock so that it can travel a fairly large distance before the inviscid postshock pressure is reached. This generally goes hand in hand with a period of relaxation in which the boundary-layer profile settles down to return to typical flat-plate conditions. Figure 3.19 shows the variation of key boundary-layer parameters through the $M = 1.3$ normal SBLI introduced in the previous images. These results were obtained with detailed nonintrusive Laser Doppler Velocimetry (LDV) measurements along a series of wall-normal traverses. Although the LDV technique can provide excellent spatial resolution, it requires the flow to be steady; thus, it is not completely able to resolve all fine-scale details. For example, although the supersonic tongue is detected in general, individual shocklets cannot be observed.

As indicated in Fig. 3.18, the interaction region as a whole can be divided into two domains: (1) the upstream supersonic region, and (2) the postshock, subsonic (except for supersonic tongues) region.

Region I (Upstream of Main Shock)

Other than very close to the surface, the flow in this region is entirely supersonic and the flow development is determined primarily by upstream flow conditions. The surface pressure increases rapidly through supersonic compression waves and the boundary layer experiences rapid decelerations. Most of the flow changes occur very close to the surface and, as a result, the skin friction decreases quickly and the boundary-layer shape factor increases rapidly. The growth of the boundary-layer displacement thickness is linked directly to the pressure rise by inviscid mechanisms. The magnitude and length scales of the flow changes are determined by the upstream boundary layer and the shock Mach number; universal solutions can be found that cover a range of practical situations as described by the free-interaction

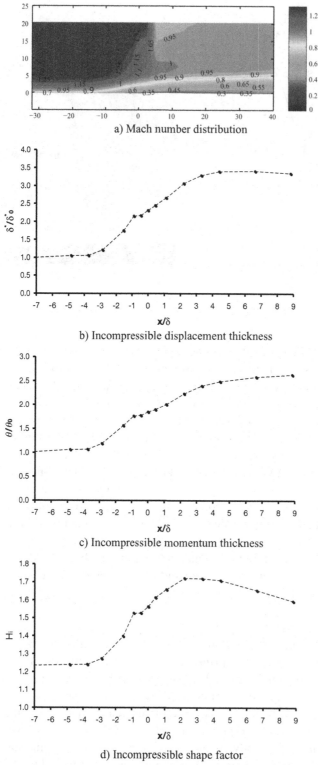

a) Mach number distribution

b) Incompressible displacement thickness

c) Incompressible momentum thickness

d) Incompressible shape factor

Figure 3.19. Flow parameters through an unseparated SBLI.

theory (see Chapter 2). Typically, Region I starts a few boundary-layer thicknesses upstream of the main shock (i.e., the upstream-influence length) and ends where the boundary-layer-edge velocity and surface pressure indicate $M = 1$.

Region II (Downstream of Main Shock)

Downstream of the shock, the flow changes much more slowly. The pressure rise is now achieved by subsonic mechanisms (supersonic tongues serve only to slow the pressure rise even further) and, as a result, there is typically a change in the slope of the pressure distribution around $p/p_0 = 0.528$ ($M = 1$) (see reference [5]). This reduction in the pressure gradient dp/dx indicates a change in the character of the flow from a supersonic to a subsonic interacting flow. Boundary-layer parameters also change much more slowly. Now that the adverse-pressure gradient has reduced significantly, the boundary layer can slowly recover towards equilibrium. This is first experienced close to the surface, where the flow accelerates and the skin friction increases. The shape factor also reduces and a reduction in displacement thickness sometimes can be observed. The flow development in this region is strongly influenced by downstream pressures and geometry effects, such as surface curvature and blockage. Region II is generally larger than Region I, and it can take many tens of boundary-layer thicknesses before equilibrium is reached.

Inflow–Shape-Factor Effects

Because of the dependency of the flow in Region I on upstream-flow properties only, it is possible to make general observations on the size of the interaction. Whereas the streamwise extent of the interaction is determined by the upstream influence, the height of the smeared shock foot above the boundary layer is determined by this length and the Mach angle of the inflow. Therefore, the upstream-influence length is the critical dimension of the interaction.

There is a clear dependence of the upstream influence on the shape factor of the incoming boundary layer. A fuller boundary-layer profile features larger velocities close to the surface, a thinner subsonic layer, and reduced upstream influence. In contrast, a less-full incoming boundary layer has a thicker subsonic layer and generally exhibits a larger upstream influence.

If we concentrate on Region I, we can define the upstream interaction length L^* as the distance between the origin of the interaction (i.e., the point where the pressure at the wall starts to rise) and the streamwise station, where the local pressure is equal to the critical value p^* corresponding to a Mach number equal to unity. The data plotted in Fig. 3.20 show the variation of the scaled interaction length L^*/δ_0^* with the upstream Mach number, Reynolds number (computed using the boundary-layer displacement thickness δ_0^*), and shape parameter H_{i_0}. It shows that for a given value of H_{i_0}, the incoming boundary-layer displacement thickness is the correct scaling for the interaction length. Moreover, the ratio L^*/δ_0^* is insensitive to the upstream Mach number. The scatter when M_0 approaches 1.3 indicates that the flow comes close to separation (see Section 3.3.2). The invariability of L^*/δ_0^* with respect to M_0 can be understood from the following argument: Raising the upstream Mach number increases the strength of the perturbation and therefore its tendency to propagate farther upstream. At the same time, the subsonic part of the boundary layer, which determines the upstream influence, becomes thinner.

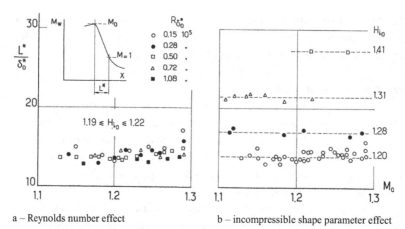

a – Reynolds number effect   b – incompressible shape parameter effect

Figure 3.20. Transonic SBLI. Influence of initial conditions on the supersonic interaction length.

Therefore, by virtue of these two compensating mechanisms, L^*/δ_0^* is nearly independent of the Mach number.

The data in Fig. 3.20 also demonstrate a strong dependence of the interaction length on the incompressible-shape parameter. As H_{i_0} increases from 1.2 to 1.4, the normalized upstream influence L^*/δ_0^* almost doubles. This confirms that the boundary-layer-velocity profile has a strong effect on the response of a boundary layer exposed to a shock.

3.3.2 The Onset of Shock-Induced Separation

As discussed previously, laminar boundary layers separate almost as soon as a sizeable shock wave is present. Turbulent boundary layers, however – and these are the most likely in practical situations – can withstand a considerable pressure increase, especially because the shock-smearing induced by the interaction reduces

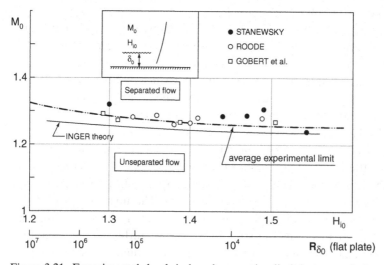

Figure 3.21. Experimental shock-induced separation limit in transonic flow.

the pressure gradients "felt" by the boundary layer. On flat surfaces in nominally two-dimensional flow, shock-induced separation of turbulent boundary layers is generally observed once the shock Mach number reaches the range of $M_s = 1.3$–1.35. There is no single exact Mach number at which separation occurs because this depends on a number of parameters.

Although we would intuitively argue that less-full boundary-layer profiles, such as those having experienced roughness or previous adverse-pressure gradients, would separate more easily, it is those same flow cases that also exhibit the largest upstream influence and, therefore, the greatest degree of shock-smearing. This causes a reduced adverse-pressure gradient that, to some extent, offsets the greater sensitivity of the flow to the pressure rise. In the case of very full boundary layers, the greater resistance to separation is offset almost exactly by the greater pressure gradients experienced due to a shortened interaction length. As a result, the onset of separation is relatively insensitive to changes in shape factor. It appears that the interaction adjusts to generate an adverse pressure gradient that suits the incoming boundary layer. Compared to other flow separations – for example, in incompressible flow – the sensitivity to separation in transonic interactions to boundary-layer shape is significantly reduced.

Much research has been undertaken to predict the onset of incipient separation in general terms. Most of it is based on the free-interaction theory outlined in Chapter 2, which states that separation occurs when the pressure rise across the shock wave exceeds the plateau value predicted by the following equation:

$$\frac{p}{p_\infty} = 1 + k \, M_\infty^2 \frac{\sqrt{C_f}}{\left(M_\infty^2 - 1\right)^{1/4}} \tag{3.3}$$

Here, the parameter k relates to the dimensionless factor F used in the original free-interaction theory, as follows:

$$k = F(x) \frac{\gamma}{\sqrt{2}} \tag{3.4}$$

For practical purposes, this makes k equivalent to F. Various researchers proposed values for k or F at separation. In extensive studies of a wide database, Zheltovodov et al. [6,7] concluded that k ranges from 6 to 7.4. They argued that the onset of intermittent separation can be predicted by comparing the theoretical pressure jump across a normal shock wave with the free-interaction theory estimate of plateau pressure. This is illustrated in Fig. 3.22, where it is shown that the intersection of the theoretical normal-shock pressure jump (ξ) with the pressure rise predicted by free-interaction theory for k values between 6 and 7.4 marks the onset of experimentally observed intermittent separation (i.e., the shock Mach numbers are around 1.25).

Conversely, the actual wall pressure across a shock wave is always smaller than the theoretical pressure jump across a normal shock due to interaction effects. The experimentally observed wall pressures immediately behind the shock for a range of shock Mach numbers are included in Fig. 3.22. Using the actual wall pressures, it is demonstrated that fully developed separation is observed when they reach the pressure rise predicted by free-interaction theory (for $k = 6$).

In practice, it is desirable to be able to predict the onset of incipient separation without having to refer to actual wall-pressure measurements. This can be

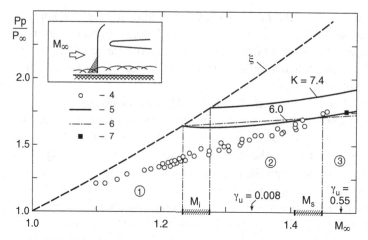

Figure 3.22. Experimentally observed pressure ratio across normal shocks: Region ① – unseparated flow; ② – intermittent separation; ③ – fully developed separated flow (adapted from Chapter 4, Fig. 4.4).

achieved by noting that the boundary-layer-edge flow just behind the shock foot is close to sonic conditions, as described previously. Figure 3.23 compares the post-shock wall pressures of Fig. 3.22 with the pressure calculated for sonic conditions ($p = 0.528p_0$), where the stagnation pressure p_0 is assumed to be unchanged from the free-stream value ahead of the shock wave. Because the main shock is smeared into isentropic compression waves, this assumption is reasonably valid and it can be seen that the predicted pressure values follow the experimental data very well. There is some departure at small Mach numbers because the assumption of a completely smeared shock foot is no longer justified and at larger shock Mach numbers, where the presence of significant separation makes this simple model invalid. Nevertheless, for shock strengths near separation (e.g., for turbulent boundary layers), it is possible to determine the separation onset by equating the theoretical pressure rise to sonic conditions with the plateau pressure from free-interaction theory. For

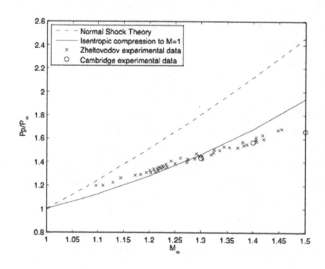

Figure 3.23. Comparison of experimentally observed pressure rise across normal SBLI (from Fig. 3.22) with pressure calculated for sonic conditions.

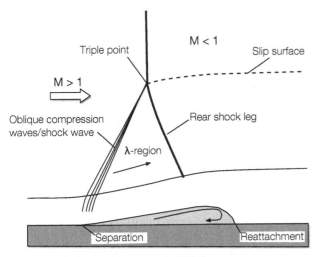

Figure 3.24. Separated transonic SBLI.

$\gamma = 1.4$ and a value of $k = 6$, this gives:

$$\frac{p}{p_\infty} = 0.528 \left(1 + 0.2 M_\infty^2\right)^{3.5} = 1 + 6 \frac{M_\infty^2 \sqrt{C_f}}{\left(M_\infty^2 - 1\right)^{1/4}} \tag{3.5}$$

To obtain an actual value for the shock strength at separation, it is necessary to express the skin-friction coefficient C_f as a function of M_∞ and Re (e.g., see reference [8]). The free-interaction theory emphasizes the viscous forces acting in an interacting boundary layer, and this dependence is represented by the skin-friction coefficient in the previous equation. However, inertial forces (i.e., momentum) also play a role and can dominate in a turbulent boundary layer. According to the free-interaction theory, resistance to separation decreases when the flow is less viscous (i.e., when the Reynolds number is increased).

This is well verified for a laminar boundary layer, which is a viscous dominated flow. Conversely, for the turbulent case, the boundary-layer profile is more full when the Reynolds number is high and therefore more resistant to separation even though the skin-friction diminishes. Thus, we observe a trend reversal, with inertia terms becoming predominant compared to viscous terms. Therefore, although the previous equation provides a reasonable (and useful) agreement for the range of Re typically observed in experiments, it does not correctly capture the influence of this parameter – in fact, as observed by Green [9], it predicts the wrong trend.

For general purposes, we may assume that shock-induced turbulent separation on a flat surface starts at shock Mach numbers in the 1.3 to 1.35 range. In practical situations, such as on transonic wings, the situation is complicated by the presence of surface curvature and the influence of three-dimensional effects. Both are discussed later in this chapter.

3.3.3 Separated SBLIs

Once shock waves are strong enough to separate the boundary layer, a different flow pattern is observed, as illustrated in Fig. 3.24. On a flat surface, the separation is usually of limited extent. In other words, there is a reattachment point at a downstream distance from the shock and the separated flow forms a bubble. As discussed later, the transonic nature of the flow actually aids the reattachment process, which keeps the size of the separation bubble relatively compact in most cases.

The first effect of separation is that the upstream-influence length is increased considerably because the reversed flow underneath the shock allows for a more effective upstream transmission of the pressure rise. At the separation point, the boundary layer is displaced from the wall, producing a relatively sharp "kink" in the boundary-layer displacement surface. This has the effect of compression waves "bunching up" around the separation point. At some distance from the boundary-layer edge, these compression waves merge into an oblique shock wave. The separated shear layer above the reversed flow cannot support significant pressure gradients; the pressure immediately after separation remains relatively uniform and the shear layer is relatively straight.

The pressure rise through these initial compression waves/oblique shock is not sufficient to incorporate the full pressure increase across the main shock; a second shock wave forms a rear "leg" of the shock structure. The rear shock leg is inclined as shown in Fig. 3.24 because it remains approximately normal to the flow behind the leading shock leg. The rear and front shock legs meet the main shock at the "triple point." This type of shock system is referred to generally as a lambda-shock structure because of its similarity with the Greek letter λ.

The flow passing through the bifurcated λ – shocks experiences a reduced entropy rise compared to the flow passing through the main shock. As a result, the flow conditions above and below the triple point differ behind the shock system and a slip surface forms to separate the two flow states. The static pressures and flow directions are equal on both sides but the velocities, densities, Mach numbers, and stagnation pressures are different.

The flow downstream of the shock system is subject to a complex interplay of subsonic and supersonic physics. The reattachment of the boundary layer and the straightening of the flow (which previously was deflected away from the surface) cause convex streamline curvature so that there is a significant vertical pressure gradient. For this reason (but also because of flow nonuniformity in the λ-region), the strength of the rear shock leg varies with wall distance. It is generally weakest (and sometimes vanishes completely) at the boundary-layer edge and strengthens towards the triple point. This goes hand in hand with a change in shock direction. At the triple point, the rear shock leg is almost a normal shock, but as the boundary-layer edge is approached and the shock strength reduces, it becomes more oblique tending towards the Mach angle.

The separated shear layer cannot sustain a strong adverse-pressure gradient and, if the rear shock leg impinges on it, an expansion fan is reflected. As a result, there is generally a supersonic tongue behind the λ region. Downstream-flow conditions, which modify the pressure field in the subsonic flow, have a strong effect on this region and also can change the shape and strength of the rear shock foot. Figure 3.25 is a sketch of this flow pattern in comparison with a schlieren photograph, experimentally observed wall pressures, and velocity measurements for a separated SBLI at $M_s = 1.5$. This shows that the wall-pressure rise across the interaction is

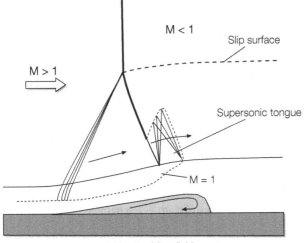

a) Sketch of flowfield

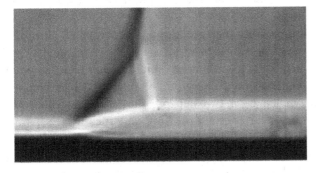

b) Schlieren photograph

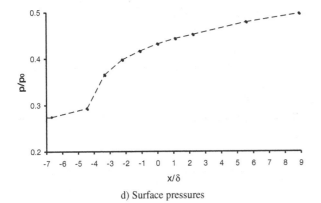

c) Mach number distribution

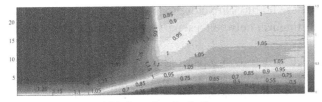

d) Surface pressures

Figure 3.25. Mildly separated transonic SBLI ($M_s = 1.5$).

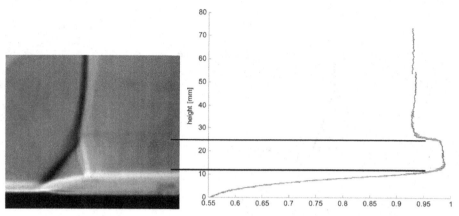

Figure 3.26. Stagnation-pressure distributions downstream of a separated SBLI.

significantly below the theoretical normal shock value, which is reached asymptotically only after a long streamwise distance.

Because of the continued pressure increase in the subsonic flow following the shock system, streamtubes expand, which in turn aids the reattachment process. If the separation bubble is relatively long, a more rapid pressure increase often is observed around the reattachment point. Therefore, a separated shock boundary-layer interaction often is typified by a two-step pressure increase with a rapid pressure rise until separation, a near-constant pressure underneath the separation bubble, and a second pressure rise (but not as steep as in the supersonic region) starting around the reattachment region. This "two-step" pressure rise distinguishes the wall-pressure profile from that of attached interactions and sometimes allows the identification of separation from pressure measurements alone. However, when the separation bubble is small, these features are close together and quite smeared such that it is often difficult to distinguish between a separated and an attached pressure distribution.

The presence of a λ-shock structure in separated transonic SBLIs has a noticeable effect on the stagnation-pressure losses, as shown in Fig. 3.26. Because multiple but weaker shocks in the shock foot incur reduced losses, there is a noticeable region of increased stagnation pressure between the boundary-layer edge and the slip surface. In practice, this region is small – typically only a few boundary-layer thicknesses in height; therefore, it has limited implications on the overall flowfield. However, the principle of achieving a reduction in losses due to a λ-shock structure forms the basis of shock control, which is discussed in Section 3.4.

Boundary-Layer Behavior in Separated Transonic Interactions

The behavior of turbulent boundary layers is best described by considering a typical case: the interaction of a normal shock wave in a slightly diverging duct. The boundary–layer-edge Mach-number distribution during the interaction is plotted in Fig. 3.27a, where the streamwise distance is scaled by the boundary-layer displacement thickness at the interaction origin. Just ahead of the shock wave, the flow reaches a peak Mach number of 1.4, giving rise to a normal SBLI that is strong enough to cause separation. Through the shock wave, the Mach number first

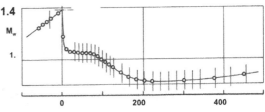

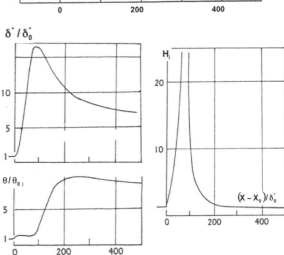

Figure 3.27. Variation of boundary-layer parameters through a separated transonic SBLI.

exhibits a rapid decrease during separation, followed by a region of almost constant M in the separated region (i.e., the pressure plateau). During reattachment, there is a further but less rapid decrease in the Mach number.

The displacement thickness δ^* (Fig. 3.27b) sharply increases at the interaction onset, generating a "viscous ramp" that is at the origin of the separation shock. Here, δ^* experiences considerable, almost linear growth during the development of the separation bubble – until it starts to decrease when reattachment begins because of the 'filling-out' of the boundary-layer-velocity profile.

Figure 3.27c shows the variation of the momentum thickness θ, which represents the momentum loss of the flow resulting in a drag increase or an efficiency loss. The momentum thickness θ undergoes an initial modest increase in conjunction with the pressure rise at separation. In the first part of the adjoining, almost isobaric separated region, θ stays constant. It then rises sharply during reattachment because of the boundary-layer thickening, the reattachment pressure rise, and the intense development of turbulence in the separated and reattaching free-shear layer. At the same time, the incompressible-shape parameter H_{inc} reaches very high values in the separated region, denoting the extreme changes to the velocity profiles that now contain reversed flow. When reattachment begins, H_{inc} decreases rapidly to return to values typical of a flat-plate boundary layer. This region characterizes the so-called rehabilitation process of the boundary layer.

The streamwise evolution of the maximum turbulent kinetic energy and shear stress inside the boundary layer is shown in Fig. 3.28. The maximum horizontal and vertical turbulence intensities (u', v') at each streamwise location are presented in Fig. 3.28b. All turbulence properties were scaled using the speed of sound a_{st} at inflow-stagnation conditions.

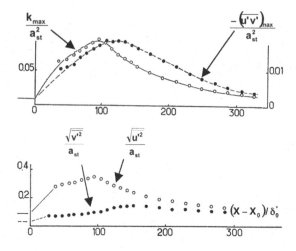

Figure 3.28. Streamwise variation of turbulence properties in a separated transonic interaction.

A feature of the turbulence behaviour is that strong anisotropy develops in the first part of the interaction, during separation, where $\overline{u'^2}$ grows more rapidly than $\overline{v'^2}$. The initially large increase in the Reynolds stress component $\overline{u'^2}$ can be explained by considering the production term in the $\overline{u'^2}$ transport equation, written here for simplicity for an incompressible flow (i.e., compressibility effects are negligible at transonic Mach numbers). This production term $P_{u'}$ is as follows (in which an overbar designates mean values, x is the streamwise coordinate, and y is the wall-normal coordinate):

$$P_{u'} = -2\overline{u'v'}\frac{\partial \overline{u}}{\partial y} - 2\overline{u'^2}\frac{\partial \overline{u}}{\partial x} \tag{3.6}$$

As a consequence of the rapid retardation of the flow in the first part of the interaction, the contribution of the streamwise derivative of $\overline{u}$ is as large as the term involving the strain rate $\partial \overline{u}/\partial y$ (and of the same sign). Thus, $P_{u'}$ is the sum of two large positive terms. Conversely, the production mechanism for the component $\overline{v'^2}$ is expressed by:

$$p_{v'} = -2\overline{u'v'}\frac{\partial \overline{v}}{\partial x} - 2\overline{v'^2}\frac{\partial \overline{u}}{\partial y} \tag{3.7}$$

The derivative $\partial \overline{v}/\partial x$ is small everywhere, whereas $\partial \overline{v}/\partial y$ is equal to $-\partial \overline{u}/\partial x$ (or nearly equal for a weakly compressible flow), so that the second term tends to decrease $\overline{v'^2}$ production in the first part of the interaction (where $\partial \overline{u}/\partial x$ is negative). Farther downstream (approaching reattachment), an ever-growing part of the viscous layer is accelerated by shear stresses, which explains the later growth of $\overline{v'^2}$.

The large turbulence anisotropy in the initial stages of the interaction tends to promote the production of the turbulent kinetic energy k because, for an incompressible flow, the production term of the k-transport equation can be written as follows (neglecting the derivative $\partial \overline{v}/\partial x$):

$$P_k = -\overline{u'v'}\frac{\partial \overline{u}}{\partial y} - (\overline{u'^2} - \overline{v'^2})\frac{\partial \overline{u}}{\partial x} \tag{3.8}$$

Measurements show that production due to normal stresses is as high as the production due to the shear stress over a streamwise distance on the order of five times the initial boundary-layer thickness. This roughly coincides with the region of steepest streamwise pressure gradient where there is a general retardation of the flow. Farther downstream, the normal stress contribution becomes negligible.

A second typical feature of turbulence behavior in strong interactions is the lag that is established between the mean and turbulent flowfields, leading to significant nonequilibrium. This can be demonstrated by considering a "phase plane" of the following two variables through the interaction and downstream recovery region:

- the square root $\sqrt{C_\tau}$ of the turbulent shear-stress coefficient defined as:

$$C_\tau = -\frac{2\,(\rho u'v')_{\max}}{\rho_e u_e^2} \qquad (3.9)$$

where $-(\rho u'v')_{\max}$ is the maximum turbulent shear-stress level measured at each streamwise location.

- the "equilibrium"-shape parameter J defined as:

$$J = 1 - \frac{1}{H_i} \qquad (3.10)$$

which has the merit of tending to a finite value when the vertical extent of the separation region increases (then H_i tends to infinity). Empirical correlation laws [10] show that the following variation of the well-known Clauser parameter:

$$G = \frac{(H_i - 1)}{\left(H_i\sqrt{0.5C_\tau}\right)} = \frac{J}{\sqrt{0.5C_\tau}} \qquad (3.11)$$

remains constant and equates to 6.55 for most equilibrium turbulent boundary-layer flows, the value of the constant being that of a "well-behaved" flat-plate boundary layer (see Chapter 2). The following equation:

$$\left(1 - \frac{1}{H_i}\right) = 6.55\sqrt{0.5C_\tau} \qquad (3.12)$$

forms a straight line in Fig. 3.29 and is representative of equilibrium boundary layers – that is, boundary layers where there is an instantaneous adjustment between turbulence and the mean-velocity field.

The data in Fig. 3.29, encompassing four transonic interactions of increasing strength including large shock-induced separation, all exhibit the following trends:

- The trajectories, or images, of the interactions in the phase plane $[J, 0.5C_\tau]$, all originate from a common point close to the equilibrium locus. Thus, the upstream boundary layer is close to equilibrium in all cases. Initially, the trajectories move below the equilibrium locus, which indicates that the boundary layer undergoes what is called a rapid interaction process in the shock-foot region. Here, there is a large departure from equilibrium characterised by a lag in the shear stress. Turbulence has no time to adjust to the rapid mean-velocity change. When the shock is strong enough, separation occurs during this part of the interaction process.

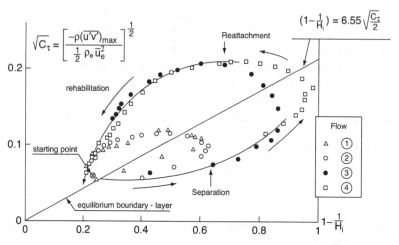

Figure 3.29. Turbulence history in transonic SBLIs.

- Thereafter, especially for separated flows, the shear stress and the shape parameter continue to increase together in a manner typical for a developing free-shear layer. This 'mixing-like' behavior could continue until $J = 1$ (i.e., $H_i \to \infty$). During the rapid-interaction process, the separating boundary layer undergoes such an overwhelming perturbation that the development of the free-shear layer is not significantly influenced by its initial conditions – that is, the initial boundary-layer properties (provided it is fully turbulent). Here, turbulence production continues in proportion to the growth of the large-scale structures forming near the separation location.
- At some location, the shape parameter J peaks and starts to decrease. This reversal signifies the onset of the reattachment process for separated flows. During this phase of the interaction, the shear-stress level continues to grow until it reaches a maximum at a streamwise station that almost coincides with the reattachment point. In the course of the reattachment process, the trajectories in the phase plane bend and cross the equilibrium locus at a point whose distance from the origin increases with the size of the separated region.
- Downstream of reattachment (and in the absence of any additional destabilizing influences), the trajectories return towards a final point located on the equilibrium line. During this relaxation process, the flow lies above the equilibrium line; at some point, a second maximum departure from equilibrium is reached. Farther downstream, the trajectories collapse onto a common trajectory tending towards the equilibrium locus.

3.3.4 Other Effects on Transonic SBLIs

Confinement Effects (Channels)

As described in the previous sections, the flow immediately behind a normal SBLI can contain secondary supersonic regions caused by wall-normal variations of flow conditions and streamline curvature. Interactions that occur in channels (or wind tunnels) in which the flow is confined by solid walls experience these effects more strongly. Because the streamline on the symmetry plane (or axis in a circular cross

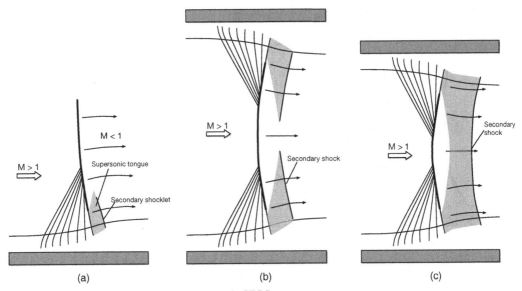

Figure 3.30. Confinement effects on transonic SBLIs.

section) is necessarily straight (ignoring asymmetric flows for the moment), the variation in streamline curvature between the boundary-layer edge and the centre is more rapid than in unconfined flows. This enhances any wall-normal flow variations and increases the size of any supersonic tongues. Furthermore, the core flow between the boundary-layer edge and the channel centre experiences a significant contraction due to the fact that the boundary layer grows rapidly through the interaction. This accelerates the subsonic flow behind the shock, which also serves to make secondary supersonic flow more likely. This is shown schematically in Fig. 3.30b.

If the blockage becomes too severe, it is possible that the supersonic tongues from all sides merge into a secondary supersonic flow region spanning the channel, as shown in Fig. 3.30c. Such a secondary supersonic flow generally features a terminating shock wave, which can be followed by further supersonic regions and shocks. When there is this close succession of supersonic regions and secondary shocks, it is called *shock train* (Fig. 3.31), which can be the cause of significant losses in transonic internal flows. These losses occur because, on the one hand, the core flow experiences a succession of (albeit weak) shock waves, each incurring a stagnation-pressure decrease; on the other hand, the boundary layer is subjected to a succession of shock waves that cause it to grow rapidly. However, the continued reacceleration to supersonic conditions reduces the pressure gradient along the wall and it can take a long streamwise distance for the pressure to reach the asymptotic value. Often more important than the losses in the flow is the nonuniform and unsteady nature of shock trains, which can have harmful consequences – for example, if the flow is to enter an engine. Therefore, it is advisable to avoid this phenomenon when possible.

Generally, the occurrence of shock trains depends on the ratio of boundary-layer displacement thickness to duct height or displacement area to duct area. Typically, whenever this ratio is greater than a few percentage points (based on displacement thickness), multiple shocks are likely. It is difficult to establish a general

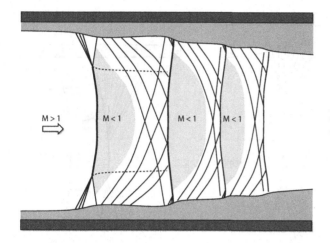

Figure 3.31. Shock train in a confined duct.

rule of thumb because the critical confinement parameter depends on the oncoming Mach number as well as the Reynolds number. For a good description of such flowfields at Mach numbers of around 1.6, readers are referred to the experimental studies by Om and Childs [11] and Carroll and Dutton [12] in circular and rectangular ducts, respectively. Both studies observed shock trains for confinement ratios δ^*/H between 4 and 6 percent, where H is the channel half-height or radius. Conversely, when the confinement ratio drops to less than 2 percent, only a single shock wave was present. Generally, it has been observed that the confinement necessary to cause shock trains reduces with increasing Mach number because the boundary-layer growth is accelerated for stronger shock waves.

In strong shock interactions in internal flows, it is possible to observe the interactions on either side of the channel interfering with one another. Strong normal-shock interactions are encountered in internal flows such as air-intakes or missile/space-launcher nozzles operating in overexpanded conditions. An example is shown in Fig. 3.32, which is a short exposure shadowgraph of strong interactions in a planar supersonic nozzle [13]. The Mach number at the interaction origins is close to 1.6, which is strong enough to cause significant separation. The λ regions of the separations on each nozzle wall are large enough to interfere with one another and, for reasons not yet fully understood, the flow adopts an asymmetric structure – despite the fact that the nozzle geometry is symmetrical. Here, the separation on

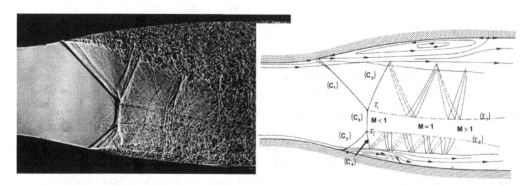

Figure 3.32. Separated transonic SBLIs in an overexpanded nozzle flow.

the upper wall occurs earlier than on the lower wall, and the intersection of the two oblique separation shocks gives rise to a Mach reflection in the centre of the channel.

The large separation on the upper wall gives rise first to a λ-shock structure, with the flow behind the rear shock leg (C_3) still supersonic (Fig. 3.32). On the lower wall, a smaller λ structure and a reduced separated region are observed. The intersection of separation shocks (C_1) and (C_2) forms an Edney Type II interference (i.e., Mach reflection) with the Mach stem being a nearly normal shock (C_5) bounded by two triple points T_1 and T_2. Slip lines (Σ_1) and (Σ_2) emanating from T_1 and T_2 form a fluidic subsonic channel between two supersonic streams. Due to conditions imposed by the adjacent supersonic flows, the flow in this channel accelerates until its reaches sonic speed at a throat. Thereafter, the fluidic channel cross section increases, with a further supersonic expansion. Where the rear legs of λ structures (C_3, C_4) impinge on the separated shear layers, expansion waves are reflected, which in turn reflect from slip lines (Σ_1, Σ_2) as converging compression waves. In turn, these waves are reflected by the nearly isobaric separated region as expansion waves, and the pattern is repeated over some distance.

This configuration demonstrates the complex flow structures caused by the coupling between the separated regions and the inviscid part of the flow in confined channels. Similar flows can be observed in other channel-type flows, such as supersonic diffusers, compressor cascades, and propulsive nozzles.

Surface-Curvature Effects

On transonic airfoils and wings, the surface underneath the interaction is not flat but rather generally features some form of convex curvature. The free-stream Mach number ahead of the shock also is not uniform but rather decreases with distance from the surface. However, apart from cases in which very large separations are present, the interaction region is relatively small compared to the size of the airfoil; the surface curvature in the shock region is also relatively mild. For this reason, most of the features described previously are as valid for attached and separated interactions on airfoils.

However, there are important consequences of surface curvature. The first effect is that the shock strength required for separation generally is found to be greater than on a flat surface. This was demonstrated theoretically [14,15,16] and in experiments [3,17]. Pearcey [3] suggested that this was due to the fact that convex curvature alone would cause a streamwise decrease in pressure that reduces the adverse-pressure gradient experienced by the boundary layer. Bohning and Zierep [16] offered an alternative explanation, arguing that convex surface curvatures lead to greater postshock expansions and that this effect can feed upstream through the subsonic portion of the boundary layer, thereby reducing the shock-induced pressure rise. Nevertheless, the curvature effect on incipient separation is relatively weak.

A more significant effect of surface curvature on the SBLI flowfield often can be observed just downstream of the interaction because the flow there is close to sonic conditions and therefore sensitive to geometric factors. Convex surface curvature in this region is likely to accelerate the flow, making secondary supersonic regions more likely or considerably increasing their size. This is generally an undesirable effect because of the increasing likelihood of secondary shocklets.

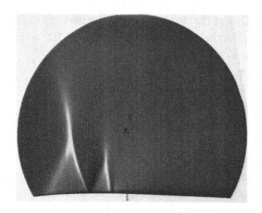

Figure 3.33. Multiple laminar SBLIs on a curved surface [18].

Thus, on well-designed transonic airfoils, shock waves are located in a region of relatively low curvature to avoid secondary supersonic regions (and any associated shock waves). However, particularly when the shock system is very weak (thus, not decelerating the flow much below $M = 1$), even mild convex curvature easily can cause significant secondary supersonic regions. This type of flowfield is seen often in low-Reynolds-number flows in which the boundary layer is laminar. In laminar interactions, even very weak shocks separate the boundary layer. These weak shock waves, however, do not decelerate the flow much below $M = 1$, and the flow curvature at reattachment combines with the effect of convex surface curvature on the airfoil to reaccelerate the flow to supersonic speeds. This can happen several times in succession, as shown in Fig. 3.33.

Sweep Effects

In transonic flows, the shock wave causing the SBLI is generally a normal or near-normal shock and the sweep angles are small. The most common example of a transonic SBLI with considerable sweep is on transonic aircraft wings where the shock follows the wing sweep, which can reach values of up to 30 degrees. Even in this case, experience seems to suggest that sweep effects are negligible as long as the flow remains attached. The main effect of sweep is on the onset of shock-induced separation. Research into the interaction of swept shock waves with flat-plate boundary layers suggests that these separate once the shock *normal* Mach number increases beyond 1.2 [19]. From this, we can conclude that cross-flow effects may bring about separation somewhat earlier than observed in the equivalent two-dimensional flow (in the absence of cross flow). Once shock-induced separation is observed, the flow becomes more three-dimensional; however, unless the flow downstream of the main shock is supersonic, the flowfield in a shock-normal plane is similar to the flow patterns shown previously.

3.3.5 Large-Scale Unsteadiness of Normal SBLIs

Unsteady effects in SBLIs are discussed in detail in Chapter 8. Shock-wave oscillations are typically triggered by disturbances in the oncoming supersonic flow, fluctuations within the incoming boundary layer, or unsteady separation bubbles underneath the shock. In transonic SBLIs, there is an additional unsteady mechanism due

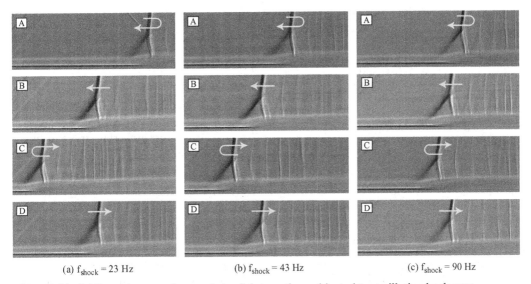

(a) f_{shock} = 23 Hz   (b) f_{shock} = 43 Hz   (c) f_{shock} = 90 Hz

Figure 3.34. Schlieren images of normal shock interaction subjected to oscillating back pressure at three different frequencies, $M = 1.4$.

to the fact that the flow downstream of the shock wave is subsonic. This allows pressure fluctuations originating downstream to travel back towards the shock wave and affect its strength and position. Typically, an increase in the back pressure causes shock waves to move upstream and vice versa.

Figure 3.34 shows a series of schlieren images from an experimental study of a normal shock wave in a constant-area channel subjected to sinusoidal back-pressure oscillations at various frequencies [20]. In response, the shock wave oscillates in a periodical motion around the steady-equilibrium position. Four images are shown per cycle: at the extremes of the shock location and halfway through each motion when the shock is near the equilibrium location but moving at maximum speed. The pressure waves originating downstream and traveling back towards the shock wave can be seen clearly in several of the images. With increasing frequency, the amplitude of the shock motion reduces (the pressure oscillation amplitude has been kept constant). The reduction of shock-wave travel is a result of the observation that the maximum shock velocity is determined by the amplitude of the pressure disturbance; because this was kept constant, a higher frequency allows the shock less time to move away from the mean position. In nonparallel ducts, the shock-motion amplitude due to pressure variations also is affected by the geometry. This is illustrated by Fig. 3.35, which compares the relationship between shock-oscillation amplitude and frequency (for a fixed-pressure amplitude) for parallel and divergent ducts. At high frequencies – where the shock oscillation amplitudes become small – there is little effect of divergence angle; however, at low frequencies, the geometry clearly has an effect.

Similar amplitude-frequency behaviour has been observed in the response of shock waves to disturbances originating from the upstream inflow or from inside the incoming boundary layer. This also explains why the response of shock waves to the relatively high frequency disturbances present in a turbulent boundary layer is

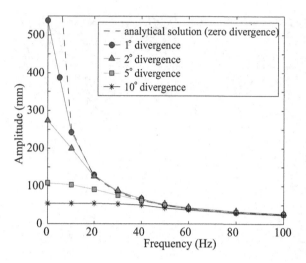

Figure 3.35. Shock oscillation amplitude in response to periodic variations of downstream pressure in constant area and divergent ducts.

relatively small in amplitude, causing shock 'rippling' rather than large-scale departures of shock location from the mean. For many technical applications in which shock oscillations are of concern, it is therefore particularly important to minimize low-frequency pressure fluctuations.

Pressure fluctuations not only cause shock motion but also affect the shock strength. A shock moving into the oncoming flowfield faces an inflow of greater relative Mach number (i.e., the free stream plus the shock motion); the opposite is true for a shock moving downstream. As a result, the pressure jump across the shock wave is increased during upstream motion and decreased when moving downstream. This is illustrated by the experimental results shown in Fig. 3.36.

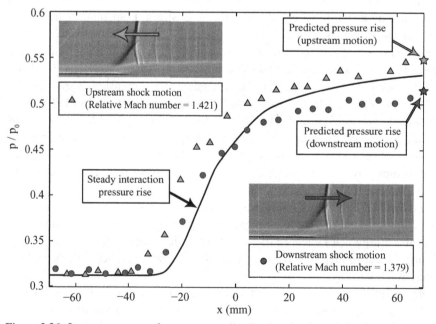

Figure 3.36. Instantaneous surface-pressure distribution for forced shock oscillation at $M = 1.4$ ($f = 43$Hz).

Even when the shock velocities are small, the differences in the shock-induced pressure rise can be considerable (e.g., a ± 1.5 percent variation in shock Mach number causes a 10 percent difference in postshock pressures), highlighting the non-linear nature of SBLIs in general. This effect is particularly noticeable at transonic Mach numbers because the relative variations are significant compared to the incident flow. Once the oncoming Mach numbers become large, however, this effect diminishes.

The sensitivity of transonic SBLIs to disturbances originating from the down-stream subsonic flow can have serious implications when the unsteady nature of the interaction causes global flowfield changes. This can be the case when there is considerable shock-induced separation. A pressure increase imposed from down-stream would cause the shock to run upstream (and increase in strength), and both may cause the separation to enlarge and change position. Depending on the overall flow geometry, this in turn can alter the downstream pressure; a dangerous feed-back mechanism may result, which greatly amplifies the pressure disturbances. An example of such an event sometimes is observed on transonic wings when large-amplitude shock oscillations occur along the upper surface (sometimes linked to a corresponding shock oscillation on the lower surface). This is one of the forms of *buffet* (there also are incompressible forms of wing buffet) that puts a severe strain on the wing structure and can destroy aircraft.

In this phenomenon, the normal shock moves over a significant portion of the airfoil-chord length, with the separated bubble disappearing and reappearing in a periodic manner. A scenario for such a periodic unsteadiness is proposed in Fig. 3.37, obtained from Navier-Stokes calculations. In the first frame, the shock occupies a downstream position but is moving upstream in response to a pressure disturbance. Thus, the flow velocity relative to the shock is increased. If the rela-tive Mach number is large enough, separation occurs and a large separation bubble forms due to the combined effect of shock-induced pressure rise and compression along the profile towards the rear of the airfoil. This separation has the effect of reducing the effective aerodynamic incidence so that the shock weakens as it con-tinues to move upstream until it stops; simultaneously, the separation bubble starts to shrink after having reached its greatest size (frame 3). This leads to an increase in the effective aerodynamic incidence, with the shock now moving downstream. The relative shock strength is then less (because the downstream movement reduces its effective Mach number) and separation at the shock foot is suppressed (frame 4). When the shock slows down, separation occurs and is then amplified when the shock starts to move upstream again and the cycle repeats.

Another example in which the flowfield downstream of an SBLI is sensitive to changes in the interaction is a supersonic engine inlet, which can suffer from severe shock unsteadiness. Inlet unstart (in which an unsteady shock is spilled out of the inlet causing engine shutdown) was previously discussed, but there also is a severe form of quasiperiodic shock unsteadiness, sometimes referred to as 'intake buzz'.

The prime purpose of an air intake is to slow down the oncoming supersonic flow before the engine to subsonic speeds through a number of shock waves. Pos-sible shock configurations are discussed in Section 3.22. For internal or mixed-compression engine inlets, the terminating near-normal shock wave is usually quite strong and flow separation is a considerable problem. When a sudden pressure

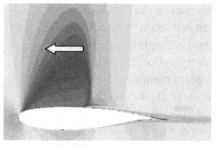

a – the shock moves upstream and induces separation

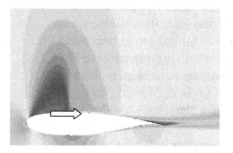

b – the shock reaches its most upstream location

Figure 3.37. A scenario for transonic buffeting over a profile. Navier-Stokes calculations (Furlano [21]).

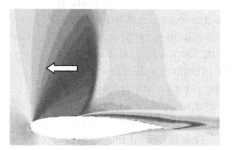

c – the shock stops and starts to move downstream

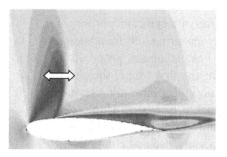

d – the shock moves downstream and does not induce separation

disturbance causes the shock to move upstream (and thus strengthen), it can increase the separated flow region and cause the downstream flow to 'choke', thereby generating an increased pressure pulse that drives the shock farther upstream. The inlet-unstart scenario is illustrated in Fig. 3.38.

A scenario similar to transonic buffet is at work in the buzz of supersonic air-intakes. In this case, a periodic large-amplitude motion of the shock system

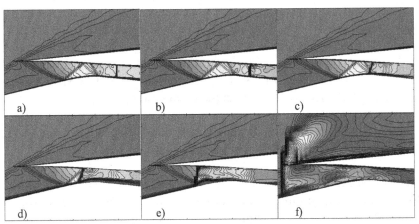

Figure 3.38. (a)–(f) Time sequence of inlet unstart developing in response to a pressure disturbance originating from downstream (RANS CFD solution provided by B. Anderson, NASA Glenn).

is triggered by unsteady SBLIs. There are two possible origins of air-intake buzz: (1) shock-induced separation, the so-called Dailey's scenario; and (2) the so-called Ferri's scenario, the swallowing of the slip line resulting from an Edney Type IV interference between the cowl shock and a shock formed by a compression ramp. Overexpanded propulsive nozzles are affected by unsteady and asymmetric shock-induced separation, which can be the source of high side loads during the start-up transient.

Large-amplitude oscillations also occur within rotating machines, such as compressors, turbines, and helicopter rotors. In these cases, the shock oscillation is forced by the device itself, which potentially can cause coupling between the body motion and the SBLI fluctuation frequencies, including those due to turbulence. A similar coupling mechanism is found in the aeroelastic response of a structure, such as compressor blades. In this case, the deformation of the structure induces shock displacements and a subsequent change in the pressure load, which can lead to a divergent process or flutter.

3.4 Control of Transonic SBLIs

From this discussion, the detrimental effects of transonic SBLIs can be summarized as follows:

1. The adverse-pressure gradient induced by the shock on the boundary layer causes significant thickening or even separation, both of which increase viscous losses. Furthermore, the changes to the boundary layer introduced by the interaction take a long time to die down, which makes the flow susceptible to separations occurring farther downstream. That said, it is theoretically possible for the interaction to have a beneficial effect on the boundary layer farther downstream because the increased turbulent mixing can promote a momentum transfer to the wall. However, the overall increase in shape factor and boundary-layer thickness generally outweighs this effect.

2. The stagnation pressure losses incurred by the shock wave introduce drag or efficiency reductions. Strictly speaking, this loss is unconnected to the shock's interaction with the boundary layer – in fact, the interaction actually helps to reduce the stagnation-pressure loss by smearing the shock wave near the surface. Nevertheless, the shock losses often are considered part of SBLI losses and changes to the SBLI structure affect their magnitude.
3. Where considerable separations occur, they lead to significant changes in the shock structure and overall flow. In some cases, this can introduce large-scale flowfield unsteadiness (i.e., buffet, inlet buzz, or engine unstart).

SBLI control aims to alleviate one or more of these effects. Depending on which type of physical mechanism is addressed, we speak of either *shock control* (which aims to change the shock-system structure to reduce stagnation-pressure losses) or *boundary-layer control* (which aims to reduce viscous losses and separation). Due to the complex nature of SBLI flows, this separation is not strict. For example, any modification of the shock wave has an effect on the boundary layer and vice versa. Depending on the exact circumstances, both methods have been shown to reduce shock-wave unsteadiness.

3.4.1 Shock Control

The principal aim of shock control is to reduce stagnation-pressure losses incurred by the presence of the shock. On transonic wings, this leads to a reduction in wave drag. In internal flows, it improves efficiency due to improved total pressures. Although shock control is applicable to both, most recent research has concentrated on the former application; therefore, we focus this discussion on transonic-wing applications of shock control. However, the underlying physical principles are just as valid in other SBLI flows.

For a normal shock wave, the stagnation-pressure decrease depends only on the incoming-flow Mach number, which we assume to be fixed. This suggests that it is impossible to reduce globally the associated losses for a given shock wave. However, as discussed, the presence of the SBLI gives rise to local reductions in stagnation-pressure losses due to the smearing of the shock foot near the wall (e.g., Fig. 3.26).

Shock control extends this principle and reduces stagnation-pressure losses *locally* by increasing the size of the smeared region and expanding the flow domain that experiences reduced losses. Instead of attempting to generate a large smeared shock foot with continuous compression waves, it is generally easier to aim for a large λ structure, replacing the shock foot with an oblique shock followed by a near-normal shock wave, as indicated in Fig. 3.39. Similar to downstream of the λ region in a separated SBLI, the flow behind a control-generated λ structure exhibits significant improvement in stagnation pressure.

Figure 3.40 shows the stagnation pressure downstream of a shock system composed of an oblique shock wave followed by a normal shock as a function of the flow deflection across the leading oblique shock wave. To illustrate the potential for stagnation-pressure savings, compare the stagnation-pressure decrease for a normal shock wave (i.e., 2.1 percent at $M = 1.3$ and 4.2 percent at $M = 1.4$) with the decrease incurred for a λ-shock system (i.e., oblique shock followed by a normal shock).

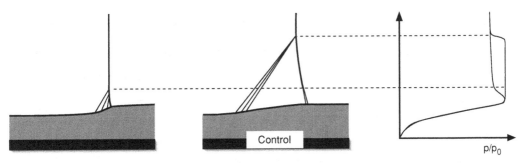

Figure 3.39. Basic mechanism of shock control.

For both Mach numbers shown, compression to subsonic speeds can be achieved with significantly reduced stagnation-pressure losses. At $M = 1.3$, an optimum lies around a deflection angle of about 5 degrees, giving an overall loss of 0.3 percent, whereas at $M = 1.4$, the optimum deflection angle is around 7 degrees with an overall loss of 0.7 percent. In both cases, the stagnation-pressure losses across the shock system are reduced by around 85 percent. However, as indicated in the flowfield sketch, the deflection angle also changes the location of the triple point and, thus,

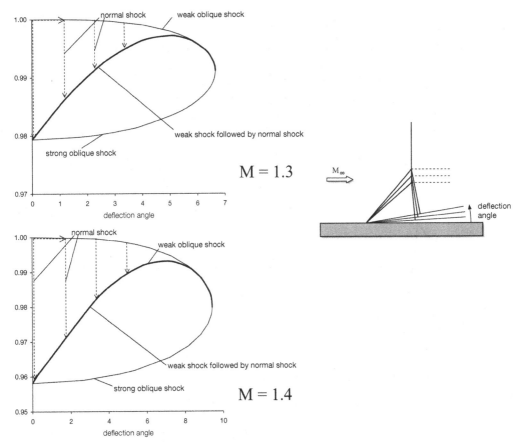

Figure 3.40. Stagnation-pressure loss across a simplified λ-shock system (oblique shock followed by normal shock wave) as a function of initial oblique-shock deflection angle for $M_\infty = 1.3$ (*left*) and $M_\infty = 1.4$ (*right*).

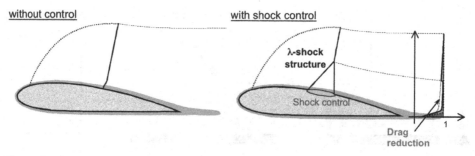

Figure 3.41. Shock control on a transonic airfoil.

the amount of fluid to experience a stagnation-pressure saving. This effect is not included in this assessment. On an actual wing, there is also the additional complexity of a varying Mach number with distance from the surface; thus, the full analysis of potential savings becomes more complex.

Although the geometric extent of the beneficial λ region is limited in wall-normal direction, this often is not a real drawback because in many applications (e.g., on transonic wings), the largest shock losses are incurred relatively close to the surface. In such cases, the shock strength diminishes rapidly with increasing distance from the wall, and a reduction relatively close to the surface is sufficient to alleviate most of the wave drag (Fig. 3.41). A theoretical study on a number of airfoils demonstrated that wave-drag reductions of about two thirds are practically possible (as long as the control device does not impose any additional viscous drag) [22].

Along the surface, the presence of shock control introduces a two-stage pressure jump, similar to that observed in separated interactions (Fig. 3.42). There is an argument that such a two-step pressure rise should be less detrimental to the boundary-layer development because the adverse-pressure gradient is effectively spread over a larger distance. In practice, however, most control mechanisms incur adverse effects on the boundary layer and the aim of good shock control is to keep

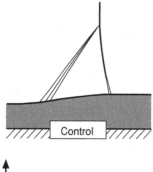

Figure 3.42. Surface pressures in a controlled SBLI.

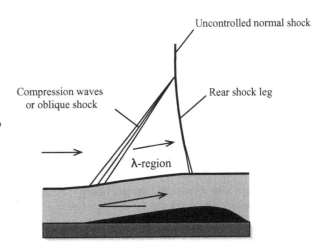

Figure 3.43. Contoured-surface bump for shock control.

them to a minimum. This means that the boundary-layer thickness and shape factor downstream of a controlled interaction should be similar to that observed behind an uncontrolled (i.e., attached) interaction. However, most control mechanisms do not fulfil this aim, and the overall benefit of shock control is a balance between the detrimental effects on the boundary layer and the advantageous effects on the shock structure. A reduction in wave drag often is counteracted by an increase in viscous drag; it is the overall balance between the two that determines whether a control can be considered effective.

An assessment of the success of a control method is complicated further by the fact that the flow downstream of the interaction is often subject to more adverse-pressure gradients. A shock control that has modified the boundary-layer development also is likely to lead to changes in the flow downstream. Such changes can be adverse (i.e., earlier separation) or beneficial (i.e., delayed separation or improved boundary-layer health) and the final balance depends on the actual geometry of the flow. Therefore, it is often difficult to fully assess the costs and benefits of a control strategy unless the complete flowfield is considered. This also means that it is difficult to take lessons learned from one situation and apply them to a different environment.

3.4.2 Methods of Shock Control

Contoured-Surface Bump

As discussed, the key to successful shock control is to generate a large λ-shock structure, which requires a leading oblique shock leg to be formed some distance ahead of the main shock. To force such a shock to develop, it is necessary for the outer inviscid flow to experience deflection away from the wall. The easiest device for achieving this is the contoured-surface bump, shown in Fig. 3.43.

The upstream portion of the contour bump deflects the flow away from the surface, thereby generating compression waves or an oblique shock wave, similar to the leading shock leg seen in separated-shock interactions. Following the initial compression waves, the bump surface is relatively flat to give a uniform flow region before reaching the rear shock leg, which constitutes the weakened portion of the

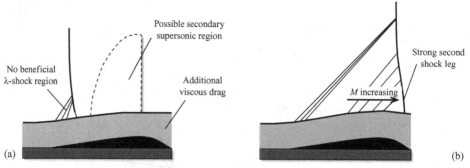

Figure 3.44. Contoured-surface bump under off-design conditions: (a) shock too far upstream; (b) shock too far downstream.

original normal shock. The maximum bump height is reached after the rear shock leg. From this point onwards, the bump height reduces and the flow is returned to the airfoil surface. This area of the bump must be shaped with care to avoid separation and reacceleration to supersonic flow (which can cause unwanted additional shock waves).

Many researchers [23] successfully demonstrated the shock-control-bump concept for transonic wings, in both computations and experiments on airfoil sections as well as on fully three-dimensional wings. A well-designed bump is capable of generating a sizeable λ-shock region while incurring only modest viscous flow penalties. In such a successful design, the boundary-layer health behind the bump is comparable to that observed without a control present.

One major difficulty of bump control is that the benefits are sensitive to the position of the shock wave relative to the bump. Under off-design conditions, with the shock wave either too far upstream or downstream from the ideal, penalties are incurred, as shown schematically in Fig. 3.44.

When the shock wave is too far upstream (Fig. 3.44a), there is virtually no beneficial smearing of the shock foot, and the additional wetted surface of the control bump incurs a viscous drag penalty. In extreme cases (i.e., for large bump heights), it is possible that the curvature of the bump causes a secondary supersonic region with a second shock that also would increase the wave drag.

When the shock wave is too far downstream (i.e., beyond the bump crest), the concave shape of the bump causes expansion waves to emerge inside the λ-shock region (Fig. 3.44b). This locally increases the flow Mach number ahead of the rear shock leg. As a result, the rear leg strengthens significantly, which incurs considerable additional wave drag. To avoid off-design penalties when using shock-control bumps, it is necessary either to use movable or deployable bumps or to employ airfoil-section shapes designed to have relatively little movement of shock location over the range of cruise Mach numbers encountered in flight.

Surface bumps also have been widely reported as capable of delaying shock-induced buffet on transonic wings [23]. This generally is observed for bumps located downstream from the optimum location for wave-drag reduction under design conditions. The exact mechanism for buffet alleviation through contour bumps is not yet well understood, but the following factors are likely to be at work:

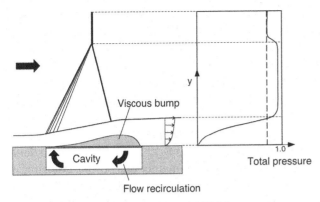
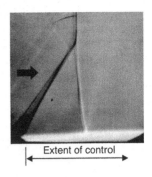

Figure 3.45. Passive control of normal SBLI.

- Buffet generally occurs when a shock wave has moved far back on a wing and grown in strength to cause significant boundary-layer separation. A contour bump placed well aft of the shock location at the design point would only come into effect under such conditions. The generation of a λ structure when a shock has moved into 'buffet territory' weakens the adverse-pressure gradients experienced by the boundary layer, which is likely to reduce the danger of buffet.
- Shock-induced buffet is driven by the large oscillating pressure loads experienced when a shock wave moves back and forth along the wing surface. When a shock moves similarly on a contour bump, the leading leg of the shock structure remains fixed to the leading edge of the bump. Only the rear leg oscillates, which reduces the overall pressure fluctuations.

'Passive' Control

Although in many ways the surface bump is a highly successful form of shock control, many researchers have looked for ways to overcome the penalties incurred when a bump is under off-design conditions. At present, active shock-bump systems (in which the bump can be retracted or moved) are considered too complex and heavy for practical applications.

An alternative concept that has been studied extensively [23] is 'passive control'. In this control method, a cavity covered with a porous surface is placed underneath the shock location, as shown in Fig. 3.45.

The pressure inside the cavity self-adjusts to a value between the preshock and postshock levels. Downstream of the shock wave, therefore, is a pressure above the porous surface that is larger than inside the cavity and vice versa ahead of the shock. As a result, boundary-layer flow is driven into the cavity behind the shock and fluid is blown out of the cavity ahead of the shock. The flow blowing out of the cavity displaces the boundary layer similar to the action of a contour bump and leading oblique compression waves are formed (often merging quickly into an oblique shock). The overall flow is similar to that observed over contour bumps and the achievable savings in wave drag are comparable. Figure 3.45 compares a sketch of this flowfield with a schlieren image of passive control applied to a normal shock wave at $M_s = 1.3$.

The 'viscous bump' formed by the recirculating flow across the porous surface self- adjusts for a wide range of shock locations. As long as the shock wave is positioned above the control, a λ structure can be observed, with an optimum when the main shock is located slightly downstream of the centre of the cavity (i.e., around two-thirds length). When the shock is outside the control region, the recirculating flow stops and the cavity has almost no aerodynamic effect, other than a slight increase in viscous drag due to the roughness of the surface. This control requires no external action or power supply – hence, the name "passive control."

Unfortunately, it has been found that introducing recirculation of boundary-layer flow incurs significant viscous-drag penalties, mainly due to the insertion of low-momentum fluid upstream of the shock wave. Although the suction region at the rear of the control improves the boundary-layer flow to some extent, it is not sufficient to avoid significant additional thickening of the boundary layer through the interaction and associated momentum loss. In most cases, it has been observed that the viscous-drag penalties incurred by passive control are equal to or greater than the achievable savings in wave drag.

Other Methods of Shock Control

In an attempt to overcome the failings of passive control but retain the beneficial attributes (e.g., the automatic tuning to the shock location), a number of variations to this concept have been proposed. Such attempts include modifications to the porous surface (i.e., forward- and backward-swept holes, variable porosity, and smart flaps/mesoflaps); separating the suction and blowing regions of the control; and application of suction to the control cavity to change the balance of suction and blowing across the porous surface. To distinguish the last concept from passive control, it is often termed *active control* because it requires an external energy input to drive the suction system. However, it should not be confused with other types of active control that respond to flow features measured by a sensor.

The search for self-activating shock-control methods remains an area of active research and, to date, no clearly successful strategy has emerged. Pure suction (without a passive-control element), however, has been found to be highly effective, but it falls under the group of boundary-layer control methods, which are discussed later in this chapter.

Three-Dimensional Shock-Control Methods

All of the control strategies discussed so far are more-or-less two-dimensional. This is understandable because the intended application – that is, the shock wave formed on a typical transonic wing – features a highly two-dimensional flowfield. However, several researchers recently suggested the use of three-dimensional controls spread in a spanwise direction along the wing, as shown in Fig. 3.46. Such localized controls are possible because to achieve a more-or-less two-dimensional λ structure, it is not necessary to use a two-dimensional control. If a shock wave is forced into a λ structure at one location, it generally takes considerable spanwise distance for it to revert to an uncontrolled single shock foot. This opens up the possibility of distributing small control devices along the shock to achieve a 'global' shock-control effect. The advantage of this strategy is that any viscous-drag penalties are confined to the

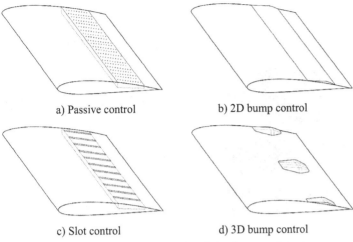

a) Passive control   b) 2D bump control

c) Slot control   d) 3D bump control

Figure 3.46. Two- and three-dimensional shock-control devices.

control locations and thus are likely to be much smaller than those observed on a comparable two-dimensional control. There also is evidence that three-dimensional controls may incur smaller off-design penalties. Furthermore, it is considerably easier to fit individual three-dimensional devices to an existing wing or even retrofit a small number of devices near 'trouble spots' where an unduly strong shock wave is observed. Finally, individual three-dimensional devices also may be easier to make 'active' (i.e., deployable or moveable).

In principle, almost all forms of two-dimensional shock control have a three-dimensional counterpart, and examples are shown in Fig. 3.46. Contour bumps have been studied in their three-dimensional 'incarnations'. Streamwise slots are similar to passive control, and various other combinations of recirculating control incorporating suction and blowing are easily imaginable. The basic flow features observed for two types of three-dimensional shock controls are illustrated in Fig. 3.47.

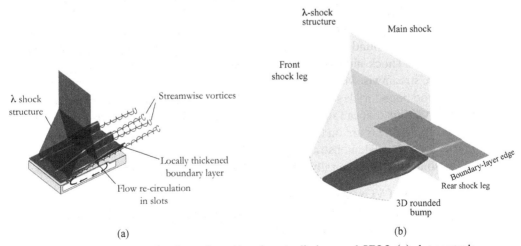

Figure 3.47. Basic features of a three-dimensional controlled normal SBLI: (a) slot control; (b) three-dimensional bump control.

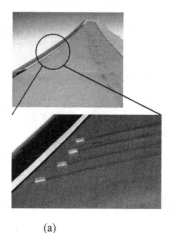

Figure 3.48. Vane-type vortex genera-tors on (a) a transonic wing, and (b) inside a supersonic inlet.

(a)   (b)

It can be seen that an array of slots (Fig. 3.47a) generates a relatively two-dimensional λ-shock structure, whereas the boundary-layer flow is highly three-dimensional (thus, the viscous penalties are highly localized). In the case of the single three-dimensional shock-control bump (Fig. 3.47b), the λ-shock structure slowly reverts back towards a single shock, but there is considerable spanwise dis-tance in which a significant λ structure remains. The boundary layer behind the three-dimensional bump is slightly thickened; however, it again is confined to the immediate wake behind the device – elsewhere, the boundary-layer profile essen-tially is unchanged. The subject of three-dimensional shock control is still under investigation but early results show promise.

3.4.3 Methods of Boundary-Layer Control

Whereas shock control aims to modify the structure of the shock foot to reduce stagnation-pressure losses, boundary-layer control alters the characteristics of the near-wall flow ahead of an SBLI to prevent or reduce shock-induced separations. The goals of this approach are to minimize viscous drag and delay or prevent the emergence of unsteady flow, such as shock-induced buffet.

The most widely used method of boundary-layer control is the vortex generator (VG), which is found on some current aircraft to control many types of separa-tion, not just shock-induced (Fig. 3.48a). As shown schematically in Fig. 3.49, VGs introduce streamwise vortices into the flow that transfer high-momentum fluid from outside the boundary layer to the near-wall regions. This generates more energetic boundary layers that can significantly enhance the resistance to shock-induced sepa-ration. Typical VGs are on the order of one boundary-layer thickness in height, and many studies have demonstrated their success in limiting shock-induced separations in transonic SBLIs. Generally, the use of VGs has the following two drawbacks:

- All types of VGs incur significant parasite drag.
- The increased fullness of the boundary-layer profile also leads to a reduc-tion in the length of the SBLI so that there is reduced shock-smearing and thus an associated (small) wave-drag penalty (i.e., increased stagnation-pressure losses).

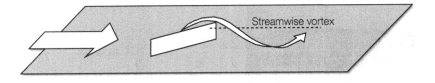

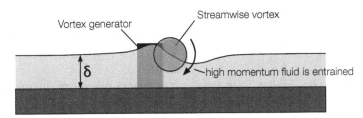

Figure 3.49. Vane-type vortex generators. *Top:* sketch of flowfield. *Bottom:* rear view.

One suggestion to reduce the parasite drag incurred by VGs is the use of sub-boundary-layer vortex generators (SBVGs), also referred to as micro-vortex generators. Such devices measure less than a boundary-layer thickness in height (some are as small as one displacement thickness) and they incur much reduced viscous drag due to the small wetted area and protrusion into much slower portions of the flow. Nevertheless, they have been shown to be surprisingly effective in reducing shock-induced separation [24, 25]. For example, Fig. 3.50 shows how pairs of counter-rotating vortices generated by micro-vanes ($h/\delta \approx 0.2$) can eradicate shock-induced separation at $M_s = 1.5$. Current research suggests that the mechanism of flow control through SBVGs is slightly different from that of more traditional VGs. Whereas both introduce streamwise vortices into the flow, traditional VGs entrain high-momentum fluid from outside the boundary layer. Typically, these devices are placed well ahead of any region of adverse-pressure gradient to allow the entrained momentum to spread throughout the boundary layer, giving a fuller velocity profile that is more resistant to separation. Conversely, micro-VGs introduce streamwise vortices that are embedded inside the boundary layer and redistribute momentum internally. Such vortices are less strong, but they are placed closer to the surface where they can be more effective. However, the location close to the wall can lead to an earlier dissipation of the vortices and it is therefore likely that such devices must be closer to the adverse-pressure gradient than traditional VGs.

Figure 3.50. Sub-boundary layer vortex generators ahead of a normal SBLI at $M_s = 1.5$.

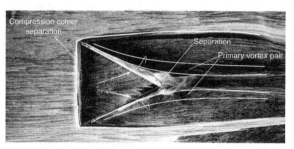 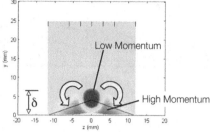

Figure 3.51. Micro-ramp vortex generator: (a) surface oil-flow visualization; (b) velocity change (relative to baseline) measured downstream of device.

In some applications (i.e., inlets), vane-type vortex generators are problematic because of their perceived mechanical fragility. For this reason, other shapes have been proposed – for example, the micro-ramp shown in Fig. 3.51. Such devices introduce a pair of counter-rotating streamwise vortices. Fig. 3.51b shows that this is effective at moving high-momentum fluid close to the wall. Device drag, in the form of a low-momentum region, is also clearly visible in the velocity-change plot.

There are other strategies for boundary-layer control through flow transpiration, such as tangential blowing, distributed suction, and discrete suction ahead of or underneath an interaction. In all of these methods, the fullness of the boundary-layer profile is enhanced significantly and resistance to separation is improved. In particular, distributed suction is widely used in supersonic-jet inlets to eliminate

Figure 3.52. Distributed surface suction in a supersonic-engine inlet.

normal shock-induced separation, even at much higher shock Mach numbers (i.e., approaching $M = 2$). Fig. 3.52 shows suction holes in the inlet of a supersonic fighter aircraft.

Surface suction, or bleed, is a highly effective method of preventing (or delaying) shock-induced separation. It is particularly attractive in inlets because it is relatively easy to implement in practice and, other than preventing separation, it also makes the boundary layer downstream of the SBLI thinner, which reduces flow distortion. However, there is a performance penalty because the mass flow reaching the engine is reduced by the suction-mass-flow rate (which may be considerable); therefore, the inlet area must be increased, which introduces additional drag. A subtler effect of bleed is that the fuller (and thinner) boundary layer reaching the SBLI causes a considerable reduction in interaction length, which leads to stronger or less smeared shocks with an associated (slight) increase in stagnation-pressure loss.

Despite the drawbacks, the need for flow control to prevent separation in supersonic inlets is so great that no current device in service operates without some form of boundary-layer bleed. Similar considerations apply to other internal flows (e.g., inside turbomachines), where bleed is also popular. For external transonic-flow applications, however, bleed is not a widely used method of boundary-layer control for SBLIs (which is not to be confused with bleed for laminar-flow control).

The area of boundary-layer control for SBLI applications is an active topic of current research. Various novel control methods have been proposed, among them zero-mass flux-pulsed jets (sometimes referred to as virtual jets) and plasma actuators, which continue to be in the early experimental phase; readers are referred to current research publications for more information.

REFERENCES

[1] A. H. Shapiro. *Dynamics and Thermodynamics of Compressible Fluid Flow* (New York: Ronald Press Co., 1954).

[2] R. C. Lock. The prediction of the drag of aerofoils and wings at high subsonic speeds. *RAE TM Aero*, 2044 (1985).

[3] H. H. Pearcey. "Shock-Induced Separation and Its Prevention by Design and Boundary Layer Control." In *Boundary Layer and Flow Control*, ed. G. V. Lachmann (Oxford: Pergamon Press, 1961), pp. 1166–344.

[4] H. Sobieczky, N. J. Yu, K. Y. Fung, and A. R. Seebass. New method for designing shock-free transonic configurations. *AIAA Journal*, 17(7) (1979), 722–9.

[5] I. Alber, J. Bacon, B. Masson, and D. Collins. An experimental investigation of turbulent transonic viscous-inviscid interactions. *AIAA Journal*, 5(11) (1973), 620–7.

[6] A. Zheltovodov, R. Dvorak, and P. Safarik. Shock wave/turbulent boundary layer interaction properties at transonic and supersonic speeds conditions. *Izvestiya of SO AN SSSR*, 6 (1990), 31–42 (in Russian).

[7] A. Zheltovodov and V. Yakovlev. Stages of development, gas dynamic structures and turbulence characteristics of turbulent compressible separated flows in the vicinity of 2-d obstacles. Preprint No. 27–86, Institute of Theoretical and Applied Mechanics, Russian Academy of Sciences, Novosibirsk (1986) (in Russian).

[8] E. R. Van Driest. Turbulent boundary layer in compressible fluids. *Journal of Aeronautical Sciences*, 18(3) (1951), 145–216.

[9] J. E. Green. Interactions between shock waves and turbulent boundary layers. *Progress in Aerospace Sciences*, 11 (1970), 235–339.

[10] L. F. East and W. G. Sawyer. An investigation of the structure of equilibrium turbulent boundary layers. AGARD Fluid Dynamics Panel Symposium, The Hague, the Netherlands, AGARD CP-271(Jan. 1980), pp. 6.1–6.19.

[11] D. Om and M. E. Childs. Multiple transonic shock-wave/turbulent boundary-layer interaction in a circular duct. *AIAA Journal*, 23(10) (1985), 1506–11.

[12] B. F. Carroll and J. C. Dutton. Multiple normal shock wave/turbulent boundary-layer interactions. *J. Propulsion and Power*, 8(2) (March–April 1982), 441–8.

[13] P. Reijasse, B. Corbel, and D. Soulevant. Unsteadiness and asymmetry of shock-induced separation in a planar two-dimensional nozzle: A flow description. 30th AIAA Fluid Dynamics Conference, 28 June–1 July 1999, Norfolk, VA, AIAA Paper 1999–3694 (1999).

[14] G. R. Inger. Transonic shock/boundary-layer interaction on curved surfaces. *J. Aircraft*, 20(6) (1983), 571–4.

[15] G. R. Inger and H. Sobieczky. Transonic shock interaction with a turbulent boundary layer on a curved wall. American Society of Mechanical Engineers (Paper), (79-WA/FE-13) (1979).

[16] R. Bohning and J. Zierep. Stoß-Grenzschichtinterferenz bei turbulenter Strömung an gekrümmten Wänden mit Ablösung. Z. Flugwiss. *Weltraumforsch*, 6(2) (1982), 68–74.

[17] G. E. Gadd. Interaction between normal shock-waves and turbulent boundary-layers. ARC R&M, No. 3262 (1961).

[18] F. Ackeret and M. Rott. Inst. Aerodyn. Zürich, No. 10; NACA Tech. Memo No. 1113, (1947). Images provided courtesy of Prof. P. Doerffer, IMP/PAN, Gdansk, Poland.

[19] R. H. Korkegi. A simple correlation for incipient turbulent boundary-layer separation due to a skewed shock-wave. *AIAA Journal*, 11(11) (1973), 1578–9.

[20] P. J. K. Bruce and H. Babinsky. Unsteady shock wave dynamics. *Journal of Fluid Mechanics*, 603 (May 2008), 463–73.

[21] F. Furlano. *Comportement de modèles de turbulence pour les écoulements décollés en entrée de tremblement* (Behaviour of turbulence models for the prediction of separated flows under buffet onset conditions). Ph.D. Thesis, Ecole Nationale Supérieure de l'Aéronautique et de l'Espace (March 2001).

[22] H. Ogawa and H. Babinsky. Evaluation of wave-drag reduction by flow control. *Aerospace Science and Technology*, 10(1) (January 2006), 1–8.

[23] See, for example: Drag Reduction by Passive Shock Control; results of the project EUROSHOCK. AER 2- CT92–0049, ed. Egon Stanewsky. Notes on numerical fluid mechanics, 56 (Braunschweig, Wiesbaden: Vieweg, 1997); and Drag reduction by shock and boundary layer control: Results of the project EUROSHOCK II. ed. Egon Stanewsky, Notes on numerical fluid mechanics and multidisciplinary design, 80 (Berlin, Heidelberg, New York, Barcelona, Hong Kong, London, Milan, Paris, Tokyo: Springer, 2002).

[24] J. Lin. Review of research on low-profile vortex generators to control boundary-layer separation. *Prog. Aerospace Sciences*, 38 (2002), 389–420.

[25] P. R. Ashill, J. L. Fulker, and K. C. Hackett. A review of recent developments in flow control. *Aeronautical J.*, 109 (1095) (May 2005), 205–32.

4 Ideal-Gas Shock Wave–Turbulent Boundary-Layer Interactions (STBLIs) in Supersonic Flows and Their Modeling: Two-Dimensional Interactions

Doyle D. Knight and Alexander A. Zheltovodov

4.1 Introduction

Effective design of modern supersonic and hypersonic vehicles requires an understanding of the physical flowfield structure of shock wave–boundary layer interactions (SBLIs) and efficient simulation methods for their description (Fig. 4.1). The focus of this chapter is two-dimensional supersonic shock wave–turbulent boundary layer interactions (STBLIs); however, even in nominally two-dimensional/axisymmetric flows, the mean flow statistics may be three-dimensional. The discussion is restricted to ideal, homogeneous gas flow wherein the upstream free-stream conditions are mainly supersonic ($1.1 \leq M_\infty \leq 5.5$). Computational fluid dynamics (CFD) simulations of two-dimensional STBLIs are evaluated in parallel with considerations of flowfield structures and physical properties obtained from both experimental data and numerical calculations.

4.1.1 Problems and Directions of Current Research

The main challenges for modeling of and understanding the wide variety of two- and three-dimensional STBLIs include the complexity of the flow topologies and physical properties and the lack of a rigorous theory describing turbulent flows. These problems have been widely discussed during various stages of STBLI research since the 1940s. In accordance with authoritative surveys [1, 2, 3, 4, 5, 6, 7] and monographs [8, 9, 10, 11], progress in understanding STBLIs can be achieved only on the basis of close symbiosis between CFD and detailed physical experiments that focus on simplified configurations (see Fig. 4.1) and that use recent advances in experimental diagnostics (e.g., planar laser scattering [PLS]; particle image velocimetry [PIV]); and turbulence modeling, including Reynolds-averaged Navier-Stokes [RANS], large eddy simulation [LES], and direct numerical simulation [DNS]).

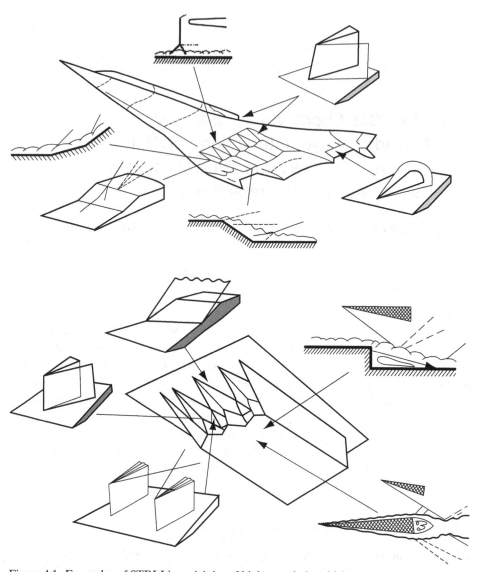

Figure 4.1. Examples of STBLI in a vicinity of high-speed air vehicle.

4.1.2 Computational Fluid Dynamics

The governing equations for STBLI are the following compressible Navier-Stokes equations:

$$\frac{\partial \rho}{\partial t} + \frac{\partial \rho u_j}{\partial x_j} = 0$$

$$\frac{\partial \rho u_i}{\partial t} + \frac{\partial \rho u_i u_j}{\partial x_j} = -\frac{\partial p}{\partial x_i} + \frac{\partial \tau_{ij}}{\partial x_j}$$

$$\frac{\partial \rho e}{\partial t} + \frac{\partial (\rho e + p)}{\partial x_j} = \frac{\partial}{\partial x_j}(u_i \tau_{ij} - q_j)$$

$$p = \rho R T \tag{4.1}$$

where x_j are the Cartesian coordinates, u_i are the corresponding velocity components, ρ is the density, p is the static pressure, e is the total energy per unit mass,

$$e = c_v T + \frac{1}{2} u_j u_i \tag{4.2}$$

R is the gas constant, τ_{ij} is the shear stress, and q_i is the heat flux. For a Newtonian fluid:

$$\tau_{ij} = \lambda \frac{\partial u_k}{\partial x_k} \delta_{ij} + \mu \left(\frac{\partial u_i}{\partial x_j} + \frac{\partial u_j}{\partial x_i} \right) \tag{4.3}$$

where μ is the dynamic molecular viscosity and, from Fourier's Law:

$$q_i = -\hat{k} \frac{\partial T}{\partial x_i} \tag{4.4}$$

where $\hat{k}$ is the thermal conductivity. These equations are valid for both laminar and turbulent flows. In the latter case, the solution of the equations is denoted as DNS, which is computationally impractical for flight-scale engineering configurations due to large values of the Reynolds number. However, DNS at lower Reynolds numbers achievable in experiments are important for a better understanding of the physical processes in STBLIs. For simulation of flight-scale engineering configurations, the compressible Navier-Stokes equations are averaged and statistical mean values are computed. In this context, two different approaches were developed. The first is the RANS equations, which are obtained by introducing the Favre ensemble average of a function $\tilde{f}$ defined by the following equation:

$$\tilde{f} = \frac{1}{\bar{\rho}} \lim_{n \to \infty} \frac{1}{n} \sum_{v=1}^{v=n} (\rho f)^{(v)}$$

$$= \frac{\overline{\rho f}}{\bar{\rho}}$$

$$f = \tilde{f} + f'' \tag{4.5}$$

in which the overbar indicates an ensemble average. The resultant equations are as follows:

$$\frac{\partial \bar{\rho}}{\partial t} + \frac{\partial \bar{\rho} \tilde{u}_j}{\partial x_j} = 0$$

$$\frac{\partial \bar{\rho} \tilde{u}_i}{\partial t} + \frac{\partial \bar{\rho} \tilde{u}_i \tilde{u}_j}{\partial x_j} = -\frac{\partial \bar{p}}{\partial x_i} + \frac{\partial T_{ij}}{\partial x_j}$$

$$\frac{\partial \bar{\rho} \tilde{e}}{\partial t} + \frac{\partial (\bar{\rho} \tilde{e} + \bar{p}) \tilde{u}_j}{\partial x_j} = \frac{\partial}{\partial x_j} (\tilde{u}_i T_{ij} - Q_j)$$

$$\bar{p} = \bar{\rho} R \tilde{T} \tag{4.6}$$

where the total stress-tensor and heat-transfer vectors are as follows:

$$T_{ij} = -\overline{\rho u_i'' u_j''} + \overline{\tau_{ij}}$$

$$Q_j = c_p \overline{\rho T'' u_j''} + \bar{q}_j \tag{4.7}$$

and where ( )'' represents the velocity fluctuations, and the first and second terms are the turbulent and mean laminar contributions, respectively.

The mathematical closure of the RANS equations requires additional equations for the turbulent stresses $-\overline{\rho u_i'' u_j''}$ and turbulent heat flux $c_p \overline{\rho' T'' u_j''}$. Models for these terms are categorized broadly in two types. The first is the eddy-viscosity model wherein the turbulent stress and heat flux are modeled analogous to molecular stress and heat flux through the use of a turbulent-eddy viscosity. Thus:

$$-\overline{\rho u_i'' u_j''} = \mu_t \left(\frac{\partial u_i}{\partial x_j} + \frac{\partial u_j}{\partial x_i} - \frac{2}{3}\Delta\delta_{ij} \right) - \frac{2}{3}\rho k \delta_{ij}$$

$$c_p \overline{\rho T'' u_j''} = -\hat{k}_t \frac{\partial \tilde{T}}{\partial x_j} \tag{4.8}$$

where μ_t is the turbulent-eddy viscosity, k is the turbulence kinetic energy, δ_{ij} is the Kronecker delta, Δ is the divergence of the velocity, and $\hat{k}_t$ is the turbulent thermal conductivity given by:

$$\hat{k}_t = \frac{\mu_t c_p}{Pr_t} \tag{4.9}$$

where $Pr_t = 0.9$ is the turbulent Prandtl number. Eddy-viscosity models may be categorized further as zero-, one-, and two-equation models based on the number of additional partial-differential equations posited to determine the turbulent-eddy viscosity. Examples of zero-equation (or algebraic) turbulence models include Cebeci and Smith [12] and Baldwin and Lomax [13]. Examples of one-equation (or algebraic) models include Baldwin and Barth [14] and Johnson and King [15]. Examples of two-equation models are the k-ε model of Jones and Launder [16] and the k-ω model of Wilcox [17]. A complete review of these and other turbulence models mentioned herein is beyond the scope of this chapter, and readers are referred to references [17, 18, 19].

The second type of turbulence models is the full Reynolds-stress equation model wherein partial-differential equations are posited for the turbulent stresses $-\overline{\rho u_i'' u_j''}$ (and, possibly, the turbulent heat flux $c_p \overline{\rho' T'' u_j''}$). In principle, this approach is less restrictive than the eddy-viscosity model because it does not imply that the principal axes of the turbulent-stress tensor are coincident with those of the mean rate-of-strain tensor. Examples of full Reynolds-stress equation models include Zhang, So, Gatski, and Speziale [20], Gnedin and Knight [21], and Zha and Knight [22].

The second approach to numerical simulation of STBLIs is LES, the equations of which are obtained by introducing the spatial average $\bar{f}$ of a function and its corresponding Favre average $\tilde{f}$ by

$$\bar{f} = \frac{1}{V} \int_V Gf dV$$

$$\tilde{f} = \frac{\overline{\rho f}}{\bar{\rho}} \tag{4.10}$$

where G is the filter function (e.g., a top hat of width Δ_G). Therefore, the LES equations represent a spatial average of the Navier-Stokes equations – unlike the RANS equations, which represent an ensemble average. The LES equations are inherently time-dependent and resolve time-scales as small as those representing the inertial sublayer scales of the turbulent motion. The RANS equations represent an average over all scales of the turbulent motion up to and including the energy-containing eddies that define the size of the turbulent boundary layer. LES equations are as follows:

$$\frac{\partial \bar{\rho}}{\partial t} + \frac{\partial \bar{\rho}\tilde{u}_k}{\partial x_k} = 0$$

$$\frac{\partial \bar{\rho}\tilde{u}_i}{\partial t} + \frac{\partial \bar{\rho}\tilde{u}_i\tilde{u}_k}{\partial x_k} = -\frac{\partial \bar{p}}{\partial x_i} + \frac{\partial T_{ik}}{\partial x_k}$$

$$\frac{\partial \bar{\rho}\tilde{e}}{\partial t} + \frac{\partial}{\partial x_k}(\bar{\rho}\tilde{e} + \bar{p})\tilde{u}_k = \frac{\partial}{\partial x_k}(Q_k + T_{ik}\tilde{u}_i)$$

$$\bar{p} = \rho R\tilde{T} \qquad (4.11)$$

where the total stress-tensor and heat-transfer vectors are as follows:

$$T_{ik} = -\bar{\rho}(\widetilde{u_i u_k} - \tilde{u}_i\tilde{u}_k) + \bar{\tau}_{ik}$$

$$Q_k = -\bar{\rho}c_p(\widetilde{Tu_k} - \tilde{T}\tilde{u}_k) - \bar{q}_k \qquad (4.12)$$

The term $-\bar{\rho}(\widetilde{u_i u_k} - \tilde{u}_i\tilde{u}_k)$ represents the Subgrid Scale (SGS) stress and the term $-\bar{\rho}c_p(\widetilde{Tu_k} - \tilde{T}\tilde{u}_k)$ represents the SGS heat flux. Two different approaches were developed to model these terms. In the first approach, explicit models for the SGS stress and heat flux are posited [23, 24, 25]. In the second approach, implicit models for the SGS stress and heat flux are posited based on the concept of mono-tone integrated large eddy simulation (MILES) [26].

Development of hybrid LES/RANS simulations of unsteady STBLIs recently appeared as a less-computationally expensive (compared to LES) approach for high Reynolds-number flows [27, 28]. The flow-dependent blending functions (similar in construction to those used in Menter's $k\text{-}\omega/k\text{-}\varepsilon$ model [29]) shift the turbulence modeling from the two-equation model near a solid surface to an LES subgrid closure away from the solid surface and in free-shear regions around a separated zone.

4.2 Two-Dimensional Turbulent Interactions

Our discussion of nominally two-dimensional STBLIs focuses on principal canonical test configurations (Fig. 4.2) – namely, a normal shock wave, an incident-oblique shock wave, a compression ramp (CR) and a compression-decompression ramp, and a forward-facing step. The flowfield structure in each case is determined by the free-stream Mach number; shock strength[1] (e.g., inviscid-static-pressure ratio $\xi = p_2/p_1$); Reynolds number[2] Re_δ based on the incoming boundary-layer thickness δ; and wall-temperature ratio T_w/T_{aw}, where T_w is the wall temperature and T_{aw} is the adiabatic-wall temperature and geometry (i.e., flat versus angled surface because a concave surface introduces additional streamline curvature that may cause Görtler vortices). For each configuration, we examine the typical STBLI regimes, discuss the flowfield structure, and present typical experimental and computational results.

---

[1] For oblique SBLIs, an equivalent specification of the shock strength is the incident inviscid flow-deflection angle α.

[2] Equivalently, the Reynolds number based on the compressible displacement thickness δ^* or momentum thickness θ may be used because the incoming turbulent boundary layer is assumed to be in equilibrium.

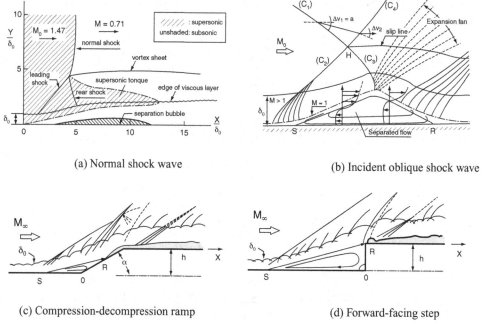

(a) Normal shock wave (b) Incident oblique shock wave

(c) Compression-decompression ramp (d) Forward-facing step

Figure 4.2. Two-dimensional shock-wave turbulent boundary layer interaction: (a) and (b) from Délery and Marvin [3]; (c) and (d) from Zheltovodov [1, 2].

Space limitations preclude an exhaustive survey of computations; the examples are selected to provide a general assessment of computational capability.

4.2.1 Normal STBLI: Flow Regimes and Incipient Separation Criteria

The schlieren photographs and surface-flow patterns in Fig. 4.3 demonstrate the different stages of transonic normal shock wave–turbulent boundary layer interaction (NSTBLI) on a flat plate [30]. The unseparated flow regime is realized at $M_\infty = 1.11$ (Fig. 4.3a). Bifurcations of the limiting streamlines upstream of the inviscid shock-wave trace (indicated by the dashed line) at $M_\infty = 1.3 \pm 0.01$ (Fig. 4.3b) are associated with small, unsteady separation zones located around the isolated saddle-type separation points along the surface span. The flow penetrates without separation between these localized zones in the downstream direction. At $M_\infty = 1.43 \pm 0.02$, a fully separated flow is realized at least in the central part of the surface (Fig. 4.3c), where three-dimensional effects are minor.

In an inviscid external flow, the normal shock-static-pressure ratio is determined by the following Rankine-Hugoniot relation:

$$\xi = p_2/p_1 = [2\gamma M_1^2 - (\gamma - 1)]/(\gamma + 1) \tag{4.13}$$

in which subscripts 1 and 2 relate to conditions immediately upstream and downstream of the shock, respectively. This pressure ratio is never observed on the surface in the immediate region of the shock due to SBLI effects. The upstream transmission of pressure through the subsonic part of the boundary layer causes a sudden local increase in the rate of change of boundary-layer displacement thickness, which in turn produces compression waves (Figs. 4.3a and 4.3b) or even oblique shock

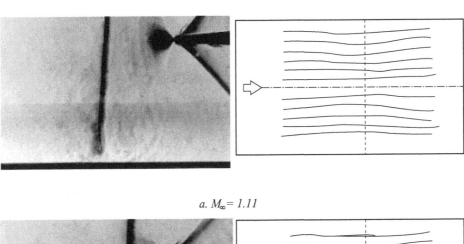

a. $M_\infty = 1.11$

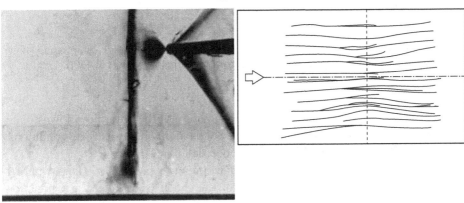

b. $M_\infty = 1.31$

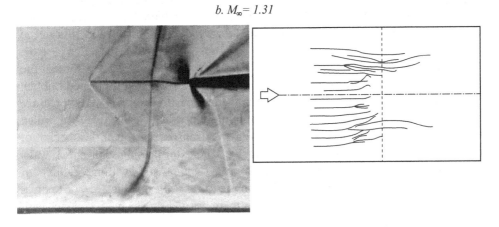

c. $M_\infty = 1.45$

Figure 4.3. Normal shock wave turbulent boundary layer interaction: Schlieren photographs (left column) and surface-flow patterns (right column) (Zheltovodov et al. [30]).

waves induced by the separation forward of the normal shock (Figs. 4.3c and 4.2a). The foot of the normal shock is transformed into compression waves or a lambda shock configuration, with the curved part of the shock in the external flow associated with the strong oblique-shock solution of the Rankine-Hugoniot equations. Consequently, the observed surface-pressure level p_2/p_1 is lower than the values predicted

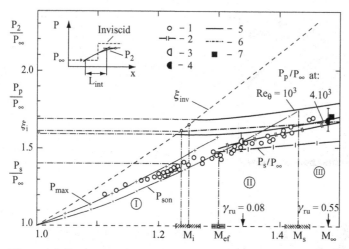

Figure 4.4. Pressure ratio across normal shock and interactions regimes: I – unseparated flow; II – intermittent separation; III – developed separation (Zheltovodov et al. [30]).

in equation (4.13). Additionally, the subsonic nature of the outer flow downstream of the normal shock renders the flow globally dependent on downstream conditions. The wave structure of NSTBLIs is discussed in detail in Chapter 3.

Figure 4.4 displays the measured surface-pressure ratio p_2/p_1 (i.e., symbol 1 [30]) in the isobaric region downstream of the shock (see sketch) as a function of the free-stream Mach number $M_\infty = M_1$, where $p_1 = p_\infty$. The normal shock-static-pressure ratio satisfying equation (4.13) is shown as a dashed line (ξ_{inv}). At unseparated-flow conditions ($M_\infty < 1.25$, regime I), the data point (1) corresponds to the maximum shock-deflection pressure curve p_{max}. A sudden reduction in the growth rate of $(dp/dx)_{max} = (p_2 - p_1)/L_{int}$ is observed at $M_\infty = M_i = 1.25 \pm 0.02$. This point corresponds to the "true" incipient-separation conditions, although the flowfield is not yet greatly altered by the existence of the small separated bubble. This conclusion is supported by the fact that the measured surface-static-pressure ratio at $M_i = 1.25$ coincides with the bare minimum value $(p_2/p_1)_{min} = 1.4$, which is required to cause true incipient separation on a transonic airfoil at the shock foot in accordance with Pearcey's [31] experiments. The external "inviscid" shock-wave strength $\xi_{inv} = f(M_\infty)$ at this stage is in good agreement (i.e., within the limits of accuracy for $M_\infty = M_i = 1.25 \pm 0.02$), with the lower critical value $\xi_i^* = 1.61$ corresponding to incipient separation in the vicinity of normal shocks in accordance with an experimental correlation by Grodzovskyi [32] (i.e., line 6).

The appearance of bifurcations in the surface-flow pattern at $M_\infty = M_{ef} = 1.3 \pm 0.01$ (Fig. 4.3b) corresponds to the beginning of "significant" (or "effective") separation, wherein a noticeable separation zone emerges in the foot of the shock. The data for p_2/p_1 confidently reach the sonic-pressure curve p_{son} at this stage (Fig. 4.4), similar to the classic experiments of Pearcey [31], and also begin to correspond to the surface-static-pressure ratio at the separation point ahead of the forward-facing step predicted by the empirical correlation [33] $p_s/p_1 = 0.365\, M_\infty + 1$ (line p_s/p_∞). Thus, the condition of equivalence between the sonic pressure and separation-point pressure ($p_{son} = p_s$) corresponding to Pearcey's [31] criterion reasonably predicts the critical Mach number value $M_{ef} \approx 1.3$ at the onset of effective incipient separation. The surface-pressure ratio p_2/p_1 in regime II at $M_\infty \geq 1.3$ begins to correspond to the plateau-pressure dependence p_p/p_1 (i.e., line 2) [33] observed for separation zones in front of forward-facing steps at low Mach numbers. LDV measurements [34] in the vicinity of the normal shock foot at $M_\infty = 1.34$ and the value p_2/p_1 shown by symbol 3 reveal the intermittent nature of a small reversed-flow zone that exists for a short fraction of time (i.e., 8 percent). This is in accordance with the intermittency coefficient $\gamma_{ru} = 0.08$, which corresponds to an "intermittent-detachment" stage [35].

The well-defined separation line forming at $M_\infty = M_s = 1.43 \pm 0.02$ (see Fig. 4.3c) corresponds to conditions when the measured p_2/p_1 level in the Reynolds-number range $Re_\theta = (1 - 4) \times 10^3$ reaches the following plateau-pressure ratio:

$$\frac{p_p}{p_\infty} = k M_\infty^2 (M_\infty^2 - 1)^{-1/4} c_f^{1/2} + 1 \tag{4.14}$$

for developed steady-separation zones predicted by the classic Free Interaction Theory (FIT) of Chapman et al. [36] (Fig. 4.4, solid line 5). The experimentally established [3] constant $k = 5.94$ for turbulent flow was used for this prediction [1, 30] with the skin-friction coefficient c_f calculated for a flat-plate turbulent boundary layer on the basis of Kutateladze and Leont'ev's theory [37]. In accordance with Fig. 4.4, the condition $p_{son} = p_p$ can be used analogously with Pearcey's [31] criterion for prediction of the critical Mach number $M_\infty = M_s$ corresponding to incipience of developed separation. As shown in experiments by Morriss et al. [38], at $M_\infty = 1.48$ (indicated by the symbol 4 for the level p_2/p_1), the reversed flow exists more than 50 percent of the time ($\gamma_{ru} = 0.55$) with a distinct separation line demonstrated by the oil-flow visualization. Thus, a developed "quasistationary" separation (or "full detachment" [35]) is realized in regime III at $M_\infty > M_s$.

Experimental and theoretical analyses of incipient-separation conditions for NSTBLI are the subject of many studies [1, 3, 8, 39]. The M_i data for true incipient separation on a flat-plate surface by Zheltovodov et al. [30] and Grodzovskyi [32] (Fig. 4.5, symbols 3 and 4) are in good agreement with the theoretical prediction by Inger [40, 41] (i.e., line 1) and close to the empirical correlation (i.e., dotted line 2), which generalizes the experiments (i.e., symbol 2) at ONERA for transonic flows past curved surfaces (e.g., airfoils and bumps on a wind-tunnel wall) in accordance with Délery and Marvin [3] and Haines [39]. Stanewsky [42] defined the limits of effective incipient separation on an airfoil surface by plotting the variation of the boundary-layer displacement-thickness δ^* kink in the vicinity of the normal shock (i.e., symbols and line 5). He also extrapolated measured separation-bubble lengths back to zero length (i.e., the solid line) and considered Pearcey's [31] criterion ($p_s = p_{son}$, symbol 6). The data (M_{ef}) for effective incipient separation on a flat-plate surface (i.e., symbols 7, 8) are in good agreement with these empirical correlations. A tendency to a small decrease in the levels of M_i and M_{ef} for the flat-plate surface (i.e., symbols 3, 4 and 7, 8), compared to experiments for slowly curved surfaces (lines 2, 5), is in agreement with a theoretical prediction [43]. All examined correlations and data demonstrate a slight increase in M_i and M_{ef} with a decreasing value of the incompressible-shape parameter $H_{io} = \delta^*_i/\theta_i$ in the incoming boundary layer, where δ^*_i and θ_i are the incompressible-displacement and momentum

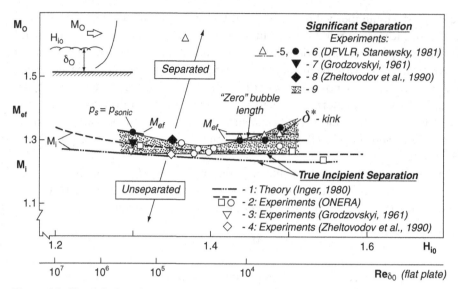

Figure 4.5. Shock-induced "true" and "effective" (or "significant") incipient-separation limits.

thicknesses [3, 39] (i.e., the increasing $Re_{\delta o}$). A lower value of H_{io} means a fuller velocity profile and, accordingly, a greater resistance of the developed ("well-behaved") turbulent boundary layer to separation. The lack of significant variation in M_i and M_{ef} is explained [3, 42] on the premise that an increase in H_{io} increased the subsonic part of the boundary layer and, consequently, the likelihood of its separation. Conversely, it also increased the length of the upstream influence, thereby alleviating the adverse pressure gradient and the tendency to separation. Therefore, the two factors tend to negate one another. Nevertheless, against a background of limited accuracy of the data presented in Fig. 4.5, they seem to demonstrate a tendency to an apparent variation with a decrease in the Reynolds numbers (at $H_{io} <$ 1.4) shown by band 9, the probable reason of which is discussed next.

A collection of numerous experimental data [30, 32, 34] for nominally two-dimensional transonic interactions with boundary-layer separation (i.e., closed symbols) and without separation (i.e., open symbols) is shown in Fig. 4.6. Mainly, the flows with no transverse curvature and with no (or, at most, mild) streamwise curvature are included. In accordance with [34], from this set of studies only the data by Schofield (1983) and the transonic diffuser experiments of Bogar et al. and Salman et al. (1983), Sajben et al. (1991) [34], and Morris et al. (1992) [38] address NSTBLIs coupled with a subsequent adverse-pressure gradient.

Considering that the empirical correlation [32] for the critical shock-wave static-pressure ratio $\xi_i^* = (p_2/p_\infty)^*$, at which an appearance of separation has been fixed at transonic Mach numbers (see Fig. 4.4, line 6), corresponds to the tendency for p_p/p_∞ predicted by equation (4.14) (i.e., solid line 5), the FIT can be applied to describe the conditions for incipient intermittent separation [1, 30]. Equating the plateau pressure (see equation 4.14) to the normal shock strength (see equation 4.13) yields the criterion for true incipient separation, as follows:

$$\frac{[2\gamma M_\infty^2 - (\gamma - 1)]}{\gamma + 1} = k M_\infty^2 (M_\infty^2 - 1)^{-1/4} c_f^{1/2} + 1 \qquad (4.15)$$

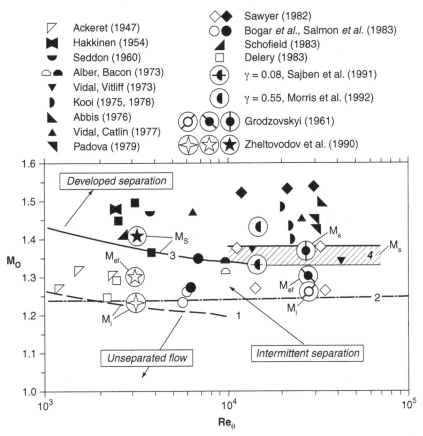

Figure 4.6. The normal shock interaction data for regimes with separation (black symbols) and without separation (opened symbols) and prediction of "true" incipient-separation conditions: 1 – the FIT (equation 4.15), 2 – theory of Inger [40, 41]; incipience of developed separation at $p_{son} = p_p$: 3 – from FIT, 4 – from experimental correlations by Grodzovskyi [32], Zukoski [33], and Zheltovodov [46].

The M_i value and corresponding normal inviscid shock-wave critical strength ξ_i are well predicted in this manner (see Fig. 4.4). Because $k = 5.94$ and c_f are functions of M_∞ and Re_δ or Re_θ (e.g., see Kutateladze and Leont'ev [37] or Van Driest II [44]), the previous equation represents a transcendental function for the Mach number M_i as a function of the Reynolds Re_θ based on the compressible momentum thickness θ; FIT (see Fig. 4.6, dashed line 1) predicts a weak decrease of the boundary-layer resistance to separation (decrease of M_i) with increasing Reynolds number (decreasing c_f); and is applicable for $Re_\delta \leq 10^5$ (or $Re_\theta \leq 10^4$). A stronger shock is required at $Re_\delta > 10^5$ ($Re_\theta > 10^4$) to separate the boundary layer in accordance with Inger's prediction (see Fig. 4.6, line 2) corresponding to experiments by Grodzovskyi (1961) [32]. Thus, an unseparated-flow regime is realized below the boundaries shown by lines 1 and 2. The stage of effective incipient separation corresponds to the critical Mach number $M_{ef} \approx 1.3$ in accordance with condition $p_{son} = p_s$. As noted previously, the condition $p_{son} = p_p$ can be used to predict the incipient stage of developed separation analogous to Pearcey's [31] criterion. The corresponding critical values of $M_\infty = M_s$ at the low Reynolds number

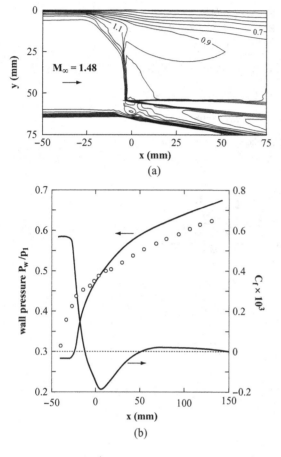

(a)

(b)

Figure 4.7. Computed Mach contours, $\Delta M = 0.1$, in two-dimensional diffuser (a), comparison of computations (solid lines) and experiment (symbols) for top wall pressure and computed surface skin friction. (b) (Blosch et al. [47]) (the total pressure p_t = 240 kPa).

$Re_\theta \leq 10^4$ using the plateau pressure p_p calculated on a basis of FIT (see equation 4.14) are shown by solid line 3. The lower boundary of the scattered data at the separation-flow regime (i.e., black symbols) corresponds to line 3 up to this value of the Reynolds number. This behavior is in agreement with a decrease in the fullness of velocity profile [3, 39, 45] and, respectively, increasing H_i in a relaxing turbulent boundary layer in the immediate region downstream of the laminar-turbulent transition region as the Reynolds number is increased up to $Re_\theta \approx 10^4$. Above these values, the tendency is reversed and the boundary-layer profile becomes fuller and H_i decreases with an increasing Reynolds number. The corresponding boundary for M_s at $Re_\theta > 10^4$ is shown by shaded band 4. The scatter of experimental data and the difference in the empirical correlations [32, 33, 46], which were used for the plateau pressure at the high Reynolds numbers, are characterized by the width of this band. As shown, this boundary is qualitatively similar to theory [40, 41] for $M_i(Re_\theta)$ shown by line 2. The demonstrated boundaries specify the regions of existence of unseparated, intermittent-separation, and developed-separation regimes in NSTBLI conditions.

4.2.2 Examples of NSTBLI Numerical Modeling

The RANS simulation [47] using the algebraic Baldwin-Lomax turbulence model for a steady, two-dimensional NSTBLI in a diffuser configuration [38] at Mach 1.48 and $Re_\delta = 2.3{\cdot}10^5$ ($Re_\theta = 1.46 \times 10^4$) is shown in Fig. 4.7. The predicted lambda

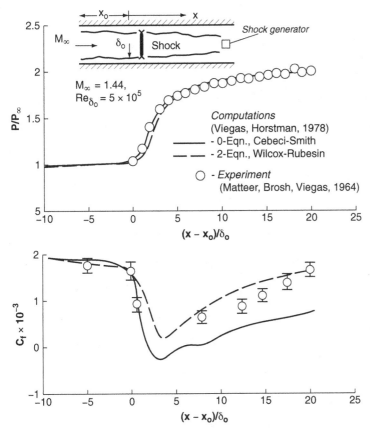

Figure 4.8. Comparison of computations and experiment for surface-pressure and skin-friction distributions in conditions of transonic normal shock interaction in axisymmetric flow (Viegas and Horstman [48]).

shock in the vicinity of the top wall (Fig. 4.7a) corresponds to the regime of developed boundary-layer separation and is in qualitative agreement with the experiment. The computed wall-pressure profile on the top surface (Fig. 4.7b, solid line) underestimates the upstream influence of the interaction, which is typical for the standard zero- and two-equation turbulence models. The computed skin-friction coefficient shows a significant region of reversed flow. The inflection point in the surface-pressure distribution is shown in the calculations and experiments in the vicinity of the minimal skin-friction region in the separation zone. The experimental and calculated static-pressure ratio in this point $p/p_\infty \approx 1.7$ (see Fig. 4.4, symbol 7) is within the limits of the generalized experimental data for the plateau-pressure value p_p/p_1 and predicted by the FIT. The computed and experimental velocity profiles at several locations upstream and downstream of the interaction are in close agreement [47].

A RANS simulation of the normal-shock interaction inside a cylindrical test section [48] is shown in Fig. 4.8. The surface pressure is accurately predicted by both algebraic (Cebeci-Smith) and two-equation (Wilcox-Rubesin) models (Fig. 4.8a). Nevertheless, the two-equation model demonstrates better agreement with the experiment in the skin-friction distribution (Fig. 4.8b) and velocity-profile development [48] downstream of the interaction. The two-equation model predicts the stage

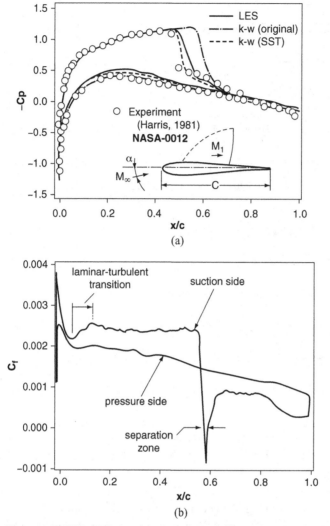

Figure 4.9. LES (Nakamori and Ikohagi [49]) and RANS (Wilcox [52]) computations of a normal shock-wave turbulent boundary layer interaction on an NACA 0012 airfoil at $M_\infty = 0.8$, $M_1 = 1.38$, $Re_c = 9 \times 10^6$, $\alpha = 2.26$ degrees.

close to incipient separation ($c_f = 0$); however, the one-equation model demonstrates a region of reversed flow. The computed and experimental profiles of turbulence kinetic energy are generally similar [48]. Nevertheless, the experimental results show significant free-stream turbulence downstream of the shock, which is not evident in the calculations. The inclusion of this effect tends to increase the calculated skin-friction downstream of reattachment relative to the values shown in Fig. 4.8b.

The capability of LES [49] to predict an NSTBLI on a NACA 0012 airfoil at Mach 0.8, angle of attack $\alpha = 2.26$ degrees, and Reynolds number $Re_c = 9 \cdot 10^6$ based on the airfoil chord is illustrated in Fig. 4.9. The Scale Similarity Model of Bardina et al. [50] was used. The normal shock appears on the upper (i.e., suction) side of the airfoil at about 60 percent of the chord length, with Mach number

$M_1 = 1.38$ in the localized supersonic region immediately upstream of the shock (Fig. 4.9a). Good agreement is observed in surface pressure (Fig. 4.9a) except for underestimation in computations at $x/c \approx 0.48 - 0.56$. The computed skin friction (Fig. 4.9b) indicates a separation zone with a reversed-flow region caused by the shock and decrease in c_f starts only at $x/c \approx 0.55$, in accordance with a later location of the region with a positive pressure gradient upstream of the separation. It is evident that the lengths of the separation zone and the upstream influence region are smaller in the computations compared to the experiment. This difference can be explained by a strong influence of the laminar-turbulent transition position on the scale effects in transonic flows [39, 51]. The predicted skin-friction coefficient increases at about 5 percent of the chord length from the leading edge on the suction side, where the transition occurs naturally. Nevertheless, its position in the experiment was not considered and compared to the predictions.

RANS computations [52] with the standard Wilcox k-ω model (see also [17]) (Fig. 4.9a) predict the shock-wave location significantly downstream compared to the experiment. The modified version of this model [52] incorporates the addition of a cross-diffusion term and a built-in stress-limiter modification that makes the eddy viscosity a function of k, ω, and the ratio of turbulence production to turbulence-energy dissipation. Adding the stress limiter yields significant improvement in predicting the normal-shock location and upstream influence of the separation zone on the upper surface of the airfoil (i.e., the dashed line identified as SST in Fig. 4.9a). Nevertheless, overestimation of the surface-pressure level appears in this zone compared to the experiment and LES.

It is important that despite the possibility of modifying RANS models to improve their predictions, the global unsteadiness of these flows is a dominant phenomenon. In accordance with recent assessments [2, 6, 7], accurate prediction of such flows therefore requires modeling of the unsteadiness; in this respect, development of LES and DNS represents the most promising direction of research.

4.2.3 Gas Dynamics Flow Structure in Compression Ramps and Compression-Decompression Ramps with Examples of Their Numerical Modeling

The compression ramp (CR) and compression-decompression ramp (CDR) interactions (see Fig. 4.2c) are characterized by a complex mean flowfield structure and various interaction regimes. For a sufficiently small angle α (i.e., where there is no or very small separation), the compression waves coalesce into a single shock (Fig. 4.10a,b). The downstream surface pressure (Fig. 4.10c) practically coincides with the inviscid-flow case. The mean surface skin-friction coefficient is everywhere $c_f > 0$ (Fig. 4.10d) and there is no mean reversed flow (Fig. 4.10e). The flowfield is accurately predicted [53, 54] with the standard two-equation Wilcox k-ω model and the Jones-Launder k-ε model.

For sufficiently large α (depending on M_∞, $Re_{\delta o}$ and T_w/T_{aw}), the boundary layer separates at point S upstream of the CR and reattaches at point R downstream (Figs. 4.11a and 4.11b). A compression-wave system forms upstream of the CR due to the deflection of the boundary layer by the separation bubble with a

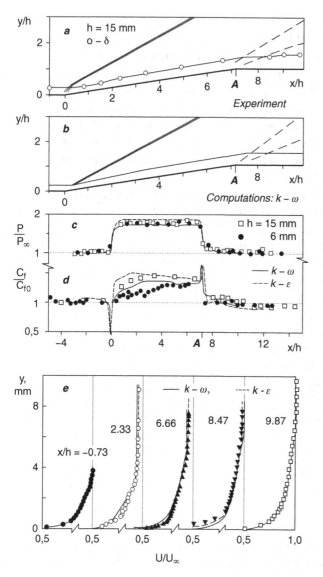

Figure 4.10. Comparison between experiment and RANS with the standard k-ω (Borisov et al. [53]) and the k-ε turbulence models (Horstman and Zheltovodov [54]) for compression/decompression ramp flow. $M_\infty = 2.9$, $\alpha = 8°$, $T_w/T_{aw} \approx 1$, $Re_{\delta o} = 1.22 \times 10^5$ at $h = 15$ mm ($h/\delta_o = 4.4$) and $Re_{\delta o} = 6.36 \times 10^4$ at $h = 6$ mm, $h/\delta_o = 2.6$ (from Zheltovodov [1], Borisov et al. [53]).

corresponding rise in the mean surface pressure (Fig. 4.11c), coalescing into a shock wave[3] (i.e., the separation shock). A "plateau" in the mean surface pressure forms in the region of reversed flow between points S and R, as shown in the velocity profiles (Fig. 4.11e). The surface-skin friction $c_f = 0$ at both points (Fig. 4.11d). A second compression-wave system forms in the vicinity of mean-reattachment point R as the flow is deflected by the corner surface and the external flow compresses to coalesce into a shock (i.e., the reattachment shock). The two shock waves intersect to form a λ-shock with a slip line (see Fig. 4.11a) and a secondary expansion fan, as shown in Fig. 4.2c, or a weak shock extending from a triple point at a low M_∞. Such

---

[3] The shock wave may form either within or outside the boundary layer depending on M_∞, $Re_{\delta o}$, and α.

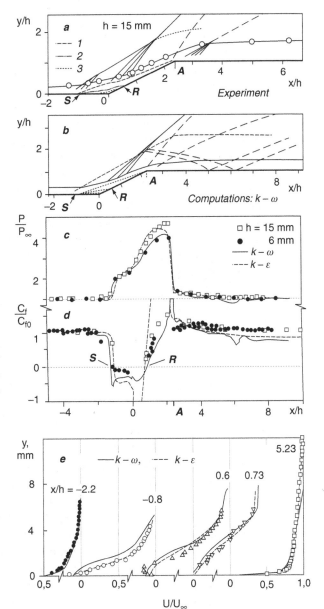

Figure 4.11. Comparison between experiment and RANS with the standard k-ω (Borisov et al. [53]) and the k-ε turbulence models (Horstman and Zheltovodov [54]) for compression-decompression ramp flow. $M_\infty = 2.9$, $\alpha = 25°$, $T_w/T_{aw} \approx 1$, $Re_{\delta_o} = 1.48 \times 10^5$ at $h = 15$ mm ($h/\delta_o = 3.7$), and $Re_{\delta_o} = 6.36 \times 10^4$ at $h = 6$ mm ($h/\delta_o = 2.6$): 1 – sonic line ($M = 1$), 2 – zero velocity line, 3 – line of maximum reversed flow velocity (from Zheltovodov [1], Borisov et al. [53]).

a secondary expansion fan is relatively weak in experiments at $M_\infty \leq 3$ but becomes noticeable at the higher Mach numbers [3, 10]. The boundary layer overexpands about the second corner with weak compression waves arising immediately downstream of the expansion fan. RANS computations [53, 54] with the k-ω and k-ε models are in good agreement with experimental data (see Figs. 4.11a–e). The flowfield structure (see Fig. 4.11b) reproduced using the computed static pressure and density panoramas correctly describes the emerging λ-shock around the CR and behavior of the sonic line (1), zero-velocity line $u = 0$ (2), and the maximum reversed-flow velocity line (3) in the separation zone, as well as the rarefaction fan around the decompression ramp. Limited experimental skin-friction data are available for $h = 15$ mm and they are supplemented with additional experimental data for $h = 6$ mm.

For configurations with a short compression-surface length, the flow properties begin to depend on the length (or the ratio h/δ_o) because the influence of the expansion of the reattaching subcritical boundary layer on the back tip of the step is transmitted in the upstream direction [55, 56]. If the reattachment point reaches the shoulder, the flow is basically the same as that ahead of a normal step, and the separation point reaches a distance approximately 4.2 step heights upstream of the shoulder [56]. These flow regimes are demonstrated in Figs. 4.12 and 4.13. The flows also exhibit a small, strongly unsteady separation zone immediately downstream of the step's top A (see Figs. 4.12a and 4.13a) and secondary separation (S') and reattachment (R') lines initiated by a localized vortex emerging in the bottom corner ahead of the step [46, 53, 57] (see Fig. 4.13a,d).

The computed flowfields (see Figs. 4.12b and 4.13b) correctly describe the shock-wave and expansion-wave structures, and they accurately predict the location of separation and reattachment as well as behavior of the sonic line (1), dividing streamline (2), zero-velocity line $u = 0$ (3), and maximum reversed-flow velocity line (3). Nevertheless, the computations do not exhibit a small, strongly unsteady separation zone near the top of the steps. The computed and experimental surface pressures (see Figs. 4.12c and 4.13c) demonstrate good agreement except in the vicinity of the decompression corner at $\alpha = 45$ degrees. The computed and experimental skin-friction coefficient is displayed in Figs. 4.12d and 4.13d. Limited experimental skin-friction data are available for $h = 15$ mm, and additional experimental data for $h = 6$ mm are shown. The agreement between computation and experiment is good; similar agreement is demonstrated in Figs. 4.12e and 4.13e for the mean-velocity profiles in the separation zone with the reversed flow.

A series of RANS computations [58] of the compression corner at Mach 2.96 for $\alpha = 15$ to 25 degrees at $Re_\delta = 1.5 \times 10^5$ is demonstrated in Fig. 4.14. The zero-equation turbulence model of Cebeci and Smith was used with a relaxation model for the turbulent-eddy viscosity $\mu_t = \mu_{t\infty} + (\mu_{t_{equil}} - \mu_{t\infty})[1 - e^{-(x-x_\infty)/\lambda}]$, where $\mu_{t\infty}$ is the eddy viscosity evaluated at location x_∞ immediately upstream of the compression-corner interaction and $\mu_{t_{equil}}$ is the standard (i.e., equilibrium) Cebeci–Smith eddy viscosity. The assumed relaxation length is $\lambda = 10\delta_o$, where δ_o is the boundary-layer thickness immediately upstream of the interaction. The computed and experimental mean-surface pressure is displayed in Fig. 4.14a, where $L = 1$ ft is the length of the upstream plate and S is the distance along the surface. The overall agreement is good; in particular, the position of upstream influence (i.e., the location of the initial mean-pressure rise) is accurately predicted as a function of the compression angle. The mean-streamwise computed and experimental velocity profiles at and downstream of the mean-separation location are shown in Fig. 4.14b. Good agreement is observed; nonetheless, the relaxation length $\lambda = 10\delta_o$ is not a universal length scale; a value twice as large is necessary to obtain close agreement with experiments for the incident-oblique shock configuration at the same free-stream Mach number and comparable Reynolds number (see the following discussion).

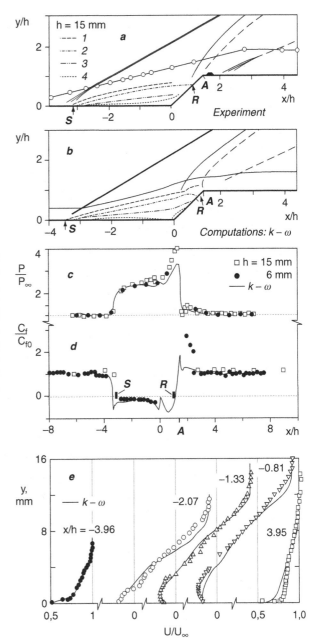

Figure 4.12. Comparison between experiment and RANS for compression/decompression ramp flow at $M_\infty = 2.9$, $\alpha = 45°$, $T_w/T_{aw} \approx 1$, $Re_{\delta o} = 1.55 \times 10^5$ at $h = 15$ mm ($h/\delta = 3.5$) and $Re_{\delta o} = 6.36 \times 10^4$ at $h = 6$ mm ($h/\delta_o = 2.6$). 1 – sonic line ($M = 1$), 2 – dividing streamline, 3 – zero velocity line, 4 – line of maximum reversed-flow velocity (from Zheltovodov [1], Borisov et al. [53]).

Extensive RANS computations [48] illustrate the possibilities of different turbulence models: zero-equation (i.e., algebraic) equilibrium, one-equation (i.e., kinetic energy) by Glushko [127], and two-equation (i.e., kinetic energy plus length scale) turbulence models by Jones-Launder and Wilcox-Rubesin (Fig. 4.15a,b). Results are presented for the surface-pressure distribution (a) and the mean-velocity profiles (b) in the vicinity of the CR, demonstrating qualitative agreement with the data. No single model shows preference for the best quantitative agreement; however, in general, overall improvement is obtained with higher-order turbulence models. This

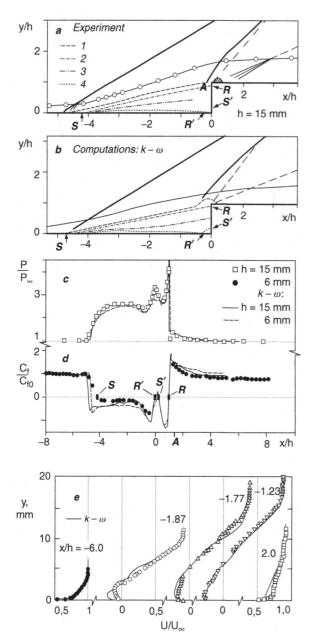

Figure 4.13. Comparison between experiment and RANS for compression/decompression ramp flow at $M_\infty = 2.9$, $\alpha = 90°$, $T_w / T_{aw} \approx 1$, $Re_{\delta o} = 1.44 \times 10^5$ at $h = 15$ mm ($h/\delta = 3.8$) and $Re_{\delta o} = 6.36 \times 10^4$ at $h = 6$ mm ($h/\delta_o = 2.6$): 1 – sonic line (M = 1), 2 – dividing streamline, 3 – zero velocity line, 4 – line of maximum reversed-flow velocity (from Zheltovodov [1], Borisov et al. [53]).

conclusion is supported by recent computations [52] (Fig. 4.15c,d) with a new version of the k-ω model both with and without the stress limiter. Most important, in the case with a stress limiter (i.e., solid lines), the computed initial pressure rise matches the measured rise and the predicted pressure-plateau level in the separation bubble is close to the measurements. The computed c_f is also in close agreement with the data, but a discrepancy with measurements downstream of the reattachment indicates a lower rate of recovery of the disturbed boundary layer to equilibrium conditions.

Despite intensive experimental and computational study, several aspects of relevant physics involved in these flows remain poorly understood and some of the

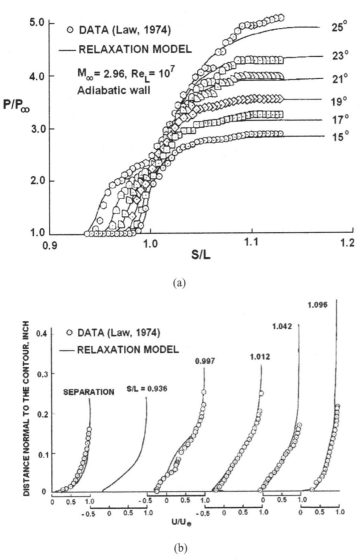

Figure 4.14. Comparison between RANS computations on a basis of relaxation version of Cebeci-Smith model with experimental data for compression-ramp flow (Shang and Hankey [58]).

physics cannot always be reproduced simply through the turbulence modifications [2, 6, 7, 11, 59]. Important physical phenomena (Fig. 4.16) include amplification of the turbulence by unsteady shock waves in the boundary layer (1) and external flow (2); suppression of turbulence by the rarefaction waves (3); formation of a new boundary layer in the near-wall region of the attached flow (4); formation of Taylor-Görtler vortices (5); and manifestation of the process, which looks like relaminarization in the separation region (6) due to the favorable pressure gradient in reverse flow and a decrease in the Reynolds number (due to the reverse-flow velocity reducing in the separation region) [46]. These elements are essential and must be considered for development of adequate mathematical models for computations of such flows.

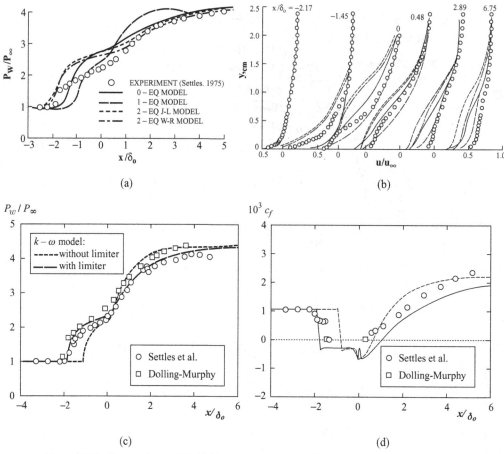

Figure 4.15. Comparison of RANS computations with experimental data for compression-ramp flow at $\alpha = 24°$, $M_\infty = 2.8$, $Re_{\delta_o} = 1.33 \times 10^6$, $T_w / T_{wa} = 0.88$: a, b – Viegas and Horstman [48]; c, d – Wilcox [52].

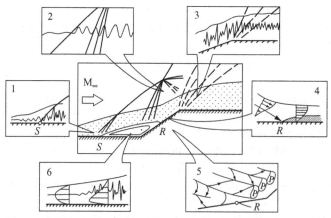

Figure 4.16. Specific physical features of flow over compression-decompression ramp configuration (Zheltovodov [1, 2, 80].

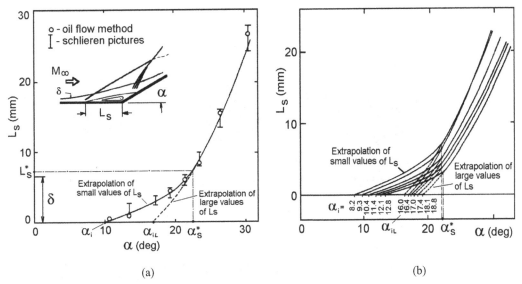

Figure 4.17. Separation length as function of the ramp angle at $M_\infty = 3.5$, $T_w/T_{aw} = 1.0$ (Appels and Richards [60]): a) $Re_{\delta o} = 2.07 \times 10^5$, b) $Re_{\delta o} = 9.34 \times 10^4$, 2.07×10^5, 3.23×10^5, 4.37×10^5, 5.38×10^5, and 6.42×10^5.

4.2.4 Incipient Separation Criteria, STBLI Regimes, and Scaling Laws for CR and CDR Flows

Separation in the CR flow is a smooth and gradual transition between attached and separated flows [60, 61, 62, 63], initially appearing in the viscous sublayer at a small value of the ramp angle $\alpha = \alpha_i$ and slowly growing with increasing α (Fig. 4.17) in the wall-interaction layer ($y/\delta \approx 0.1 - 0.2$). At sufficiently large α, the outer supersonic portion of the boundary layer becomes involved and the separation length increases with increasing α at a higher rate. These observations define two separate critical values for α – namely: (1) the first value α_i corresponding to the onset of "small" separation; and (2) the larger value α_{iL} characterizing the onset of the high growth rate or "large-scale" separation region [60, 61]. Additionally, the third value α_s^* corresponding to the bifurcation point for the extrapolated curves (Fig. 4.17) can be recommended in parallel with α_{iL} to define the onset of large-scale separation.

Data for the critical CR angles α_i, α_{iL}, and α_s^* are presented in Fig. 4.18a,b. The measurements of α_s^* by Settles et al. [62, 63] at $M_\infty = 2.9$, $T_w/T_{aw} \approx 1$ performed with the use of different experimental techniques (Fig. 4.18a, open symbols) fall within a band of 15 to 18 degrees and are independent of $Re_{\delta o}$ for fully turbulent flow ($Re_{\delta o} > 10^5$). As concluded, this behavior of the data refutes a tendency to increase with the Reynolds number displayed by previous measurements of α_s^* by Law, and Roshko, and Tomke [56] shown in Fig. 4.18a in accordance with [62]. Similarly, the data for α_{iL} in accordance with measurements by Appels and Richards [60] at different M_∞ display similar independence of $Re_{\delta o}$ and increasing α_{iL} with M_∞. Holden's [64] correlation (Fig. 4.18a, solid line), developed for hypersonic flows but applied for $M_\infty = 2.9$, predicts a decrease in α_s^* with increasing $Re_{\delta o}$. This trend is typical for viscous-dominated flows and agreement with the supersonic data is limited to low $Re_{\delta o}$. Elfstrom's [65] prediction (Fig. 4.18a, dashed line), applied for $M_\infty = 2.9$,

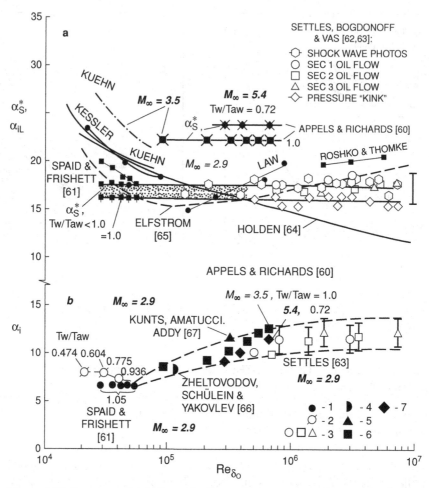

Figure 4.18. Summary of data for critical compression-ramp angles α_i and α_S^* corresponding, respectively, to incipience of "small" and "large-scale" separation: $1 - \delta_o = 8.38 \div 8.0$ mm (Spaid and Frishett [61]); $2 - 9.96 \div 8.08$ mm (Spaid and Frishett [61]); $3 - 30 \div 10$ mm (Settles [63]); $4 - 3.4$ mm (Zheltovodov et al. [66]); $5 - 8.27$ mm (Kuntz et al. [67]; $6 - 6.5 \div 5.9$ mm (Appels and Richards [60]); $7 - 19.8 \div 18.2$ mm (Appels and Richards [60]).

shows good agreement with experiments except for a doubtful tendency for increasing α_s^* at higher $Re_{\delta o}$.

The experimental data for α_i (Fig. 4.18b) represent the best efforts in the exceedingly difficult task of measuring the onset of small separation (see, e.g., Appels and Richards [60], Spaid and Frishett [61], and Settles et al. [62, 63]). It is clearly evident, however, that α_i is smaller than α_s^* for a given M_∞ and $Re_{\delta o}$. The α_i values for the adiabatic wall at $M_\infty = 2.9$ (symbols $1, 3 - 5$) range from 6.5 to 12 degrees for $10^4 < Re_{\delta o} \leq 10^7$ and increase with $Re_{\delta o}$ for $Re_{\delta o} > 10^5$. This behavior is consistent with the decreasing fullness of velocity profiles (i.e., increasing H_i) in a relaxing boundary layer immediately downstream of the laminar-turbulent transition up to $Re_{\delta o} \approx 10^5$ [39, 45], which becomes less resistant to separation. Also, the increased thickness of the interacting viscous-wall layer and its subsonic portion at

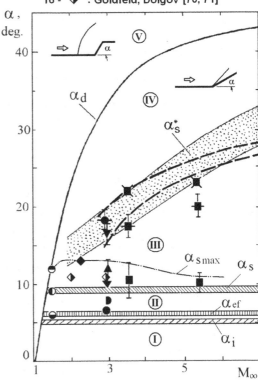

1- ●, 2-- ●: Spaid, Frishett [61]
3- ■, 4-- ■, 5- ▒: Appels, Richards [60]
6 - ▼, 7- ▼: Settles, Bogdonoff, Vas [62,63]
8 - ◗ : Zheltovodov, Schülein, Yakovlev [66]
9 - ▲ : Kuntz, Amatussi, Addy [67]
10 - ◓, 11- ◒, 12- ◒: Thomas, Putnam, Chu [68]
13 - ◆: Ardonceau, Lee, Alziary de Roquefort, Goethals [69]
14 - ▒▒▒ : Korkegi [72]
15 - ≡≡≡ : Elfstrom [65]
16 - ◈ : Goldfeld, Dolgov [70, 71]

Figure 4.19. The stages of turbulent-separation development in the vicinity of compression and compression/decompression ramps (Zheltovodov [1, 2]) I – unseparated flow; II – intermittent separation; III – developing small-scale separation; IV – large-scale separation in compression ramp; V – maximum-scale separation in front of the forward-facing step.

low $Re_{\delta o}$ promotes an earlier onset of small separation. The increasing fullness of velocity profiles (i.e., decreasing H_i) with reduced shear stress at $Re_{\delta o} > 10^5$ makes the boundary layer more resistant to separation. Also, the movement of the sonic line closer to the wall and the decrease in viscous-layer thickness with increasing $Re_{\delta o}$ reduces the thickness of the interacting wall layer, which defines the scale of the interaction and causes a rise in α_i. It has been noted (see, e.g., Spaid and Frishett [61]) that wall-cooling suppresses incipient separation (Fig. 4.18b, symbol 2).

The five stages of turbulent-separation development for adiabatic-wall CR and CDR (Fig. 4.19) are I: unseparated flow; II: intermittent separation; III: developing small-scale separation; IV: large-scale separation in CR; and V: maximum-scale separation in front of the forward-facing step. The features of separation development in the vicinity of a normal shock explain the stages of the separated-flow development near a two-dimensional CR. Following Elfstrom's [65] concept,

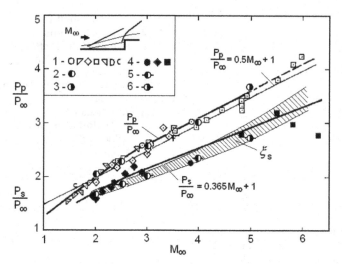

Figure 4.20. Comparison of pressure ratio in the "plateau" region p_p/p_∞ (symbols 1–3) and in separation point p_s/p_∞ (4–6) with critical shock-wave strength ξ_s in compression ramp: 1, 4 – Zukoski [33]; 2, 5 – Zheltovodov [46]; 3, 6 – Zheltovodov and Schülein [73].

Zheltovodov [1] proved that incipient small separation and the early stages of its development at low to moderate Reynolds numbers ($Re_{\delta o} \leq 10^5$) are characterized by the CR critical angles α_i, α_{ef}, and α_s (Fig. 4.19), corresponding to detachment shock-wave conditions realized at different distances from the surface where the local critical Mach numbers achieve the values M_i, M_{ef}, and M_s, respectively. These critical Mach numbers characterize the stages of true incipient separation, beginning effective intermittent and developed separation in the vicinity of the normal shock, respectively (see Fig. 4.4). The data [60, 61, 62, 63, 66, 67] for incipient-separation conditions in CRs with adiabatic walls (see Fig. 4.18b) are concentrated close to these boundaries (Fig. 4.19, symbols 1, 3, 6, 8, and 9) with additional data [68, 69, 70, 71] (i.e., symbols 10–13, 16) and support the proposed considerations. Decreasing the external-flow Mach number increases the thickness of the transonic near-wall portion of the boundary layer and produces earlier onset of resolved small separation, even at the higher Reynolds number. In this connection, the additional data [68] at $M_\infty = 1.5$, $Re_\delta = 1.79 \cdot 10^5$ ($Re_{\theta o} = 17,500$) are indicative because a weak intermittent separation was found at these conditions at $\alpha = 6$ degrees (i.e., symbol 10) and more prominently at $\alpha = 9$ and 12 degrees (i.e., symbols 11 and 12, respectively).

The strength (measured by the pressure ratio) of the oblique shock wave for the CR $\xi_{inv} = \xi_s$ corresponding to $\alpha = \alpha_s$ (see Fig. 4.19) for a range of Mach numbers $M_\infty \approx 2 - 5$ practically coincides with experimental data and an empirical correlation by Zukoski [33] for the static-pressure ratio in the separation point p_s/p_∞ (Fig. 4.20). The following simple expression:

$$\xi_s = p_s/p_\infty = 0.365 M_\infty + 1 \qquad (4.17)$$

is a convenient condition for practical determination of incipient effective separation, which is close to true separation. By equating the oblique-shock-wave strength

ξ_{inv} with the plateau-pressure ratio $\xi_p = p_p/p_\infty$ realized in the separation region [33]:

$$\xi_{inv} = \xi_p = p_p/p_\infty = 0.5M_\infty + 1 \qquad (4.18)$$

another value of the critical CR $\alpha = \alpha_{s\,max}$ can be predicted. As shown in Fig. 4.19, this prediction (i.e., line $\alpha_{s\,max}$) is in good agreement with the upper limit of the CR values revealed in experiments at high Reynolds numbers $Re_{\delta o} > (1-2) \cdot 10^5$.

The next boundary marked as α_s^* in Fig. 4.19 is represented by band 14 corresponding to the incipient large separation [72] at $Re_{\delta o} > 10^5$. The tendency for increasing α_s^* with Mach number was predicted by Elfstrom [65] (i.e., band 15 between two dashed lines) for $2 \cdot 10^5 < Re_{\delta o} \leq 10^7$. The experimental data for α_s^* (i.e., symbols 2, 5, 7) are also in good agreement with this correlation. Thus, area III in Fig. 4.19 corresponds to developing small-scale separation and area IV characterizes large-scale separation when the attached oblique shock should form for CRs in inviscid flow. The last boundary marked as α_d corresponds to the appearance of a detached shock ahead of the step with an inclined face of limited length at inviscid-flow conditions. The maximum-scale separation zones are realized above this boundary in region V, in fact, in NSTBLI conditions ahead of the forward-facing steps.

The stages and features of separation development explain the different behaviors of the upstream influence distance L_u/δ_o (i.e., the location of the rise in surface pressure, measured from the corner) and the separation-line location L_s/δ_o (i.e., measured from the corner) with increasing α for adiabatic and cold walls [61] at $M_\infty \approx 2.9$ and at low Reynolds numbers $Re_{\delta o} < 10^5$ (Fig. 4.21). In the case of an adiabatic wall (Fig. 4.21a), the linear behavior of L_u/δ_o corresponds to the unseparated flow stage I, and its sudden reduction or constant value occurs at intermittent-separation conditions ($\alpha_{ef} \leq \alpha \leq \alpha_s$, stage II). The development of a small-scale separation in the near-wall portion of the boundary layer at $\alpha_s \leq \alpha \leq \alpha_s^*$ (i.e., stage III) is accompanied by an increase in L_u/δ_o. The rate of increase is significantly greater in the next stage of large-scale separation development (i.e., stage IV) in the external portion of the boundary layer. The separation-line position L_s/δ_o is practically independent of the Reynolds number in conditions of intermittent and small-scale separation (i.e., stages II and III), but a tendency to increase with rising $Re_{\delta o} = (3.63 - 5.92) \cdot 10^4$ emerges on the stage of large-scale separation development (i.e., stage IV). This can be explained by a decreasing mean velocity-profile fullness in the external portion at such conditions. It is remarkable that the rise of L_u/δ_o at $\alpha_s < \alpha \leq \alpha_s^*$ (i.e., stages III and IV) is caused mainly by increasing separation distance L_s/δ_o. The wall-cooling promotes the degeneration of subsonic and viscous near-wall portions of the interacting boundary layer and the suppression of separation (Fig. 4.21c). As a result, an obvious reduction of the upstream-influence distance L_u/δ_o with decreasing wall-temperature ratio T_w/T_{aw}, especially pronounced at $T_w/T_{aw} = 0.775 - 0.474$, occurs at all stages I–IV (Fig. 4.21b) compared to the adiabatic-wall case. The value $\alpha_s^* \approx 16 - 17$ degrees determined from Figs. 4.21a,c is approximately the same for adiabatic and cooling walls and independent of the Reynolds number in the considered range (see the band of data by Spaid and Frishett [61] for $T_w/T_{aw} = 1.0$ and $T_w/T_{aw} < 1.0$; Fig. 4.18a) as in the data by Settles et al. [62] (i.e., open symbols) at $Re_{\delta o} > 10^5$. This value corresponds to

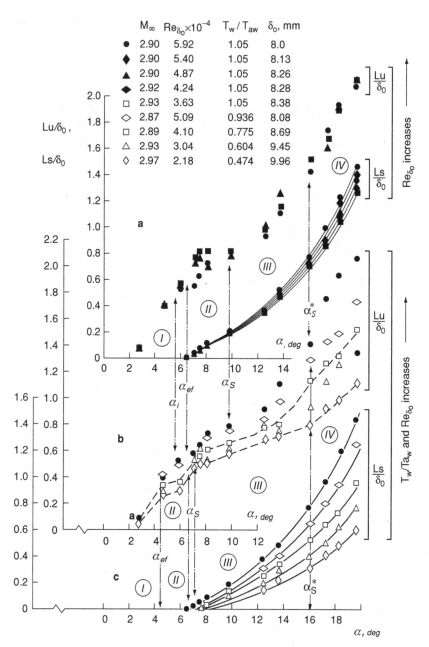

Figure 4.21. Dimensionless upstream influence distance (L_u/δ_o) and separation position (L_S/δ_o) at adiabatic (a) and cooling wall (b, c) conditions (Spaid and Frishett [61]) and specific stages of separation development (Zheltovodov [1, 2]): I – unseparated flow; II – intermittent separation; III – developing small-scale separation; IV – large-scale separation.

experimental correlation 14 by Korkegi [72] (see Fig. 4.19, symbol 2) at $M_\infty = 2.9$. Discussed previously, the data for α_s^* by Appels and Richards [60] for an adiabatic wall at $M_\infty = 3.5$ and a cooling wall at $M_\infty = 5.4$ (see Fig. 4.18a) shown in Fig. 4.19 by symbol 5 demonstrate better agreement with this correlation than α_{iL} values (i.e., symbol 4).

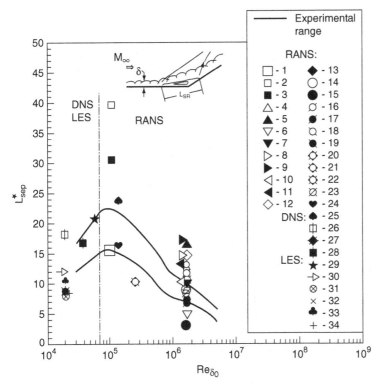

Figure 4.22. Comparison of separation length for RANS, DNS, and LES with experiment (from Knight et al. [7]). RANS: 1 – 3 – $\beta = 25°$ (Shang and Hankey [58]); 4 – 7 – $\beta = 20°$ (Horstman and Hung [74]); 8 – 11 – $\beta = 24°$ (Shang and Hankey [58]; 12 – 15 – $\beta = 16°$, 16 – 19 – $\beta = 20°$, 19 – 20 – $\beta = 24°$ (Visbal and Knight [75]); 21, 22 – $\beta = 16°$, 23 – (Ong and Knight [76]); 24 – $\beta = 25°$, k–ε model (Horstman and Zheltovodov [54]); 25 – $\beta = 25°$, k-ω model (Borisov et al. [53]). DNS: $M_\infty = 3$, $\beta = 18°$: 26 – (Adams [98]), 27 – (Rizzetta and Visbal [111]); 28 – $M_\infty = 2.9$, $\beta = 24°$ (Ringuetta, Wu, and Martin [112]). LES: M_∞ =2.95: 29 – $\beta = 25°$ (Loginov et al. [103, 105]), 30 – $\beta = 25°$ (Yan et al. [100]), 31, 32, 33 – $\beta = 18°$, 20° and 24° (Rizzetta and Visbal [111]), 34 – $\beta = 18°$ (El-Askary [96]).

An empirical scaling law [73] for the total extension of large-scale separation in CRs in regime IV (see Fig. 4.19) at adiabatic-wall conditions $L^*_{sep} = L_{SR}/L_c = f(Re_{\delta o})$ (Fig. 4.22; the region between solid lines) incorporates the effects of Reynolds number $Re_{\delta o}$, Mach number M_∞, compression angle α (through the inviscid-pressure rise p_2/p_∞), and boundary-layer thickness δ_0. Here, L_{SR} is the shortest distance between separation S and reattachment R points, $L_c = (\delta_0/M_\infty^3)(p_2/p_p)^{3.1}$, and the plateau pressure p_p/p_∞ is described by the empirical formula (4.18) [33]. There is a distinct change in trend at $Re_\delta \cong 10^5$ that corresponds to the variation of the fullness of the mean-velocity profile with $R_{\delta o}$ noted previously. Fig. 4.22 is a summary [2, 7] of several RANS predictions of separation length for two-dimensional CR configurations. Results of different researchers (i.e., symbols 1–25) who used various algebraic turbulence models (including the Baldwin-Lomax model and its modification) [58, 74, 75, 76], the k-ε and k-ω models [53, 54], and various ad hoc modifications correspond to the range $\beta = 16$–25 degrees, $M_\infty = 1.96$–3.0 at $Re_{\delta o} \geq 10^5$. As shown, in general, the simulations do not predict accurately the separation length in comparison with the experimental correlation

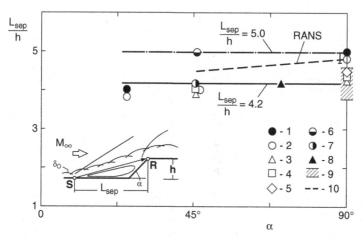

Figure 4.23. Correlation for extension of the maximum-scale separated zone ahead CDR in regime V *(at $h/\delta_o > 1.5$)*. $\alpha = 25°$, $45°$ and $90°$: 1 – $M_\infty = 2.0$, 2 – 2.25, 3 – 3.0, 4 – 4.0 (Zheltovodov [46]; Dem'yanenko and Zheltovodov [57]; Zheltovodov, Schülein, and Yakovlev [66]); 5 – $\alpha = 90°$, $M_\infty = 5.0$ (Schülein and Zheltovodov [77]); $\alpha = 24°$: 6 – $M_\infty = 1.6$, 7 – 2.5 (Coe, from [33]); 8 – $\alpha = 70°$, $M_\infty = 3.0$ (Hahn [78]); 9 – scatter of data for $\alpha = 90°$ at $M_\infty = 2.1$– 6.5 (Zukoski [33]); 10 – RANS with the k-ω model at $M_\infty = 2.9$ (Borisov et al. [53]).

(i.e., solid lines). The behavior of calculations on the basis of LES and DNS is discussed later in this chapter.

Another simple empirical correlation $L_{sep}/h \cong 4.2$ can be used to predict the extent of the maximum-scale separation in front of CDR (Fig. 4.23) in regime V (see Fig. 4.19) corresponding to the situation with a detached normal shock in inviscid-flow conditions. This correlation was demonstrated by Zukoski [33] for normal ($\alpha = 90$ degrees) steps at $M_\infty = 2.1 - 6.5$ and can be used at $h/\delta_o > 1$. Shown in Fig. 4.23 are additional data [46,57] for $\alpha = 90$ degrees at $M_\infty = 3.0$ and 4.0 (i.e., symbols 3 and 4) and at $M_\infty = 5.0$ (i.e., symbol 5 [77]) correspond to this correlation and are located in the limits of scatter band 9 of the data [33]. Experimental data by Zheltovodov et al. [66] for $\alpha = 25$ degrees at $M_\infty = 2.0$ and 2.25 (i.e., symbols 1 and 2), $\alpha = 45$ degrees at $M_\infty = 2.25$, 3.0, and 4.0 (i.e., symbols 2, 3, and 4), and by Coe (i.e., symbol 7, see [33]) at $M_\infty = 2.5$, as well as by Hahn [78] at $\alpha = 70$ degrees at $M_\infty = 3.0$ (i.e., symbol 8) are also in good agreement with this correlation. The RANS computations [53] with the k-ω model for $\alpha = 45$ and 90 degrees at $M_\infty = 2.9$ demonstrate good agreement with the experiments (i.e., line 10) (see also Figs. 4.12 and 4.13), although a small overprediction of L_{sep}/h is seen at $\alpha = 90$ degrees. The data [46, 57] for $\alpha = 90$ degrees at $M_\infty = 2.0$ and 2.25 (i.e., symbols 1 and 2) and by Coe at $\alpha = 45$ degrees at $M_\infty = 1.6$ (i.e., symbol 6) demonstrate the increase of L_{sep}/h up to the value $L_{sep}/h \cong 5.0$. This tendency corresponds to the data considered by Zukoski at $1.4 \leq M_\infty < 2.1$. It can be explained by a significant rise of the normal-shock detachment distance ahead of the step in inviscid-flow conditions at low Mach numbers (see, e.g., [79]).

4.2.5 Heat Transfer and Turbulence in CR and CDR Flows

The importance of various physical processes (see Fig. 4.16) can be confirmed by analysis of the surface-heat transfer and turbulence modification for CR and CDR

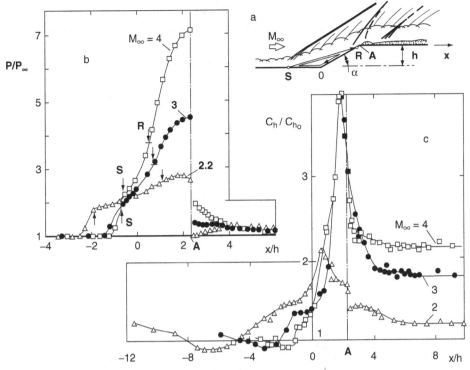

Figure 4.24. Surface pressure (b) and heat-transfer coefficient (c) distributions in the vicinity of CDR configuration (a) at $\alpha = 25°$: $b - h/\delta_0 = 3.3$–3.9, $Re_{\delta_0} = 1.25 \times 10^5$, 1.48×10^5, and 1.76×10^5, respectively, at $M_\infty = 2.2$, 3.0 and 4.0, $T_w/T_{aw} \approx 1$; $c - h/\delta_0 = 2.3 - 2.4$, $Re_{\delta_0} = 8.06 \times 10^4$, 9.3×10^4, and 1.16×10^5, respectively, at $M_\infty = 2.0$, 3.0, and 4.0, $T_w/T_{aw} = 1.04 - 1.05$ (from Zheltovodov [80] and Zheltovodov et al. [81]).

configurations [80, 81]. Figure 4.24 shows the variation of the surface-pressure p/p_∞ (Fig. 4.24b) and the heat-transfer coefficient C_h/C_{ho} distributions (Fig. 4.24c) for a CDR at $\alpha = 25$ degrees and different Mach numbers M_∞, where C_{ho} is the heat-transfer coefficient value immediately upstream of the initial surface-pressure-rising region. Some of the observed variations in the surface-heat transfer correlate qualitatively with the surface-pressure distributions, including (1) the increase in length of the separation zone upstream of the CR with a decreasing Mach number; and (2) the appearance of a localized region of positive surface-pressure and heat-transfer gradients immediately downstream of the top of A at $M_\infty = 2.2$ and 2.0 (Fig. 4.24b,c), with the advent of an additional small separation zone (Fig. 4.24a) at $\alpha \geq \alpha_d$ in regime V (see Fig. 4.19). However, there are notable differences between the surface-heat transfer and pressure distributions, including (1) the maximum heat transfer is the same for $M_\infty = 3.0$ and 4.0 upstream of the top of A (Fig. 4.23c), whereas the maximum pressure increases with M_∞ (Fig. 4.24b); (2) the significant rise of the heat transfer on the step's top surface downstream of point A with an increasing M_∞ value despite the same pressure levels at least at $x/h > 3.5$; and (3) the decrease in the heat-transfer level in the separation zone on the CR with an increasing Mach number opposite the rising surface-pressure level between separation S and reattachment R lines.

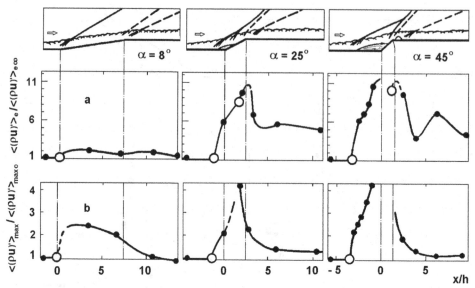

Figure 4.25. Longitudinal mass-flux fluctuations in the external flow (a) and along a line of maximal value in the external portion of boundary layer (b): $M_\infty = 2.95$, $Re_{\delta_0} = 6.36 \times 10^4$, $T_w/T_{aw} \approx 1$, $h/\delta_o = 2.5 - 2.64$ (from Zheltovodov [80]).

It is not difficult to understand the similar surface-heat transfer peaks on the compression surface ahead of point A (Fig. 4.24c). The step heights h/δ_0 in the heat-transfer experiments were $h/\delta_0 = 2.3 - 2.4$, whereas in the surface-pressure experiments, $h/\delta_0 = 3.3 - 3.9$ [81]. As a result, the length of the CR's surface (i.e., the distance x/δ_0 between points 0 and A; see Fig. 4.23a) was limited in the former case and prevented the heat transfer from rising in qualitatively similar way as the surface pressure at $M_\infty = 3.0$ and 4.0. For a CR with a sufficiently long compression surface, the maximum heating level correlates with the maximum pressure rise in the interaction region [3, 10, 82] and can be described by the simple power-law relationship proposed by Holden [82], $q_{max}/q_o = (p_{max}/p_o)^{0.85}$, which generalizes experimental data for a wide range of supersonic and hypersonic Mach numbers. The heightened surface-heat-transfer level on the top surface (Fig. 4.24c) can be explained by the formation of a thinner, new boundary layer downstream of the disturbances (i.e., shock waves, separation zone, and rarefaction fan) that develops in the near-wall portion of the relaxing "old" boundary layer as in an external flow with the heightened turbulence level [80, 81, 83] (see Fig. 4.16, sketch 4, and Fig. 4.24a).

Figure 4.25 shows the variation of the root mean square (rms) mass-flux fluctuations $<(\rho u)'>_e$ in the external flow above the boundary-layer edge (Fig. 4.25a) and along the line of their maximum value in the outer region ($y/\delta \geq 0.4$) of the boundary layer (Fig. 4.25b), which were measured by the constant-current hot-wire anemometry [80, 84, 85] at $M_\infty = 2.95$ and the low Reynolds number $Re_{\delta_0} = 6.34 \cdot 10^4$. Fluctuations in the disturbed external flow above the boundary layer are normalized by their value $<(\rho u)'>_{e\infty}$ in the undisturbed flow, and $<(\rho u)'>_{max}$ in the boundary layers are normalized by the maximum value $<(\rho u)'>_{max0}$ in the undisturbed boundary layer upstream of the CR. The regions shown by open circles correspond to locations where the mass flux jumps across the unsteady shock, thereby producing a false turbulence peak in the measured profiles. It is evident that the SBLI in the CR significantly enhances the mass-flow fluctuations in the disturbed external flow (Fig. 4.25a) and in the outer region of the boundary layer on the compression surface (Fig. 4.25b). The turbulence amplification increases with the rising CR inclination angle α and, despite partial suppression by the expansion waves in the vicinity of the decompression ramp, the heightened level remains downstream in the external flow over a long distance, whereas a more rapid relaxation of the maximum fluctuations to their initial value is observed in the boundary layer. Predominant acoustic fluctuations were revealed in the external flow [84, 85] that are associated with moving weak shocklets generated by large-scale vortices in the external part of the disturbed boundary layer.

Figure 4.26 compares RANS computations with different turbulence models and experimental data for surface-heat-transfer coefficient distributions supporting previous conclusions regarding the limitations of the RANS eddy-viscosity models. As shown, Horstman's computations [86] with the standard Jones-Launder k–ε turbulence model and its two-layer modification by Rodi [86] did not obtain an acceptable agreement with experimental data by Zheltovodov et al. [81, 83]. (The scatter bars in Fig. 4.26a indicate the level of spanwise heat-transfer variation in the experimental data in accordance with additional measurements by Trofimov and Shtrekalkin [87] caused by Görtler-type vortices [see Fig. 4.16, sketch 5].) The two-layer k-ε model by Rodi shows improvement in the prediction of wall-heat transfer in the CR vicinity but underestimates the level downstream of the expansion corner similar to the first model. Later attempts [88] were directed toward improving the prediction of different parameters in the separation zones ahead of CR and CDR test configurations by controlling the balance between the turbulence production and dissipation in the framework of the Wilcox k–ω model. Significant improvements were achieved by limiting the specific turbulent kinetic dissipation rate $\omega_e \sim \omega_w 10^{-3}$, where ω_w is the maximum value on the wall. These computations improved the heat-transfer prediction in the CR vicinity (Fig. 4.26a), as well as surface-pressure, skin-friction, and velocity profiles by modeling the phenomenon of the reverse-flow relaminarization in the separation zone in accordance with Zheltovodov's [46] experiments (see also [89]). However, as shown in Fig. 4.26a, significant underestimation of surface-heat transfer remained on the top surface downstream of the expansion fan, as in other turbulence models.

4.2.6 Unsteadiness of Flow Over CR and CDR Configurations and Its Numerical Modeling

The limitations of RANS turbulence models typically are blamed for the discrepancies between computations of STBLIs and experiments. However, it is clear that global unsteadiness is one of the dominant phenomena of STBLIs (see Chapter 9) and without its modeling, accurate predictions of the time-averaged wall pressure, velocity profiles, heat transfer, and other parameters are not likely to be possible, irrespective of the turbulence model [6, 59]. As shown in Fig. 4.19, the second through fifth regimes of separation development in the CR and CDR vicinity are not strictly steady due to shock-wave unsteadiness and the consequent intermittent nature of flow in the vicinity of the separation and reattachment lines. The

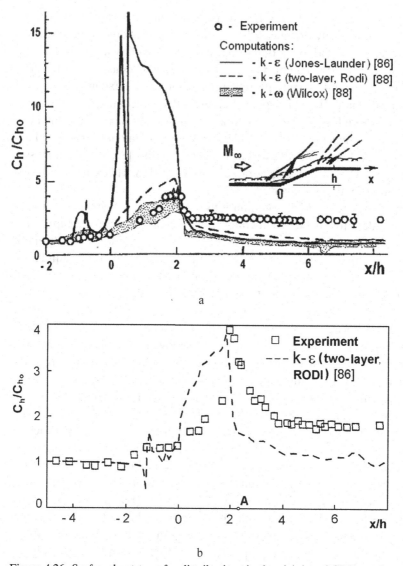

Figure 4.26. Surface-heat-transfer distributions in the vicinity of CDR configuration $\alpha = 25°$, $h/\delta_0 = 2.4 - 2.5$, $T_w/T_{aw} \approx 1.04 - 1.05$. (a) $M_\infty = 4.0$, $Re_{\delta_0} = 1.16 \times 10^5$; (b) $M_\infty = 3.0$, $Re_{\delta_0} = 9.3 \times 10^4$. Experiment [81, 83]; k-ε computation [86], k-ω computations [88] (from Zheltovodov [2]).

separation shock is unsteady and exhibits both low-frequency (i.e., small compared to the frequency U_∞/δ of the energy-containing eddies in the incoming turbulent boundary layer) and high-frequency motions. This is illustrated by experimental results [68] for a CR at the low Mach number $M_\infty = 1.5$. Figure 4.27 presents the variation in rms wall-pressure fluctuation intensity σ_p (in the 0–65 kHz bandwidth) normalized by the mean wall pressure upstream of the interaction for angles $\alpha = 6$, 9, and 12 degrees. The fluctuation intensity increases with the length of the interaction region upstream of the ramp ($x/\delta_0 < 0$), with rising α (i.e., inviscid oblique shock strength ξ) under the influence of both shear-layer flapping and decays downstream of the interaction ($x/\delta_0 > 0$) with turbulent stress relaxation. The positions

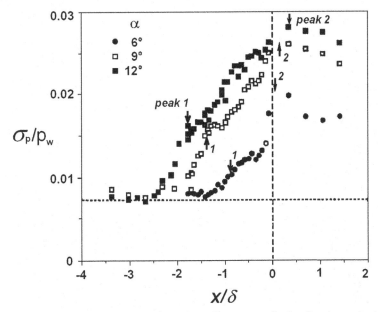

Figure 4.27. Streamwise variation in wall-pressure fluctuation intensity in the vicinity of CR at $M_\infty = 1.5$, $Re_{\delta o} = 1.79 \times 10^5$ (Thomas et al. [68]).

of the low-frequency peaks (in the 0–4 kHz bandwidth) – analyzed in detail herein – are indicated by arrows. The appearance of a small intermittent-separation bubble at $\alpha = 6$ degrees in the experiments at $M_\infty = 1.5$ (see Fig. 4.19, symbol 10) is in agreement with the predicted critical value $\alpha = \alpha_{ef}$ corresponding to the beginning of the intermittent-separation regime. More prominent unsteady separation zones were observed at $\alpha = 9$ and 12 degrees (i.e., symbols 11 and 12).

The power spectra [68] for the wall-pressure fluctuations p' presented in the terms of $f \cdot G(f)$ versus $\ln(f)$, where $G(f)$ is the power spectral density and f is a temporal frequency, are characterized by many common features; the 12-degree case is presented as an example in Fig. 4.28. Two characteristic regimes are evident. The fluctuations with frequency on the order of the peak value $f_p \approx U_\infty/\delta \approx 30$ kHz correspond to the characteristic frequency of turbulence in the incoming boundary-layer flow. These fluctuations exhibit continuous amplification through the interaction region, and a decrease in the spectral peak f_p value is caused by a reduction in convection velocity during the interaction process. The development of a secondary central peak centered near a much lower characteristic frequency on the order $f_{sh} \approx 0.1 U_\infty/\delta$ is indicated. These low frequencies are characteristic of shock oscillations and also were observed in several CR experiments at higher Mach numbers $M_\infty = 2$–5 (see Chapter 9). The ratio of rms pressure fluctuations in the 0–4 kHz bandwidth to the total fluctuation intensity (in the entire 0–65 kHz band) is shown in Fig. 4.29. The streamwise variation of this ratio exhibits two well-defined peaks upstream and downstream of the CR apex that are associated with localized separation-shock oscillations and separation-bubble motion (i.e., unsteady flow reattachment), respectively. The locations of these peaks are indicated by arrows in Fig. 4.27. The existing upstream peak in the low-frequency wall-pressure fluctuation intensity (within the 0–4 kHz bandwidth) corresponds to the location of the

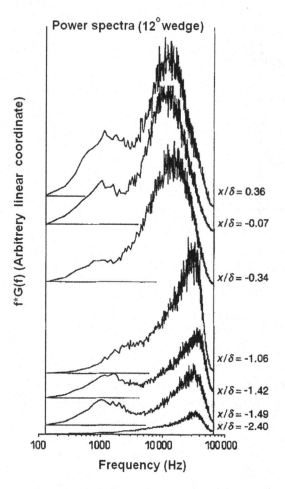

Figure 4.28. Pressure fluctuations spectra at selected streamwise locations for $\alpha = 12°$ at $M_\infty = 1.5$ (Thomas et al. [68]).

maximum value of the gradient of the measured intermittency function, thereby confirming the conclusion that this peak is associated with the spatial oscillation of the shock system [68]. The rise of the two peaks with increasing α (Fig. 4.29) corresponds to regimes II–IV and the onset of regime V at $\alpha = \alpha_d$ (see Fig. 4.19, symbols 10–12).

The behaviour of the wall-pressure fluctuations is closely correlated to the different flow regimes in accordance with variation of the corresponding scaling laws for extension of separation (see Figs. 4.22 and 4.23; see also, e.g., [90, 91, 92]). For example, in accordance with data by Bibko et al. [91] for CR and CDR at the fixed Mach and Reynolds numbers $M_\infty = 2.0$, $Re_{\delta*} \approx 3.2 \cdot 10^4$ presented in terms of sound-pressure level (SPL), wall-pressure-fluctuation distributions (Fig. 4.30a) and their corresponding spectra in the first maximum point ($x = x_1$) (Fig. 4.30b) in regimes III and IV ($\alpha = 10 - 20$ degrees) depend significantly on the angle α and differ from those in regime V ($\alpha \geq 25$ degrees), where the inviscid detached normal shock appearing upstream of the CDR creates a separation zone of maximal size with a free-separation point on the plate ahead of the step and a fixed-reattachment point on the face near the top of the decompression ramp. Considered wall-pressure-fluctuation parameters are independent from α in regime V and their behavior corresponds to correlations shown in Fig. 4.23, which demonstrate the independence of

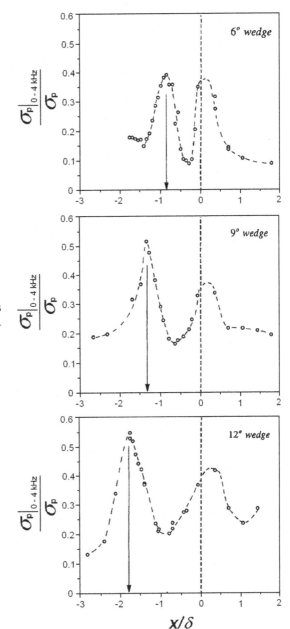

Figure 4.29. Bands limited relative intensities $\sigma_p(f)|_{0-4\,kHz}/\sigma_p$ at $M_\infty = 1.5$ (Thomas et al. [68]).

separation extent from α. At the same time, significant influence of the step height h (or the ratio h/δ^*) is demonstrated in regime V, which does not exist in regime IV [90, 91]. Similarly, the influence of the Reynolds number is different in regimes IV and V. In regime IV, an increase in the Reynolds number causes a broadening of the range of pressure fluctuations at high frequencies and a decrease in the intensity at low frequencies associated with the formation of free-separation zones (with free separation and reattachment points). However, in conditions of fixed reattachment (i.e., regime V), an increase in the Reynolds number causes an increase in the wall-pressure fluctuation intensity throughout the entire spectrum. The influence of the

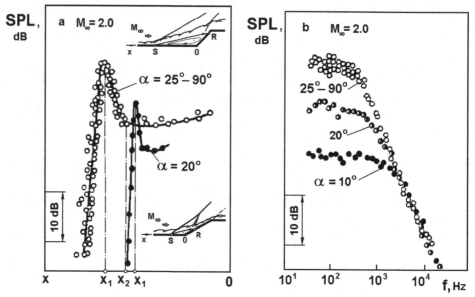

Figure 4.30. Wall-pressure fluctuations in terms of SPL distributions at $f = \omega/2\pi = 250$ Hz (a) and corresponding spectra of surface-pressure fluctuations in the first maximum (at $x = x_1$) in the frequency band $\Delta f = 1$ Hz (b) at $M_\infty = 2.0$, $\mathrm{Re}_{\delta*} \approx 3.2 \cdot 10^4$ (Bibko et al. [91]).

Mach number on pressure fluctuations is virtually independent of the regime, and an increase in Mach number causes a decrease in pressure fluctuations throughout the entire spectrum [90].

The mechanism responsible for shock oscillations is the focus of considerable interest, and controversy in understanding it has existed for a long time (see, e.g., Dolling [6, 59], Smits and Dussauge [11], Thomas et al. [68], Efimtsov and Kuznetzov [90], Bibko et al. [91, 92], and the detailed discussion in Chapter 9). The unsteady motions of the shock in conditions of turbulent separation are characterized by a wide range of frequencies and length scales. The small-scale shock fluctuations usually are related to boundary-layer and separated-shear-layer turbulence, whereas the large-scale, low-frequency motions are associated mainly with pulsations of the separation zone. The question of causality (i.e., whether the separated bubble or the shock is the source of the unsteady motion) is the subject of wide discussions and further research. In accordance with the early concept by Plotkin [93], the turbulence in the upstream boundary layer (i.e., the burst-sweep cycle of events) is considered the mechanism that triggers the shock-front unsteadiness. However, the high frequency of the energy-containing eddies in the incoming turbulent boundary layer (i.e., $f = U_\infty/\delta = 30$–50 kHz) cannot explain the dominant low frequency $f_{sh} = 0.4$–2 kHz of shock motion. Another explanation is that the separation zone undergoes a relatively low-frequency expansion/contraction from about $(2$–$4)\delta_0$ in streamwise extent; several investigations indicated that the shock oscillation is related to such motion at the foot of the shock [6]. The separated shear layer can amplify the low-frequency separation-zone motion; these perturbations then penetrate upstream due to a feedback mechanism through the large subsonic region in the separation zone (see also [92]). Recent experiments by Ganapathisubramani, Clemens, and

Dolling [94] also demonstrated that the low-frequency unsteadiness of the separation region/shock foot can be explained by a turbulent mechanism because of the presence of elongated (i.e., lengths greater than 40δ) spanwise coherent structures (i.e., "strips") of uniform momentum in the upstream supersonic boundary layer. Thus, as suggested by the authors, a physical model that includes the presence of such elongated strips probably can coexist with Plotkin's [93] mathematical model by reevaluation of the constant that defines the restoring mechanism to account for the presence of large-scale turbulence structures.

LES and DNS enable the investigation of unsteady effects in parallel with experimental research. Figure 4.31a shows the flowfield structure in a CR at $\alpha = 8$ degrees, $M_\infty = 2.9$ $Re_{\delta o} = 2 \times 10^4$ in accordance with LES by Urbin et al. [95] using a static Smagorinsky model. An instantaneous image of the shock wave is visualized by the isosurface $p/p_\infty = 1.25$, and large-scale spanwise turbulent structures are seen clearly in the undisturbed boundary layer in the horizontal cut of the streamwise velocity field at a distance from a wall $y^+ = 10$ (in wall units). The oblique shock wave is significantly disturbed due to the interaction with these structures, and the individual localized intermittent-separation zones with instantaneous negative velocity appearing in the shock-foot vicinity correspond to intermittent-separation conditions (see Fig. 4.19, regime II). The computed timewise and spanwise averaged mean wall-pressure distribution indicated by line 4 (Fig. 4.31b), as well as a similar prediction by El-Askary [96] using MILES (i.e., line 5), are in good agreement with the experiment [61] at the low Reynolds number $Re_{\delta o} = 2.18 \cdot 10^4$ (i.e., symbol 1) – which in this regime is close to the data [53, 66, 84, 97] at the higher $Re_{\delta o}$ values (i.e., symbols 2 and 3). RANS calculations performed for the higher Reynolds number using the Jones-Launder k-ε turbulence model [54] (i.e., line 6) and Wilcox k-ω model [53, 97] (i.e., line 7) also show good agreement with experimental data. However, the LES predicts the mean wall-skin-friction coefficient $C_f \approx 0$ (Fig. 4.31c) corresponding to intermittent-separation conditions (Fig. 4.31a) in contrast with RANS, which predict $C_f > 0$. At that time, a small incipient-separation zone with the size $L_s \approx (0.2$–$0.3)\delta_o = (0.44$–$0.66)$ mm was revealed by the oil-flow visualization (see Fig. 4.19, symbol 8) at the higher $Re_{\delta o}$, which was not realized in RANS. As also shown in Fig. 4.31c, various LES give higher skin-friction levels downstream of the CR ($x/\delta_o > 0$) compared to the experiment and RANS predictions at the higher Reynolds numbers. At the same time, predictions with the k-ω model (i.e., symbol 7) correspond better to experimental data than computations with the k-ε model (i.e., symbol 6).

DNS of a supersonic CR flow [98] for $\alpha = 18$ degrees at Mach 3 and $Re_{\theta o} = 1,685$ ($Re_{\delta o} = 2.1 \cdot 10^4$) provides insight to the flowfield structure and turbulence behavior (Fig. 4.32) at conditions corresponding to the boundary between regimes III and IV (see Fig. 4.19, band 14). The predicted mean wall-skin-friction coefficient distribution (Fig. 4.32b) displays a small quasistationary separation zone. A time trace of the spanwise-averaged locations of separation and reattachment is shown in Fig. 4.32c. The computed shock motion is limited to less than about 10 percent of the mean boundary-layer thickness δ_o and it oscillates slightly around the mean location with a frequency of similar magnitude to the bursting frequency of the incoming boundary layer $f = U_\infty/\delta_o$. Schlieren visualization of the instantaneous density-gradient magnitude averaged in the spanwise direction demonstrates the emergence of

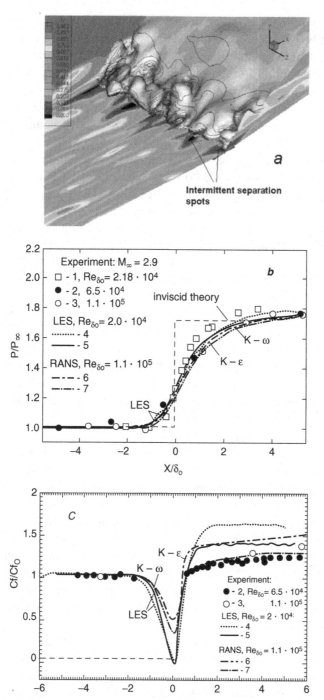

Figure 4.31. LES and RANS of interaction in CR at $\alpha = 8°$, $M_\infty = 2.9$, $T_w/Tw_a \approx 1$ (regime II, from Zheltovododov [2]), a – instantaneous flowfield visualization [95], b – mean wall-surface pressure, c – mean wall skin-friction coefficient distribution. Experimental data: 1 – [61], 2 – [53, 66, 84], 3 – [53, 97]; LES: 4 – [95], 5 – [96]; RANS: 6 – k-ε model [54], 7 – k-ω model [53, 97].

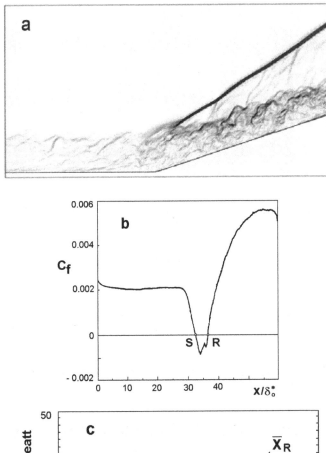

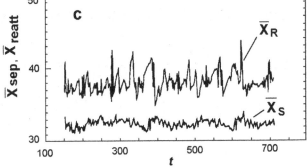

Figure 4.32. DNS of interaction in CR at $\alpha = 8°$, $M_\infty = 2.9$, $Re_{\theta o} = 1685$ ($Re_{\delta o} = 2.3 \times 10^4$), $T_w/Tw_a = 1$ (the boundary between regimes III and IV): a – flowfield Schlieren imitation; b – mean wall skin-friction coefficient distribution; c – time trace of spanwise averaged separation and reattachment line (Adams [98]).

distinct coherent vortex structures in the boundary layer downstream of the interaction region (Fig. 4.32a) and compression waves emanating from its external edge that stimulate undulations of the oblique shock front.

Figure 4.33 presents numerical results by Urbin et al. [99] and Yan et al. [100] for the flow over a CR at $\alpha = 25$ degrees, $M_\infty = 2.9$, $Re_{\delta o} = 2 \times 10^4$ using MILES. The flow conditions correspond to regime IV (see Fig. 4.19). An instantaneous image of the λ-shock (specifically, the isosurface $p/p_\infty = 1.4$ and $p/p_\infty = 2.0$) and streamwise cut of the velocity (at $y^+ = 20$) are shown in Fig. 4.33a. The interaction of the

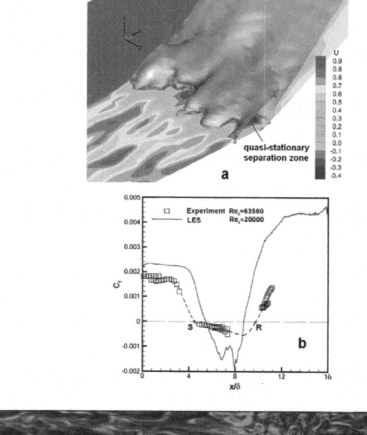

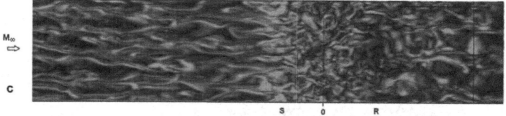

Figure 4.33. LES of interaction in CR at $\alpha = 25°$, $M_\infty = 2.9$, $Re_{\delta o} = 2 \times 10^4$, $T_w/T_{w_a} = 1$ (regime IV): a – instantaneous flowfield visualization (Urbin et al. [99]); b – mean wall skin-friction coefficient distribution (Yan et al. [100]); c – instantaneous vorticity (Urbin et al. [99]) at $y^+ = 15$.

shock wave with the streaks in the undisturbed boundary layer causes intensive rippling of the separation shock and undulation of the unsteady separation line close to the foot, as also observed in the experiments [94]. The separation-shock rippling penetrates along the front of the external oblique shock wave. The modulus of the vorticity (Fig. 4.33c) indicates streaks in the boundary layer close to the wall (at $y^+ = 15$) and their behavior during the interaction with the shock wave is reminiscent of the vortex-breakdown phenomenon [101]. Downstream of the interaction, the size of the eddies has increased but small-scale vortices penetrate the wall in the reversed flow between lines R and S. A quasisteady separation zone appears with an intermittency region upstream of the mean separation as typical interaction

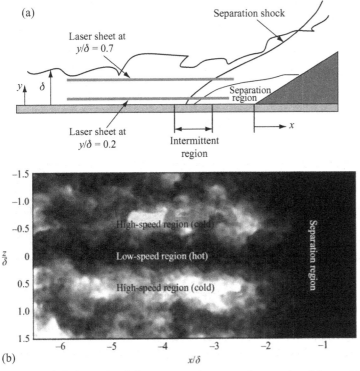

Figure 4.34. Schematic of the compression-ramp interaction (a) $\alpha = 20°$, $M_\infty = 2.0$, $Re_{\delta o} \approx$ $4.86 \cdot 10^5$ (regime IV), and planar laser scattering image (b) at $y/\delta = 0.2$. The ramp corner is at $x/\delta = 0$ (Ganapathisubramani et al. [94]).

features in this regime. As shown in the predicted spanwise and time-averaged wall-skin-friction distribution (Fig. 4.33b), the predicted separation zone at the low Reynolds number $Re_{\delta o} = 2 \cdot 10^4$ is smaller compared to the experiments [66, 84] at the higher value $Re_{\delta o} = 6{,}35 \times 10^4$.

The observed turbulence evolution in the interaction region (Fig. 4.33a,c) can be described as consisting of two main stages – namely: (1) turbulent-bursting events in the oncoming boundary layer because an ejection of low-speed, high-temperature, near-wall fluid appears as low-Mach-number spots [98] and streaks [94] within the ambient low-temperature and high-speed outer fluid; and (2) the interaction of moving spots and streaks with the downstream-located shock waves. Figure 4.34b shows a sample PLS image [94] of flow in the vicinity of the CR ($\alpha = 20$ degrees at $M_\infty = 2.0$) at a wall-normal location $y/\delta = 0.2$ (Fig. 4.34a). The figure demonstrates the presence of large-scale spanwise strips of low- and high-momentum regions in the incoming boundary layer and their interaction with separation shock ahead of the CR in regime IV. The instantaneous separation line (defined in Ganapathisubramani et al. [94] as the spanwise line at any given wall-normal location) penetrates farther upstream through the low-momentum (i.e., hot) streaks and recedes downstream of high-momentum (i.e., cold) streaks. LES demonstrates qualitatively similar flow features (Fig. 4.33a,c) and specifies the vortex-scales transformation in the interaction region and downstream. Some of the events observed and predicted by LES can be explained by interaction of a traveling temperature (or density)

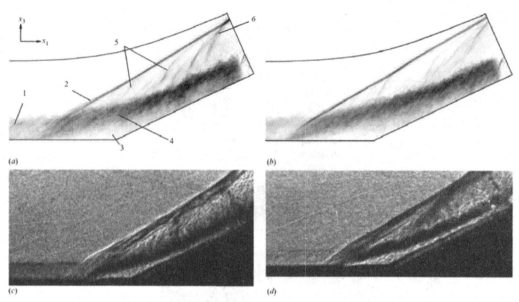

Figure 4.35. Computed (a, b) and experimental (c, d) schlieren images at two different instants of time of interaction in CR at $\alpha = 25°$, $M_\infty = 2.9$, $Re_{\delta o} = 6.3 \times 10^4$, $T_w/Tw_a = 1$ (regime IV). (The computed and experimental instants are not related.) 1 – incoming boundary layer, 2 – forward (separation) shock, 3 – reverse-flow region, 4 – detached shear layer, 5 – traveling compression waves (shocklets), 6 – unsteady rearward shock waves in reattachment region (Loginov et al. [105]).

inhomogeneity with the shock wave similar to the heated air-bubbles/shock-wave interaction considered by Schülein et al. [102]. The instability of an interface between gases with different temperatures (or densities) caused by the impulsive impact of the passing of a shock wave is known as Richtmeyer-Meshkov (R-M) instability. In the case of spanwise-located periodic hot and cold streaks, the shock wave causes the growth of initial periodic perturbations, which begin to increase in amplitude with formation of spike-like and bubble-like structures. Finally, the vortex sheet rolls up and accumulates into periodic vortex cores in the postshock flow. The mechanism of this baroclinic-vorticity generation is based on misalignment of the existing temperature (i.e., density) gradient across the bubble or streak surface and the shock-induced gradient according to the two-dimensional, compressible vorticity equation:

$$\rho \frac{D}{Dt} \frac{w}{\rho} = \frac{\nabla \rho \times \nabla p}{\rho^2}$$

where ω is the vorticity vector directed normal to the plane, including the pressure- and density- (temperature) gradient vectors.

LES of a CR [103, 104, 105] and a CDR [106] were performed for $M_\infty = 2.9$ at $\alpha = 25$ degrees and $Re_{\delta o} = 6.3 \times 10^4$ (regime IV) corresponding to experiments [66,84, 97, 107]. The Approximate Deconvolution Model by Stolz and Adams [108] was used. Computed and experimental schlieren images are shown in Fig. 4.35 for the CR flow. The computed (Fig. 4.35a,b) and experimental images (Fig. 4.35c,d) are taken at two different times and, although the times are not related in the computations and experiment, all the images demonstrate a qualitatively similar unsteady shock system. The forward-separated shock 2 (Fig. 4.35a), reverse flow region 3, detached shear layer 4, moving shock waves (shocklets) 5 above the detached shear layer, and rear compression shock waves 6 in the reattachment region are evident. Along with high-frequency fluctuations of the shock system, a large-scale shock motion was confirmed by the simulation, and the streamwise length of the separation shock excursion was estimated as $1.3\delta_o$. Compression waves are formed behind the forward shock around the coherent vortex structures and propagate downstream together with them at velocities ranging from 0.1 to $0.4U_\infty$. The higher-speed waves form shocklets, thereby accounting for the high level of turbulence in the external flow between the separation shock and the detached shear layer. The rearward shocks in the reattachment region are highly unsteady, as in the experiments, and become invisible at irregular time intervals.

The CR flow was realized in experiments [66, 84, 97, 107] using a CDR test configuration (Fig. 4.36b) and LES of the flow over the entire test configuration (Fig. 4.36a) was performed by Loginov et al. [106]. In these conditions, the boundary-layer turbulence interacts downstream of reattachment with the expansion fan 7, and compression waves arise immediately downstream of the expansion due to over-expansion of the near-wall portion of the boundary layer around the decompression ramp (see Figs. 4.11a and 4.36b, arrow 8). (The flowfield features indicated by symbols 1–7 correspond to those described in Fig. 4.35.) The computed and experimental mean surface-pressure and mean-skin-friction coefficient are displayed, respectively, in Fig. 4.36c,d. As shown, the LES accurately predicts the mean surface pressure and skin friction.

In accordance with surface oil-flow visualization, specific three-dimensional effects reminiscent of periodic Görtler-like vortices were observed in experiments (Fig. 4.37a). Such periodic vortices usually are associated with a boundary-layer instability due to the influence of centrifugal forces that try to remove the gas from the external part of the boundary layer to a concave wall. The signs of these vortex structures are seen especially well in the vicinity of the reattachment region, where distinct saddle (Sd) and nodal (N) points are located periodically along divergence line R; periodic longitudinal convergence and divergence lines develop downstream from these points (Fig. 4.37b). These longitudinal convergence and divergence lines also tend to spread with the reversed flow to separation line S; however, by then, they become rather blurred. Undulation of upstream separation line S also is observed, indicating the probable existence of periodic saddle and nodal points along the line as on reattachment line R. The periodic spanwise disturbances can be amplified significantly by initiating periodic vortex perturbations upstream of the CR using thin foils in the form of a zigzag band [109], which stimulates a more distinct flow topology corresponding to the plots in Fig. 4.37b. The three-dimensional flowfield structure, as well as the periodic spanwise wall-skin-friction and heat-transfer variation caused by artificially stimulated intensive periodic vortexes, was predicted in the framework of RANS with the Spalart-Allmaras turbulence model [126].

LES results [103, 104, 105, 106] exhibit Görtler-like vortices formed naturally at the SBLI and demonstrate agreement with the experiments [66, 84]. The predicted mean-wall-skin-friction coefficient panorama (Fig. 4.38a) corresponds to an experimental oil-flow image (see Fig. 4.37a). In accordance with the computations,

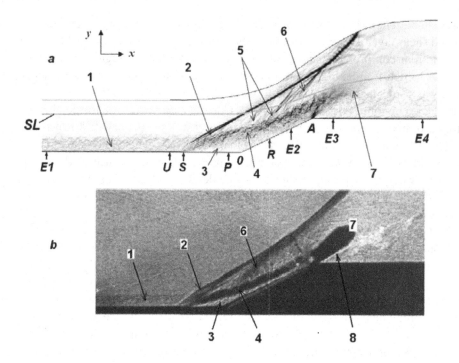

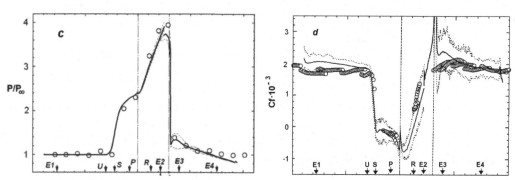

Figure 4.36. LES of interaction in CDR at $\alpha = 25°$, $M_\infty = 2.9$, $Re_{\delta o} = 6.36 \times 10^4$, $T_w/Tw_a = 1$ (regime IV): a – LES schlieren-type visualization, b – schlieren photograph [66], c – wall-pressure distribution, d – wall skin-friction coefficient distribution. Computations: solid line – LES averaged in time and over spanwise direction; dotted lines – LES averaged in time, min and max values over the spanwise direction (Loginov et al. [106]).

undulating solid convergence and divergence lines S and R correspond to the mean-flow separation and reattachment at $c_f = 0$. The darker color corresponds to the lower c_f value. Two pairs of flow convergence and divergence lines are observed downstream of reattachment on the CR surface and their period is about $2\delta_o$ in the computations, as in the experiments. This period corresponds to the distance between the streaks observed in the undisturbed boundary layer [94] and between the maxima and minima appearing on the predicted undulating separation line S. This indicates that the interaction of the streaks with successive separation and reattachment shock waves and the baroclinic-vorticity generation can be considered the predominant mechanism that stimulates the appearance of Görtler-like vortices.

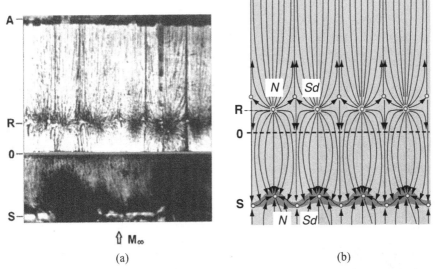

Figure 4.37. Oil-flow visualization pattern (a) and its interpretation (b) on CR surface for the configuration shown in Fig. 4.36 (Zheltovodov et al. [66]).

This conclusion is supported by the experiments of Adelgren et al. [110], in which the emergence of large-scale and periodic small-scale vortices in conditions of heated-bubble interaction with successive shock waves was explained by Richtmeyer-Meshkov instability and a baroclinic-vorticity generation mechanism [102]. A spanwise variation of the time-averaged wall-skin-friction coefficient with maxima and minima begins to be observed distinctly at separation line S (Fig. 4.38b), at reattachment line R, and downstream at section E_2 on the CR surface; whereas, in the reversed flow (section P), an obvious decrease of periodic disturbances amplitude is observed. (Positions of the considered different sections are indicated in Fig. 4.36a,c,d.) In accordance with LES, the flow acceleration in the expansion fan around the concave wall in the vicinity of the step top A significantly suppressed the appearance of periodic structures (Fig. 4.38b, sections E_3 and E_4). The oil-flow visualization revealed only a continuation of the periodic longitudinal oil-coalescence lines without any signs of divergence between them.

A summary [7] of RANS, LES, and DNS predictions for the mean-separation length for a broad range of compression corners is shown in Fig. 4.22 with the experimental correlation of Zheltovodov and Schülein [73]. Results for DNS [98, 111, 112] (i.e., symbols 26–28) and LES [96, 103, 100, 105, 111] (i.e., symbols 29–34) are available only at low Reynolds numbers; nevertheless, they show good agreement with experimental trends. Results for RANS simulations (i.e., symbols 1–25) show a wide scatter depending on the turbulence model used. A comparison of LES results with experimental data for the CR [103, 104, 105] and CDR [106] also demonstrated good agreement for the mean-flow velocity profiles as well as mass-flux-, density-, and velocity-fluctuation profiles and surface-pressure fluctuations (see also [2]). For example, Fig. 4.39a shows good correspondence with experiments of the predicted heightened rms mass-flux fluctuations in the external disturbed flow (i.e., line 1) along streamline SL (see Fig. 4.36a) and along a line of the fluctuation maximum in the external part of the boundary layer (i.e., symbol 2) (see [84]) (i.e., symbols 3 and 4, respectively). The data are normalized with their

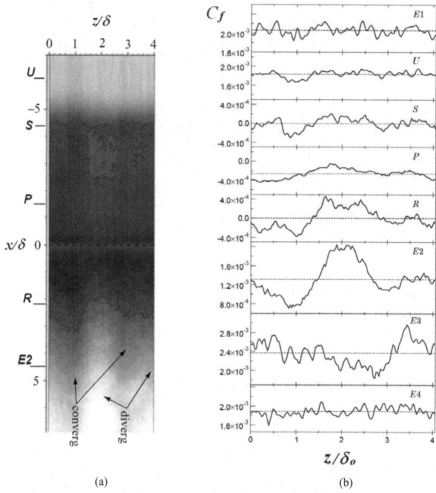

(a)   (b)

Figure 4.38. Predicted by LES the mean wall skin-friction coefficient panorama on CR surface (a) and C_f distributions in different CDR spanwise sections (b): averaged in time – solid lines, averaged in time and over the spanwise direction, z – dotted lines (Loginov et al. [103, 105, 106]).

respective values at initial station E_1 in the undisturbed flow. Also shown are preliminary results [106] from processing additional statistical data for the flow fluctuations downstream of decompression ramp A. Peaks I and II (Fig. 4.39a) correspond to unsteady separation and reattachment shock system (see Figs. 4.35a and 4.36a), and additional peak III is caused by an unsteady shock emerging downstream of the expansion fan due to the flow overexpansion around the decompression ramp. Traveling shocklets above the shear layer downstream of the separation line support the higher turbulence level, which persists for a long distance after the step top A. The predicted distribution of surface-pressure fluctuations (Fig. 4.39b) correlates with the turbulence behaviour and demonstrates three similar peaks (i.e., I, II, and III), as well as the higher level in the separation zone ahead of the CR and on the upper surface of the step downstream of top A. The wall-pressure-fluctuation distribution on the CR (up to point A) is in good agreement with experimental data at the higher Reynolds numbers (see [2, 103, 104, 105]).

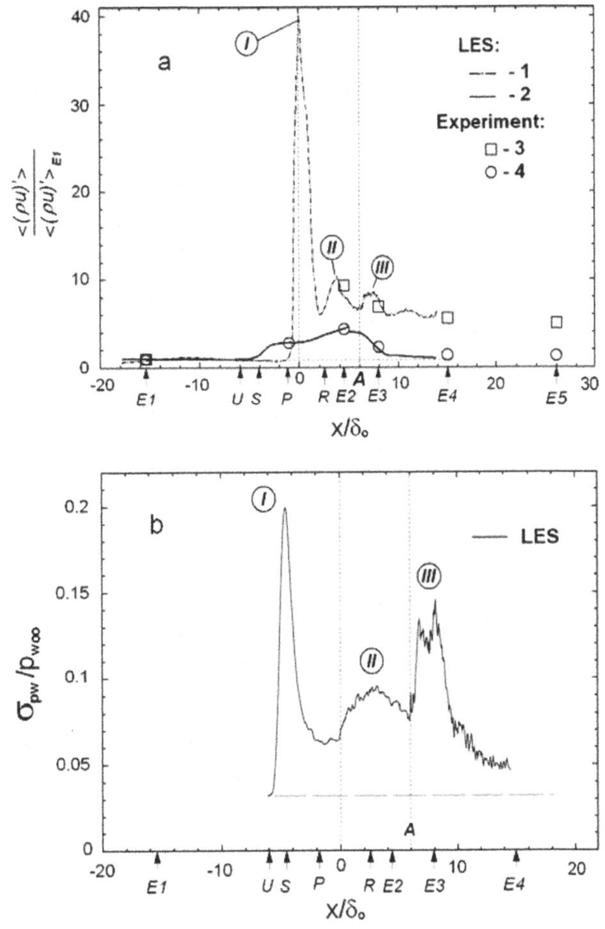

Figure 4.39. LES of CDR flow ($\alpha = 25°$, $M_\infty = 2.9$, $Re_{\delta o} = 6.36 \cdot 10^4$, $T_w/Tw_a = 1$). RMS of mass-flux fluctuations variation with streamwise direction (a): 1, 3 – in the external flow; 2, 4 – along a line of maximum fluctuations in the external part of the boundary-layer profile, and surface-pressure fluctuations distribution (b) (Loginov et al. [106]).

Numerical simulation [27, 28] using a hybrid LES/RANS approach demonstrated the potential of this more economical approach compared to LES and DNS for the prediction of high-Reynolds-number SWTBLIs. In particular, recent simulation [28] of the CR flow at $\alpha = 28$ degrees, $M_\infty = 5.0$, and $Re_{\delta o} = 8.77 \times 10^5$ with Menter's hybrid k-ω/k-ε SST model [29] demonstrated rather good agreement with the experiments. The computations captured the mean-flow and time-dependent-flow structure and support the presence of streaks with lower and higher momentum in the boundary layer, which induces a low-frequency undulation of the separation front. The simulations also capture the turbulence transformation in the interaction region and the three-dimensional mean-flow vortical structures that produce significant variations in the surface-skin friction in the reattaching boundary layer. As concluded previously, the existence of counter-rotating vortices may explain why RANS models fail to capture the correct recovery rate of the boundary layer downstream of reattachment.

4.2.7 Oblique Shock Wave–Turbulent Boundary-Layer Interaction

The mean flowfield structure of the incident oblique shock wave–turbulent boundary-layer interaction (IOSTBLI) is shown in Fig. 4.2b for a shock strength sufficient to cause separation. An incident shock with flow-deflection angle α

interacts with an equilibrium-turbulent boundary layer on a flat plate. The separation of the boundary layer causes a deflection of the flow, which generates a compression-wave system that coalesces into the separation shock. The incident shock intersects the separation shock and the refracted incident shock interacts with the free shear layer, deflecting the free shear layer and generating a reflected expansion fan [3, 113, 114]. The boundary layer is recompressed at the attachment point, generating a compression-wave system. If the incident shock strength is sufficiently modest, it does not cause boundary-layer separation and it curves in; its intensity weakens as it penetrates the boundary layer up to the sonic line. The pressure rise behind the incident shock tends to propagate upstream through a thin subsonic region in the boundary layer, causing this part to thicken and generating outgoing compression waves instead of a separation shock that coalesce to form a reflected shock. The refracted incident shock and the secondary weak outgoing waves formed due to the outgoing compression-wave refraction in the supersonic portion of the boundary layer effectively "reflect" as expansion waves from the sonic line within the boundary layer.

In accordance with the fundamental Free Interaction Concept (FIC) [36, 115, 116], the flow properties in the vicinity of the separation line for moderate and large-scale separation must be independent of the type of disturbance (i.e., the test configuration) that causes the boundary-layer separation, and the flowfield in the vicinity of the separation line can be predicted by the FIT. The similarities between the CR and incident oblique-shock interactions were noted [3, 113] and, as pointed out, a CR of angle 2α produces the same series of compression interactions at separation and reattachment as an incident oblique shock with an initial flow-deflection angle α. Thus, it can be supposed that the interaction regimes for CR flows (see Fig. 4.19) also are realized in IOSTBLI cases with critical flow-deflection angles α_i, α_{ef}, α_s, α_s^*, and α_d, which are two times less than the corresponding values for CR flows. Such critical deflection angles are indicated for the IOSTBLI case in Fig. 4.40a with experimental data [113] for separation-length variation L_{sep}/δ_o versus α in accordance with surface-oil-flow visualization at $M_\infty = 2.5$. The incipient-separation stage at $\alpha_{ef} \leq \alpha \leq \alpha_s$, $Re_{\delta o} = 10^5$ and 5×10^5 revealed by this technique corresponds to regime II of intermittent-separation appearance, as in the CR interaction case (see Fig. 4.19). A tendency toward an increasing critical flow-deflection angle with Reynolds number also corresponds to the behavior of the CR data (see, e.g., Fig. 4.18b). The development of small-scale separation at regime III (Fig. 4.40a) and the higher growth rate of the large-scale separation at $\alpha > \alpha_s^*$ (regime IV) also are similar to CR flow tendencies. A significant change in the evolution of the separation zone in IOSTBLI at a critical value α_d corresponding to the detachment of the inviscid shock ahead of the CDR (see Fig. 4.19) is demonstrated in Fig. 4.40b by experimental data [115] for the separation-shock-wave detachment distance L_o at $M_\infty = 2.15$. This distance was measured from the trace B (at $x = x_B$) of the incident oblique shock-wave prolongation on a flat plate surface (shown by the dotted line in the sketch corresponding to regime IV). The length of separation zone at $\alpha \geq \alpha_d$ achieves the maximum size and begins to be independent of α in the same way as the maximum-scale separation in the vicinity of forward-facing steps in regime V (see Fig. 4.19). In accordance with shadowgraphs [115], a complex flow structure with a nearly normal central shock (or Mach stem) is realized in regime V (see the sketch in Fig. 4.40b).

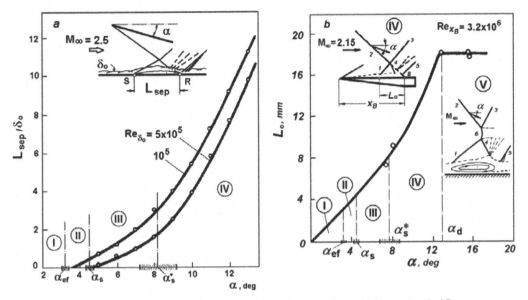

Figure 4.40. Variation of separation length (a) beneath an incident oblique shock (Green [113]) and separation shock-wave detachment distance (b) (Petrov et al. [115]) at different interaction regimes.

The similarity between the IOSTBLI and the CR interactions [3, 113] also is confirmed by good agreement of experimental surface-pressure distributions and their numerical (i.e., RANS) predictions [117] at $M_\infty = 2.96$, $Re_\delta \approx 1.5 \cdot 10^5$ ($Re_x = (1.0-1.2) \times 10^7$) (Fig. 4.41c). The incident shock was induced by a 12.7-degree shock-wave generator (Fig. 4.41a) so that the inviscid pressure rise is identical to the 25-degree CR (Fig. 4.41b). RANS calculations were performed with the zero-equation turbulence model of Cebeci and Smith with a relaxation model for the turbulent-eddy viscosity that was used successfully to predict CR interactions [58] (see Fig. 4.14). The predicted density-contour plots (Fig. 4.41a,b)

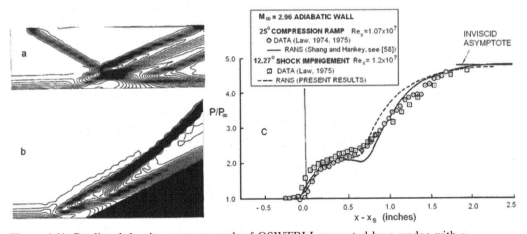

Figure 4.41. Predicted density contour graph of OSWTBLI generated by a wedge with $\alpha_w = 12.27°$ (a) and compression ramp $\alpha = 25°$ flow (b) (regime IV); comparison of surface-pressure distribution (c) (Shang et al. [117]).

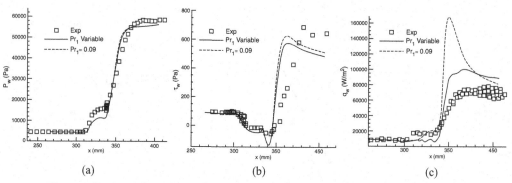

Figure 4.42. Computed and measured wall pressure (a), wall-shear stress (b) and wall-heat flux for IOSTBLI at $\alpha = 14°$, $M_\infty = 5$, $Re_{\delta o} = 1.72 \times 10^5$, $T_w/T_{aw} = 0.8$ (regime IV) (Xiao et al. [118]).

are in general agreement with the holographic interferogram images for shock generators varying from $\alpha = 7.93$ to 12.17 degrees. However, a parametric study of the relaxation-length scale λ [117] revealed that this parameter is not universal, and the value of $\lambda = 20\delta_o$ is superior to the value $\lambda = 10\delta_o$ previously used for CR flows to achieve better agreement with experiments in the flowfield structure, surface-pressure distribution, and locations of separation and attachment points.

Figure 4.42 is a comparison of RANS calculations [118] with experimental data [119, 120] at $\alpha = 14$ degrees, $M_\infty = 2.9$ (i.e., regime IV). A fresh approach was used to calculate the turbulent Prandtl number Pr_t as part of the solution to ensure the incorporation of relevant physics at higher Mach numbers into the model equations. The approach is based on a two-equation model for the enthalpy variance and its dissipation rate [118]. As indicated in Fig. 4.42a,b, the prediction of surface-pressure and wall-shear stress is almost identical for both constant and variable Pr_t. Both calculations underpredict the pressure in the separation zone, but the extent of separation is well predicted. However, the wall-shear stress is overpredicted in the relaxing flow downstream of reattachment. As shown in Fig. 4.42c, although the variable Pr_t results demonstrate improvement in the prediction of the wall-heat flux over constant Pr_t in the relaxing-flow downstream of reattachment including peak heating, the discrepancy with experiments remains significant. Moreover, the variable Pr_t results underpredict wall-heat flux in the separation zone as compared to a constant Pr_t.

These significant errors in RANS predictions of wall pressure and heat flux with different modern turbulence models are illustrated by many examples [59] for a wide range of Mach numbers and oblique shock-wave strengths; therefore, an understanding of unsteadiness effects is important. Unique measurements of fluctuating heat fluxes in IOSTBLIs were carried out [121] at $M_\infty = 3.85$ (± 0.04), $T_w/T_{aw} \approx 0.6$, $Re_{\delta o} \approx 1.76 \times 10^5$. Two representative oblique shock angles $\alpha_{sw} = 18.5$ and 22.3 degrees corresponding to flow-deflection angles $\alpha \approx 4.15$ and 8.35 degrees, respectively, were selected to analyze unseparated and separated flow cases. In accordance with the interaction regimes shown in Fig. 4.19 for CR interactions at adiabatic-wall conditions, incipient intermittent separation (i.e., regime II) is expected at the double CR angle value $2\alpha = 8.3$ degrees and the stage

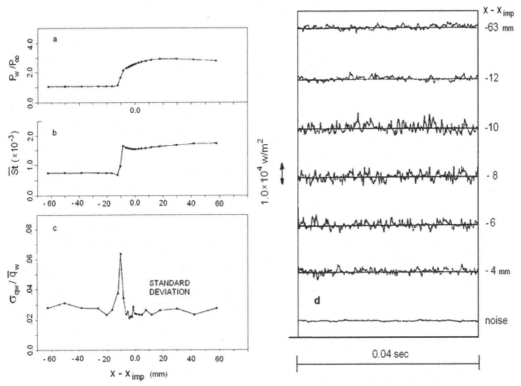

Figure 4.43. Wall pressure (a), mean heat-transfer distribution (b), rms heat-transfer fluctuations (c), and signals of the heat transfer in different sections for IOSTBLI (d), at $\alpha \approx 4.15°$, $M_\infty = 3.85$, $\delta_o \approx 7$ mm (± 0.04) (unseparated flow regime) (Hayashi et al. [121]).

of developed small-scale-separation regime III, which is very close to the onset of large-scale separation, corresponds to $2\alpha = 16.7$ degrees. However, unseparated-flow and small-scale-separation regimes were realized at such conditions due to a wall cooling. The inviscid-shock-wave impingement point x_{imp} is taken as the origin of the streamwise coordinate for presentation of the mean-wall pressure, the mean-heat-transfer distributions, and the standard deviation of heat-transfer fluctuations normalized by mean heat flux (σ_{qw}/q_w) (Figs. 4.43a–c). The actual impingement point of the shock wave is about 10 mm upstream of the inviscid-impingement point and corresponds to $(x - x_{imp}) = -10$ mm. In the unseparated flow regime, the amplitude of the heat-flux fluctuations becomes large near the impingement point of the shock wave (Fig. 4.43c), but σ_{qw}/q_w decays rapidly downstream. No intermittent signals of heat flux in such conditions are observed (Fig. 4.42d) and the fluctuations follow the Gaussian probability distribution that is qualitatively similar to the wall-pressure fluctuations [121]. A small mean-heat-flux peak also is observed in the mean-heat transfer near the shock-impingement point (Fig. 4.43b).

In the case of separated flow, a peak appears in the mean-heat-transfer distribution near separation point S and the level increases noticeably downstream of reattachment point R (Fig. 4.44b). Intermittency is observed in the heat-transfer fluctuations near the separation point (Fig. 4.44d); after the separation point, turbulent fluctuations are observed. The σ_{qw}/q_w distribution demonstrates a sharp peak

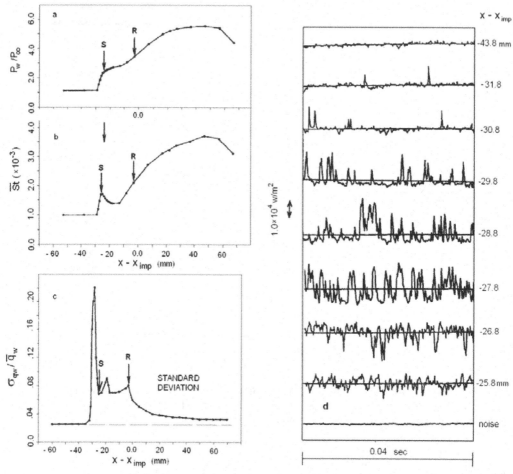

Figure 4.44. IOSTBLI (small-scale separation regime) at $\alpha \approx 8.35°$, $\delta_o \approx 7$ mm (Hayashi et al. [121]).

near the separation point (Fig. 4.44c) due to the large oscillatory motion of the separation shock that is similar to that of the pressure fluctuations. A second small peak in the separation zone is associated with the incident shock wave, and a third peak corresponds to the reattachment point. These additional three peaks also are observed in the wall-pressure fluctuation measurements [121]. The heat-flux fluctuations are amplified through the boundary-layer separation and reattachment, and the values of σ_{qw}/q_w in the relaxing flow downstream of reattachment point R are higher than those upstream for $(x - x_{imp}) > 10\delta_o$ (Fig. 4.44c). The heat flux fluctuates intermittently and more frequently up to $x - x_{imp} \approx -25.8$ mm (Fig. 4.44d), and no intermittency is observed downstream of this section. The wall pressure also demonstrates comparable intermittency in the same region with nearly the same frequency at the same locations [121]. This analogy of the behavior of wall-heat-flux and pressure fluctuations also is observed in the flow over the CDR configuration. In particular, the higher level of the predicted surface-pressure fluctuations not only in the separation zone ahead of the CR but also on a surface downstream of top A of the decompression ramp (see Fig. 4.39b) corresponds to the observed higher level of heat transfer (see Fig. 4.26). Thus, the unsteady nature of the complex flows is a serious limitation for their adequate modeling on the basis of RANS turbulence models.

DNS of an impinging oblique shock-wave interaction [122] at $\alpha_{sw} = 33.2$ degrees, $\alpha \approx 8$ degrees, $M_\infty = 2.25$, $Re_{\theta o} = 3,725$ ($Re_{\delta o} = 5.15 \times 10^4$), $T_w/T_{aw} = 1$ demonstrates the principal unsteady effects and explains fundamental physical mechanisms. The flow corresponds to the boundary between the regimes of small-scale separation (II) and large-scale separation (III) (see Fig. 4.19, $2\alpha \approx 16$ degrees). In accordance with the computed instantaneous density field shown in a vertical plane (Fig. 4.45a) and pressure and velocity fields (not shown), the existence of a complex organized motion in the outer part of the boundary layer is revealed. It is characterized by the occurrence of turbulent three-dimensional bulging coherent structures inclined at an acute angle with respect to the wall and also observed in experiments. This figure confirms the enhancement of mixing, which is typical for STBLI. The velocity isocontours reveal a small separation zone of reverse flow confined between the upstream influence point x_0 and oblique shock-impingement point x_I, which exhibit a highly intermittent nature. The computed mean-wall-pressure and average-skin-friction coefficient show rather good agreement with the experiments of Deleuze [123] and Laurent [124] (Fig. 4.45b,c). The faster relaxation of the calculated skin friction to an equilibrium state was explained by the higher-momentum thickness in the experiments $Re_{\theta o} = 4,808$, which also caused a 20 percent difference in the skin friction upstream of interaction. The existence of elongated, intermittent, streaky vortical structures upstream of the interaction and the appearance of larger, coherent structures in the mixing layer above the separation zone (Fig. 4.46a) are qualitatively similar to CR interactions, as discussed previously. The flow in the separation zone is highly three-dimensional and characterized by scattered spots of flow reversal in the interaction zone (Fig. 4.46b). However, the existence of the Görtler-like vortices was not observed in the flow, although they have been observed distinctly in Brazhko's [125] experiments by optical (i.e., PLS) visualization and surface-heat-transfer measurements at $M_\infty = 5$ and 6, $\alpha = 15$ degrees (i.e., regime IV). The average separation and reattachment points exhibited an rms displacement (normalized by δ_o) on the order of 5 and 16 percent, respectively. As concluded on the basis of analysis of the calculated mean-velocity profiles, the mean-flow relaxation distance after the interaction is close to $10\delta_o$; however, in accordance with turbulence statistics, complete equilibrium is attained in the inner part of the boundary layer, whereas in the outer region, the relaxation process is incomplete. This conclusion is in agreement with the behavior of wall-heat-flux fluctuations shown in Fig. 4.44c.

Figure 4.47 illustrates several important dynamic features of IOSTBLI flow as shown by the flowfield structure at a time interval of $0.22\delta_o/u_\infty$. The complex organized motion in the outer part of the boundary layer is characterized by the occurrence of large coherent vortical structures that are shed close to the average separation point, resulting in a mixing layer that grows in the streamwise direction and is considered a major cause of large-scale unsteadiness. The turbulence-amplification mechanism is associated primarily with the formation of the mixing layer. The figure reveals the intense flapping motion of the reflected shock (and, to a lesser extent, the incident shock) and its branching. The foot of the incident shock experiences an

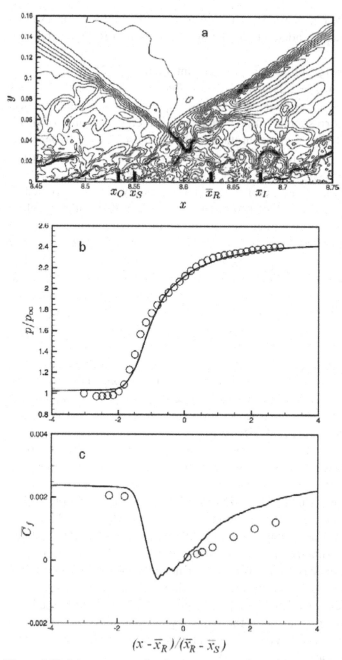

Figure 4.45. Instantaneous density isocountour lines (a) and comparison of the DNS with experimental measurements by Deleuze [123] and Laurent [124] for average wall pressure (b) and skin-friction coefficient (c) (Pirozzoli and Grasso [122]).

oscillatory motion that is related to the passage of the coherent structures through the tip. As concluded on the basis of a detailed analysis of calculated wall-pressure fluctuations of power-spectral density, the large-scale, low-frequency unsteadiness is sustained by an acoustic-resonance mechanism established in the interaction region [122]. The power-spectral density at various locations exhibits a nearly constant

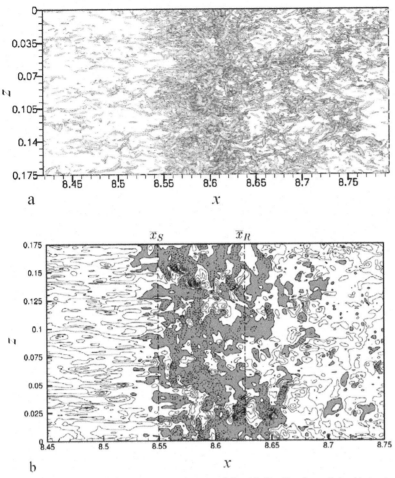

Figure 4.46. Projection of vortex structures (a) and distribution of the instantaneous "stream-wise" skin-friction coefficient (b) on flat plate surface (Pirozzoli and Grasso [122]).

distribution in the low-frequency range and a power-law decay at higher frequencies. In the interaction zone, these spectra show a dominant low-frequency energy content and peaks at discrete frequencies.

4.3 Summary

Our knowledge and understanding of two-dimensional STBLIs is virtually complete. Extensive experimental investigations revealed accurate and useful correlations for determining the conditions for fully attached, incipiently separated, and fully separated flows in STBLIs for canonical geometries including normal shock, CR, compression-expansion ramp, and oblique-shock interactions. The correlations are based on fundamental physical knowledge of gas dynamics and turbulence structure, and they provide useful data for engineering design. The inherent similarities in two-dimensional SBLIs also are evident and provide a better understanding of flowfield physics. It is now entirely feasible to predict not only peak aerothermodynamic loadings associated with two-dimensional STBLIs but also details of the

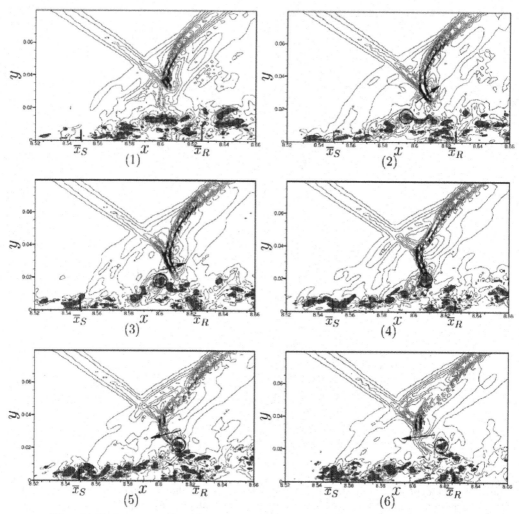

Figure 4.47. Sequence of x-y flowfield visualization at time intervals of $0.22\delta_o/u_\infty$. Solid lines depict isocontour levels of the pressure-gradient modulus, $20 \leq |\nabla p| \leq 400$; gray patches correspond to vortex structures. The arrow indicates the instantaneous direction of the incident shock foot motion, and the circle tracks the evolution of specific vortex (Pirozzoli and Grasso [122]).

aerothermodynamic interactions on the basis of the synthesis of the experimental data and the correlation based on physical understanding.

Our ability to predict two-dimensional STBLIs progressed dramatically in recent decades; however, a complete model capable of predicting all relevant details at flight conditions is lacking. RANS models (in particular, two-equation turbulence models) continue to remain the standard for engineering applications. Significant improvement in predictive capability of RANS models was achieved in the past two decades; however, their accuracy in predicting complex (e.g., multiple) shock-expansion interactions over a wide range of Reynolds numbers, pressure ratios, and Mach numbers remains to be fully explored. LES methods demonstrated dramatic accuracy with minimal (or no) requirement for calibration of model constants; however, LES methods are limited to low Reynolds numbers due to the requirement for resolution of the viscous sublayer, wherein the dynamics of turbulence production are of paramount importance. Further research is needed in the development of approximate models of the viscous-sublayer dynamics to enable LES at higher Reynolds numbers. DES models combining RANS approaches in the near-wall region (to avoid the Reynolds-number limitation of LES) with LES in the outer region represent an area of current research interest. Further progress in this methodology can be expected, principally in the assimilation of accurate RANS models in the near-wall region to LES in the outer portion of the flow. DNS provides accurate predictions of SBLIs without introduction of modeling parameters; however, it is limited to low Reynolds numbers for the same reason as LES. Additional research is needed in the development of faster algorithms for DNS to raise (or, it is hoped, remove) the Reynolds-number barrier.

REFERENCES

[1] A. A. Zheltovodov. Shock waves/turbulent boundary-layer interactions: Fundamental studies and applications. *AIAA Paper* 96–1977 (1996).

[2] A. A. Zheltovodov. Some advances in research of shock wave turbulent boundary-layer interactions. *AIAA Paper* 2006–0496 (2006).

[3] J. Délery and J. G. Marvin. Shock-wave boundary layer interactions. *AGARDograph No.* 280 (1986).

[4] G. S. Settles and D. S. Dolling. Swept shock-wave/boundary-layer interactions. Tactical missile aerodynamics. *Prog. Astronautics and Aeronautics*, ed. M. Hemsch and J. Neilsen, Vol. 104 (New York: AIAA, 1986), 297–379.

[5] D. D. Knight and G. Degrez. Shock wave boundary layer interactions in high Mach number flows: A critical survey of current CFD prediction capabilities. *AGARD Report* 319, 2 (1998), pp. 1-1–1-35.

[6] D. S. Dolling. Fifty years of shock-wave/boundary-layer interaction research: What next? *AIAA J.*, 39 (2001), 8, 1517–31.

[7] D. Knight, H. Yan, A. G. Panaras, and A. Zheltovodov. "Advances in CFD prediction of shock wave turbulent boundary layer interactions." In *Progress in Aerospace Sciences* (Oxford: Pergamon Press, 2003), 39, pp. 121–84.

[8] P. K. Chang. Separation of flow. *International Series of Monographs in Interdisciplinary and Advanced Topics in Science and Engineering*, Vol. 3 (New York: Pergamon, 1970), xviii, 778.

[9] L. V. Gogish and G. Yu. Stepanov. *Turbulent Separated Flows* (Moscow: Nauka, Central Edition of Physical-Mathematics Literature, 1979) (in Russian).

[10] V. Ya. Borovoi. *Gas Flow Field and Heat Exchange in the Zones of Shock Waves Interactions with a Boundary Layer* (Moscow: Mashinostroenie, 1983) (in Russian).

[11] A. J. Smits and J. P. Dussauge. *Turbulent Shear Layers in Supersonic Flow*, 2nd Edition (Berlin Heilderberg: Springer Science + Business Media, 2006).

[12] T. Cebeci and A. Smith. *Analysis of Turbulent Boundary Layers* (New York: Academic Press, 1974).

[13] B. Baldwin and H. Lomax. Thin layer approximation and algebraic model for separated flows. *AIAA Paper* 78–257 (1978).

[14] B. Baldwin and T. Barth. A one-equation turbulence transport model for high Reynolds number wall-bounded flows. *AIAA Paper* 1991–610 (1991).

[15] D. Johnson and L. King. A new turbulence closure model for boundary layer flows with strong adverse pressure gradients and separation. *AIAA Paper* 1984–175 (1984).

[16] W. Jones and B. Launder. The prediction of laminarization with a two-equation model of turbulence. *Int. J. Heat and Mass Transfer*, 15 (1972), 301–4.

[17] D. Wilcox. *Turbulence Modeling for CFD*. 2nd Edition (La Canada, CA: DCW Industries, 2002).

[18] J. Marvin. Turbulence modeling for computational aerodynamics. *AIAA J.*, 21 (1983), 7, 941–55.

[19] D. D. Knight. "Numerical simulation of compressible turbulent flows using the Reynolds-averaged Navier-Stokes equations: Turbulence in compressible fluids. *AGARD Report* 819, pp. 5–1 to 5–52 (1997).

[20] H. S. Zhang, R. M. C. So, T. B. Gatski, and C. G. Speziale. "A Near-Wall 2nd-Order Closure for Compressible Turbulent Flows." In *Near-Wall Turbulent Flows* (1993), pp. 209–218.

[21] M. Gnedin and D. Knight. A Reynolds stress turbulence model for compressible flows. Part I: Flat-plate boundary layer. *AIAA Paper* 95–0869 (1995).

[22] G. Zha and D. Knight. Three-dimensional shock boundary layer interaction using Reynolds stress equation turbulence model. *AIAA J.*, 34 (1996), 7, 1313–20.

[23] G. Erlebacher, M. Hussaini, C. Speziale, and T. Zang. Toward the large eddy simulation of compressible turbulent flows. *J. Flui. Mech.*, 238 (1992), 1550–85.

[24] M. Lesieur and P. Comte. Large eddy simulations of compressible turbulent flows: Turbulence in compressible fluids. *AGARD Report* 819 (1997), 4-1–4-39.

[25] J. A. Domaradzki, T. Dubois, and A. Honein. A subgrid-scale estimation model applied to large eddy simulations of compressible turbulence. In *CTR Proceedings, 1998 Summer Program*, Center for Turbulence Research, Stanford University and NASA Ames Research Center, Stanford, CA (1998).

[26] F. Grinstein, L. Margolin, and W. Rider (eds.) *Implicit Large Eddy Simulation* (New York: Cambridge University Press, 2007).

[27] X. Xiao, J. R. Edwards, H. A. Hassan, and R. A. Baurle. Inflow boundary conditions for hybrid large eddy–Reynolds averaged Navier-Stokes simulation. *AIAA J.*, 41 (2003), 8, 1418–89.

[28] J. R. Edwards, J.-L. Choi, and J. A. Boles. Large-eddy/Reynolds-averaged Navier-corner interaction. *AIAA J.*, 46 (2008), 4, 977–91.

[29] F. R. Menter. Two-equation eddy-viscosity turbulence model for engineering applications. *AIAA J.*, 32 (1994), No. 8, 1598–605.

[30] A. Zheltovodov, R. Dvorak, and P. Safarik. Shock waves/turbulent boundary layer interaction properties at transonic and supersonic speeds conditions. *Izvestiya SO AN SSSR, Seriya Tekhnicheskih Nauk*, 6, (1990), 31–42 (in Russian).

[31] H. H. Pearcey. Some effects of shock-induced separation of turbulent boundary layers in transonic flow past airfoils. *ACR R&M*, No. 3108 (1959).

[32] L. G. Grodzovskyi. Experimental research of shock waves/boundary layer interaction at the Mach number range M = 1.0 − 1.8. *Izvestiya AN SSR, Energetika i Avtomatika*, No. 4 (1961), 20–31 (in Russian).

[33] E. E. Zukoski. Turbulent boundary-layer separation in front of a forward-facing step. *AIAA J.*, 5 (1967), 10, 1746–53.

[34] M. Sajben, M. J. Morris, T. J. Bogar, and J. C. Kroutil. Confined normal–shock/turbulent–boundary-layer interaction followed by an adverse pressure gradient. *AIAA J.*, 29 (1991), 22, 2115–23.

[35] S. J. Kline, J. L. Bardina, and R. C. Strawn. Correlation of the detachment of two-dimensional turbulent boundary layers. *AIAA J.*, 21 (1983), 1, 68–73.

[36] D. Chapman, D. Kuehn, and H. Larson. Investigation of separated flows in supersonic and subsonic streams with emphasis on the effect of transition. *NACA Report* 1356 (1958).

[37] S. S. Kutateladze and A. I. Leont'ev. *Turbulent Boundary Layer of Compressible Gas* (Novosibirsk: SO AN SSSR, 1962) (in Russian).

[38] M. J. Morris, M. Sajben, and J. C. Kroutil. Experimental investigation of normal–shock/turbulent–boundary-layer interaction with and without mass removal. *AIAA J.*, 30 (1992), 2, 359–66.

[39] A. B. Haines. 27th Lanchester Memorial Lecture: Scale effects in transonic flow. *Aeronauutical J.*, 91 (1987), 907, 291–313.

[40] G. R. Inger. Upstream influence and skin friction in non-separating shock/turbulent boundary-layer interactions. *AIAA Paper* 80–1411 (1980).

[41] G. R. Inger. Some features of a shock/turbulent boundary-layer interaction theory in transonic fields. *AGARD-CP*-291 (1980).

[42] E. Stanewsky. Interaction between the outer inviscid flow and the boundary layer on transonic airfoils. Doctoral Engineer Dissertation, TU Berlin (D83) (1981), pp. 242–52.

[43] R. Bohning and J. Zieper. Normal shock/turbulent boundary-layer interaction at a curved wall. *AGARD-CP*-291 (1981).

[44] E. Hopkins and M. Inouye. An evaluation of theories for predicting turbulent skin friction and heat transfer on flat plates at supersonic and hypersonic Mach numbers. *AIAA J.*, 9 (1971), 6, 993–1003.

[45] C. D. Johnson and D. M. Bushnell. "Power-law velocity-profile-exponent variations with Reynolds number, wall cooling, and Mach number in a turbulent boundary layer. *NASA TN D*-5753 (1970).

[46] A. A. Zheltovodov. "Analysis of Two-Dimensional Separated Flows at Supersonic Speeds Conditions." In *Investigations of the Near-Wall Flows of Viscid Gas*, ed. Academician N. N. Yanenko (Novosibirsk, 1979), pp. 59–94.

[47] E. Blosch, B. Carroll, and M. Morris. Numerical simulation of confined transonic normal shock wave/turbulent boundary-layer interactions. *AIAA J.*, 31 (1993), 12, 2241–6.

[48] J. R. Viegas and C. C. Horstman. Comparison of multiequation turbulence models for several shock separated boundary-layer interaction flows. *AIAA J.*, 17 (1979), 8, 811–20.

[49] I. Nakamori and T. Ikohagi. Large eddy simulation of transonic turbulent flow over an airfoil using a shock capturing scheme with zonal embedded mesh. *Proceedings of the Third AFOSR International Conference on DNS/LES*, (Columbus, OH: Greyden Press, 2001), 743–50.

[50] J. Bardina, J. Ferziger, and W. Reynolds. Improved subgrid scale models for large eddy simulation. *AIAA Paper* 80–1357 (1980).

[51] H. H. Pearcey, J. Osborne, and A. B. Haines. The interaction between local effects at the shock and rear separation a source of significant scale effects in wind-tunnel tests on airfoils and wings. *AGARD CP* 35 (1968).

[52] D. Wilcox. Formulation of the k-ω turbulence model revised. *AIAA J.*, 46 (2008), 11, 2823–38.

[53] A. V. Borisov, A. A. Zheltovodov, A. I. Maksimov, N. N. Fedorova, and S. I. Shpak. Experimental and numerical study of supersonic turbulent separated flows in the neighborhood of two-dimensional obstacles. *Flui. Dyn.*, 34 (1999), 2, 181–9.

[54] C. C. Horstman and A. A. Zheltovodov. Numerical simulation of shock waves/expansion fans-turbulent boundary layer interaction. *International Conference on the Methods of Aerophysical Research*, Russia, Novosibirsk, Proc., Part 2, 118–22 (1994).

[55] L. G. Hunter and B. L. Reeves: Results of a strong interaction, wake-like model of supersonic separated and reattaching turbulent flows. *AIAA J.*, 9 (1971), 4, 703–12.

[56] A. Roshko and G. L Thomke. Flare-induced interaction length in supersonic, turbulent boundary layers, *AIAA J.*, 14 (1976), 7, 873–9.

[57] V. S. Dem'yanenko and A. A Zheltovodov. Experimental investigation of turbulent boundary layer separation in the vicinity of step. *Mekhanika Zhidkosti i Gaza* (Fluid Dynamics), No. 5, 73–80 (1977) (in Russian).

[58] J. Shang and W. Hankey. Numerical solution for supersonic turbulent flow over a compression ramp. *AIAA J.*, 13 (1975), 10, 1368–74.

[59] D. S. Dolling. High-speed turbulent separated flows: Consistency of mathematical models and flow physics. *AIAA J.*, 36 (1998), 5, 725–32.

[60] C. Appels and B. E. Richards. Incipient separation of a compressible turbulent boundary layer. *AGARD CP – 168, Flow Separation*, 21-1–21-12 (1975).

[61] F. W. Spaid and J. L. Frishett. Incipient separation of a supersonic, turbulent boundary layer, including effect of heat transfer. *AIAA J.*, 10 (1972), 7, 915–22.

[62] G. S. Settles, S. M. Bogdonoff, and I. E. Vas. Incipient separation of a supersonic turbulent boundary layer at high Reynolds numbers. *AIAA J.*, 14 (1976), 1, 50–6.

[63] G. S. Settles. "An experimental study of compressible turbulent boundary layer separation at high Reynolds numbers." Ph.D. Dissertation (Princeton, NJ: Aerospace and Mechanical Sciences Department, Princeton University, 1975).

[64] M. S. Holden. Shock wave-turbulent boundary layer interaction in hypersonic flow. *AIAA Paper* 72–74 (1972).

[65] G. M. Elfstrom. Turbulent hypersonic flow at a wedge-compression corner. *J. Flui. Mech.*, 53 (1972), Pt. 1, 113–27.

[66] A. A. Zheltovodov, E. Kh. Schülein, and V. N. Yakovlev. Development of turbulent boundary layer at the conditions of mixed interaction with shock waves and expansion fans. Preprint No. 28–83, ITAM USSR Academy of Sciences, Novosibirsk, (1983) (in Russian).

[67] D. W. Kuntz, V. A. Amatucci, and A. L. Addy. Turbulent boundary-layer properties downstream of the shock-wave/boundary-layer interaction. *AIAA J.*, 25 (1987), 5, 668–75.

[68] F. O. Thomas, C. M. Putnam, and H. C. Chu. On the mechanism of unsteady shock oscillation in shock wave/turbulent boundary layer interactions. *Experiments in Fluids*, 18 (1994), 1/2, 69–81.

[69] P. Ardonceau, D. H. Lee, T. Alziary de Roquefort, and R. Goethals. Turbulence behavior in a shock wave/boundary layer interaction. In *AGARD CP*-271, 8-1–8-14 (1999).

[70] M. A. Goldfeld and V. N. Dolgov. Investigations of turbulent boundary layer on test model of compression surfaces of supersonic diffuser. In *Aerophysical Research*, ITAM SB USSR Acad. Sci., Novosibirsk, 2, 98–9 (1972) (in Russian).

[71] M. A. Goldfeld and V. N. Dolgov. Experimental research of turbulent boundary layer on delta plate with wedge. *Izv. SO AN SSSR, Ser. Tekhn. Nauk* 8 (1973), 2, 16–22 (in Russian).

[72] R. H. Korkegi. Comparison of shock-induced two- and three-dimensional incipient turbulent separation. *AIAA J.*, 13 (1975), 4, 534–5.

[73] A. Zheltovodov and E. Schülein. Peculiarities of turbulent separation development in disturbed boundary layers. *Modelirovanie v Mekhanike (Modeling in Mechanics)*, 2, 1, 53–8, 1988 (in Russian).

[74] C. Horstman and C. Hung. Reynolds number effects on shock-wave turbulent boundary-layer interaction: A comparison of numerical and experimental results. *AIAA Paper* 77–42 (1977).

[75] M. Visbal and D. Knight. The Baldwin–Lomax turbulence model for two-dimensional shock-wave/boundary-layer interactions. *AIAA J.*, 22 (1984), 7, 921–8.

[76] C. Ong and D. Knight. Hybrid MacCormack and implicit Beam-Warming algorithms for a supersonic compression corner. *AIAA J.*, 25 (1987), 3, 401–7.

[77] E. Schülein and A. A. Zheltovodov. Development of experimental methods for the hypersonic flows studies in Ludwieg tube. International Conference on the Methods of Aerophysical Research, Part 1, Russia, Novosibirsk, 191–9 (1998).

[78] J. S. Hahn. Experimental investigation of turbulent step-induced boundary-layer separation at Mach numbers 2.5, 3 and 4. *AEDC-TR*-69–1 (1969), 31.

[79] H. Liepman and A. Roshko. *Elements of Gasdynamics* (New York: John Wiley & Sons, 1957).

[80] A. A. Zheltovodov. Peculiarities of development and modeling possibilities of supersonic turbulent separated flows. *Separated Flows and Jets IUTAM Symposium Novosibirsk, USSR*, eds. V. V. Kozlov and A. V. Dovgal (Berlin Heilderberg: Springer-Verlag Berlin, 1991), 225–36.

[81] A. A. Zheltovodov, E. G. Zaulichnyi, and V. M. Trofimov. Development of models for calculations of heat transfer at supersonic turbulent separated flows conditions. *Zhurnal Prikladnoi Mekhaniki i Tekhnicheskoi Fiziki (J. of Applied Mechanics and Technical Physics)*, No. 4, 96–104 (1990) (in Russian).

[82] M. S. Holden. Shock wave-turbulent boundary layer interaction in hypersonic flow. *AIAA Paper* 77–45 (1977).

[83] A. Zheltovodov, E. Zaulichniy, V. Trofimov, and V. Yakovlev. The study of heat transfer and turbulence in compressible separated flows. Preprint No. 22–87, ITAM, USSR Academy of Sciences, Siberian Branch, Novosibirsk (1987), 48 (in Russian).

[84] A. A. Zheltovodov and V. N. Yakovlev. Stages of development, gas dynamic structure and turbulence characteristics of turbulent compressible separated flows in the vicinity of 2-D obstacles. Preprint No. 27–86, ITAM, USSR Academy of Sciences, Novosibirsk, 55 (1986) (in Russian).

[85] A. A. Zheltovodov, V. A. Lebiga, and V. N. Yakovlev. Measurement of turbulence parameters in compressible boundary layers in the vicinity of separation zones. *Zhurnal Prikladnoi Mechaniki i Tekhnicheskoi Fiziki (J. of Applied Mechanics and Technical Physics)*, No. 3, 108–13 (1989) (in Russian).

[86] A. Zheltovodov, A. Borisov, D. Knight, C. Horstman, and G. Settles. The possibilities of numerical simulation of shock waves/boundary layer interaction in supersonic and hypersonic flows. *International Conference on the Methods of Aerophysical Research, Part 1*, Novosibirsk, Russia, 164–70 (1992).

[87] V. Trofimov and S. Shtrekalkin. Longitudinal vortices and heat transfer in reattached shear layers. *Separated Flows and Jets*, IUTAM Symposium, Novosibirsk, Russia, July 9–13, 1990, eds. V. V. Kozlov and A. V. Dovgal (Berlin Heilderberg: Springer-Verlag, 1991), 417–20.

[88] I. Bedarev, A. Zheltovodov, and N. Fedorova. Supersonic turbulent separated flows numerical model verification. *International Conference on the Methods of Aerophysical Research – Part 1*, Novosibirsk, Russia, 30–5 (1998).

[89] A. V. Borisov, A. A. Zheltovodov, A. I. Maksimov, N. N. Fedorova, and S. I. Shpak. Verification of turbulence models and computational methods of supersonic separated flows. *International Conference on the Methods of Aerophysical Research, Part 1*, Novosibirsk, Russia, pp. 54–61 (1996).

[90] B. M. Efimtsov and V. B. Kuznetsov. Spectrums of surface pressure pulsations at the supersonic flow over forward-facing step. *Uchenie Zapiski TSAGI (Scientific Notes of TSAGI)*, 20 (1989), 3, 111–15 (in Russian).

[91] V. N. Bibko, B. M. Efimtsov, and V. B. Kuznetsov. Spectrums of surface pressure pulsations ahead of inside corners. *Uchenie Zapiski TSAGI (Scientific Notes of TSAGI)*, 20 (1989), 4, 112–17 (in Russian).

[92] V. N. Bibko, B. M. Efimtsov, V. G. Korkach, and V. B. Kuznetsov. About oscillations of shock wave induced by boundary layer separation. *Mekhanika Zhidkosti i Gaza (Fluid Dynamics)*, No. 4, 168–70 (1990) (in Russian).

[93] K. J. Plotkin. Shock wave oscillation driven by turbulent boundary layer fluctuations. *AIAA J.*, 13 (1975), 8, 1036–40.

[94] B. Ganapathisubramani, N. T. Clemens, and D. S. Dolling. Effects of upstream boundary layer on the unsteadiness of shock-induced separation. *J. Flui. Mech.*, 585 (2007), 369–94.

[95] G. Urbin, D. Knight, and A. Zheltovodov. Compressible large eddy simulation using unstructured grid: supersonic boundary layer in compression corner. *AIAA Paper* 99–0427 (1999).

[96] W. El-Askary. Large eddy simulation of subsonic and supersonic wall-bounded flows. Abhundlungen aus dem Aerodynamishen Institut, der Rhein.-Westf. Technischen Hoschule Aachen (Proceedings), Heft 34. Institute of Aerodynamics Aachen University, ed. Prof. Dr. W. Schröder, Aachen, 12–27 (2003).

[97] A. V. Borisov, A. A. Zheltovodov, A. I. Maksimov, N. N. Fedorova, and S. I. Shpak. Verification of turbulence models and computational methods of supersonic separated flows. *International Conference on the Methods of Aerophysical Research, Part 1*, Russia, Novosibirsk, 54–61 (1996).

[98] N. A. Adams. Direct simulation of the turbulent boundary layer along a compression ramp at M = 3 and Re_o = 1,685. *J. Flui. Mech.*, 420 (2000), 47–83.

[99] G. Urbin, D. Knight, and A. Zheltovodov. Large eddy simulation of a supersonic compression corner: Part I. *AIAA Paper* 2000–0398 (2000).

[100] H. Yan, D. Knight, and A. Zheltovodov. Large eddy simulation of supersonic compression corner using ENO scheme. *Third AFOSR International Conference on DNS and LES, August 5–9,* 2001. Arlington: University of Texas, 381–8 (2001).

[101] A. A. Zheltovodov, E. A. Pimonov, and D. D. Knight. Numerical modeling of vortex/shock wave interaction and its transformation by localized energy deposition. *Shock Wave,* No. 17, 273–90 (2007).

[102] E. Schülein, A. A. Zheltovodov, E. A. Pimonov, and M. S. Loginov. Study of the bow shock interaction with laser-pulse-heated air bubbles. *AIAA Paper* 2009–3568 (2009).

[103] M. Loginov, N. Adams, and A. Zheltovodov. Large-eddy simulation of turbulent boundary layer interaction with successive shock and expansion waves. *International Conference on the Methods of Aerophysical Research, Part I.* Publishing House Nonparel, Novosibirsk, 149–57 (2004).

[104] M. Loginov, N. Adams, and A. Zheltovodov. LES of shock wave/turbulent boundary layer interaction. *Proc. High Performance Computing in Science and Engineering'04,* eds. E. Krause, W. Jäger, and M. Resch (Berlin Heilderberg: Springer-Verlag, 2005), 177–88.

[105] M. Loginov, N. Adams, and A. Zheltovodov. Large-eddy simulation of shock-wave/turbulent-boundary-layer interaction. *J. Flui. Mech,* 565 (2006), 135–69.

[106] M. Loginov, N. A. Adams, and A. A. Zheltovodov. Shock-wave system analysis for compression-decompression ramp flow. *Fifth International Symposium on Turbulence and Shear Flow Phenomenon.* Eds. R. Friedrich, N. A. Adams, J. K. Eaton, J. A. C. Humprey, N. Kasagi, and M. A. Leschziner. TU München, Garching, Germany, 27–29 August 2007, I, 87–92.

[107] A. Zheltovodov, V. Trofimov, E. Schülein, and V. Yakovlev. An experimental documentation of supersonic turbulent flows in the vicinity of forward- and backward-facing ramps. Report No. 2030, Institute of Theoretical and Applied Mechanics, USSR Academy of Sciences, Novosibirsk (1990).

[108] S. Stolz and N. A. Adams. An approximate deconvolution procedure for large-eddy simulation. *Phys. Fluids,* 11 (1999), 1699–1701.

[109] H. Lüdeke, R. Radespiel, and E. Schülein. Simulation of streamwise vortices at the flaps of reentry vechicles. *AIAA Paper* 2004–0915 (2004).

[110] R. G. Adelgren, H. Yan, G. S. Elliott, D. D. Knight, T. J. Beutner, and A. A. Zheltovodov. Control of Edney IV interaction by pulsed laser energy deposition. *AIAA J.,* 43 (2005), 2, 256–69.

[111] D. Rizzetta and M. Visbal. Large eddy simulation of supersonic compression ramp flows. *AIAA Paper* 2001–2858 (2001).

[112] M. Ringuette, M. Wu, and M. P. Martin. Low Reynolds number effects in a Mach 3 shock/turbulent-boundary-layer interaction. *AIAA J.,* 46 (2008), 7, 1883–6.

[113] J. Green. Interactions between shock waves and turbulent boundary layers. *Prog. Aerospace Sci.,* 11, 235–340 (1970).

[114] L. Landau and E. Lifshitz. *Fluid Mechanics* (Oxford: Pergammon Press, 1959).

[115] G. I. Petrov, V. Ya. Likhushin, I. P. Nekrasov, and L. I. Sorkin. Influence of viscosity on a supersonic flow with shock waves. *Trudi CIAM (Proceedings of CIAM),* No. 224, 28 (1952) (in Russian).

[116] S. Bogdonoff and C. Kepler. Separation of a supersonic turbulent boundary layer. *J. Aero. Sci.,* 22 (1955), 414–30.

[117] J. S. Shang, W. L. Hankey Jr., and C. H. Law. Numerical simulation of shock wave–turbulent boundary-layer interaction. *AIAA J.,* 14 (1976), 10, 1451–7.

[118] X. Xiao, H. A. Hassan, J. R. Edwards, and R. L. Gaffney Jr. Role of turbulent Prandtl numbers on heat flux at hypersonic Mach numbers. *AIAA J.,* 45 (2007), 4, 806–13.

[119] E. Schülein. Optical skin-friction measurements in short-duration facilities. *AIAA Paper* 2004–2115 (2004).

[120] E. Schülein, P. Krogmann, and E. Stanewsky. Documentation of two-dimensional impinging shock/turbulent boundary layer interaction flows. DLR, German Aerospace Research Center, Paper 1, B 223–96 A 49 (1996).

[121] M. Hayashi, S. Aso, and A. Tan. Fluctuation of heat transfer in shock wave/turbulent boundary-layer interaction. *AIAA J.*, 27 (1989), 4, 399–404.

[122] S. Pirozzoli and F. Grasso. Direct numerical simulation of impinging shock wave/ turbulent boundary layer interaction at $M = 2.25$. *Phy. Flui.*, 18 (2006), 6, Art. 065113, 1–17.

[123] J. Deleuze. *Structure d'une couche limite turbulente soumise á une onde de choc incidente*. Ph.D. Thesis, Université Aix-Marseille II (1995).

[124] H. Laurent. *Turbulence d'une interaction onde de choc/couche limite sur une paroi plane adiabatique ou chaufée*. Ph.D. Thesis, Université Aix-Marseille II (1996).

[125] V. N. Brazhko. Periodic flowfield and heat transfer structure in the region of attachment of supersonic flows. *Uchenie Zapiski TSAGI (Scientific Notes of TSAGI)*, X (1979), 2, 113–18 (in Russian).

[126] P. R. Spalart and S. R. Allmaras. A One-Equation Turbulence Model for Aerodynamic Flows. *AIAA Paper 92-0439* (1992).

[127] G. S. Glushko. Turbulent Boundary Layer on a Flat Plate in an Incompressible Fluid. *Bulletin of Academic Sciences USSR, Mechanical Series*, No. 4 (1965), 13–23.

5 Ideal-Gas Shock Wave–Turbulent Boundary-Layer Interactions in Supersonic Flows and Their Modeling: Three-Dimensional Interactions

Alexander A. Zheltovodov and Doyle D. Knight

5.1 Introduction

This chapter continues the description of supersonic turbulent shock wave–boundary layer interactions (STBLIs) by examining the flowfield structure of three-dimensional interactions. The capability of modern computational methods to predict the observed details of these flowfields is discussed for several canonical configurations, and the relationships between them and two-dimensional interactions (see Chapter 4) are explored.

5.2 Three-Dimensional Turbulent Interactions

To aid in the understanding of three-dimensional STBLIs, we consider a number of fundamental geometries based on the shape of the shock-wave generator – namely, sharp unswept (Fig. 5.1a) and swept (Fig. 5.1b) fins, semicones (Fig. 5.1c), swept compression ramps (SCRs) (Fig. 5.1d), blunt fins (Fig. 5.1e), and double sharp unswept fins (Fig. 5.1f). More complex three-dimensional shock-wave interactions generally contain elements of one or more of these basic categories. The first four types of shock-wave generators are examples of so-called dimensionless interactions [1] (Fig. 5.1a–d). Here, the shock-wave generator has an overall size sufficiently large compared to the boundary-layer thickness δ that any further increase in size does not affect the flow. The blunt-fin case (Fig. 5.1e) is an example of a dimensional interaction characterized by the additional length scale of the shock-wave generator (i.e., the leading-edge thickness). The crossing swept-shock-wave interaction case (Fig. 5.1f) represents a situation with a more complex three-dimensional flow topology. We briefly discuss the most important physical properties of these three-dimensional flows and provide examples of numerical simulations.

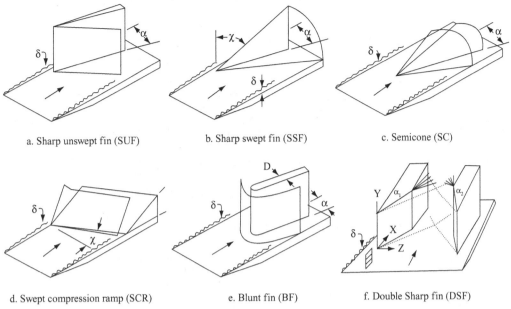

a. Sharp unswept fin (SUF)   b. Sharp swept fin (SSF)   c. Semicone (SC)

d. Swept compression ramp (SCR)   e. Blunt fin (BF)   f. Double Sharp fin (DSF)

Figure 5.1. Swept shock wave generators.

5.3 Three-Dimensional Nature of Separated Flows

5.3.1 Introduction

In accordance with the fundamental work of Prandtl [2], a two-dimensional steady flow separates from a no-slip boundary at the point where the wall shear vanishes and exhibits a negative gradient. The concept of two-dimensional separation was useful in the early stages of development of theoretical approaches for various engineering applications. However, in reality, the two-dimensional STBLI examples considered in Chapter 4 frequently demonstrate an obvious three-dimensional structure that is especially pronounced at conditions of incipient and developed separation. For example, the two-dimensional compression corner (see Fig. 4.37) displays Görtler-like vortices and a set of singular (i.e., critical) node–saddle-point combinations located along the convergence (i.e., separation) and divergence (i.e., attachment) lines. Consequently, the two-dimensional separation concept cannot be strictly applied to achieve an adequate understanding of the organization (i.e., topology) of complex separated flows and to obtain a correct interpretation of their physics.

The Critical Point Theory (CPT) – developed by Legendre [3, 4, 5] to analyze wall shear lines in a steady flow in topological terms based on Poincare's theory of two-dimensional vector fields – is the appropriate tool for interpreting experimental and computational results and for understanding the topology of three-dimensional separated flows (see, e.g., Délery [6, 7]). The trajectories of the wall shear vector field almost coincide with the limiting streamlines (or surface streamlines) that can be observed in experiments using the oil-soot-film visualization technique (i.e., oil-flow visualization). CPT can be used to interpret the surface streamlines

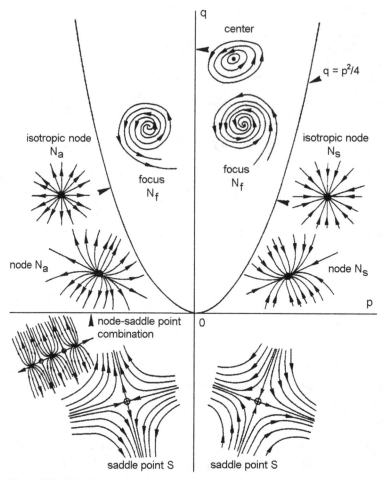

Figure 5.2. Classifications critical points in the $[p, q]$ plane (Délery [6, 7]).

generated by three-dimensional steady laminar and time-averaged turbulent-flow Navier-Stokes computations.

A time-averaged phase portrait of the surface streamlines comprises a continuous vector field with a number of singular points. It can be classified in terms of the trace $p = \lambda_1 + \lambda_2$ and determinant $q = \lambda_1 \cdot \lambda_2$ of the Jacobian matrix of the skin-friction-line partial derivatives [6, 7]. Depending on the value of p and q, the phase portraits can be classified as saddles, nodes, and foci, as well as centers, star-shaped sinks, and sources (Fig. 5.2). Only two skin-friction lines (i.e., separators) pass through saddle point S, whereas all of the other lines miss the origin and take directions consistent with those of the adjacent lines. Nodes are subdivided into nodal points and foci. There is one tangent line at a nodal point. All skin-friction lines are directed either outward away from the node (i.e., a nodal point of attachment N_a) or inward toward the node (i.e., a nodal point of separation N_s). The other type of node (i.e., the focus N_f) has no common tangent line. All skin-friction lines spiral around the origin, either out of it (i.e., attachment focus) or into it (i.e., separation focus). When the location and type of singular points are known, the qualitative properties of the vector field are determined. The location of convergence and

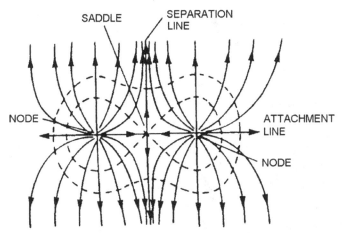

Figure 5.3. Adjacent nodes and saddle-point combination (Lighthill [9]).

divergence lines on the surface, which present the boundaries of secondary flows in the boundary layer or separation zones, are determined by the location of such singular points that significantly affect the three-dimensional flowfield structure. The separation and attachment surfaces are attached to these lines, which emanate from the saddle points and end at a node or focus [6]. The following relationship exists between the number of nodes/foci and saddle points on a closed-surface obstacle [6, 8]

$$\Sigma \text{ nodes} - \Sigma \text{ saddle points} = 2 \tag{5.1}$$

and can be used to check the consistency of a three-dimensional flow description in terms of critical points.

In accordance with Fig. 5.2, an infinite set of node–saddle-point combinations distributed along common separation or attachment lines (at $q = 0$, see left side of the diagram) corresponds to a particular case of three-dimensional separated flow, which historically was termed *two-dimensional*. Lighthill [9] (see also Tobak and Peake [8]) described the topological scheme (Fig. 5.3) that characterizes the three-dimensional effects observed experimentally on the attachment line in a nominal two-dimensional separation and predicted by Large Eddy Simulation (LES) (see Figs. 4.37 and 4.38). The particular convergence line emerging from the saddle point was labeled a line of separation (or, conversely, a line of attachment) because it separates skin-friction lines emerging from different nodal points of attachment.

5.3.2 STBLI in the Vicinity of Sharp Unswept Fins

5.3.2.1 Flow Regimes and Incipient Separation Criteria
A vertical sharp unswept fin with deflection-angle α generates a planar swept shock wave that interacts with an equilibrium turbulent boundary layer on a flat plate (see Fig. 5.1a). The transverse pressure gradient generated by the swept shock wave causes the development of a secondary cross flow in the near-wall part of the boundary layer with a higher deflection of the slower-moving fluid close to the wall (Fig. 5.4). Consequently, convergence and divergence lines can arise on the surface

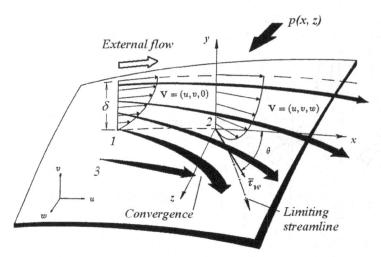

Figure 5.4. Development of secondary crossflow in three-dimensional boundary layer.

associated with the places where the boundary layer can separate or attach. Figure 5.5a illustrates the boundaries of the interaction regimes for STBLI in the vicinity of a vertical unswept fin depending on the deflection-angle α and the inviscid static-pressure ratio in the swept shock ($\xi = p_2/p_1$), as defined by changes in the limiting streamline patterns [10, 11, 12, 13] (Fig. 5.5b). The experimental data characterize the stages of formation of the primary convergence (S_1) and divergence (R_1) lines

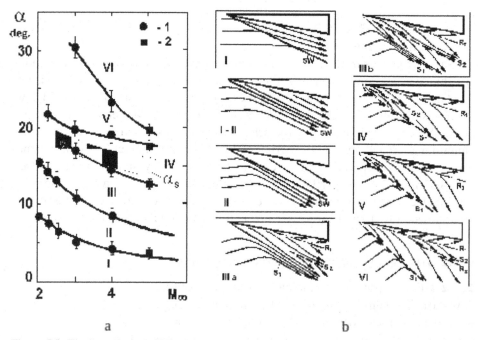

a   b

Figure 5.5. The boundaries of STBLI regimes (a) and corresponding surface-flow patterns (b) for *SUF*. $1 - M_\infty = 2 - 4$, $Re_{\theta o} = (1.1 - 2.3) \cdot 10^4$, $T_w/T_{aw} \approx 1$ (Zheltovodov [11], Zheltovodov et al. [12]); $2 - M_\infty = 5$, $Re_{\theta o} = (6.2 - 7.1) \cdot 10^3$, $T_w/T_{aw} \approx 0.8$ (Schülein and Zheltovodov [13]).

as well as the secondary lines (S_2, R_2) that are associated, respectively, with primary and secondary separation and attachment.

We identified six specific regimes [11, 12] depending on the deflection-angle α with increasing shock-wave strength. In weak unseparated interaction regime I (Fig. 5.5b), the limiting streamlines veer underneath the inviscid shock wave SW without forming a convergence line. At an increased angle α (i.e., regime II), a zone with approximately parallel limiting streamlines moves outward of the inviscid shock-wave trace. At an intermediate stage (i.e., I–II) corresponding to the boundary between regimes I and II (Fig. 5.5a), the limiting streamlines on either side of the SW trace are almost parallel to it. Stanbrook [14] arbitrarily considered this stage to be incipient swept separation of the flow and Korkegi [15, 16], following the arguments of McCabe [17], proposed a practical criterion for incipient swept separation of the following form:

$$M_\infty \cdot \alpha_i = 0.3 \qquad (5.2)$$

for $M_\infty > 1.6$ and $\gamma = c_p/c_v = 1.4$, where the fin deflection angle at incipient-separation α_i is measured in radians. The boundary between regimes I and II is shown in Fig. 5.5a in accordance with this relationship.

As shown in Fig. 5.5b (stages IIIa and IIIb), a further increase in α stimulates the appearance of a convergence zone with asymptotic-convergence line S_1 upstream of the shock-wave trace as well as primary divergence line R_1 near the fin–plate junction. Nevertheless, asymptotic-convergence line S_1 exists on a plate even at stage I and lies well inside the zone bounded by the fin and the inviscid shock [18, 19]. This was demonstrated by placing surface tracer material only on the flat plate upstream of the fin prior to the experiment. Such an asymptotic-convergence line spreads from a saddle point that always exists on a plate close to the sharp-fin leading edge. In this connection, it was concluded that the flow, in fact, is always separated; Stanbrook's regime can be considered a stage of the emergence of separation but not a strict criterion of incipient separation. At the same time, an explicit separation line S_1 begins to form with increasing shock strength and intensification of the convergence of the limiting streamlines (see stage IIIb). Another important feature of regime III is the appearance of secondary convergence line S_2. With increasing α, this line extends farther toward the fin leading edge, as seen in stages IIIa and IIIb. Regimes IV and V are characterized, respectively, by a gradual suppression of line S_2 and eventual disappearance (apart from a limited region in the vicinity of the fin leading edge). In regime VI, secondary convergence line (S_2) reappears accompanied by a divergence line (R_2) considerably closer to the fin (approximately in the vicinity of the inviscid shock-wave trace) than in previous regimes. The following empirical correlation [20]:

$$M_\infty \cdot \alpha_{i2} = 0.6, \qquad (5.3)$$

where α_{i2} is the specific fin deflection angle measured in radians, can be used to determine the appearance of secondary convergence line S_2 – hence, the boundary between regimes II and III (Fig. 5.5a).

In accordance with Lighthill's [9] arguments, the convergence of the limiting streamlines in the vicinity of separation line S_1 (Fig. 5.5b; see stages IIIa, IIIb, and IV) is an important factor resulting in the abrupt departure from the surface (see

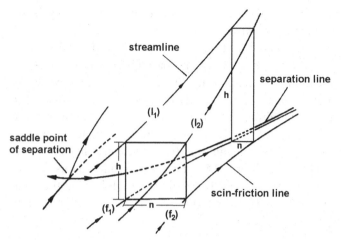

Figure 5.6. Flow in the vicinity of a separation line in accordance with Lighthill [9] (from Délery [6]).

also Délery [6] and Tobak and Peake [8]). Consider an infinitely small streamtube with a rectangular cross section located close to the wall between the two adjacent limiting streamlines l_1 and l_2 and two skin-friction lines f_1 and f_2 (Fig. 5.6). From the conservation of mass, the height of streamtube $h = C \, (\mu/n\rho\tau_w)^{1/2}$. Consequently, the height can increase greatly not only in the vicinity of a singular point where τ_w becomes very small but also in the vicinity of the convergence (i.e., separation) line where n significantly decreases. The latter describes the emergence of three-dimensional separation far from the singular point where the limiting streamlines start to leave the surface with a significant vertical-inclination angle. The symmetrical attachment process can be defined, *mutatis mutandis*, in identical terms, with the difference that the skin-friction lines in such a situation diverge from the separator, which becomes an attachment (or reattachment) line [5, 6]. The velocity along the attachment line tends to increase with increasing distance from the saddle point where the attachment line originates. The distance n between diverging skin-friction lines increases, and distance h of the outer-flow streamlines from the surface can decrease in the attachment region.

An understanding of the factors that determine the type of flow symmetry in the vicinity of swept shock waves is important to gain insight into the three-dimensional flowfield structure and to develop scaling laws. Many investigators concluded that the swept-shock interaction is quasiconical, except for an initial region in the immediate vicinity of the juncture of the fin leading edge and the flat plate (see, e.g., [1, 18, 21, 22, 23, 24, 25, 26, 27, 28, 29]). Consequently, the spherical polar-coordinate system is a proper frame for analysis of these flows. The flowfield features (Fig. 5.7a) and surface streamlines (Fig. 5.7b), outside of an inception region of length L_i, can be considered in two dimensions by projection from the vertex of the Virtual Conical Origin (VCO) onto the surface of the sphere. A commonly used simplification is the replacement of such a spherical segment with a plane tangent to the sphere and normal to the inviscid shock wave. In this case, the pertinent specific Mach number describing the interaction depends on the shock-angle β_0 and the component of M_∞ in this plane according to $M_n = M_\infty \sin \beta_0$, rather than M_∞. The

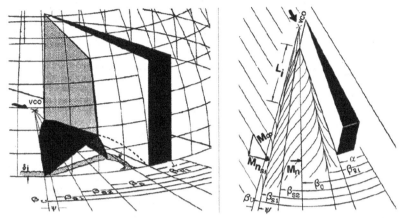

Figure 5.7. Projection of quasiconical interaction flowfield onto spherical polar coordinate surface (a) and sharp-fin interaction footprint (b) (Settles [18]).

flow in the cross section normal to a swept shock wave is similar to the equivalent two-dimensional normal shock wave–turbulent boundary layer interaction (NST-BLI) case. Fig. 5.8 demonstrates (a) a typical surface oil flow pattern, (b) a photograph of flow visualization by Planar Laser Scattering (PLS) in the vertical section normal to a swept shock wave, (c) a sketch of the flowfield in this cross section, with (d) a qualitative surface pressure distribution. Flow conditions correspond to the boundary between regimes I and II (i.e., Stanbrook's flow regime) at $M_\infty = 3$, $\alpha = 5$ degrees (see Fig. 5.5a). As shown in Fig. 5.8b–d, there is gradual compression between the upstream-influence line U (corresponding to beginning deflection

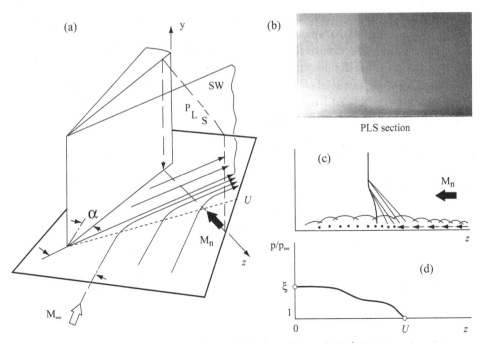

Figure 5.8. *SUF* interaction at $M_\infty = 3$, $\alpha = 5$ degrees, $Re_\theta = 2 \cdot 10^4$ (the boundary between regimes I and II) (Zheltovodov [21], Zheltovodov et al. [23]).

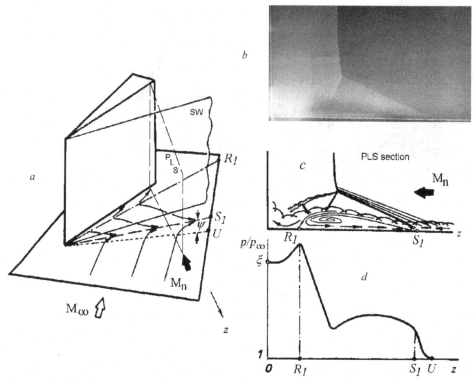

Figure 5.9. SUF interaction at $M_\infty = 3$, $\alpha = 19$ deg., $Re_\theta = 2 \cdot 10^4$ (regime IV) (Zheltovodov et al. [23]).

of limiting streamlines in the interaction region) and surface trace of the deformed shock wave; the pressure "plateau" region forms in this region where the limiting streamlines are almost parallel. An additional pressure rise is observed after the shock wave in the direction of the fin. The flowfield structure (Fig. 5.8b,c) is qualitatively similar to two-dimensional NSTBLI (see Fig. 4.3a,b). At greater deflection-angles α, a λ-shock foot is observed (Fig. 5.9b,c) as in the separated flow regime of a two-dimensional NSTBLI (see Figs. 4.2a and 4.3c) and a distinct surface pressure plateau is formed downstream of explicit coalescence line S_1 (Fig. 5.9a,c,d). However, in contrast to the two-dimensional interaction, the three-dimensional flow is capable of developing transverse-velocity components, allowing fluid to escape laterally, and the streamlines exhibit a helix structure (Fig. 5.9c). Thus, the flow shown in a cut (i.e., PLS) plane characterizes the behavior of "pseudo-streamlines" appearing as a result of projection of the velocity-vector field in this plane. In these conditions, the streamline emanating from separation line S_1 is distinct from the streamline penetrating reattachment line R_1, and the separation zone is opened for the external flow penetrating it.

Within the framework of the two-dimensional analogy, the plateau pressure data $p_p/p_\infty = f(M_n)$ for a fin-generated three-dimensional STBLI (Fig. 5.10a, symbols 5–8) are in good agreement with empirical correlations for two-dimensional interactions [24, 30] (i.e., lines 2 and 3). The Free Interaction Theory (FIT) [31] expression (4.14) in Chapter 4 (i.e., solid line 1) agrees well with experimental data (i.e., symbols 5–7) for an adiabatic wall and the Reynolds numbers

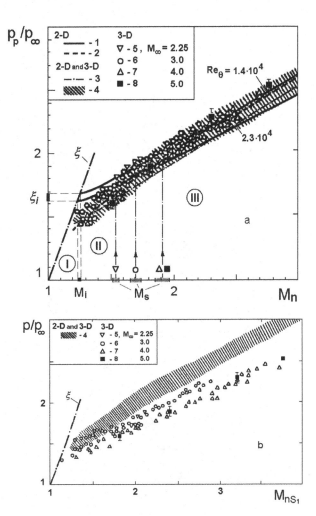

Figure 5.10. Plateau pressure comparison for 2-D and 3-D STBLI. 2-D: 1 – FIT (Chapman et al. [31]), 2 – Zukoski [30]; 2-D and 3-D: 3 – Dem'yanenko and Igumnov [24]; 4 – collected data (Dem'yanenko and Igumnov [24], Hayes [25]); 3-D (sharp fins): 5, 6, 7 – $\alpha = 4° - 31°$, $\chi = 0°$, 15°, 30°, 45° and 60°, $Re_\theta = (1.4 - 2.3) \cdot 10^4$, $T_w/T_{wa} \approx 1$ (Zheltovodov et al. [11, 12, 20, 22]); 8 – $\alpha = 8 - 23°$, $\chi = 0°$, $Re_\theta = 6.9 \cdot 10^3$, $T_w/T_{wa} = 0.8$ (Schülein and Zheltovodov [13, 26]).

$Re_\theta = (1.4 - 2.3) \times 10^4$ with $k = 7.4$ recommended for these conditions [10, 32]. The value of c_f was determined by Kutateladze and Leont'ev's [33] method. Similar to two-dimensional normal-shock-wave interactions (see, e.g., Fig. 4.4), the plateau pressures tend to lie below the FIT predictions for small normal Mach numbers, although here, the spread is somewhat larger. The scatter in the collected two- and three-dimensional data [24, 25] is within the limits of the experimental results shown by band 4 (Fig. 5.10a). This correspondence of two- and three-dimensional data follows from the validity of the FIC within the framework of the two-dimensional analogy for different three-dimensional STBLI flows with cylindrical and conical symmetry [34, 35]. In accordance with other recommendations [28, 29], the normal component of the Mach number $M_{nS1} = M_\infty \sin \beta_{S1}$ to separation line S_1 (see Fig. 5.7b) also can be used to generalize the plateau pressure data of three-dimensional interactions. However, the Mach-number component M_n normal to the swept-shock trace, rather than M_{nS1}, provides the best quantitative agreement with existing two-dimensional correlations [34]. This is shown clearly in Fig. 5.10b, where M_{nS1} values are used to generalize the data.

In the framework of the two-dimensional analogy by assuming that $M_\infty = M_n$ and ξ_i is equal to the plateau pressure ratio for separated two-dimensional normal

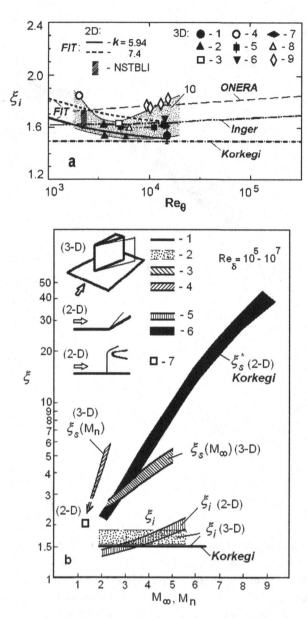

Figure 5.11. Incipience-separation pressure rise, a – incipient-separation conditions (Stanbrook's regime) for fin-generated 3-D STBL: 1 – $M_\infty = 2.95$, 3 and 4 (Zheltovodov et al. [12]); 2 – $M_\infty = 2.0, 3.0$ and 4 (Dem'yanenko and Igumnov [24]); 3 – $M_\infty = 5.0$ (Schülein and Zheltovodov [26]); 4 – $M_\infty = 2.3$ (Kubota and Stollery [27]); 5 – $M_\infty = 2.95$ (Zubin and Ostapenko [28, 29]); 6 – $M_\infty = 2.95$ (Oskam et al. [37]); 7 – $M_\infty = 2.5$ (Lu and Settles [38]); 8 – $M_\infty = 2.45$ (Leung and Squire[39]); 9 – $M_\infty = 2.47, 2.95$, and 3.44 (Lu and Settles [19]); b – incipience-separation pressure rise: 3-D: $1 - \xi_i = 1.5$ (Korkegi [15]), $2 - \xi_i$, $3 - \xi_S$; 2-D: $4 - \xi_i$, $6 - \xi^*_S$ (Korkegi [15]).

shock-wave interactions, equation (4.15) can be used to predict the critical Mach number M_{ni} and corresponding pressure rise ξ_i required to produce fin-generated incipient swept separation at $Re_\theta = (1.4 - 2.3) \times 10^4$, $T_w/T_{aw} = 1$. The predicted values $M_{ni} = 1.24 - 1.26$ and $\xi_i = 1.62 - 1.69$ (Fig. 5.10a) correspond to Stanbrook's interaction regime on the boundary between the regimes I and II (see Fig. 5.5b, case I–II and Fig. 5.8a) and agree well with conditions of emerging "true" incipient separation for two-dimensional NSTBLIs (see Fig. 4.4). Fig. 5.11a shows the predicted values $\xi_i(Re_\theta)$ using $k = 7.4$ (following recommendations [10, 32]) and $k = 5.94$ (as suggested by Charwat [36]), demonstrating the tendency to decrease as Re_θ increases from 10^3 to 1.5×10^4. Experimental data gathered by a number of researchers [12, 24, 26, 27, 28, 29, 37, 38, 39] (i.e., symbols 1–8) are in agreement with the predicted tendency. This view of incipient-separation conditions is an improvement over Korkegi's criterion [15, 16] (i.e., $\xi_i = 1.5$), which ignores the Reynolds-number and shape-factor dependence inherent in the properties of the incoming boundary layer. The data by Lu and Settles [19] (i.e., symbol 9) draw attention by their higher values as compared to their previous result (i.e., symbol 7) and the tendency to increase with Reynolds number. The behavior of the entire data band (i.e., band 10) is qualitatively similar to two-dimensional NSTBLIs, as indicated in Fig. 4.5 (i.e., band 9). The tendency for increasing ξ_i with Reynolds number – in accordance with Inger's theory [40, 41] and the ONERA experimental correlation [42] – is demonstrated in Fig. 5.11a, which corresponds to the respective dependencies for M_i considered in Chapter 4 (see Fig. 4.5, lines 1 and 2). The ξ_i levels predicted by Inger are within the band of the data scatter and calculations on the basis of FIT with different k at $2.5 \times 10^3 \leq Re_\theta \leq 2.4 \times 10^4$. However, the Lu and Settles data (i.e., symbol 9) are in better agreement with the ONERA correlation at $Re_\theta \geq 10^4$.

Analysis demonstrated that the emergence of Stanbrook's three-dimensional swept-shock-wave interaction regime corresponds to conditions of true incipient-intermittent separation at the two-dimensional normal-shock-wave interaction. Moreover, the values ξ_i for the corresponding regime for the three-dimensional interaction case (Fig. 5.11b, band 2) also correspond to the critical pressure rise ξ_i in the regime of true incipient intermittent separation over two-dimensional compression ramps (CRs) (i.e., band 5) at deflection-angles $\alpha_i \leq \alpha \leq \alpha_{ef}$ (see Fig. 4.19). Because the three-dimensional flow is capable of developing transverse-velocity components, allowing fluid to escape laterally (see Fig. 5.4), a secondary flow is formed in the near-wall part of the boundary layer in such a regime with higher horizontal streamline inclinations. Analogous to two-dimensional interactions, the onset of developed separation with distinct reversed flow should be expected at a higher interaction strength with increasing fin-deflection-angle α and corresponding inviscid shock-inclination-angle β_0. In accordance with Kubota and Stollery [27], this regime is associated with the formation of convergence line S_1 upstream of the inviscid shock trace (see Fig. 5.5b, stages IIIa–IIIb). However, even after the emergence of line S_1, a further increase of α and β_0 is accompanied by a reduction of the angle Ψ between this line and the upstream-influence line U (shown in Fig. 5.7) up to a constant limiting value that is specific for each Mach number M_∞ [10, 12]. This is plotted in Fig. 5.12 as $\Delta\Psi_i$ versus $\Delta\beta_{0i}$, where $\Delta\Psi_i = \Psi - \Psi_i$ and $\Delta\beta_0 = \beta_0 - \beta_{0i}$. The values Ψ_i and β_{0i} correspond to the incipient intermittent-separation conditions (i.e., Stanbrook's regime at $\xi = \xi_i$ or $M_n = M_{ni}$). The black symbols at $\Delta\Psi_i \approx 2 - 8$ degrees correspond to the developing intermittent-separation regime in which a gradual process of convergence-line S_1 formation is observed at different values of Mach number M_∞. The dashed region indicates the domain in which the limiting constant values $\Delta\Psi_S = \Psi_S - \Psi_i$ were achieved at some $\Delta\beta_{0S} = \beta_{0S} - \beta_{0i}$ for various Mach numbers M_∞. The values β_{0S} correspond to the fin critical-inclination-angles $\alpha = \alpha_S$, which are shown in Fig. 5.5a by the band α_S located in the limits of regime IV. Once $\alpha = \alpha_S$ (and, respectively, $\Delta\Psi = \Delta\Psi_S$), it can be argued that the interaction has reached the regime of a fully developed, large-scale swept separation. Indeed, in accordance with this interpretation, the plateau pressure in the vicinity of the swept shock reaches the value typical for large-scale, two-dimensional separation at the corresponding critical normal-to-the-shock-wave Mach number $M_S = M_\infty \sin$

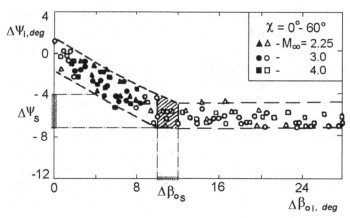

Figure 5.12. Extent of upstream-influence region vs. swept-shock-trace deflection angle (Zheltovodov et al. [12]).

β_{0S} (see Fig. 5.10a). A distinct λ-shock foot is formed around the separation zone at $\alpha \geq \alpha_S$ (see Fig. 5.9b,c). Analogous to the regimes of two-dimensional NSTBLIs (see Fig. 4.4), the respective regimes of unseparated flow (I), intermittent separation (II), and developed large-scale separation (III) can be identified for the three-dimensional swept-shock-wave interaction (see Fig. 5.10a). In accordance with the indicated critical Mach numbers $M_n = M_S$, the third regime in the three-dimensional case is realized at the higher pressure ratio in shock wave (ξ_S) with an increase in the Mach number M_∞. The shock-pressure-ratio dependency ξ_S versus Mach number M_∞ for conditions of fully developed, large-scale swept separation emerging at $\alpha = \alpha_S$ or $M_n = M_S$ (see Fig. 5.11b, band 3) almost coincides with Mach numbers $M_\infty \approx 2.2 - 3.6$, with the correlation $\xi_S^*(M_\infty)$ (i.e., band 6) proposed by Korkegi [15] for the onset of the large-scale separation stage in two-dimensional compression ramp (CR) flows corresponding to $\alpha = \alpha_S^*$ (see band 14 in Fig. 4.19). However, within the framework of a two-dimensional analogy, the swept-shock-pressure ratio ξ_S variation must be considered versus the normal Mach number M_n (see Fig. 5.11b, band 4) in the same way as the data for the plateau pressure (see Fig. 5.10a). With decreasing M_n, the dependence 4 logically tends to the value ξ_S (i.e., symbol 7) for two-dimensional NSTBLIs in accordance with values of M_S being considered for this case (see Fig. 4.6) at $Re_\delta \geq 10^5$ (or $Re_\theta \geq 2 \times 10^3$). Comparison of the generalized data for the critical shock-wave strengths ξ_i and ξ_S at the stages of incipient true and large-scale separation in conditions of two- and three-dimensional SWTB-LIs demonstrate their close correspondence at the supersonic Mach numbers for the respective interaction regimes. This is opposite to the conclusion by Korkegi [15] – namely, that three-dimensional swept-shock-wave–turbulent boundary-layer interaction always leads first to separation and possible flow breakdown in rectangular diffusers and inlets and the critical shock-pressure ratios are well below the incipient values for two-dimensional cases. This conclusion, however, cannot be recognized as totally reliable because it was based on the comparison of empirical dependencies 1 ($\xi_i = 1.5$) and 6 (see Fig. 5.11b) that characterize different three- and two-dimensional interaction regimes.

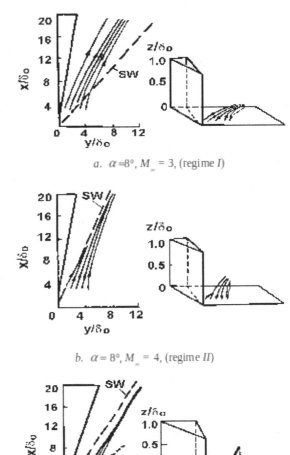

a. $\alpha = 8°$, $M_\infty = 3$, (regime I)

Figure 5.13. Computed streamline trajectory (Horstman and Hung [43]).

b. $\alpha = 8°$, $M_\infty = 4$, (regime II)

c. $\alpha = 16°$, $M_\infty = 4$, (regime IV)

5.3.2.2 Flow Structure and Its Numerical Prediction

Flow behavior in the vicinity of convergence line S_1 for different stages of formation was examined by RANS simulation with a simple algebraic eddy-viscosity turbulence model [43]. The streamlines originating near the surface at $z = 0.05\delta_0$ are shown in Fig. 5.13. For each case, a top view of the streamlines and a three-dimensional view are shown on the left and right sides of the figure, respectively. The projection of the right-hand streamline on the $z = 0$ surface for each case also is demonstrated. In accordance with the boundaries of the different regimes presented in Fig. 5.5a, the first case ($M_\infty = 2$, $\alpha = 8$ degrees) corresponds to fully attached flow (i.e., regime I) and the streamline trajectories do not converge and show little liftoff from the surface (up to $z/\delta_0 \sim 0.1$). The second case ($M_\infty = 4$, $\alpha = 8$ degrees, Fig. 5.13b) corresponds to regime II, where a distinct primary-coalescence line has not

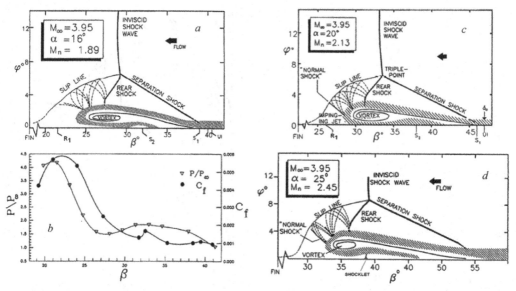

Figure 5.14. Transformation of the flowfield structure with increase of fin-inclination angle at $M_\infty = 3.95$: a, b – $\alpha = 16°$ (regime IV), c – $\alpha = 20°$ (the boundary between regimes IV and V), d – $\alpha = 25°$ (regime VI) (Settles [18] and Alvi and Settles [44, 45]).

yet formed (see Fig. 5.5b) and there is only little liftoff ($z/\delta_0 \sim 0.2$) of the streamlines. Similar flow behavior was observed in experiments [37] in conditions of the same interaction regime II at $M_\infty = 2.95$, $\alpha = 10$ degrees. In accordance with the experiments, large horizontal and small vertical inclinations of the streamlines are typical for the formation of secondary flow in the near-wall portion of the boundary layer. The third case, shown in Fig. 5.13c ($M_\infty = 4$, $\alpha = 16$ degrees), corresponds to the band of critical values $\alpha = \alpha_S$ (shown in the limits of regime IV in Fig. 5.5a) at which the large-scale separation appears, and the streamlines converge and lift off the surface to reach a significant height of $z/\delta_0 = 0.55$.

Figure 5.14 depicts the transformation of the flowfield structure for an interaction case studied experimentally by Alvi and Settles [18, 44, 45]. The flowfield features in different interaction regimes were obtained by optical visualization using conical shadowgram and interferogram as well as PLS methods. In conditions of regime IV at $M_\infty = 3.95$, $\alpha = 16$ degrees (Fig. 5.14a) corresponding to RANS calculations shown in Fig. 5.13c, the separated free-shear-layer is deflected away from the wall by the separation shock and then returns toward it near the rear leg of the λ-shock. The separated free-shear-layer rolls into a tight vortex with a distinct core. Aft of the separation bubble, the flow impinges on the surface, forming attachment line R_1, where some flow is directed back upstream. This reverse flow then encounters secondary-coalescence line S_2. The impinging jet structure is bounded by the slip line originating from the triple point of the λ-shock (i.e., the outer boundary) and the separation bubble (i.e., the inner boundary). The turning of this jet is accomplished via a Prandtl-Meyer expansion fan that reflects from the slip line as a compression fan. The compression waves occasionally coalesce and form "shocklets," which are drawn as solid lines. The structure of this jet impingement [18, 44, 45] is

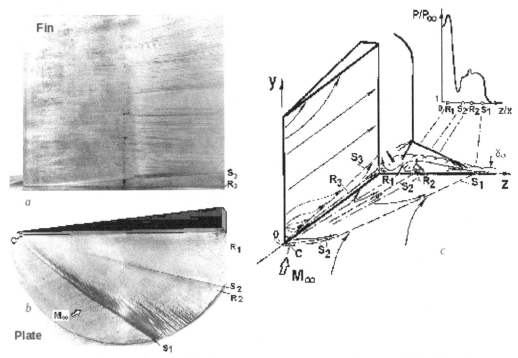

Figure 5.15. Surface-flow visualization (a, b) and flowfield structure (c) at $\alpha = 30.5°$, $M_\infty = 4$ (regime VI); Zheltovodov [11] and Zheltovodov et al. [12]).

reminiscent of the Edney type-IV leading-edge shock impingement [46] and thus explains the reason for the peak pressure, skin friction (see Fig. 5.14b), and heating in the rear part of the fin interaction. The surface skin friction is finite (i.e., nonzero) in the vicinity of primary- and secondary-coalescence lines S_1 and S_2 in the section that is distant from the singular point located ahead of the fin leading edge.

As the interaction strength increases further (Fig. 5.14c,d), the shocklets in the impinging jet become more intensive; eventually, the final shocklet becomes a "normal shock" (i.e., to the flow in the conical projection) that terminates supersonic flow in the jet prior to impingement. Secondary-coalescence lines S_1 and S_2 are located at a distance upstream from the inviscid shock-wave trace during regimes III and IV up to regime V (see, e.g., Fig. 5.14a,c). The conical shadowgrams reveal an obvious "bulge" in the reversed flow in the vicinity of line S_2, but the authors concluded that it is not a clear indication of secondary separation. For even greater interaction strengths (starting from regime V), the "reverse" flow inside the separation zone accelerates to supersonic speeds (see also [24, 28]), which can lead to the formation of a shocklet located approximately underneath the inviscid shock wave (Fig. 5.14d).

The surface-flow-pattern visualization with the corresponding qualitative flow-field structure and surface pressure distribution at strong interaction conditions (i.e., regime VI, $\alpha = 30.5$ degrees, $M_\infty = 4$) [11, 12] are shown in Fig. 5.15. Here, the flow is characterized by the appearance not only of primary-coalescence and divergence lines S_1 and R_1 but also of distinct secondary lines (i.e., S_2 and R_2), which are

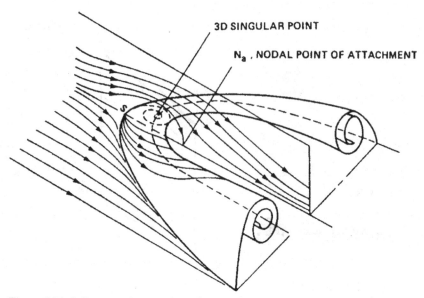

Figure 5.16. 3-D separation topology (Tobak and Peak [8]).

associated with secondary separation and reattachment (Fig. 5.15b,c). Surface-pressure maxima occur in the vicinity of primary as well as secondary reattachment lines R_1 and R_2 (Fig. 5.15c). There is also a λ-shock structure in the vicinity of the secondary-separation zone. Convergence line S_3 and a divergence line R_3 were observed on the fin side (Fig. 5.15a,c), which suggests the presence of a longitudinal vortex in the vicinity of the fin and plate junction. The existence of saddle-point C with a vanishing shear stress located on primary-separation line S_1 upstream of the fin leading edge (Fig. 5.15b,c) is evidence of the predominant singular nature of separation at this location. Farther downstream, along convergence line S_1, the character of separation changes to ordinary (or global) separation and the fluid leaves the surface in accordance with Lighthill's [9] concept (see Fig. 5.6), as illustrated by the topological schema in Fig. 5.16. Detailed optical measurements of the surface-skin friction [47] for flow conditions in regime VI (at $\alpha = 23$ degrees, $M_\infty = 5$) are shown in Fig. 5.17. It can be seen that the skin friction at primary (S_1) and secondary (S_2) separation remains nonzero, as expected for a three-dimensional ordinary separation.

Overall, the surface-streamline pattern in the vicinity of the leading edge of a sharp fin with a significant inclination angle (Fig. 5.15) is reminiscent of a typical three-dimensional separated flow in the vicinity of a blunt fin or a vertical cylinder [48, 49] as shown in Fig. 5.18. Here, a singular separation at saddle-point C is accompanied by the appearance of nodal-reattachment-point N on the cylinder surface. The intensive flow divergence from this point on the cylinder and the appearance of a vortex between lines S_3 and R_3 are qualitatively similar to the flow along the bottom part of the fin-side surface close to the leading edge (see Fig. 5.15a,c). The blunt-cylinder flow includes secondary-separation and reattachment lines S_2 and R_2, which originate ahead of cylinder saddle-point C_2 and node N_2, respectively. The development of similar additional secondary-separation lines – often seen near the leading edge of the fin at regimes V and VI (i.e., the dotted line denoted $S_2{'}$ in

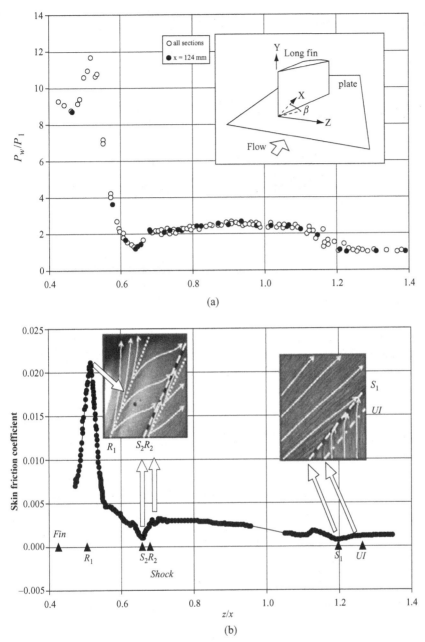

Figure 5.17. Surface pressure (Schülein and Zheltovodov [26]) (a) and skin-friction distributions (Schülein [47]) (b) at $x/\delta_0 = 32.6$, $\alpha = 23°$, $M_\infty = 5$, $Re_{\delta 0} = 1.4 \times 10^5$ (regime VI).

Fig. 5.15c) – may be caused by similar effects. However, these secondary lines are of limited extent because they are shifted quickly by the "reversed" flow to primary line S_1 in contrast to similar lines S_2 and R_2 emerging downstream close to the inviscid shock-wave trace.

RANS simulations are capable of resolving important details of the flow structures described herein. Fig. 5.19 shows the flowfield predicted by calculations [50] using the Baldwin-Lomax turbulence model for Mach $M_\infty = 4$, $\alpha = 20$ degrees at

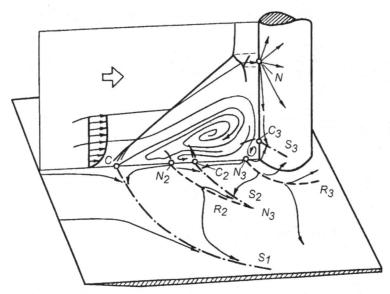

Figure 5.18. 3-D separation in the vicinity of vertical cylinder (Voitenko et al. [48, 49]).

Reynolds number $R_{\delta 0} = 2.3 \times 10^5$ corresponding to regime V. Two surfaces were derived from many streamline traces: Surface No. 1 separates the initial boundary layer into two parts. The lower part is entrained in a vortical structure. The outer part flows up and over the vortical structure, turning toward the inviscid downstream angle. The line of divergence represents the intersection of this surface with the wall. It is a line of separation according to Lighthill's terminology [9] and it defines the origin of the three-dimensional separation sheet (i.e., Surface No. 2), which consists of streamlines originating off the surface at different lateral positions. To clarify the complex flow, some streamline traces (i.e., A, B, C, and D) are included in the illustration. The flowfield is compressed in the spanwise direction. In reality, streamlines

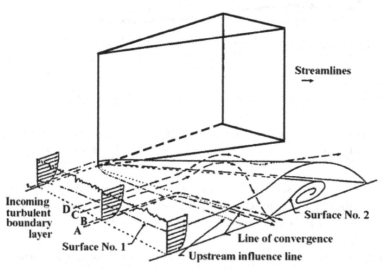

Figure 5.19. Predicted by RANS calculations flowfield model for $\alpha = 20°$, $M_\infty = 4$ (regime V) (Knight et al. [50]).

like B and C may take more than 100 upstream boundary-layer thicknesses in the spanwise direction to complete a single revolution of the vortical structure. Streamline A, on the surface, deflects at the upstream-influence line and flows along the surface, parallel to the line of convergence. Streamline B, slightly above streamline A, passes over the convergence line, turns to the approximate inviscid shock-wave direction, and flows along Surface No. 2. Streamline C, just below Surface No. 1, turns and rises initially but then descends, continuing to turn slowly. Streamline D, just above Surface No. 2, follows a similar path but, as it descends and approaches the floor, it continues downstream as the surface streamline for the downstream flow. Bogdonoff [51] concluded that the primary feature of this type of flowfield is a flattened vortical flow. Nevertheless, if the rotational velocities or static pressures are analyzed, a fully developed vortex cannot be detected and the structure is primarily a supersonic flow along the shock direction with a slow rotational component.

5.3.2.3 Secondary-Separation Phenomenon and Its Prediction

The presence of secondary separation and its evolution with increasing swept-shock- wave strength in regimes III–VI (see Fig. 5.5) was described previously. Following experimental research [11, 12, 22, 28], two important factors determine the appearance, disappearance, and reappearance of the secondary separation in regimes III–VI: (1) the state (laminar or turbulent) of "reverse" flow between primary-reattachment and separation lines R_1 and S_1; and (2) the acceleration of the near-wall cross flow up to supersonic speeds ($M_n > 1$). Zubin and Ostapenko [28] suggested that the formation of a limited secondary separation in the downstream direction in regime III is related to the subsonic laminar near-wall cross flow ($M_n < 1$) and that secondary-convergence line S_2 can disappear downstream when this flow becomes transitional and turbulent along the longer streamlines extending from divergence line R_1 to line S_1. The reason for suppression of the secondary separation was associated with achievement of the supersonic cross-flow state ($M_n > 1$) inside the laminar reversed flow in regime V. These logical considerations were examined experimentally by Zheltovodov et al. [12] by applying sand-grain roughness along primary-attachment line R_1 to trigger transition of the reversed flow. It was possible to suppress the secondary separation in regimes III and IV but not in regime VI. The reappearance of secondary separation in regime VI occurs in the vicinity of an imbedded normal shock wave with the critical strength ($\xi_i \geq 1.5$–1.6), which is typical for turbulent reversed flow. In addition to the experiments, this conclusion was supported by theoretical analysis within the framework of the two-dimensional analogy and the FIT [12, 22]. Thus, secondary separation in regimes III–V disappears when the flow transitions to turbulence, whereas its reappearance in regime VI occurs when the cross flow is turbulent, as was assumed in [11]. At this stage, the embedded normal shock wave has reached the critical strength required to force turbulent separation.

RANS predictions of secondary separation and other flow parameters in the interaction region under different states of reversed flow (i.e., laminar, transitional, and turbulent) is a complex problem. Traditional turbulence models (e.g., the algebraic Baldwin-Lomax model and the two-equation k-ε model) have only limited success. This is demonstrated in Fig. 5.20, where different eddy-viscosity models are used in computations of sharp-fin interactions. The computed surface pressure

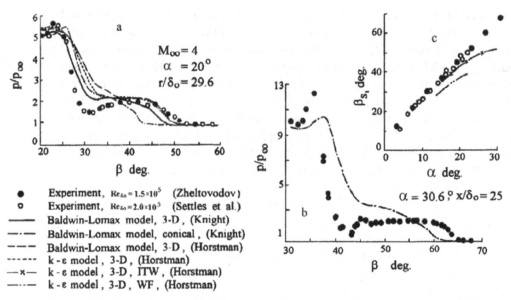

●	Experiment, $Re_{\delta_0}=1.5 \times 10^5$ (Zheltovodov)
○	Experiment, $Re_{\delta_0}=2.0 \times 10^5$ (Settles et al.)
——	Baldwin-Lomax model, 3-D, (Knight)
—·—	Baldwin-Lomax model, conical, (Knight)
— — —	Baldwin-Lomax model, 3-D, (Horstman)
······	k-ε model, 3-D, (Horstman)
—×—	k-ε model, 3-D, ITW, (Horstman)
—··—	k-ε model, 3-D, WF, (Horstman)

Figure 5.20. Comparison of RANS computations with surface pressure data (a, b) and primary-separation-line angle variation vs. fin-inclination angle (c) (Zheltovodov [10]).

shows the closest agreement with experiments for the weaker interaction at $\alpha <$ 20 degrees (see, e.g., [10, 52, 53]). As the interaction strength increases, differences between the computations and experiments become more severe with a comparable level of accuracy for different turbulence models (Fig. 5.20a,b). None of the computations predicted secondary separation in the regimes in which it was observed experimentally. A similar tendency is seen when comparing primary-separation-line-angle β_{S1} versus fin-angle α (Fig. 5.20c). The integrated-to-the-wall (ITW) results are in better agreement than the wall-function (WF) results. For $\alpha < 20$ degrees, these computations are definitely within the uncertainties of the measurements; however, for $\alpha > 20$ degrees, the computations and data diverge.

Figure 5.21 shows a RANS computation [54] for conditions close to the boundary between regimes III and IV (see Fig. 5.5a), using a modified Baldwin-Lomax turbulence model. Several cross-flow cuts and skin-friction lines are presented in Fig. 5.21b in accordance with their positions marked in Fig. 5.21a. It can be seen that this calculation can capture the secondary separation. It also can be observed that the inner layers of the undisturbed boundary layer, where the eddy viscosity is high, wind around the core of the vortex. However, the outer layers (which have low turbulence) move above the vortex and penetrate the separation bubble at the reattachment region, forming a low-turbulence "tongue" close to the wall underneath the vortex (Fig. 5.21b, sections VI and VIII). Because the intermittency of the flow inside this tongue and the outer layers of the vortex is small, the flow is almost laminar in this region. At the initial stage of development, the conical separation vortex is completely composed of turbulent fluid (section II); however, as it grows in downstream direction, the low-turbulence tongue forms (sections IV, VI, and VIII). Panaras [53, 54] observed that an increase of the interaction strength causes more of the low-turbulence outer flow to be entrained in the vortex. Conversely, in a weak interaction, a low-turbulence tongue is not formed. Figure 5.22 demonstrates that an

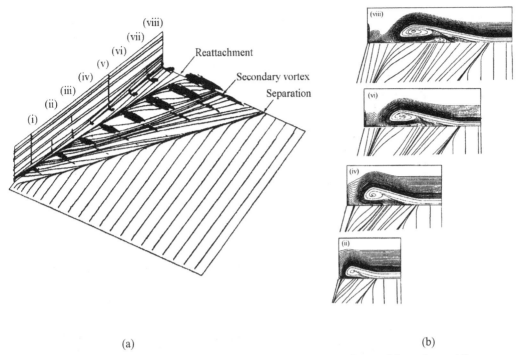

(a) (b)

Figure 5.21. Skin-friction line patterns and cross sections of the vortices at $M_\infty = 4$, $\alpha = 16°$, $Re_{\delta_0} = 2.04 \times 10^5$ (the boundary between regimes III and IV) (Panaras [54]).

obvious improvement in the accuracy of the numerical prediction can be achieved by using an alternative equation for the eddy-viscosity coefficient [55]. Here, the calculations are compared with experimental data [56] for $M_\infty = 4$ and $\alpha = 16$ degrees and calculations by other researchers [57].

RANS simulations [58] examined the linear Wilcox k–ω – model and its weakly nonlinear version (denoted $WD+$ in Fig. 5.23), which represents an extension to compressible flows of the nonlinear correction of Durbin [59]. As shown in Fig. 5.23a, the $WD+$ model (i.e., line 4) shows significant improvement over the linear k–ω model (3) and standard Baldwin-Lomax (B-L) model (line 2). It gives a good prediction of the wall-pressure distribution as well as capturing the secondary separation in the vicinity of the embedded normal shock wave (Fig. 5.23b) in regime VI in accordance with experiments (see Fig. 5.15b,c). In contrast to other standard turbulence models (i.e., B-L, k-ω, k-ε), the $WD+$ model predicts secondary separation in regimes III and IV and the wall-skin friction in the interaction region (Knight et al. [52], Thivet et al. [58], and Thivet [60]). This improvement is associated with a significant reduction in the peak turbulent kinetic energy (TKE) in the flow that penetrates from outside the shear layer to the wall near the secondary-separation line (Fig. 5.24a) in contrast to calculations with a standard Wilcox (WI) model (Fig. 5.24b), which is characterized by high turbulence levels in the near-wall "reversed" flow. Small underpredictions of the plateau pressure and a more severe overprediction of the interaction-region length are evident in the calculations using the $WD+$ turbulence model (see Fig. 5.23a, line 4). Various physical factors can cause these features, one of which may be due to unsteady effects that cannot be

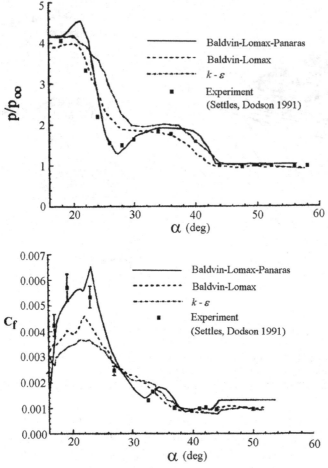

Figure 5.22. Surface pressure and skin-friction prediction at $M_\infty = 4$, $\alpha = 16°$, $Re_{\delta o} = 2.04 \times 10^5$ (Panaras [55]).

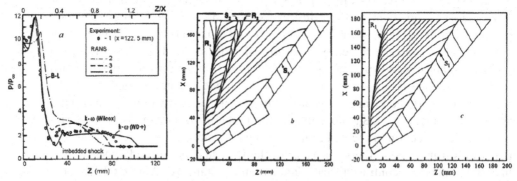

Figure 5.23. Surface pressure (a) and surface-streamlines prediction for $\alpha = 30.6°$ and $M_\infty = 4$, $Re_{\delta o} = 1.6 \times 10^5$ (regime VI); (b) weakly nonlinear k-omega model (WD +); (c) linear Wilcox's k-ω model (Thivet et al. [58]).

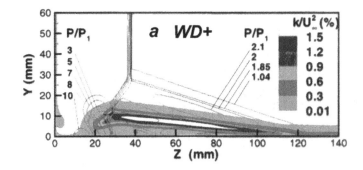

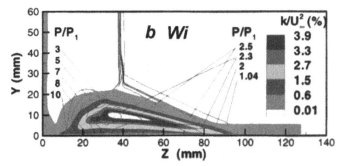

Figure 5.24. Turbulence kinetic energy in cross section ($x = 122.5$ mm): a – weakly nonlinear k-ω model (WD +), c – linear Wilcox's k-ω model (Thivet et al. [58]).

correctly captured in the steady RANS computations. Another difficulty may be the accurate prediction of turbulence evolution and amplification in the outer part of the separated boundary layer during the interaction with the λ-shock system, as well as with unsteady shocklets and the normal shock in the jet-impingement region above the three-dimensional separation zone. As described previously, this outer part of the boundary layer penetrates the near-wall reverse flow, and the turbulence level in this region is critical for development of the secondary separation.

5.3.3 Sharp Swept Fin and Semi-Cone: Interaction Regimes and Scaling Laws

The nonplanar swept shock-wave–turbulent boundary-layer interactions in the vicinity of sharp swept fins [12, 22, 61] and semi-cones [12, 62] represent more general cases than the unswept-fin configuration. Other than the fact that the main shock is now curved (Fig. 5.25a), the flowfield structure for the swept-fin case (Fig. 5.25b) and the interaction regimes with the boundaries indicated by solid lines (Fig. 5.26) are qualitatively similar to the unswept-fin case. As shown in Fig. 5.25c, an increase of χ at fixed α and M_∞ values weakens the shock wave, thereby decreasing (1) the surface pressure levels at the maximum peak on reattachment line (R_1) close to the fin; (2) the plateau pressure level; and (3) the angular position β_U of the upstream-influence line where the surface pressure rise is observed. The λ-shock foot appears only at a sufficient value of the swept-fin inclination-angle α, and the

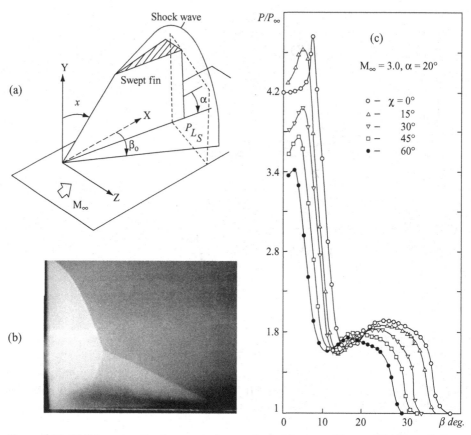

Figure 5.25. PLS flow visualization around swept fin (a, b) at $\chi = 60°$, $\alpha = 20°$, $M_\infty = 3$, $Re_\theta = 2 \cdot 10^4$ and surface pressure distributions (b) at different χ values (Zheltovodov and Schülein [61]).

attachment of separated flow occurs closer to the fin with increasing leading-edge sweep-angle χ. However, the plateau pressure ratio p_p/p_∞ versus M_n (see Fig. 5.10a), the critical normal Mach number $M_n = M_i$, and shock-wave critical strength $\xi = \xi_i$ corresponding to Stanbrook's interaction regime – as well as the second critical Mach number $M_n = M_S$ and shock-wave critical value $\xi = \xi_S$ corresponding to the emergence of developed large-scale separation – are the same for unswept- and swept-fin interactions (Fig. 5.26).

Fig. 5.27a demonstrates the evolution of the surface-flow pattern near a semi-cone for interaction regimes I–VI that is similar in some details to the regimes observed for unswept and swept sharp fins. The corresponding cross-sectional flow structures are shown in Fig. 5.27b with increasing half-angle α and shock-strength ξ. The sketches are based on observations made from PLS and surface-flow visualizations as well as surface pressure measurements [62]. The boundaries of these regimes (see Fig. 5.26, dotted lines) correspond to extrapolated boundaries for swept fins. The experiments with the semi-cone half-angles $\alpha = 8, 10, 15, 20, 25$, and 30 degrees are indicated by open symbols. Closed symbols correspond to experiments with asymmetric flows when the cone axis on the plate was turned by an additional angle α_0 relative to the undisturbed-flow direction. The flowfields in

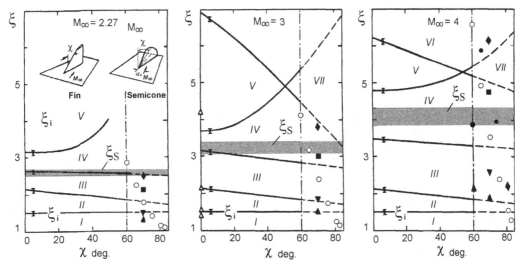

Figure 5.26. The boundaries of STBLI regimes for sharp unswept and swept fins and semi-cone at $Re_{\delta 0} = 1.45 - 2.76) \times 10^5$, $T_w/T_{ad} = 1$: open symbols – symmetric flow over semi-cone ($\alpha_0 = 0$), closed symbols – asymmetric flow ($\alpha_0 > 0$) (Zheltovodov et al. [12]).

regimes I–II (Fig. 5.27) are characterized by an increasing penetration of the semi-cone surface boundary layer onto the plate with increasing α and ξ. This stimulates the appearance of the distinct secondary-convergence line S_2 on the plate surface in regime III. A localized vortex appears close to the junction of the cone surface and the plate between lines S_3 and R_3. Similar flow features were observed in the early stages of interactions in the vicinity of an unswept sharp fin [27].

The flow in regime IV is characterized by the disappearance of secondary-convergence line S_2 at a downstream distance from the semi-cone tip and by the appearance of a distinct primary-convergence line S_1 that is associated with a stage when the fluid starts actively to leave the plate surface. The compression-wave system above the plate has changed into a λ-shock configuration. The onset of developed boundary-layer separation on the plate appears once the shock strength exceeds a second critical value $\xi = \xi_S$ (indicated by black bands in Fig. 5.26a–c). As shown in this figure, ξ_S is constant within experimental uncertainty throughout the range of χ for Mach numbers $M_\infty = 2.27$, 3, and 4. At higher Mach numbers, however, larger ξ_S values are necessary to cause fully developed three-dimensional separation.

As shown in Fig. 5.27b (i.e., regimes IV–VI), divergence line R_1 appears on the semi-cone surface for $\xi > \xi_S$, which is associated with flow attachment. A surface-pressure maximum appears on this line as in the reattachment point below the crest of a two-dimensional forward-facing step (see Fig. 4.13a,b) and on the reattach-ment line below the top of a swept, forward-facing step [20, 62]. The appearance of secondary separation and reattachment (S_2, R_2) underneath the reversed flow is characteristic for regime VI. Regime VII, observed at $M_\infty = 3$ and 4 (see Fig. 5.26), corresponds to conditions when secondary lines S_2 and R_2 emerging at regime VI and located closer to the semi-cone coexist with secondary-convergence line S_2 aris-ing in the vicinity of its top at regime III.

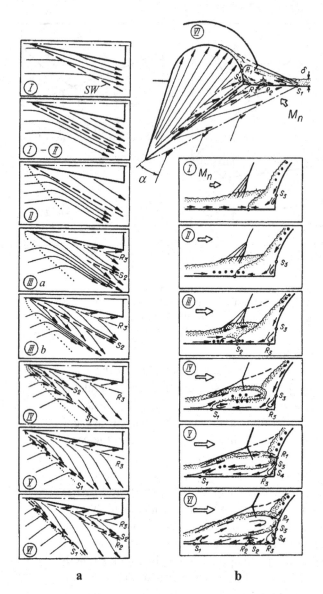

Figure 5.27. Surface-flow pattern and flowfield structure in the vicinity of semi-cone at various interaction regime (Zheltovodov and Maksimov [62]).

Many investigators have suggested suitable scaling laws for quasiconical-swept SWTBLIs to correlate locations of the upstream-influence line, the primary- and secondary-separation lines, and the primary-attachment line (see, e.g., [1, 11, 12, 22, 18, 25, 29, 63, 64, 65, 66]). The key parameters are the free-stream Reynolds number and the incoming boundary-layer thickness at the fin leading-edge location. An inception-length L_i can be defined, beyond which the interaction becomes purely conical (see Fig. 5.7). As shown [63, 64, 65, 66], the upstream influence of swept interactions accurately scales with coordinates tangential and normal to the shock wave if the following nondimensionalization is used:

$$\left(l_n / \delta \, Re_\delta^{1/3} \right)_U / M_n = f \left[\left(l_S / \delta Re_\delta^{1/3} \right)_U \right] \tag{5.4}$$

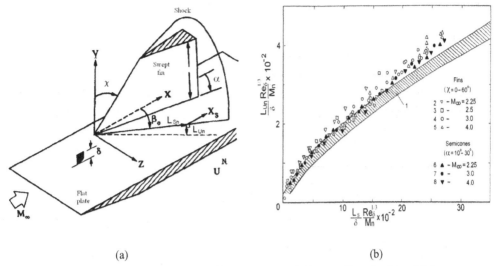

(a)   (b)

Figure 5.28. Scaling of the spanwise extent of the upstream influence line: $1 - M = 2.94$, $\alpha = 12 - 24°$, $\chi = 0$ (Wang and Bogdonoff [65]); 2–5 – fins, 6–8 – semi-cones (Zheltovodov et al. [12]).

or

$$(L_n)_U / M_n = f[(L_S)_U] \qquad (5.5)$$

Here, the scaling $(1/\delta)Re_\delta^{1/3}$ is applied uniformly in two orthogonal directions (Fig. 5.28a) (i.e., along the inviscid shock-wave trace l_s and in the normal direction l_n). M_n is the Mach-number component normal to the shock and Re_δ is the Reynolds number based on the boundary-layer thickness δ at the fin leading edge. Because sharp-fin interactions have no other characteristic length, equations (5.4) and (5.5) apply only to the inception length. Downstream of this inception zone, the flowfield is invariant to any scale transformation. Thus, these equations simplify to the following:

$$(L_i/\delta)_U / M_n \propto 1/Re_\delta^{1/3} \qquad (5.6)$$

Figure 5.28b demonstrates generalized experimental data for unswept and swept fins $(1 - 5)$ and semi-cones $(6 - 8)$ at different Mach numbers. As shown [66] (Fig. 5.29), the inception length to conical symmetry for sharp unswept-fin interaction is weakly dependent on the Mach number and strongly dependent on shock-angle β_0.

The conical angles for upstream-influence β_U, primary-separation β_{S1}, and primary-attachment β_{R1} lines can be described for unswept- and swept-fin interactions by the following empirical relationships [22]:

$$\beta_U - \beta_{Ui} = 1.53(\beta_0 - \beta_{0i}) \qquad (5.7)$$
$$\beta_{S1} - \beta_{S1i} = 2.15(\beta_0 - \beta_{0i}) - 0.0144(\beta_0 - \beta_{0i})^2 \qquad (5.8)$$
$$\beta_{R1} - \beta_{R1i} = 1.41(\beta_0 - \beta_{0i}) - 0.0139(\beta_0 - \beta_{0i})^2 \qquad (5.9)$$

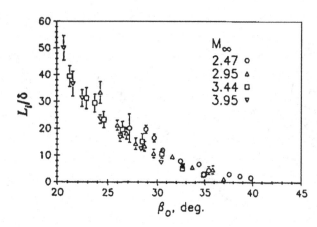

Figure 5.29. Inception length scaling for sharp unswept fins in terms of inviscid shock angle (Lu and Settles [66]).

where β_{U1}, β_{S1i}, and β_{R1i} are the inclination angles of these lines corresponding to incipient-intermittent separation ($\xi = \xi_i$ and, respectively, $\beta_0 = \beta_{0i}$). They can be predicted by the following approximate relationships:

$$\beta_{Ui} = 1.22\beta_{0i} + 3.4 \tag{5.10}$$

$$\beta_{S1i} = \beta_{0i} \tag{5.11}$$

$$\beta_{R1i} = 0.55\beta_{0i} \tag{5.12}$$

The angle of the secondary-separation line for subsonic reversed flow (i.e., regimes III–IV) can be described by the following equation:

$$\beta_{S2} - \beta_{S2i} = 1.6(\beta_0 - \beta_{02i}) \tag{5.13}$$

and for supersonic reversed flow (i.e., regime VI) by the following equation:

$$\beta_{S2} = \beta_0 \tag{5.14}$$

The value β_{S2i} in equation (5.13) corresponds to the first appearance of the secondary-separation line at $\alpha = \alpha_{i2}$ and, respectively, $\beta_0 = \beta_{02i}$ in accordance with condition (5.3) on the boundary between regimes II and III (see Fig. 5.5a). These relationships proposed on a basis of generalization of wide experimental data at $M_\infty = 2-4$ also provide good agreement with experimental data [13, 26] for sharp-unswept fins at $M_\infty = 5$ up to fin-inclination angles $\alpha = 27$ degrees, as shown in Fig. 5.30a. The empirical correlations also can be used with some success for flow over semi-cones [12, 62], as shown in Fig. 5.30b–d.

5.3.4 Swept Compression Ramp Interaction and Its Modeling

The SCR configuration (see Fig. 5.1d) is another widely studied example of three-dimensional SWTBLIs [1, 10, 56]. At small sweep-angles χ, these flows display cylindrical symmetry (Fig. 5.31a) whereas conical symmetry appears with increasing sweepback [67, 68] (Fig. 5.31b). The upstream-influence U, primary-separation S, and reattachment R lines are shown in these figures with the swept-corner line C. The inception zone L_i for the SCR interaction is analogous to the fin case but with a length L_i that varies with α and χ and increases without limit as the cylindrical–conical boundary is approached. In accordance with detailed parametric experimental studies [35, 68], at $M_\infty = 3$, the boundary between the cylindrical and

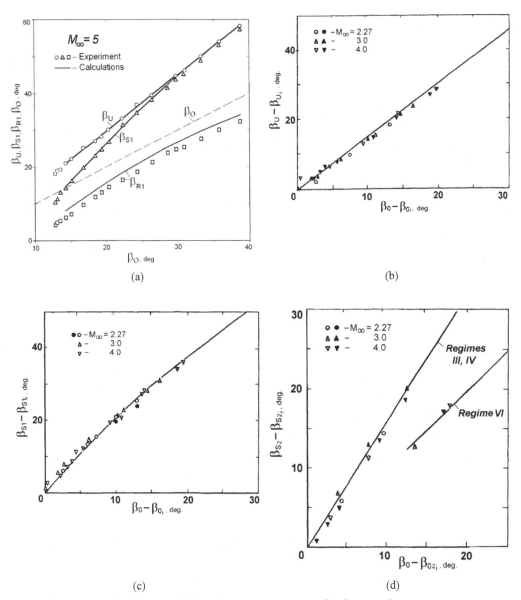

Figure 5.30. Upstream influence (β_U), primary convergence (β_{S1}), secondary convergence (β_{S2}), and primary divergence (β_{R1}) lines scaling for unswept fin (a) (Schülein and Zheltovodov [26]) and semi-cones (b–d) (Zheltovodov and Maksimov [62]).

conical regimes is determined by the inviscid shock detachment stage from a SCR (Fig. 5.32a, solid line); this boundary is invariant with δ_0 and $Re_{\delta 0}$. Thus, cylindrical and conical symmetries correspond to attached and detached shocks, respectively. Different regimes of separation with cylindrical (1) and conical (2, 3) symmetry with reattaching flow on the ramp or the test (plate) surface can be realized in these conditions depending on the values of χ and α (see sketches 1, 2, and 3, respectively, in Fig. 5.32b). The inviscid shock shapes for different test configurations can be correlated by a detachment-similarity parameter [35], as follows:

$$\zeta = (\alpha_n + 38.53)/[M_n(1 - 0.149 M_n)] \tag{5.15}$$

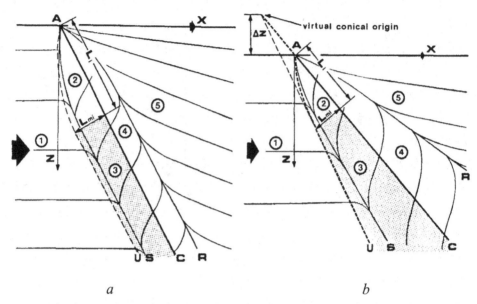

Figure 5.31. Schematic of surface streak lines for asymptotic cylindrical (a) and conical (b) symmetry of flow in the vicinity of swept compression ramp (Settles and Teng [68]).

which is a simplified form of the detachment relationship in oblique-shock theory; it has a constant value of 43.6 for shock detachment from swept wedges at Mach numbers up to $M_\infty = 3$. (Here, α_n and M_n are the shock-generator angles and the Mach number normal to the leading edge of the shock generator.) Using the detachment parameter, Settles [69] generalized the normalized inception lengths in the

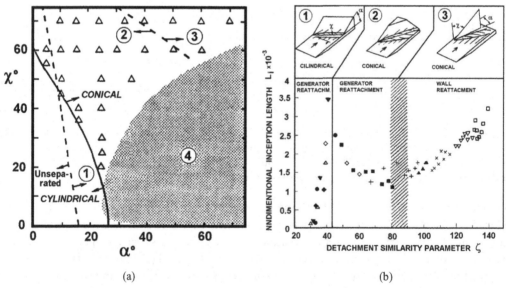

Figure 5.32. Swept compression ramp flow regimes (a) (Settles and Kimmel [35]) and inception length similarity plot (b) (Settles [69]) at $M_\infty = 3$. Cylindrical flow symmetry: 1 – separation with reattachment on the ramp surface; conical flow symmetry: 2 – separation with reattachment on the ramp surface; 3 – separation with reattachment on the wall surface; 4 – inaccessible due to tunnel blockage.

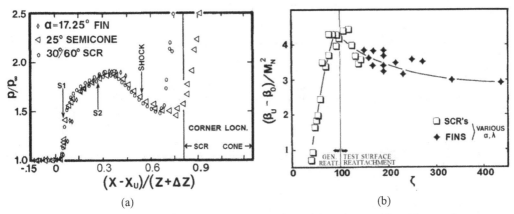

Figure 5.33. Free-interaction surface pressure comparison (a) and normalized interaction response $(\beta_U - \beta_U)/M_n$ vs. detachment similarity parameter ζ (b) for quasiconical interactions (Settles and Kimmel [35]).

form $(L_i/\delta_0)Re_{\delta o}^a = f(\zeta)$, with the empirical constant $\alpha = 1/3$, for quasiconical fins and SCR interactions. A similarity plot (Fig. 5.32b) compares all three members of the fin/swept-corner shock-interaction family in accordance with symmetry type and position of the primary-reattachment line. Transition from a cylindrical (1) to a conical (2) flow regime corresponds to $\zeta = 43.6$; above a critical value $\zeta \approx 100$, it was found that both fin and swept-corner generators produced the same generic flow topology with reattachment on the wall surface.

The same surface-flow patterns for SCRs, fins, and semi-cones are produced when the inviscid shock and upstream-influence-line angles β_0 and β_U have similar values at these conditions [35]. The similarity of surface topography is a consequence of close agreement in surface pressure distributions for these different cases (Fig. 5.33a), which can be considered to illustrate the classic FIC for three-dimensional STBLI cases. This similarity is in agreement with the conclusions of Zheltovodov and Kharitonov [34], who demonstrated the validity of FIC for various three-dimensional STBLIs with cylindrical and conical symmetry by generalizing experimental data for surface pressure distributions in the framework of the two-dimensional analogy in the form $(p - p_\infty)/(p_p - p_\infty) = F(x - x_{S1})/(x_{S1} - x_{R1})$ at different Mach numbers M_∞ and their comparison with similar generalized data for different two-dimensional separated flows. Based on experimental data at $M_\infty = 2.95$, Settles and Kimmel [35] proposed the following empirical relationships for quasiconical interactions:

$$\beta_U = 1.59\,\beta_0 - 8.3 \quad \text{(for semi-cones)} \tag{5.16}$$

$$\beta_U = 1.59\beta_0 - 10.0 \quad \text{(swept and unswept fins and swept corners, } \zeta \ll 100\text{)} \tag{5.17}$$

$$(\beta_U - \beta_0)/M_n^2 = f(\zeta) \,\text{(swept corners, } \zeta \leq 500\text{)} \tag{5.18}$$

The similarity among the flows produced by different shock generators is demonstrated in Fig. 5.33b, which also reveals the functional dependency f in equation (5.18).

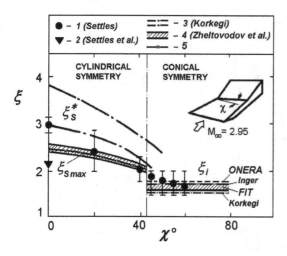

Figure 5.34. Prediction of incipient separation in swept compression ramp at $M_\infty = 2.95$, $Re_{\delta o} = 2.88 \times 10^5$, $T_w/T_{wa} \approx 1$. Experiment: 1 – Settles (in accordance with Korkegi [70]), 2 – Settles et al. [71], Settles [72]; predictions: 3 – Korkegi [70], 4 – with using FIT, Zheltovodov et al. [12], 5 – with using correlation by Zukoski [30].

A theoretical prediction of incipient separation in SCRs is of considerable practical interest. Such a prediction can be made using two-dimensional analogies by considering the cross section normal to the inviscid shock trace and the corresponding normal Mach number M_n. Figure 5.34 compares experimental data of Settles (1) for incipient separation on SCRs at $M_\infty = 2.95$, which were used by Korkegi [70] for comparison with prediction (i.e., band 3) on the basis of his empirical correlation [15] for two-dimensional CRs (see band 14 in Fig. 4.19). As shown previously, this correlation characterizes the critical angle α_S* and, respectively, the critical oblique shock-wave strength ξ_S* corresponding to large-scale incipient separation in a two-dimensional CR flow. Although the predicted trend (i.e., band 3) in the framework of the two-dimensional analogy is correct in conditions of cylindrical symmetry (see Fig. 5.34), data point 1 is located below this band. Moreover, such a prediction cannot be applied in conditions of conical symmetry. As suggested by Zheltovodov et al. [12], a similar prediction can be improved significantly in the framework of the classic FIT by considering the type of flow symmetry (i.e., conical or cylindrical) depending on the value of χ. Here, equation (4.15) was used with the empirical constant $k = 7.4$ to predict the critical Mach number M_i and the critical normal-shock-wave strength ξ_i in the range of the Reynolds numbers $Re_\theta = (1.4 - 2.3) \cdot 10^4$ (see Fig. 5.10a) corresponding to incipient-intermittent-separation conditions (i.e., Stanbrook's regime) in the region of conical symmetry (see Fig. 5.34, band 4). In the region of cylindrical symmetry, the critical strength of oblique shock wave $\xi_{s\,max}$ corresponding to the onset of developing small-scale separation in two-dimensional CRs at $\alpha = \alpha_{s\,max}$ on the boundary between regimes 2 and 3 (see Fig. 4.19) was calculated using condition (4.18) with FIT equation (4.14) (see Chapter 4) for the plateau pressure ratio for the corresponding values $M_n = M_\infty \cos\chi$. (The plateau pressure ratio versus M_n predicted by the FIT is denoted in Fig. 5.10a by two solid lines 1). As shown in Fig. 5.34, such calculations yield a better fit to the experiments than the value $\xi = \xi_S*$ according to Korkegi [15]. Indeed, in the region of conical symmetry, the experiments demonstrate a tendency toward constant critical shock strength when considering the measurement accuracy, which also is predicted by ONERA's correlation, Inger's theory, and the empirical condition proposed by Korkegi [15] ($\xi_i = 1.5$) for swept shock waves in the vicinity of sharp fins

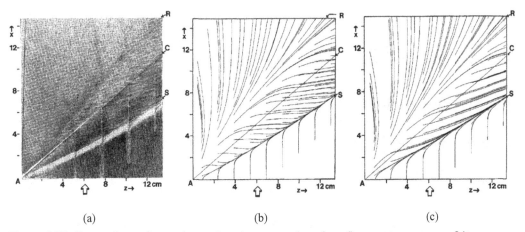

(a)   (b)   (c)

Figure 5.35. Comparison of experimental and computed surface flow pattern at $\alpha = 24°$, $\chi = 40°$, $M_\infty = 2.95$, $Re_{\delta_0} = 9.39 \times 10^5$, $T_w/T_{wa} \approx 1$: b – algebraic (Cebeci-Smith) model, c – two-equation (Jones-Launder) model (Settles et al. [73]).

(see Fig. 5.11). In the region of cylindrical symmetry at $\chi > 0$, the calculated values $\xi_{S\,max}$ also coincide with the experiment (1) (Fig. 5.34, band 4). The calculated values $\xi_{inv} = \xi_{S\,max}$ at the higher Reynolds numbers $Re_\theta > 2.3 \cdot 10^4$ with using equation (4.18) with empirical correlation by Zukoski [30] for the "plateau" pressure ratio $p_p/p_\infty = 0.5M + 1$ (line 5) correspond to the data (1) and predictions (4) on a basis of FIT. At $\chi = 0$, the top experimental point 1 corresponds to two-dimensional, large-scale incipient-separation conditions at $\alpha = \alpha_S^*$ (i.e., point 7; see Fig. 4.19), which agrees with Korkegi's predictions. Also shown in Fig. 5.34 is an additional point 2 of Settles et al. [71, 72] that corresponds to the incipient, steady small-scale separation, which is in good agreement with the predicted $\xi_{S\,max}$ value.

Figure 5.35 demonstrates a comparison of experimental (a) and RANS-predicted surface flow patterns [73] with the algebraic (b) and the k-ε turbulence models in a SCR vicinity in conditions of conical-interaction regime 2 (i.e., $\alpha = 24$ degrees, $\chi = 40$ degrees, $M_\infty = 2.9$; see Fig. 5.32) when separated flow attaches to the ramp surface. The calculations with both turbulence models are in good agreement with the experiments. Flowfield surveys were conducted for two δ_0 values differing by 3:1 to check the accuracy of the following Reynolds-number similarity law proposed by Settles and Bogdonoff [63]:

$$(x_w/\delta_0)Re_{\delta_0}^a = f\left[(z_w/\delta_0)Re_{\delta_0}^a\right] \quad (\alpha, \lambda, M_\infty \text{ fixed}) \tag{5.19}$$

Equation (5.19) was formulated in terms of x and z for footprint scaling; by replacing z with y, it also can be applied for flowfield scaling. This comparison is shown in Fig. 5.36a, which demonstrates excellent agreement between elements of the flowfield structure at different boundary-layer thicknesses. The calculations demonstrated qualitative agreement with the similarity law; however, some measured features of the flow were poorly predicted. In accordance with Fig. 5.36b, the scaling law works for the surface flow pattern and is supported by experiments and computations, although the computations tend to underpredict the conical asymptote of upstream influence.

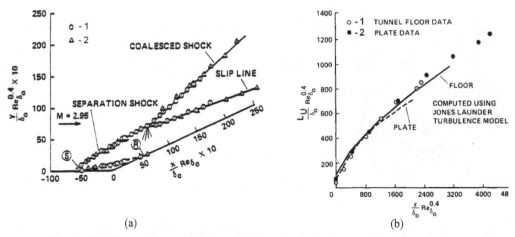

(a)   (b)

Figure 5.36. Reynolds number scaling (Settles et al. [73]): a – of floor and plate interaction structure, b – of measured and computed upstream influence line. $1 - \delta_0 = 1.494$ cm, $Re_{\delta_0} = 9.39 \times 10^5$, $2 - \delta_0 = 0.429$ cm, $Re_{\delta_0} = 2.7 \times 10^5$.

Figure 5.37 shows results of experimental and computational investigations [74] of STBLI flows in the vicinity of SCR corners $(\alpha, \chi) = (24$ and 40 degrees$)$ and $(24$ and 60 degrees$)$ at $M_\infty = 3$, $Re_{\delta_\infty} \approx 9 \times 10^5$. RANS computations with four different turbulence models were used (i.e., B-L, Cebeci-Smith, Jones-Launder with WF, and Jones-Launder ITW). The calculated flowfields show general agreement with experimental data for surface pressure and good agreement with experimental flowfield pitot pressure and yaw-angle profiles.

A general flowfield model for SCR corners is presented in Fig. 5.38 and shows a time-averaged streamline pattern. The actual flowfield is unsteady. The line of coalescence defines a boundary of three-dimensional separation-surface 1, which spirals around the main vortical structure. The flow in the separation-surface vicinity exhibits a large yaw angle, and the streamlines are strongly skewed in the spanwise

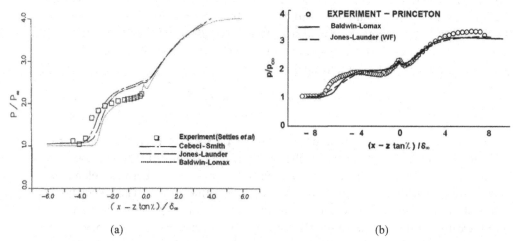

(a)   (b)

Figure 5.37. Computed and experimental surface pressure for $\alpha = 24°$ at $Re_{\delta_0} = 9.39 \times 10^5$, $T_w/T_{wa} \approx 1$: a – $\chi = 40°$, b – $\chi = 60°$, $M_\infty = 2.95$, $Re_{\delta_0} \approx 9 \times 10^5$: showing Cebeci-Smith turbulence model, and Jones-Launder model (Knight et al. [74]) respectively.

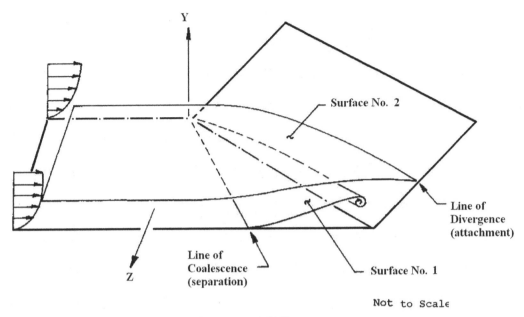

Figure 5.38. Mean streamline model (Knight et al. [74]).

direction. The line of divergence defines the intersection of the second surface (2) with the SCR. The fluid contained between the wall and surface 2 is entrained in the vortical flow. The fluid above surface 2 passes over the vortical structure and continues up the CR. As shown, the general features are similar to those observed for the sharp fin (see Fig. 5.19).

5.3.5 Double Sharp-Fin Interaction

The crossing STBLIs generated by double symmetric ($\alpha_1 = \alpha_2$) and asymmetric ($\alpha_1 \neq \alpha_2$) sharp fins (see Fig. 5.1f) were widely studied experimentally and using RANS simulations (see, e.g., Zheltovodov [10, 21], Knight et al. [52], Knight and Degrez [57]). The primary aims of this research are to gain insight into various physical properties and to develop turbulence models capable of providing accurate predictions of flowfield structure and aerothermodynamic loads.

Figure 5.39 compares an experimental oil-flow visualization on the bottom plate [75] (a) with computed skin-friction lines using the Jones-Launder k-ε model [76, 77] (b) and the standard Wilcox k-ω model [78, 79, 80] (c) for a symmetric $\alpha_1 \times \alpha_2$ $= 7 \times 7$-degree double-fin configuration at adiabatic-wall conditions. In this weak-interaction case, the experimental and numerical surface-flow patterns are in good agreement. The limiting streamlines begin to curve in the vicinity of upstream-influence line U (Fig. 5.39a) and asymptotically approach one another to form a narrow region of parallel flow in the direction toward the channel axis (i.e., throat middle line [TML]). Some streamlines penetrate downstream in the symmetry-axis vicinity and converge to form a characteristic throat. The flow passes through this fluidic throat without any signs of separation. The interaction of intense secondary flows, propagating from divergence lines R_1 and R_2 and the flow passing through the throat, results in the formation of secondary-convergence lines S_3 and S_4.

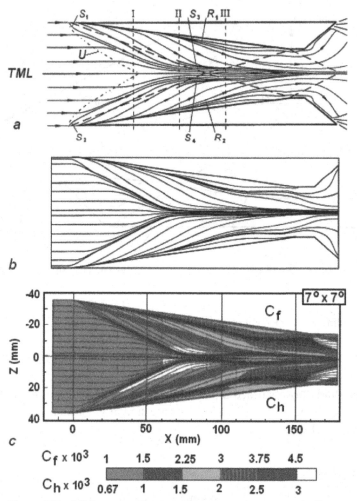

Figure 5.39. Oil-flow visualization (a) (Zheltovodov et al. [75]), and computed skin-friction lines (b) $-k - \varepsilon$ model (Zheltovodov et al. [76, 77]) and (c) $-k - \omega$ model (Thivet et al. [78, 79, 80] for $7° \times 7°$ case at $M_\infty = 4$, $Re_{\delta\infty} = 3.1 \times 10^5$.

Figure 5.40 compares the computed distributions of surface pressure and heat-transfer coefficients using the Wilcox k-ω model with the experimental data along the axis of symmetry (i.e., TML) and in three cross sections denoted, respectively, as I, II, and III (see Fig. 5.39a). The heat-transfer measurements were carried out at $T_w/T_{aw} = 1.04 - 1.05$. Again, the computations on different grids are in good agreement with the measurements.

In accordance with the experiment at $\alpha_1 \times \alpha_2 = 11 \times 11$ degrees (Fig. 5.41a), the throat formed between lines S_1 and S_2 becomes narrower. The flow expanding behind this throat and directed along the centerline is terminated above the point of the first intersection of the inviscid shocks (denoted by dotted lines), with a subsequent formation of central-longitudinal-divergence line R_3 of finite length along the flow. This divergence line gradually degenerates to a dividing streamline in the region of transverse-flow intersection. Secondary-convergence lines S_3 and S_4 propagate more upstream toward the fin vertices than in the previous case. The flow

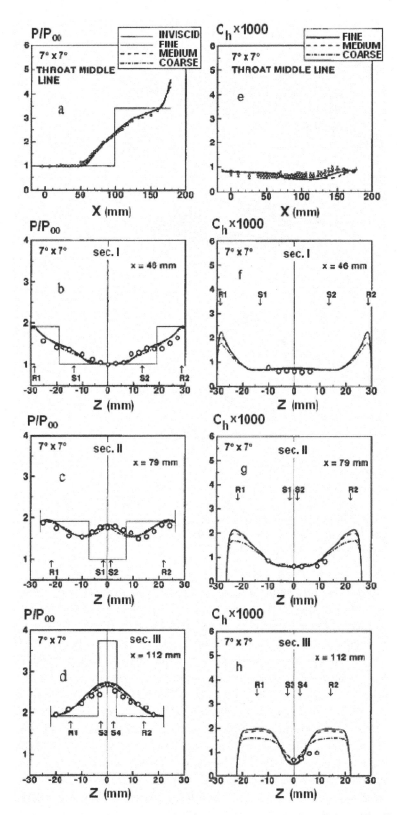

Figure 5.40. Surface pressure (a–d) and heat-transfer coefficient distributions (e–h) for $7° \times 7°$ case and their prediction with the k-ω model at $M_\infty = 4$, $Re_{\delta\infty} = 3.1 \times 10^5$ (Thivet et al. [78, 79, 80]).

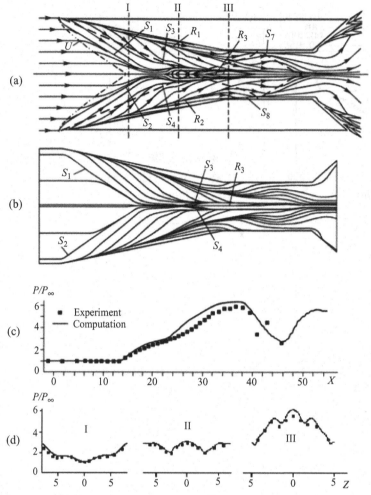

Figure 5.41. Oil flow visualization (a) (Zheltovodov et al. [75]) and computed with the k-ε model (Zheltovodov et al. [76, 77]) skin-friction lines (b) and surface pressure distributions (c, d) for $11° \times 11°$ case at $M_\infty = 4$, $Re_{\delta\infty} = 3.1 \times 10^5$.

character at the constant-width channel entrance is partly determined by the influence of expansion waves propagating from the points of inflection at the fin-side surfaces. The influence of these expansion waves favors a more intense motion of the near-wall flow toward the fin surfaces. The crossing shock waves reflected from the fins' surfaces initiate an opposing pressure gradient and cause a deflection of the limiting streamlines toward the channel centerline with the formation of convergence lines S_7 and S_8. The shock wave reflected from the fin surfaces initiates an opposing pressure gradient and causes a deflection of the limiting streamlines to the channel centerline. The limiting streamlines calculated with the k-ε model (Fig. 5.41b) reproduce qualitatively the characteristic features of the flow considered. At the same time, similar to the previously considered case, a significantly smaller width of the flow penetrating through the throat in the vicinity of the centerline and the regions of the flow diverging from divergence line R_3 are worth noting. Secondary-convergence lines S_3 and S_4 appear in the calculations farther downstream than in the experiment. Comparison of numerical and experimental data for pressure

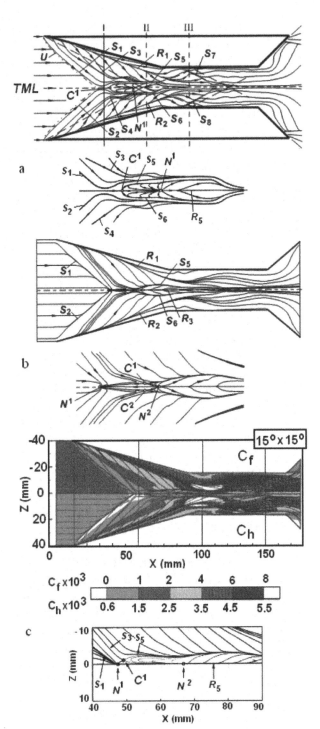

Figure 5.42. Oil-flow visualization [75] (a) and computed skin-friction lines (b) with the k-ε model [76, 77]) and (c) k-ω model [78, 79, 80]) for $15° \times 15°$ case at $M_\infty = 4$, $Re_{\delta\infty} = 3.1 \times 10^5$.

distribution along the channel centerline (Fig. 5.41c) and in cross sections I, II, and III (Fig. 5.41d) shows good agreement for $X = x/\delta_0 < 24$. For $X > 24$, however, the calculations tend to overestimate the value of the pressure level near the centerline at $24 < X < 40$. The calculated and measured pressure distributions in cross sections I, II, and III (Fig. 5.41d) are in good agreement.

Experimental and computed surface flow patterns for a stronger-shock, 15×15–degree fin-angle case are shown in Fig. 5.42. The experiment (Fig. 5.42a) shows the appearance of definite primary-separation and attachment lines (S_1 and S_2 and R_1 and R_2, respectively). Secondary-separation lines S_3 and S_4 become more distinct under the influence of strong convergence of the secondary flow propagating from lines R_1 and R_2. A large-scale separation region is formed in the throat behind the centerline singular saddle-point C^1, bounded by separation lines S_5 and S_6 (see enlargement in Fig. 5.42a). Centerline nodal-point N^1 was observed downstream of saddle-point C^1. Secondary-convergence line S_3 is almost merging with S_5 (as is S_4 with S_6 on the opposite side) approximately in the middle of the central separated zone. The computed-flow topologies around the separated region are different for this case (Fig. 5.42b,c), which is shown more clearly in the enlargements. Instead of the combination of saddle and node points (i.e., C^1 and N^1), another combination of the centerline singular points appears including two nodes (N^1 and N^2) and two saddle points (C^1 and C^2) located symmetrically about the centerline. The rule (i.e., equation 5.1) governing singular points requiring equality of nodes and saddle points is satisfied in these conditions. Moreover, the calculations do not show any secondary-separation lines S_3 and S_4; the central separated zone is more compressed by the flows propagating from the fins; and it is noticeably smaller in width compared to the experiment. When comparing the two turbulence models, the Wilcox k-ω model predicts a similar combination of singular points but exhibits a better prediction of the width of this zone (Fig. 5.42c). As shown in Fig. 5.43a–d, the computations using the Wilcox k-ω model provide good quantitative agreement with surface pressure data except for a region in the vicinity of the channel centerline (60 mm $< x < 110$ mm), where the computed surface pressure is noticeably overestimated. A similar tendency was observed in computations using the k-ε model [76, 77]. A comparison of surface-heat-transfer coefficients is shown in Fig. 5.43e–h. As shown in Fig. 5.43e, the heat-transfer maximum on the TML is overpredicted by the calculations (using the k-ω model) up to 2.5 times above the experimental data. Substantial differences are seen also in cross sections I–III (Fig. 5.43f–h), the locations of which are shown in Fig. 5.42a.

Thivet et al. [79] suggested that two-equation turbulence models overpredict the growth of TKE in the outer part of the boundary layer as it crosses the shock wave. This excessive TKE is then convected to the wall by the vortices downstream of the shock wave, causing an overprediction of the surface-heat transfer. Various attempts to decrease the TKE level by limiting coefficients α_ν and C_μ in the viscosity equation $\mu_t = \alpha_\nu \, C_\mu \, \rho \, (k/\omega)$ were suggested [81]. Four models were tested [79] on crossing shock-wave interactions – namely, Wilcox–Moore (WM), Wilcox–Durbin (WD), and their modifications (WM$^+$) and (WD$^+$). As shown in Fig. 5.44, in the case of the WM$^+$ solution on the fine grid, the correction significantly changes wall-pressure and heat-transfer distributions. There is obvious improvement in the heat-transfer prediction, and the width of the central separation has increased. Nevertheless, considerable discrepancy remains between computations and experiments in surface pressure and heat transfer.

The computations of Panaras [53] using his modification of the B-L turbulence model demonstrate good agreement with experimental data by Settles et al. [56] in the surface flow pattern (Fig. 5.45a,b), surface pressure distribution (c), and

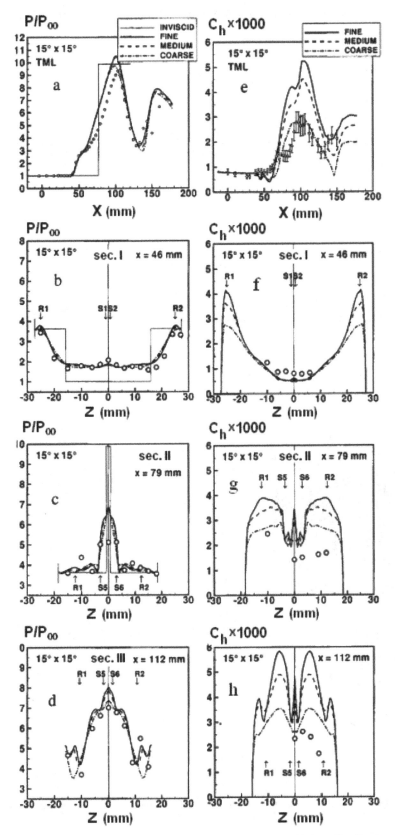

Figure 5.43. Surface pressure (a–d) and heat-transfer coefficient distributions (e–h) and their prediction using the k-ω model for $15° \times 15°$ case at $M_\infty = 4$, $Re_{\delta\infty} = 3.1 \times 10^5$ (Thivet et al. [78, 79, 80]).

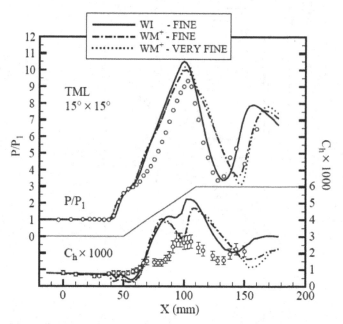

Figure 5.44. Comparison of experiment and computations for surface pressure and heat-transfer coefficient along TMP for $15° \times 15°$ case at $M_\infty = 4$, $T_w/T_{adw} = 1.05$ (Thivet et al. [79]).

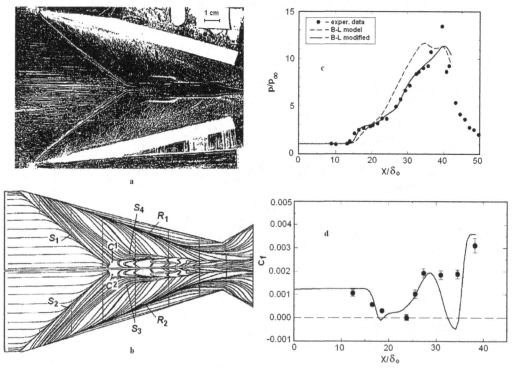

Figure 5.45. Prediction of double-fin interaction for $15° \times 15°$ case at $M_\infty = 3.98$, $Re_{\delta o} = 2.66 \times 10^5$ (Panaras [53]).

skin-friction coefficient distribution (d) along the centerline as compared to the standard B-L model (see Fig. 5.45c). The model provides a mixed-type separation flow in the vicinity of every isolated fin: turbulent in the region of separation lines (S_1 and S_2; Fig. 5.45b) and almost laminar between reattachment lines (R_1 and R_2) and corresponding secondary-separation lines (S_3 and S_4). This yields significant improvement in predicting crossing shock-wave-interaction flow between two fins. However, despite the improvement, the unsteady nature of the crossing SBLI flow is not captured by steady RANS computations and may be an additional reason for the observed discrepancies in the predicted flow topology, surface pressure, and heat-transfer level compared to the experiments. In accordance with the experimental research of Batcho et al. [82], in conditions of a weak 7×7–degree interaction at $M_\infty = 2.95$, the rms surface pressure fluctuations increase along the centerline, reach a constant value before the theoretical shock-wave crossing point, and are relatively constant downstream of this point in the remainder of the investigated region. In contrast, for the 11×11–degree case, after the initial rise, the large rms surface pressure fluctuation peak lies behind the theoretical shock-wave crossing point and is followed by a decrease and then another increase. The unique measurements of surface temperature fluctuations in the vicinity of the large rms pressure-fluctuations peak support the assumption that the pressure fluctuations are connected to surface temperature and heat-transfer fluctuations. The unsteady nature of these interactions is supported by additional surface pressure-fluctuation measurements performed by Poddar and Bogdonoff [83] at $M_\infty = 2.95$ and by Davis and Hingst [84] at $M_\infty = 2.95$. Thus, LES and Direct Numerical Simulation (DNS) computations of crossing shock-wave interaction are important for the analysis of unsteadiness effects at such conditions on surface heat-transfer, skin-friction, and pressure-distribution predictions.

Figure 5.46 is a three-dimensional perspective of the complex crossing shock-wave structure realized in the vicinity of a 15×15–degree double-fin configuration at $M_\infty = 3.83$ in accordance with the experimental research of Garrison and Settles [85] using PLS images. Considering the inherent symmetry of the crossing shock-wave interaction, only half of the flowfield structure is shown. Initial cross section I represents a flow structure typical of the single-fin interaction case. The incident-separation (1), rear (2), and inviscid (3) shocks with a slip line (4) penetrating from the triple point to a fin-side/plate cross line, as well as a separation vortex (5) under the bifurcated shock system, are visible in this cross section. To understand the intersection of the two separate single-fin interactions, the vertical plane of symmetry is considered an inviscid-reflection plane. For the symmetrical crossing shock-wave interaction, shock waves that intersect this plane must reflect from it to satisfy continuity. As shown, in cross section II, the incident-separation shock-wave reflection from the symmetry plane is an irregular (i.e., Mach) reflection. The Mach reflection results in a straight shock-wave segment, the Mach stem (7), which spans the interaction centerline, a reflected portion of the separation shock (6), and the newly formed triple point (10). In accordance with the perspective view and cross section III, the entire incident λ-shock structure reflects from the center plane in an irregular manner and remains intact (although somewhat distorted), propagating away from the center toward the fin surface. As shown in cross section III, two additional shock-wave segments and two triple points are observed to form as a result of this

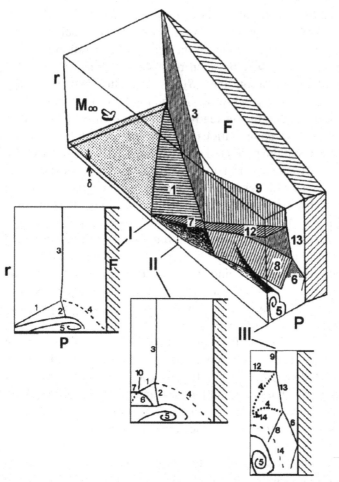

Figure 5.46. Double-fin flow structure at $\alpha_1 \times \alpha_2 = 15° \times 15°$, $M_\infty = 3.83$ (Garrison and Settles [85]).

crossing. One centerline segment (12) spans the interaction centerline between the reflected inviscid shock waves (9) and the second "bridge" segment (13), and then connects the reflected inviscid shock (9) to the reflected separation (6) and rear (8) shocks. A "mushroom-shaped" separation structure (5) is observed in the centerline vicinity in the last-shown cross section of the twin, single-fin separations that begin the crossing shock-wave interaction.

Elements of the effect of interaction strength on the flowfield structure are illustrated in Fig. 5.47a–d in accordance with different interaction regimes (see Figs. 5.39b, 5.41b, and 5.42b) predicted at $M_\infty = 4$ on the basis of the k-ε model [76, 77, 86, 87, 88]. Select streamlines shown in the symmetry plane normal to the plate may be interpreted as intersections of stream surfaces with this plane. In the case of a weak interaction (i.e., 7×7–degree, Fig. 5.47a), the fluid in the incoming thickening boundary layer moves away from the surface under the action of the opposite pressure gradient, which occurs without any sign of separation and reverse flow. With increasing interaction strength (11×11–degree, Fig. 5.47b), line S_2 is strengthened. However, near the plane of symmetry, it turns partially parallel to a line of

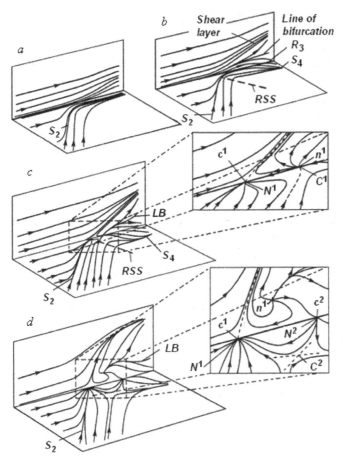

Figure 5.47. Evolution of the computed separation topology with increase in interaction strength (Zheltovodov et al. [77]; Schmisseur et al.[87]).

symmetry and all upstream lines are squeezed into the narrow channel between line S_2 and the line of symmetry. On the downstream side of Line S_2, each line is turned streamwise prior to reaching the symmetry plane under the influence of the reflected separation shock (RSS). These lines then form secondary-separation line S_4. On the symmetry plane, the pattern shows the lifting of the incoming boundary layer and the formation of a shear layer. Beneath this forms a line of divergence/bifurcation (LB), which can be viewed as denoting the approach of fluid from the sides toward the symmetry plane. Fluid below the LB attaches to the plate at the line of symmetry at R_3 and forms the longitudinal centerline vortices.

Increased interaction strength produces the first pair of critical points on the line of symmetry with zero shear stress on the surface (Fig. 5.47c). The primary line of separation S_2 terminates at node N^1 on the line of symmetry. Saddle-point C^1 is formed downstream. All lines of separation and attachment have numerical designators subscripted, critical points have designators superscripted, and critical points on the symmetry plane are designated by lowercase letters. On the symmetry plane, critical point N^1 corresponds to saddle point c^1, whereas saddle-point C^1 is associated with node n^1, which acts as a source of all fluid beneath the separated boundary layer. The LB now emanates from this node and again can be interpreted as denoting the approach of fluid from the sides toward the symmetry plane. The bifurcation corresponding to the 15×15–degree case is shown in Fig. 5.47d. Four critical points arranged in a diamond-shaped pattern are observed on the plate. Downstream-saddle-point C^1 observed in the previous regime bifurcates into saddle-point C^2 off the line of symmetry; its mirror image C^1 on the other side under the influence of intensive reversed flow penetrates from node N^2 to node N^1. This reversed flow replaces node n^1 on the symmetry plane, which forms an interior stagnation point as a focus/node above the centerline. Correspondingly, on the symmetry plane, point n^1 is replaced by saddle-point c^2.

Schülein and Zheltovodov [13, 26] performed experiments with crossing shock-wave interactions at $M_\infty = 5$ to identify important details of the topology evolution in the real flow. A very liquid mixture of mineral oil with lampblack and oil paint was used for this visualization. The movement of the mixture particles along the limiting streamlines was recorded using a video camera, which provided the possibility of analyzing the dynamics of the surface flow topology development. Increased fragments of the surface flow pattern and their corresponding schemes in the vicinity of the central separated zone "apex" specify the stages of the flow-topology development at different double-fin inclination angles (Fig. 5.48). The formation of two symmetric primary-separation lines S_1 and S_2, two secondary-coalescence lines S_3 and S_4, and the centerline divergence line R_3 was fixed at the 16×16–degree case (Fig. 5.48a). Centerline saddle-point C^1 exists downstream of the liquid throat between lines S_1 and S_2 in the region shaded in the sketch, with node N^1 located downstream (not shown), as in the 15×15–degree case at $M_\infty = 4$ (see Fig. 5.42a). The appearance of central–cross-separation line S_0 closing the liquid throat and located downstream of the small central separation zone was registered in the 17×17–degree case (Fig. 5.48b). This local zone is limited both upstream and downstream by two centerline saddle-points, C_0^1 and C_0^2. Two additional symmetric nodes, N_0^1 and N_0^2, are located off the centerline, and the rule (i.e., equation 5.1) governing critical points requiring equality of nodes and saddle points is satisfied under this configuration considering downstream node N^1. The size of the central-separation zone increases in the 17.5×17.5–degree interaction case, but the general surface flow topology does not change (Fig. 5.48c).

The next photograph and scheme for the 18×18–degree crossing-shock-wave interaction case (Fig. 5.48d) displays the appearance of reversed flow from downstream-located centerline-node N^1 to saddle-point C_0^1 in the center of cross-separation line S_0. To show all significant features including nodal point N_1, the scale of this figure is decreased approximately twice compared to previous figures. Centerline saddle-point C^1 bifurcates into a pair of symmetric saddle points C^1 and C^2 under the influence of such active reversed flow directed along central-convergence line S_c. In a similar way, centerline saddle-point C_0^2 bifurcates into a pair of symmetric saddle points, C_0^2 and C_0^3. These points are drawn into the separation zone and located between node N_0^2 and focus N^2 as well as between node N_0^1 and focus N^3, respectively. Similar large-scale topological structures are observed in experiments with crossing SWTBLIs in conditions of flow around two parallel, conically sharpened bodies of revolution located above a plate [89] (Fig. 5.49). The figure specifies details of the surface flow pattern in the vicinity of saddle-points C_0^2 and C_0^3. Some

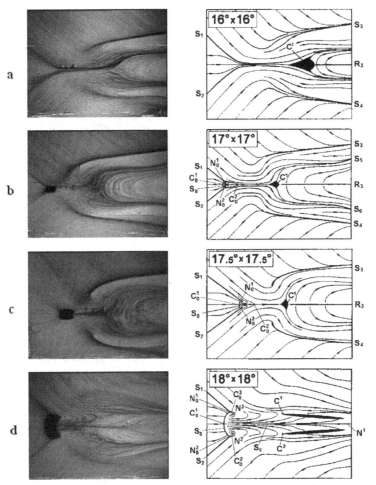

Figure 5.48. Surface flow-pattern transformation with increase in interaction strength at $M_\infty = 5$, $Re_{\delta_0} = 1.4 \times 10^5$, $T_w/T_{adw} = 0.76$ (Schülein and Zheltovodov [13, 26]). (Image d about half the scale of a–c).

Figure 5.49. Surface flow topology on a plate under two conically sharpened parallel bodies of revolution at $M_\infty = 4$ (Derunov et al. [89]).

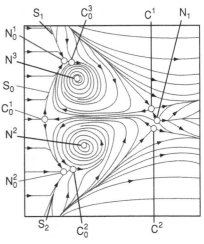

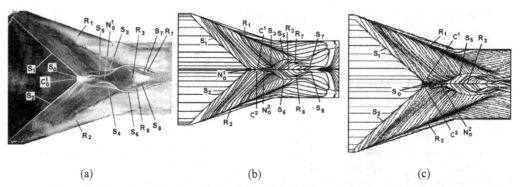

(a)   (b)   (c)

Figure 5.50. Surface flow pattern for $18° \times 18°$ at $M_\infty = 5$, $Re_{\delta o} = 1.5 \times 10^5$, $T_w/T_{wa} = 0.8$: a – experiment (Schülein and Zheltovodov [13, 26]), b – k-ε model (Gaitonde and Schmisseur [76, 87, 88]), c – B–L model modified by Panaras [55].

notation is changed in the figure from that used in [89] in accordance with Fig. 5.48d. Topological rule (i.e., equation 5.1) is satisfied again under this new configuration.

A comparison between the experiments of Schülein and Zheltovodov [13, 26] and the RANS computations of Gaitonde and Schmisseur using the k-ε model (see [76, 87, 88]), as well as with the computations of Panaras with his modification of the B-L model [55], is shown in Fig. 5.50 for strong 18×18–degree CSTBLIs at $M_\infty = 5$. The predicted surface flow patterns (Fig. 5.50b,c) are in general good agreement with the experimental portrait (Fig. 5.50a). However, the first center-line nodal point N_0^1 is predicted using the k-ε model in the apex of the separation zone (Fig. 50b) instead of the central saddle-point C_0^1 and two symmetric nodes N_0^1 and N_0^2 on cross-separation line S_0 observed in the experiment (see Figs. 5.48d and 5.50a), as well as symmetric saddle-points C_0^2, C_0^3 and nodes N^2, N^3 emerging in the confluence of small forward and large-scale downstream separation zones. This difference is associated with an overpredicted turbulence level by this model, which stimulates more active compression of the central separation zone by the contrary-directed near-wall flows that spread from primary-reattachment lines R_1 and R_2 to the centerline. Computations using the modified B-L model demonstrate better prediction of the flow details observed in the experiments (Fig. 5.50c). Both models predict symmetric saddle-points C^1 and C^2 in agreement with the experiments (see Fig. 5.48d). The secondary-separation lines are not observed in experiments and in computations in the flows directed to the centerline between primary-reattachment (R_1, R_2) and separation (S_1, S_2) lines in agreement with regime V (see Fig. 5.5) for the isolated fin. Computations using the k-ε model [76, 87, 88] and the modified B-L model of Panaras [55] predict well the surface pressure distributions in different cross sections. However, they overpredict the level in the centerline vicinity (Fig. 5.51), although the computations using the modified B-L model are again in better agreement with the experiments (Fig. 5.51b).

The comparison between the experiments and RANS computations is shown in Fig. 5.52 for a strong 23×23–degree CSTBLI at $M_\infty = 5$. The surface-oil-flow photograph (Fig. 52a) demonstrates a significant growth of the central separation zone and increased width of the cross separation line at the apex with a middle saddle-point C_0^1. The flow properties downstream of this line are qualitatively similar to the 18×18–degree case (see Fig. 5.50a) and the RANS computations using the k-ε

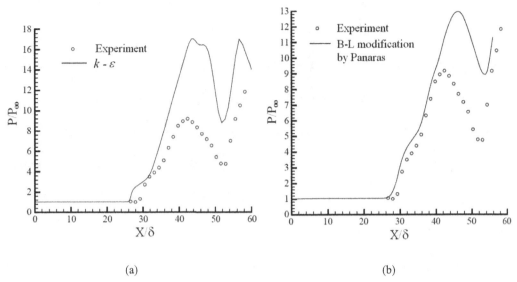

Figure 5.51. Surface pressure centerline distribution for $18° \times 18°$ interaction at $M_\infty = 5$.

and modified B-L models (Fig. 5.52b,c) to reproduce centerline saddle-point C^1_0, side-node N^1_0 and its mirror image, and downstream saddle-point C^1 with its mirror image, as well as centerline node N_1. In accordance with regime VI (see Fig. 5.5), secondary-separation S_2 and reattachment R_2 lines emerge between the primary-separation and reattachment lines S_1 and R_1; this phenomenon is predicted by both of the turbulence models. Downstream, lines S_2 and R_2 terminate in a node–saddle combination (i.e., N^4–C^4) as shown in the enlargement in Fig. 5.53a with additional details of the flow topology in the symmetry plane. The three-dimensional structure of the flow in the separation- and attachment-line vicinity, as well as different singular points, was characterized through analysis of paths of theoretical particles released in the flowfield [87, 88]. By joining the paths of carefully selected particles into ribbons identifying stream surfaces (i.e., ribbon plots), the complex flowfield is clearly described (Fig. 5.53b). The incoming boundary layer separates along the

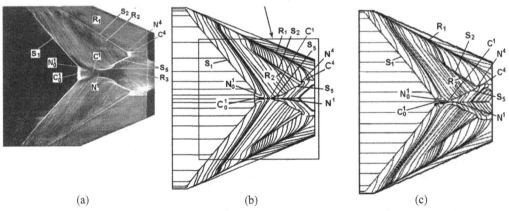

Figure 5.52. Oil-flow visualization [13, 26] (a) and computed surface streamlines with k-ε model [76, 87, 88] (b), and modified B–L model [55] (c) for $23° \times 23°$ CSW interaction at $M_\infty = 5$.

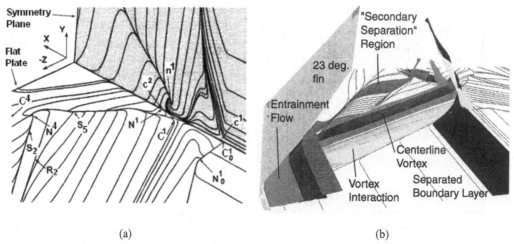

Figure 5.53. Flow topology [76, 87] (a) and ribbon plot [87, 88] (b) prediction on a basis of the k-ε model, for $23° \times 23°$ CSW interaction at $M_\infty = 5$.

primary-separation line and does not reattach. The vortex-interaction region contains the fluid that attaches near the fin and sweeps spanwise to fill the void left by separated boundary-layer fluid. The boundary-layer and vortex-interaction flows separate on either side of the primary-separation line. Two vortex filaments (only one is shown in the figure) leave the surface at the two foci on either side of the central cross-separation line. Fluid originating near the fin leading edge, which is swept spanwise to the symmetry plane before turning downstream, comprises the central vortex region. The high-energy entrainment flow from the inviscid region near the fin leading edge attaches downstream and sweeps toward the centerline before separating. Separation identified with line S_5 (Fig. 5.53a) is associated with a shock-wave feature that can be traced to the system originating at the opposite fin. The entrained flow penetrating as a wall-jet-like structure to the symmetry plane is separated by this shock wave. Secondary-separation line S_2 terminates at downstream foci N_4, where a vortex filament leaves the surface. A similar second vortex filament forms on the opposite side of the interaction region.

As shown in Fig. 5.54 and similar to previous examples, surface pressure along the TML again is overpredicted by calculations of the 23×23–degree interaction case. Although some improvement can be achieved using the modified B-L turbulence model, the discrepancy apparently is not accidental but rather the result of limitations of the RANS approach. As in other types of SBLIs, the unsteadiness inherent in the flow may be of considerable importance. Such unsteadiness was observed experimentally, but it cannot be reproduced by RANS computations. For example, computations using the k-ε model demonstrated the regular interaction of external crossing shock waves [87, 88], whereas in experiments, a strongly unsteady interaction regime was observed [76].

Experimental research and RANS computations of the asymmetric ($\alpha_1 \neq \alpha_1$) double-fin interactions demonstrated more complex and manifold flowfield structures and topology. Interested readers are directed to papers in which these interactions are analyzed in more detail [10, 52, 57, 78, 79, 90, 91, 92, 93, 94].

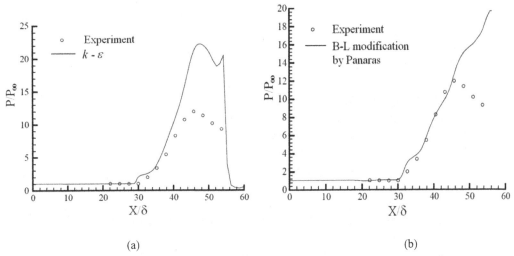

Figure 5.54. Centerline surface pressure distribution for $23° \times 23°$ interaction at $M_\infty = 5$.

5.4 Summary

Our knowledge and insight of three-dimensional STBLIs has progressed dramatically in the past three decades; nevertheless, it is far from complete and further research is needed to provide a comprehensive understanding. Numerous experimental investigations yielded detailed descriptions of the flowfield structure for several canonical three-dimensional configurations, including the single unswept/swept fin, SCR corner, and double-fin configurations. Correlations for determining conditions for fully attached, incipiently separated, and fully separated flows for several simple canonical three-dimensional configurations were developed analogous to and with knowledge gleaned from two-dimensional STBLIs. Despite the knowledge gained from the canonical configurations, their usefulness in predicting critical aerothermodynamic loads (i.e., surface pressure and heat transfer) and flowfield structure (in particular, the location and extent of separation) for realistic air-vehicle configurations is limited by the sheer complexity of typical shock-wave interactions in three dimensions.

The potential to predict complex flows using computational methods advanced dramatically in recent years. Concerning three-dimensional STBLIs, the RANS models (the standard for engineering applications) demonstrated a general capability to predict the mean aerothermodynamic loads for *weak* three-dimensional interactions. Furthermore, these models defined the flowfield structure for the canonical configurations discussed in this chapter. However, they failed to predict peak aerothermodynamic loading for *strong* three-dimensional STBLIs except when configuration-specific (i.e., nonuniversal) modifications to the models are utilized. To date, it has not been possible to apply the more sophisticated LES and Detached Eddy Simulation (DES) numerical methods to even the specially chosen canonical configurations due to high computational costs. The application of the even more demanding DNS to these flows remains a distant hope. Progress in using LES, DES, and (ultimately) DNS methods for the three-dimensional STBLIs will be possible only with dramatic improvements in the efficiency of numerical algorithms and reduced-order modeling of near-wall turbulence dynamics.

Abreviations used in this chapter

B-L   Baldwin-Lomax
CPT   Critical Point Theory
CR   Compression ramp
CSTBLI   Crossing shock wave–turbulent boundary-layer interaction
DES   Detached Eddy Simulation
DNS   Direct Numerical Simulation
FIC   Free Interaction Concept
FIT   Free Interaction Theory
LES   Large eddy simulation
LES   Large Eddy Simulation
LB   Line of divergence/bifurcation reflected
NSTBLI   Normal shock wave–turbulent boundary-layer interaction
PLS   Planar Laser Scattering
RANS   Reynolds average Navier Stokes
RSS   Reflected separation shock
SCR   Swept compression ramps
STBLI   Shock wave–turbulent boundary-layer interaction
SW   Shock wave
TKE   Turbulent kinetic energy
TML   Throat middle line
WI   Wilcox model
VCO   Virtual Conical Origin

REFERENCES

[1] G. S. Settles and D. S. Dolling. "Swept Shock-Wave/Boundary-Layer Interactions. Tactical Missile Aerodynamics." In *Progress in Astronautics and Aeronautics*, eds. M. Hemsch and J. Neilsen, Vol. 104 (New York: AIAA, 1986), 297–379.

[2] L. Prandtl. *Über Flüssigkeitsbewegung bei Sehr Kleiner Reibung. Verhandlungen des 3 Internationalen Mathematiker-Kongresses* (Leipzig, Germany: Teubner, 1904), 484–91.

[3] R. Legendre. *Ecoulement au voisinage de la pointe avant d'une aile á forte fleche aux incidences moyeness* (Flow in the vicinity of the apex of a wing with large sweep angle at moderate incidences). *La Recherche Aéronautique*, 30 (1952), 3–8.

[4] R Legendre. *Séparation de l'écoulement laminaire tridimensionnel* (Separation of a laminar three-dimensional flow). *La Recherche Aéronautique*, 54 (1956), 3–8.

[5] R. Legendre. *Lignes de courant d'un écoulement permanent: Décollement et séparation* (Streamlines of a steady flow: separation and separators). *La Recherche Aéronautique*, 6 (1977), 327–35.

[6] J. M. Délery. Physics of vortical flows. *J. of Aircraft*, 29 (1992), 5, 856–76.

[7] J. M. Délery. Robert Legendre and Henri Werlé: Toward the elucidation of three-dimensional separation. *Annual Review of Fluid Mechanics*, 33 (2001), 129–54.

[8] M. Tobak and D. J. Peake. Topology of three-dimensional separated flows. *Annual Review of Fluid Mechanics*, 14 (1982), 61–85.

[9] J. M. Lighthill. "Attachment and Separation in Three-Dimensional Flow." In *Laminar Boundary-Layer Theory*, ed. L. Rosenhead (Oxford, UK: Oxford University Press, 1963), Sec. II, Chap. 2.6, pp. 72–82.

[10] A. A. Zheltovodov. Shock waves/turbulent boundary-layer interactions: Fundamental studies and applications. *AIAA Paper 96–1977* (1996).

[11] A. Zheltovodov. Regimes and properties of three-dimensional separation flows initiated by skewed compression shocks. *J. Applied of Mechanics and Technical Physics*, 23 (1982), 3, 413–18.

[12] A. Zheltovodov, A. Maksimov, and E. Schülein. "Development of Turbulent Separated Flows in the Vicinity of Swept Shock Waves." In *The Interactions of Complex 3-D Flows*, ed. A. Kharitonov (Novosibirsk, 1987), 67–91 (in Russian).

[13] E. Schülein and A. A. Zheltovodov. Development of experimental methods for the hypersonic flows studies in Ludwieg tube. *Proc. International Conference on the Methods of Aerophysical Research – Pt.* 1. Novosibirsk, Russia (1998), 191–9.

[14] A. Stanbrook. An experimental study of the glancing interaction between a shock wave and a turbulent boundary layer. *British ARC, C.P.*, No. 555 (1960).

[15] R. H. Korkegi. Comparison of shock-induced two- and three-dimensional incipient turbulent separation. *AIAA J.*, 13 (1975), 4, 534–5.

[16] R. H. Korkegi. A simple correlation for incipient turbulent boundary-layer separation due to a skewed shock wave. *AIAA J.*, 11 (1973), 1, 1575–9.

[17] A. McCabe. A three-dimensional interaction of a shock wave with a turbulent boundary layer. *The Aeron. Quart.*, XVII (1966), Pt. 3, 231–52.

[18] G. S. Settles. Swept shock/boundary-layer interactions: Scaling laws, flowfield structure, and experimental methods, special course on shock-wave/boundary-layer interactions in supersonic and hypersonic flows. *AGARD Report* 762 (1993), 1-1–1-40.

[19] F. Lu and G. Settles. Color surface-flow visualization of fin-generated shock wave boundary-layer interactions. *Experiments in Fluids*, 8 (1990), 6, 352–4.

[20] A. Zheltovodov. Physical features and properties of two- and three-dimensional separated flows at supersonic velocities, *Izvestiya AN SSSR, Mekhanika Zhidkosti i Gaza (Fluid Dynamics)*, 3 (1979), 42–50 (in Russian).

[21] A. A. Zheltovodov. Some advances in research of shock wave turbulent boundary-layer interactions. *AIAA Paper* 2006–0496 (2006).

[22] A. A. Zheltovodov and E. Schülein. Three-dimensional swept shock waves/turbulent boundary layer interaction in angle configurations. *Preprint No. 34–86, ITAM*, USSR Academy of Sciences, Novosibirsk, 49, 1986 (in Russian).

[23] A. Zheltovodov, R. Dvorak, and P. Safarik. Shock waves/turbulent boundary layer interaction properties at transonic and supersonic speeds conditions. *Izvestiya SO AN SSSR, Seriya Tekhnicheskih Nauk*, 6 (1990), 31–42 (in Russian).

[24] V. S. Dem'yanenko and V. A. Igumnov. Spatial shock wave–turbulent boundary layer interactions in the interference region of intersecting surfaces. *Izvestiya Sibirskogo Otdeleniya Akademii Nauk SSSR (Proceedings of the USSR Academy of Sciences, Siberian Branch), Seriya Tekhnicheskh Nauk*, 8 (1975) (248), No. 2, 56–62 (in Russian).

[25] J. R. Hayes. Prediction techniques for the characteristics of fin-generated three-dimensional shock wave turbulent boundary layer interactions. *Technical Report AFFDL–TR–77–*10 (1976).

[26] E. Schülein and A. A. Zheltovodov. Documentation of experimental data for hypersonic 3-D shock waves/turbulent boundary layer interaction flows. *DLR Internal Report*, IB 223–99 A 26, 95 (2001).

[27] H. Kubota and J. L. Stollery. An experimental study of the interaction between a glancing shock and a turbulent boundary layer. *J. Fluid Mech.*, 116 (1982), 431–58.

[28] M. Zubin and N. Ostapenko. Structure of the flow in the region of separation for interaction of a normal shock wave with a boundary layer. *Izvestiya AN SSSR, Mekhanika Zhidkosti i Gaza*, 3 (1979), 51–8 (in Russian).

[29] M. A. Zubin and N. A. Ostapenko. Geometrical characteristics of turbulent boundary layer separation at interaction with normal shock wave in conical flows. *Izvestiya AN SSSR, Mekhanika Zhidkosti i Gaza*, No. 6 (1983), 43–51 (in Russian).

[30] E. E. Zukoski. Turbulent boundary-layer separation in front of a forward-facing step. *AIAA J.*, 5 (1967), 10, 1746–53.

[31] D. Chapman, D. Kuehn, and H Larson. Investigation of separated flows in supersonic and subsonic streams with emphasis on the effect of transition. *NACA Report* 1356 (1958).

[32] A. A. Zheltovodov and V. N. Yakovlev. Stages of development, gas dynamic structure and turbulence characteristics of turbulent compressible separated flows in the vicinity of 2-D obstacles. *ITAM*, USSR Academy of Sciences, Novosibirsk, Preprint No. 27–86 (1986), 55 (in Russian).

[33] S. S. Kutateladze and A. I. Leont'ev. *Turbulent Boundary Layer of Compressible Gas. SO AN SSSR* (Novosibirsk, 1962) (in Russian).

[34] A. A. Zheltovodov and A. M. Kharitonov. About the analogy of 2-D and 3-D separated flows. *Fizicheskaya Gazodinamika (Physical Gas Dynamics)*. Novosibirsk, 1 (1976), 6, 130–3 (in Russian).

[35] G. S. Settles and R. L. Kimmel. Similarity of quasiconical shock wave/turbulent boundary layer interactions. *AIAA J.*, 24 (1986), 1, 47–53.

[36] A. F. Charwat. Supersonic flows imbedded separated regions. *Advances in Heat Transfer*, 6, (1970), 1–32.

[37] B. Oskam, I. E. Vas, and S. M. Bogdonoff. Mach 3 oblique shock wave/turbulent boundary layer interactions in three dimensions, *AIAA Paper* 76–336, p. 19 (1976).

[38] F. Lu and G. Settles. Conical similarity of shock/boundary layer interaction generated by swept fins, *AIAA Paper* 83–1756 (1983).

[39] A. W. C. Leung and L. C. Squire. Reynolds number effects in swept-shock-wave/turbulent-boundary-layer interaction. *AIAA J.*, 33 (1995), 5, 798–803.

[40] G. R. Inger. Upstream influence and skin friction in non-separating shock/turbulent boundary-layer interactions. *AIAA Paper* 80–1411 (1980).

[41] G. R. Inger. Some features of a shock/turbulent boundary-layer interaction theory in transonic fields. *AGARD-CP-291* (1980).

[42] J. Délery and J. G. Marvin. Shock-wave boundary layer interactions. AGARDograph *No.* 280 (1986).

[43] C. C. Horstman and C. M. Hung. Computation of three-dimensional turbulent separated flows at supersonic speeds, *AIAA Paper* 1979–0002, p. 41 (1979).

[44] F. S. Alvi and G. S. Settles. Structure of swept shock wave/boundary layer interactions using conical shadowgraphy. *AIAA Paper* 90–1644 (1990).

[45] F. S. Alvi and G. S. Settles. A physical model of the swept shock/boundary-layer interaction flowfield. *AIAA Paper* 91–1768 (1991).

[46] B. Edney. Anomalous heat transfer and pressure distributions on blunt bodies at hypersonic speeds in the presence of an impinging shock. *Aeronautical Research Institute in Sweden, FFA Report* 115 (1968).

[47] E. Schülein. Skin-friction and heat-flux measurements in shock/boundary-layer interaction flow. *AIAA J.*, 44 (2006), 8, 1732–41.

[48] D. M. Voitenko, A. I. Zubkov, and Yu. A. Panov. Supersonic gas flow past a cylindrical obstacle on a plate. *Izvestiya Akademii Nauk SSSR, Mekhanika Zhidkosti i Gaza (Fluid Dynamics)*, 1 (1966), 120–5 (in Russian).

[49] D. M. Voitenko, A. I. Zubkov, and Yu. A. Panov. About existence of supersonic zones in three-dimensional separated flows. *Izvestiya Akademii Nauk SSSR, Mekhanika Zhidkosti i Gaza (Fluid Dynamics)*, 1 (1967), 20–4 (in Russian).

[50] D. D. Knight, C. C. Horstman, B. Shapey, and S. M. Bogdonoff. Structure of supersonic turbulent flow past a sharp fin. *AIAA J.*, 25 (1987), 10, 1331–7.

[51] S. M. Bogdonoff. The modeling of a three-dimensional shock wave turbulent boundary layer interaction: The Dryden Lecture. *AIAA Paper* 90–0766 (1990).

[52] D. Knight, H. Yan, A. G. Panaras, and A. Zheltovodov. Advances in CFD prediction of shock wave turbulent boundary layer interactions. *Progress in Aerospace Sciences*, 39 (2003), 121–84.

[53] A. G. Panaras. Algebraic turbulence modeling for swept shock-wave/turbulent boundary-layer interactions, *AIAA J.*, 35 (1997), 3, 456–63.

[54] A. G. Panaras. The effect of the structure of swept-shock-wave/turbulent-boundary-layer interactions on turbulence modeling. *J. Fluid Mech.*, 338 (1997), 203–30.

[55] A. G. Panaras. Calculation of flows characterized by extensive cross-flow separation, *AIAA J.*, 42 (2004), 12, 2474–81.

[56] G. S. Settles and L. J. Dodson. Hypersonic shock/boundary-layer interaction database, *NASA CR*-177577 (1991).

[57] D. D. Knight and G. Degrez. Shock wave boundary layer interactions in high Mach number flows: A critical survey of current CFD prediction capabilities. AGARD Report *319*, 2, 1-1–1-35 (1998).

[58] F. Thivet, D. Knight, A. Zheltovodov, and A. Maksimov. Importance of limiting the turbulence stresses to predict 3D shock wave boundary layer interactions. *Proc. 23rd International Symposium on Shock Waves* (Ft. Worth, TX, July 2001), Paper No. 2761, p. 7.

[59] P. Durbin. On the *k-ε* stagnation point anomaly. *Int. J. Heat Fluid Flow*, 17 (1996), 1, 89–90.

[60] F. Thivet. Lessons learned from RANS simulations of shock-wave/boundary-layer interactions. *AIAA Paper* 2002–0583, p. 11 (2002).

[61] A. A. Zheltovodov and E. Schülein. Problems and capabilities of modeling of turbulent separation at supersonic speeds conditions. *Proc. The Seventh All-Union Congress on Theoretical and Applied Mechanics, Reports Annotations* (Moscow, 1991), 153–4 (in Russian).

[62] A. A. Zheltovodov and A. I. Maksimov. Development of three-dimensional flows at conical shock-wave/turbulent boundary layer interaction. *Sibirskiy Fiziko-Technicheskiy Zhurnal (Siberian Physical-Technical Journal)*, 2 (1991), 88–98 (in Russian).

[63] G. S. Settles and S. M. Bogdonoff. Scaling of two- and three-dimensional shock/turbulent boundary-layer interactions at compression corners. *AIAA J.*, 20 (1982), 6, 782–9.

[64] D. S. Dolling and S. M. Bogdonoff. Upstream influence in sharp fin-induced shock wave turbulent boundary layer interaction, *AIAA J.*, 21 (1983), 1, 143–5.

[65] S. W. Wang and S. M. Bogdonoff. A re-examination of the upstream influence scaling and similarity laws for 3-D shock wave/turbulent boundary layer interaction, *AIAA Paper* 83–0347 (1986), 7.

[66] F. K. Lu and G. S. Settles. Upstream-influence scaling of sharp fin interactions, *AIAA J.*, 29 (1991), 1180–1.

[67] G. S. Settles, J. J. Perkins, and S. M. Bogdonoff. Investigation of three-dimensional shock/boundary-layer interaction at swept compression corners, *AIAA J.* 18 (1980), 779–85.

[68] G. S. Settles and H. Teng. Cylindrical and conical flow regimes of three-dimensional shock/boundary-layer interactions. *AIAA J.*, 22 (1984), 2, 194–200.

[69] G. S. Settles. On the inception lengths of swept shock-wave/turbulent boundary-layer interactions. *Proc. IUTAM Symposium on Turbulent Shear-Layer/Shock-Wave Interactions*, Palaiseau, France, ed. J. Délery (Springer Verlag, 1985), 203–13.

[70] R. H. Korkegi. A lower bound for three-dimensional turbulent separation in supersonic flow. *AIAA J.*, 23 (1985), 3, 475–6.

[71] G. S. Settles, S. M. Bogdonoff, and I. E. Vas. Incipient separation of a supersonic turbulent boundary layer at high Reynolds numbers. *AIAA J.*, 14 (1976), 1, 50–6.

[72] G. S. Settles. An experimental study of compressible turbulent boundary layer separation at high Reynolds numbers. Ph.D. Dissertation, Princeton, NJ: Aerospace and Mechanical Sciences Department, Princeton University (1975).

[73] G. S. Settles, C. C. Horstman, and T. M. McKenzie. Experimental and computational study of a swept compression corner interaction flowfield. *AIAA J.*, 24 (1986), 5, 744–52.

[74] D. D. Knight, C. C. Horstman, and S. M. Bogdonoff. Structure of supersonic turbulent flow past a swept compression corner, *AIAA J.*, 30 (1992), 4, 890–6.

[75] A. A. Zheltovodov, A. I. Maksimov, and A. M. Shevchenko. Topology of three-dimensional separation under the conditions of symmetric interaction of crossing shocks and expansion waves with turbulent boundary layer. *Thermophysics and Aeromechanics*, 5 (1998), 3, 293–312.

[76] A. A. Zheltovodov, A. I. Maksimov, E. Schülein, D. V. Gaitonde, and J. D. Schmisseur. Verification of crossing-shock-wave/boundary layer interaction computations with the

k-ε turbulence model. *Proc. International Conference on the Methods of Aerophysical Research*, Novosibirsk, Russia, 9–16 July 2000. Part 1. 231–41.

[77] A. A. Zheltovodov, A. I. Maksimov, D. Gaitonde, M. Visbal, and J. S. Shang. Experimental and numerical study of symmetric interaction of crossing shocks and expansion waves with a turbulent boundary layer. *Thermophysics and Aeromechanics*, 7 (2000), 2, 155–71.

[78] F. Thivet, D. D. Knight, A. A. Zheltovodov, and A. I. Maksimov. Numerical prediction of heat-transfer in supersonic inlets. *Proc. European Congress on Computational Methods in Applied Sciences and Engineering (ECCOMAS 2000)*, Barcelona, CD Contents, p. 1 (September 2000).

[79] F. Thivet, D. D. Knight, A. A. Zheltovodov, and A. I. Maksimov. Insights in turbulence modeling for crossing shock wave boundary layer interactions. *AIAA J.*, 39 (2001), 7, 985–95.

[80] F. Thivet, D. D. Knight, A. A. Zheltovodov, and A. I. Maksimov. Analysis of observed and computed crossing-shock-wave/turbulent-boundary-layer interactions. *Aerospace Sci. and Technology*, 6 (2002), 3–17.

[81] J. F. Moore and J. Moore. Realizability in two-equation turbulence models. *AIAA Paper* 99–3779 (1999).

[82] P. F. Batcho, A. C. Ketchum, S. M. Bogdonoff, and E. M. Fernando. Preliminary study of the interactions caused by crossing shock waves and a turbulent boundary layer. *AIAA Paper* 89–359 (1989).

[83] K. Poddar and S. Bogdonoff. A study of unsteadiness of crossing shock wave turbulent boundary layer interactions. *AIAA Paper* 90–1456 (1990).

[84] D. O. Davis and W. R. Hingst. Surface and flowfield measurements in a symmetric crossing shock wave/turbulent boundary layer interaction. *AIAA Paper* 92–2634, (1992).

[85] T. J. Garrison and G. S. Settles. Flowfield visualization of crossing shock-wave/boundary-layer interactions. *AIAA Paper* 92–0750, p. 10 (1992).

[86] D. V. Gaitonde, J. S. Shang, T. J. Garrison. A. A. Zheltovodov, and A. I. Maksimov. Evolution of the separated flowfield in a 3-D shock wave/turbulent boundary layer interaction. *AIAA Paper* 97–1837 (1997).

[87] J. D. Schmisseur, D. V. Gaitonde, and A. A. Zheltovodov. Exploration of 3-D shock turbulent boundary layer interactions through combined experimental/computational analysis. *AIAA Paper* 2000–2378, p. 11 (2000).

[88] J. D. Schmisseur and D. V. Gaitonde. Numerical investigation of strong crossing shock-wave/turbulent boundary-layer interactions. *AIAA J.*, 39 (2001), 9, 1742–49.

[89] E. K. Derunov, A. A. Zheltovodov, and A. I. Maksimov. Development of three-dimensional turbulent separation in the neighborhood of incident crossing shock waves. *Thermophysics and Aeromechanics*, 15 (2008), 1, 29–54.

[90] A. A. Zheltovodov, A. I. Maksimov, A. M. Shevchenko, and D. D. Knight. Topology of three-dimensional separation under the conditions of asymmetric interaction of crossing shocks and expansion waves with turbulent boundary layer. *Thermophysics and Aeromechanics*, 5 (1998), 4, 483–503.

[91] D. Gaitonde, J. Shang, T. Garrison, A. Zheltovodov, and A. Maksimov. Three-dimensional turbulent interactions caused by asymmetric crossing shock configurations. *AIAA J.*, 37 (1999), 12, 1602–8.

[92] D. Knight, M. Gnedin, R. Becht, and A. Zheltovodov. Numerical simulation of crossing-shock-wave/turbulent-boundary-layer interaction using a two-equation model of turbulence. *J. Fluid Mech.*, 409 (2000), 121–47.

[93] D. D. Knight, T. J. Garrison, G. S. Settles, A. A. Zheltovodov, A. I. Maksimov, A. M. Shevchenko, and S. S. Vorontsov. Asymmetric crossing-shock-wave/turbulent-boundary-layer interaction. *AIAA J.*, 33 (2001), 12, 2241–58.

[94] D. V. Gaitonde, M. R. Visbal, J. S. Shang, A. A. Zheltovodov, and A. I. Maksimov. Sidewall interaction in an asymmetric simulated scramjet inlet configuration. *J. Propulsion and Power*, 17 (2001), 3, 579–84.

6 Experimental Studies of Shock Wave–Boundary-Layer Interactions in Hypersonic Flows

Michael S. Holden

6.1 Introduction

Some of the most serious and challenging problems encountered by the designers of hypersonic vehicles arise because of the severity of the heating loads and the steepness of the flow gradients that are generated in shock wave–boundary layer interaction (SBLI) regions. The characteristics of these flows are difficult to predict accurately due in no small measure to the significant complexity caused by shear-layer transition, which occurs at very low Reynolds numbers and can lead to enhanced heating loads and large-scale unsteadiness. Even for completely laminar flows, viscous interaction can degrade appreciably the performance of control and propulsion systems. It is interesting that both of the two major problems encountered with the U.S. Space Shuttle program were associated with SBLI. The first was the so-called Shuttle Flap Anomaly that nearly resulted in disaster on the craft's maiden flight due to a failure in the design phases to account correctly for the influence of real-gas effects on the shock-interaction regions over the control surfaces. During the flight, a significantly larger flap deflection was required to stabilize the vehicle than had been determined from ground tests in cold-flow facilities. Miraculously, it was possible to achieve the necessary control, and disaster was narrowly averted. The second problem was the leading-edge structural failure caused by the impact of foam that had been fractured and released from the shuttle tank as a result of the dynamic loads caused by a shock interaction. Figure 6.1 is an example of the shock structures that are generated among the shuttle, the main tank, and the solid reusable boosters. The contour plot illustrates the corresponding computer-predicted pressure distribution. Aerothermal loads generated by shock waves in the region of the bipod that supports the shuttle nose caused the foam glove to be fractured and released. Unfortunately, the damage this caused resulted in a tragic accident.

Even on the Apollo capsule, which is the simplest of hypersonic reentry vehicles, significant problems arose due to SBLIs that occurred in the base region as a consequence of the interplay between the reaction control system jets and the near-wake. As in previous examples, the occurrence of transition in the separated flow, coupled in this case with the complexity of combustion in the recirculated region, makes these flows – including shock-shock interactions as well as SBLIs – extremely

Figure 6.1. Shock interactions on OTS shuttle configuration.

difficult to predict. Some of the largest heating loads experienced on hypersonic vehicles are associated with flows of this type and occur where the interaction is between an oblique shock wave and a near-normal shock wave formed over a leading edge (Fig. 6.2a). The resultant intense surface heating can be orders of magnitude larger than the stagnation-point heating to the leading edge on its own. Shock-interaction heating of this type caused the dramatic failure of the pylon support to a ramjet carried beneath the X-15 research vehicle, as illustrated in Fig. 6.2b. Because transition occurs at very low Reynolds numbers ($Re_D \approx 100$) within the shock-shock interaction regions, the magnitude and severity of the heating loads, as in previous examples, are difficult to predict. Also, the multiple expansion and compression regions generated in these flows result in a sensitivity to real-gas effects; this feature was exploited in the design of "double-cone" models used in

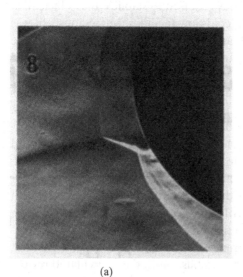

(a)   (b)

Figure 6.2. (a) Type IV interaction; (b) X15 pylon.

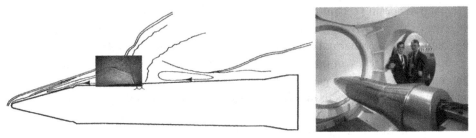

Figure 6.3. Schematic/Schlieren photograph of divert/thruster flow over interceptor interceptor with control thruster.

experimental studies of real-gas effects, which are discussed in this chapter and elsewhere in this book.

The effect on the control forces, acoustic loads, and aero-optic phenomena associated with jet interaction and the accompanying SBLIs are of major concern to designers of hypersonic vehicles that are maneuvered with large thrusters. These problems are particularly challenging at high altitudes where jet interaction produces a separated region (Fig. 6.3), which can extend to envelop the vehicle almost to the nose tip. Infrared images of this flowfield confirm that burning gases from the thrusters are entrained in the recirculating flows, potentially blinding the sensors placed behind windows even close to the nose tip. When laminar-boundary layer separation occurs with transition in the free-shear layer, large-scale instabilities occur in the recirculating flow upstream of the transverse jet. The fluctuating pressures generated by the unsteady interaction region can create serious acoustic loads with major consequences on the performance of the optical sensors. Here, the combined effects of shear-layer transition, air and combustion chemistry, and large-scale unsteadiness render these flows virtually impossible to predict with any degree of accuracy.

All of these shock-interaction phenomena and others confront designers of ram/scramjet-powered boost and cruise vehicles. In scramjet-powered hypersonic vehicles such as the X-43, the flow in the inlet and isolator sections of the engine contain laminar, transitional, and turbulent regions of viscous interaction and shock-shock interaction. The interaction regions generated by multiple reflected and swept shock waves control the efficiency of the inlet and the characteristics of the flow entering the injector/combustor section of the engine. Although the multiple shock interactions in the engine (Fig. 6.4) promote mixing and can act as sites for ignition and flame-holding, they also are responsible for pressure losses and drag penalties, which can significantly reduce engine performance. Accurately predicting shock interaction in the presence of transition, compressible nonequilibrium turbulence, and air and combustor chemistry remains a significant task.

Figure 6.4. Shock interactions through scramjet-powered research engine.

Active cooling techniques must be used in hypersonic vehicles and scramjets to ensure the survival of leading edges and internal components of an engine. Predicting the aerothermal loads resulting from the interaction of a shock wave with a cooling layer poses an important design requirement. These flows, which involve the mixing of dissimilar gases, render the calculations far more complex and – as in many other instances in these flows – experimental measurements are extremely important to provide or validate the design of the vehicle component.

So far in this introduction, some of the more challenging problem areas associated with SBLIs within the hypersonic regime have been highlighted. Nevertheless, there has been extensive research on this topic and, in many instances, the flows are understood and can be predicted with reasonable certainty. To provide insight, the remainder of this chapter is devoted to outlining this knowledge. Section 6.1 discusses experimental studies to define the characteristics of regions of laminar, transitional, and turbulent SBLIs in hypersonic flow. We also evaluate the semi-empirical and numerical techniques developed to predict these flows. Section 6.2 discusses the salient features of laminar-interaction regions and traces the development of validation studies to evaluate prediction techniques initially based on solutions to the first- and second-order boundary-layer equations and, more recently, on solutions to the Navier-Stokes equations or by using Direct Simulation Monte Carlo (DSMC) techniques.

Section 6.3 describes the experimental evaluation of the characteristics of SBLI regions in turbulent and transitional flows. Correlations of measurements made in studies of turbulent-hypersonic flows provide simple prediction methods to describe the major aerothermal characteristics of these flows. Comparisons between Reynolds-Averaged Navier-Stokes (RANS) solutions and measurements in fully turbulent interaction regions yield disappointing results. Swept-shock interaction regions were studied extensively in adiabatic supersonic flow and similar measurements are presented for hypersonic interacting flows over highly cooled surfaces. Transitional shock-interaction regions are discussed briefly in the context of their importance to nose-shaping on ablated surfaces.

Section 6.4 discusses experimental studies that evaluated the extreme heating loads developed in regions of shock-shock interaction. In these studies, measurements were obtained for laminar, transitional, and turbulent interaction regions. Although Navier-Stokes and DSMC methods succeeded in describing these flows when they are laminar, for transitional- and turbulent-interaction regions, only correlations of the experimental data can be relied on to predict correctly the aerothermal loads in regions of shock-shock interaction. Section 6.5 discusses experimental studies in the important area of protecting surfaces in regions of shock interaction using transpiration- and film-cooled or ablating surfaces. Here, experimental studies reveal the difficulty of using these techniques to protect surfaces subjected to shock interaction. Section 6.6 concludes by discussing experimental studies that evaluated the influence of real-gas effects on the aerothermal characteristics of the flow structures and heating levels developed in regions of SBLI and shock-shock interaction. The failure to predict the influence of these effects on the control-surface characteristics of the U.S. Space Shuttle – which were responsible for the narrowly averted catastrophe on its first flight – highlights the importance of understanding those flows.

6.2 SBLI in Laminar Two-Dimensional and Axisymmetric Hypersonic Flows

6.2.1 Introduction

The intense aerothermal heating loads that arise within regions of shock-wave inter-action with turbulent-boundary layers are regarded generally as presenting the most significant problems for hypersonic flight. However, it may well be the *laminar* viscous–inviscid interaction and flow separations that are associated with these flows at greater altitudes that pose the greatest fundamental limitations on the per-formance of maneuvering and air-breathing hypersonic vehicles. The effectiveness of intakes and flap-control systems, as well as the performance of vehicles using jet interaction, may be compromised seriously by the occurrence of shock-induced laminar separation. For the intakes and flaps, compression surfaces essentially can be "faired in," and the separated regions formed in front of the transverse jet can change the force or moment characteristics of the vehicle or – more seriously for some applications – result in the obscuration of optical-seeker devices. The range in performance of scramjet engines on a single stage to orbit vehicles also may be limited basically by the occurrence of laminar-flow separation on the sidewall and cowls of the engine, which eventually can result in "engine unstart." The use of boundary-layer controls to alleviate these problems is difficult in hypersonic flow because the major portion of the mass and momentum in the layer over a cooled wall is contained at its outer edge.

In the past twenty years, there has been a massive increase in computational capabilities with which to make a direct assault on using the Navier-Stokes equa-tions to solve flows that include SBLIs with embedded recirculating regions. These methods have had great success in describing the flows for hypersonic Mach num-bers, and certain outstanding questions for cases with large embedded separated regions were resolved by three-dimensional computation.

6.2.2 Salient Characteristics for Laminar Regions of SBLI in Hypersonic Flows

To discuss the salient characteristics of hypersonic SBLIs in laminar flows, this chap-ter focuses on two configurations. The first occurs on compression surfaces on which the shock waves are developed "internally" and the upstream influence and sepa-ration are produced by "free interaction" between the viscous and inviscid flow. The second configuration is induced by the impingement of a shock wave onto the boundary layer.

Most of the previous SBLI studies focused on flows over two-dimensional con-figurations because the experiments were more straightforward to conduct and interpret, and a boundary-layer–based analysis was more tractable. Flow char-acteristics of wedge-induced and externally generated shock-induced attached and separated interaction regions are illustrated in the schlieren photographs in Fig. 6.5.

Figure 6.5a,b shows wedge-induced attached and separated laminar interact-ing flows. For the attached flow, the interaction region occurs principally on the

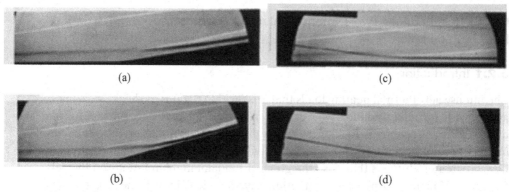

Figure 6.5. Wedge- and shock-induced laminar interaction regions in hypersonic flow. $M_\infty = 11.7$ and $\mathrm{Re}_L = 2.5 \times 10^5$. (a) attached laminar interaction region; (b) separated laminar interaction region; (c) attached laminar interaction region; (d) separated laminar interaction region.

wedge surface downstream of the corner (Fig. 6.5a); it is only when a large separated region is created (Fig. 6.5b) that the interaction region moves upstream. In these flows, the strong curvature in the reattachment region results in a significant normal-pressure gradient, which must be modeled correctly to obtain accurate predictions. A similar situation occurs for interaction regions generated by an external shock (Fig. 6.5c,d). Here, we observe that for a less-intense shock, most of the interaction region occurs downstream of the shock-impingement point. Only when extensive separation occurs does the separation shock appear ahead of the incident shock (Fig. 6.5d). The interaction takes place in only four or five boundary-layer thicknesses, and there is significant flow curvature during the reattachment-compression process.

Typical skin-friction distributions measured in wedge-induced interaction regions for attached and separated flows are shown in Fig. 6.6a [1]. Incipient separation occurs in these flows for wedge angles of just over 10 degrees. Measurements of the pressure rise needed to induce incipient separation, and the value of the plateau pressure from a number of sources for both wedge- and shock-induced interaction regions are compared in Fig. 6.6b [1]. The figure shows that the measurements correlate remarkably well when plotted in terms of the following viscous-interaction parameter:

$$\bar{\chi} = M^3 \sqrt{C} \Big/ \sqrt{\mathrm{Re}_L} \qquad \text{where} \quad C = \frac{\mu}{\mu_r} \frac{T_r}{T}. \tag{6.1}$$

Incipient separation also can be correlated in terms of hypersonic similitude parameters, as shown in Fig. 6.7a. The correlations suggest that even in high-Mach-number flows, flow separation is controlled principally by free interaction between viscous and inviscid flows. Also, it is observed that the peak-heat transfer in the reattachment regions nondimensionalized by the undisturbed value ahead of the interaction can be correlated simply in terms of the pressure rise through the interaction region (Fig. 6.7b).

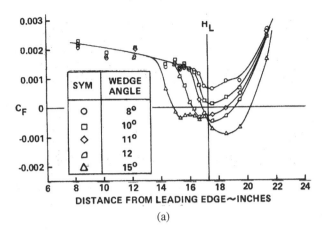

(a)

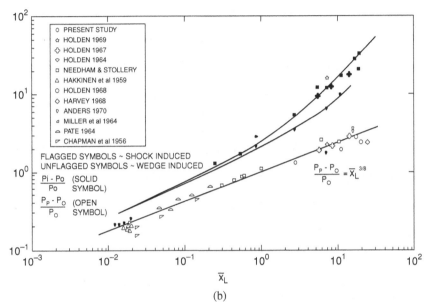

(b)

Figure 6.6. (a) Skin-friction measurements to define incipient separation and correlations of incipient separation for wedge- and shock-induced flows [1]. $M_\infty = 11.7$ and $Re/m = 5.2 \times 10^4$ ($Re/ft = 1.7 \times 10^5$); (b) Correlation of pressure rise to induce incipient separation and the plateau pressure in shock- and wedge-induced interactions taken from a variety of sources [1].

6.2.3 Boundary-Layer Models of Shock Wave–Laminar Boundary-Layer Interaction

Before the development of massive computer capability – which has enabled direct Navier-Stokes and DSMC solutions to be obtained for complex flows – the efforts to understand and predict the characteristics of SBLI regions were focused on numerical solutions to the first- or second-order boundary-layer equations. These approaches, in which separation was postulated to occur through the free interaction between the growth of the viscous layer and the outer supersonic-inviscid flow, had their roots in the early studies of Lighthill [2] and Oswatitsch and Weighardt [3] to model the mechanism of upstream influence. This phenomenon originally was

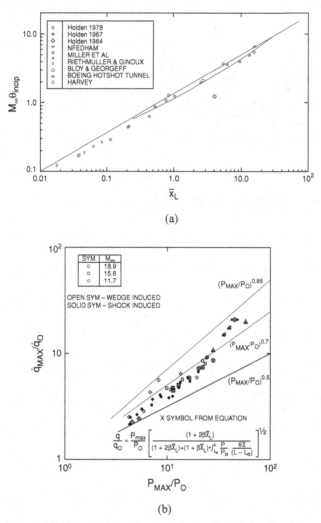

Figure 6.7. Correlations of separation and peak heating in wedge- and shock-induced laminar shock interaction regions: (a) correlation of incipient separation conditions for wedge-induced interactions; (b) comparison between simple correlation and prediction method.

believed to be associated with upstream propagation through the subsonic region of the boundary layer; however, the model was inconsistent with measurements, indicating a large upstream extent of the separation region in low-Mach-number flows. Rather, laminar separation in supersonic flows could be described better by a model in which the viscous layer grew rapidly by mutual interaction between the viscous and inviscid flow [4]. Glick [5] formulated free-interaction models using modifications to the Crocco-Lees mixing theory that were capable of describing the pressure rise and boundary-layer growth leading to boundary-layer separation. Glick, Honda [6], and Lees and Reeves [7] all used momentum-integral techniques, adding the moment-of-momentum equation to develop prediction techniques capable of describing a shock-induced separated region from separation through to reattachment, similar to what is shown in Fig. 6.8a.

In their method, Lees and Reeves [7] used the compressible form of Stewartson's reverse-flow profiles (Fig. 6.8b) to describe the structure of the flow in

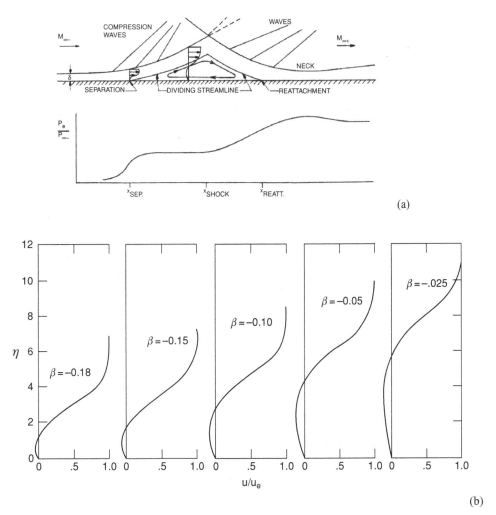

Figure 6.8. "Free interaction" flow model used to describe shock-induced separated regions with momentum integral techniques: (a) schematic of flowfield and pressure distribution in shock-induced laminar separated interaction; (b) reverse-flow profiles used in integral methods to predict separated flow.

and above the reverse-flow region. Figure 6.9a is an example of the good agreement between the Lees and Reeves prediction and the measurements of Chapman, Kuehn, and Larson [8] in laminar regions of SBLI. A similar approach by Nielsen et al. [9] that used the power-law profiles to describe the velocity distribution across the interaction regions is compared with measurements of Lewis, Kubota, and Lees [10] and the Lees and Reeves predictions for wedge-induced separated flow shown in Fig. 6.9b. Again, there is reasonable agreement between predictions and measurement.

The momentum-integral analysis, although adequate for supersonic flows over adiabatic walls, begins to break down in hypersonic flows where regions of viscous–inviscid interaction occur over fewer boundary-layer thicknesses (see Fig. 6.5). One of the last sets of analysis to use integral techniques to describe the heat-transfer, pressure, and skin-friction distribution in SBLI regions over cold walls was developed by Holden, who added the energy equation [11] and then incorporated

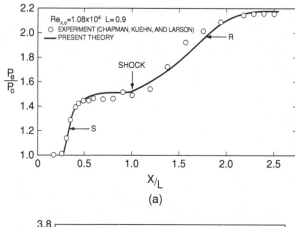

(a)

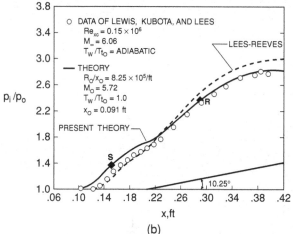

(b)

Figure 6.9. Comparison between experiments in supersonic flow and momentum integral predictions by Lees and Reeves, and Nielsen, Goodwin and Kuhn for adiabatic-wall conditions: (a) Lees and Reeves prediction for adiabatic wall; (b) Nielsen, Goodwin, and Kuhn predictions for adiabatic wall.

a normal-pressure gradient [12] in integral solutions for the first- and second-order boundary-layer equations. Heat-transfer and pressure predictions made in laminar-wedge-induced separated regions are compared with Holden's predictions [11] in Fig. 6.10, which shows that although the length of the interaction and the pressure distribution are calculated with reasonable accuracy, the method overpredicts heat transfer in the separated region.

Comparisons with measurements in higher-Mach-number flows with the theory incorporating normal-pressure gradient (Fig. 6.11) [13] show good agreement between theory and experiment for the pressure and skin-friction distributions. However, heat transfer in the separated region is overpredicted.

6.2.4 Early Navier-Stokes Validation Studies

In the same period that integral techniques showed serious limitations in calculating laminar separated regions in high-Mach-number flows, the first accurate numerical-solution technique to the Navier-Stokes equations was being developed by MacCormack [14]. One of the first sets of calculations using this code predicted the characteristics of attached and separated regions in the corner flow generated over

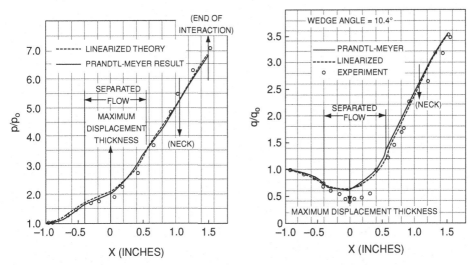

Figure 6.10. Comparison between pressure heat and transfer measurements in wedge-induced separated flows and predictions by Holden ([11]). $M_\infty = 10$ and $\mathrm{Re}_L = 1.4 \times 10^5$.

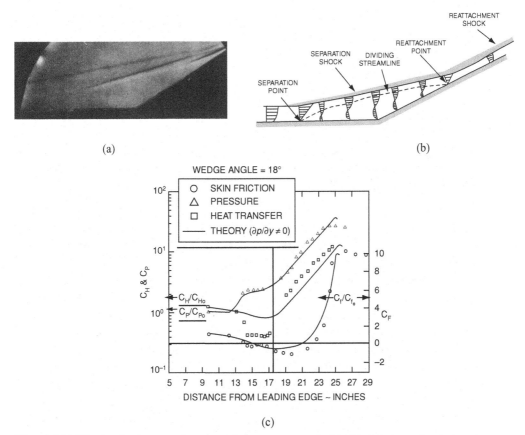

Figure 6.11. Comparison among heat transfer, pressure, and skin friction in wedge-induced separated flows and predictions by Holden incorporating normal-pressure gradient. $M_\infty = 16$ and $ReL = \chi L = 19.8$.

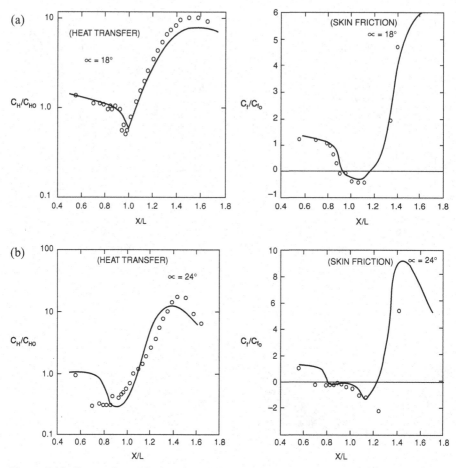

Figure 6.12. Comparisons between Navier-Stokes computations and measurements in attached and separated regions of laminar shock wave–boundary layer interaction: (a) incipient separation; (b) well separated.

a sharp-flat-plate–wedge configuration. The calculations of Hung and MacCormack [15] are compared with the heat-transfer and skin-friction measurements in a Mach 14 flow obtained by Holden [13] in Fig. 6.12a,b.

The predictions were in remarkably good agreement with the experiment for attached and small separated regions. However, for the large separated flows induced by a 24-degree wedge, the prediction significantly underestimated the measured size of the separated region. Several years later, using the Computational Fluids Laboratory 3-Dimentional flow solver (CFL3D), Rudy et al. [16] revisited this dataset including three-dimensional flow effects and obtained excellent agreement for the measurements in both wedge- and shock-induced separated flows (Figs. 6.13 and 6.14). At that time, there was no further reason to question the capabilities of well-performed Navier-Stokes computations to predict the characteristics of two- and three-dimensional laminar separated flows induced by complex regions of shock–boundary layer and shock-shock interaction – at least, in the absence of real-gas effects.

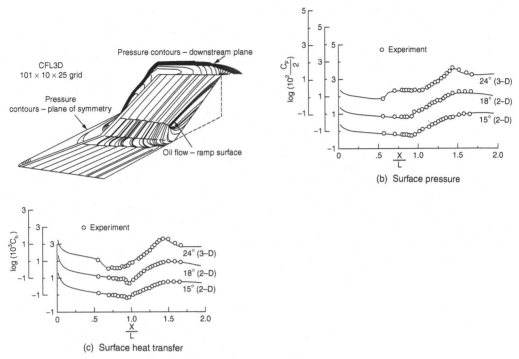

(b) Surface pressure

(c) Surface heat transfer

Figure 6.13. Comparisons between measurements and Navier-Stokes solutions with the CFL 3D code for wedge-induced separated flows.

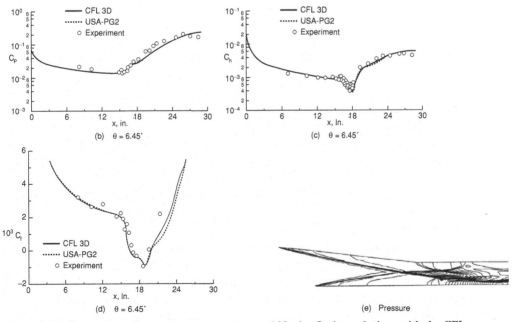

(b) $\theta = 6.45°$

(c) $\theta = 6.45°$

(d) $\theta = 6.45°$

(e) Pressure

Figure 6.14. Comparisons between measurements and Navier-Stokes solutions with the CFL 3D code for shock-induced separated flows.

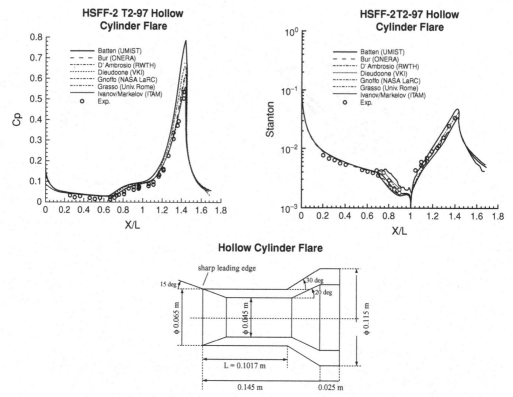

Figure 6.15. Comparison between Navier-Stokes predictions and pressure and heat-transfer measurements on hollow-cylinder–flare configuration.

During development of the Hermes spacecraft, the European Space Agency (ESA) actively promoted workshops to evaluate the performance of numerical techniques with which to predict aerothermal characteristics of the Hermes vehicle. In particular, those efforts focused on accurately calculating the hypersonic flow around the flap controls, which were found to be of major concern in the U.S. Space Shuttle program. A series of experimental studies was conducted in which suitable test cases were identified that would generate flow phenomena of key interest in the Hermes program. One experimental case devised and tested at ONERA [17] was a hollow-cylinder–flare configuration that provided measurements in separated regions of shock wave–laminar boundary-layer interaction. Results for this configuration were compared with computations by well-established scientists in Europe and America. It is surprising that there were significant disagreements among the different computational results and the pressure and heat-transfer measurements (Fig. 6.15). These discrepancies were manifested in the length of the separation region and in the pressure upstream and downstream of the interaction. These results were unanticipated in the wake of the successes with Navier-Stokes computations that were achieved in the earlier studies discussed previously, and they signaled concerns associated with the gridding and the dissipated nature of the various numerical schemes. Comparisons also were made between measurements in separated flows over a hyperboloid–flare configuration (Fig. 6.16) [18]. Similar

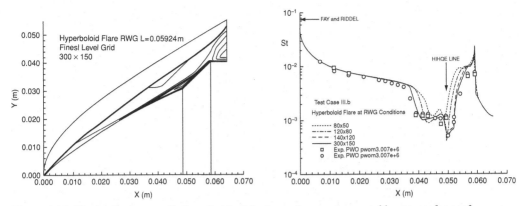

Figure 6.16. Navier-Stokes predictions for the Mach number contours and heat transfer coefficient compared with measurements on a hyperboloid flare configurations [18].

conclusions were reached for these comparisons; again, differences associated with numerical dissipation in the different schemes were considered to be important.

6.2.5 Recent Navier-Stokes and DSMC Code-Validation Studies of Hypersonic SBLIs

In an attempt to resolve the experimental or numerical problems that arose in the ESA studies, a new series of code-validation measurements was conducted by Holden and Wadhams [19]. The measurements were obtained in regions of shock wave–laminar boundary-layer interaction over a cone–flare configuration (Fig. 6.17) similar to that used in the European studies. However, the flare was extended to allow the flow to become fully reattached with a period of constant pressure before reaching the end, thus providing well-defined downstream-boundary conditions.

The initial comparisons between theory and experiment were conducted "blind" and computations again were performed by experienced scientists from America and Europe using Navier-Stokes and DSMC techniques. In general, there was good agreement between theory and experiment; however, questions arose

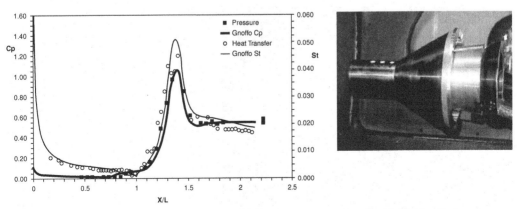

Figure 6.17. Hollow-cylinder–flare model and measurements compared with Navier-Stokes predictions.

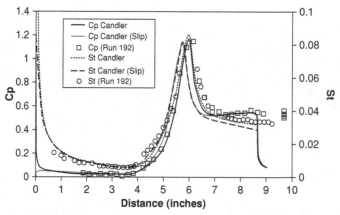

Figure 6.18. Comparisons between measurements with Navier-Stokes solutions incorporating surface slip for pressure coefficient and Stanton number.

because of grid fidelity and differences in computational schemes related to numerical dissipation. In a second round of activities [20], additional measurements were made including at lower densities, and flow-calibration studies were conducted to refine characterization of the test conditions. At that time, major improvements to the DSMC rarefied-flow numerical schemes improved their performance toward the high-density limit of their applicability. Advances also had been made with the Navier-Stokes schemes to improve surface-accommodation models for low-density flows. Sample results from those studies, shown in Fig. 6.18, demonstrate that for this configuration, the Navier-Stokes method is fully capable of accurately describing the laminar regions of separated flows. Although not shown here, the same conclusion can be drawn for solutions obtained using the DSMC method.

A second configuration selected for this code-validation exercise was the double cone shown in Fig. 6.19. The flow for this geometry is similar to that over the indented nose shapes in which combined shock-wave–boundary-layer and shock-shock-interaction phenomena controlled the flowfield, surface pressure, and heating. The wind-tunnel studies conducted on this shape provided precise validation-quality measurements for a flowfield that also combined regions of SBLI

Figure 6.19. The double-cone model.

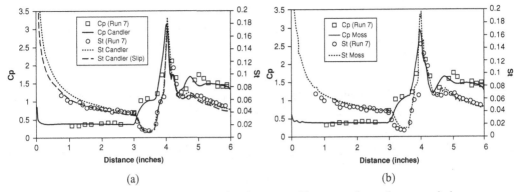

Figure 6.20. Comparison between measured and computed heat transfer and pressure in low-density flows over the cone/cone body using (a) Navier-Stokes, and (b) DSMC codes.

and shock-shock interaction and it was more challenging to compute than the relatively benign hollow-cylinder–flare flow. The new studies were conducted under high-enthalpy conditions using nitrogen as the test gas to eliminate the more complex nonequilibrium and chemistry effects present in air flows. Low-Reynolds-number conditions were selected for all of the studies to ensure that the flow remained fully laminar at all times, and low-density cases were added to enable DSMC methods to be used.

Typical results from this study are shown in Fig. 6.20. For denser flows, Navier-Stokes solutions were obtained by Candler and Gnoffo [21] and were in excellent agreement with one another and the experiment. As the density is reduced, boundary-layer slip effects begin to be significant; however, it was demonstrated that by accounting for them, the Navier-Stokes method could still be used to good effect. Figure 6.20a is a typical comparison between prediction and the measured heat-transfer and pressure distributions on the double cone; again, it exhibits excellent agreement between prediction and experiments. Recent improvements by Bird [22] and others to the DSMC codes also resulted in computations that are in excellent agreement with the measurements over a large range of Reynolds numbers. An example of such a comparison is shown in Fig. 6.20b. The clear conclusion from these and other comparative studies is that when appropriately and expertly applied, both the Navier-Stokes and DSMC methods can describe accurately the flowfield and the pressure and heat-transfer distributions in these complex flows, *proving that they are fully laminar and free from flow chemistry or real-gas effects*. It is apparent that within these limitations, both computational methods are capable of predicting even the most complicated interaction regions likely to develop over hypersonic vehicles.

6.3 SBLI in Turbulent and Transitional Flows

6.3.1 Introduction

Predicting the size and distribution of flow properties through a region of shock wave–*turbulent* boundary-layer interaction for supersonic and hypersonic Mach numbers remains a major challenge for numerical simulation, which, thus far, has had little success. The major problem encountered is the difficulty in finding adequate ways to model the turbulence through the separated-interaction regions in which reverse flow, turbulent nonequilibrium, compressibility, and shock–turbulence interaction effects must be accurately reproduced. Intuitively, it is expected that modeling the macroscopic and major unsteadiness of turbulent-separated regions necessitates the use of the more complicated and time-consuming Large Eddy Simulation (LES) and Direct Numerical Simulation (DNS) techniques. However, so far, they also have been relatively unsuccessful in determining the position of separation, possibly principally due to the difficulties associated with the "wall-layer" modeling. One saving feature for vehicle designers is that turbulent-boundary layers are difficult to separate in hypersonic flows and, when separation occurs, the major relevant characteristics (i.e., peak heating and pressure) can be estimated with simple methods combined with correlations from the experiments. As yet, no definitive methods have been developed to describe the length of the separated interaction region or the distribution of skin friction and heat transfer in the separated region. Earlier detailed reviews of shock-induced separated-turbulent flows and the methods to predict their characteristics by Greene [23], Stollery [24], Délery and Marvin [25], and Knight and Degrez [26] concentrated principally in the supersonic-flow regime for adiabatic walls. The large Reynolds numbers required to generate fully turbulent regions of SBLI in hypersonic flows are difficult to attain; hence, there are significantly fewer data from experimental studies available as well as validated models to characterize the size and detailed structure of turbulent-separated flows induced by SBLIs.

6.3.2 Characteristics of Turbulent SBLI in Two-Dimensional Configurations

Many of the earlier studies about the characteristics of shock wave–turbulent boundary-layer interaction in hypersonic flow were conducted with two-dimensional models, as in the case for laminar flows (see Section 6.2.2). The major flow features of attached and separated wedge- and shock-induced interaction regions are illustrated in the schlieren photographs in Figs. 6.21 and 6.22. A distinct characteristic of these flows close to when boundary-layer separation occurs is that the interaction region takes place at the base of the boundary layer. In contrast to laminar flows, the shock wave emanating from this region must traverse nearly all of the boundary layer because the sonic point is much closer to the wall. These features are evident in the schlieren photographs of the corner-interaction regions generated by 27- and 30-degree wedges on a flat plate in a Mach 8 flow (Fig. 6.21a,b). The imbedded separated regions induced in these flows do not generate a flowfield that can be described by a mutual interaction that results from the growth of the boundary layer responding to an adverse-pressure gradient. Only when the turning angle of the wedge approaches 33 degrees can the interaction begin to propagate upstream of the corner (Fig. 6.21c). For a wedge angle of 36 degrees (Fig. 6.21d), a well-separated region is formed with a clearly defined shear layer and constant-pressure-plateau region. A similar situation is observed for externally generated shock-induced flows. Separation ahead of the incident

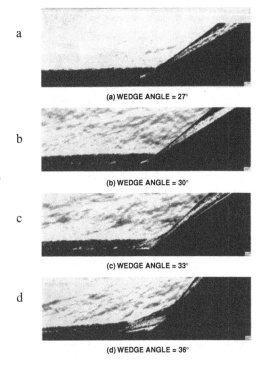

(a) WEDGE ANGLE = 27°

(b) WEDGE ANGLE = 30°

Figure 6.21. Turbulent separation over flat-plate wedge compression surfaces at Mach 8.

(c) WEDGE ANGLE = 33°

(d) WEDGE ANGLE = 36°

shock does not occur in the Mach 8 flow until the turning angle exceeds 17 degrees (Fig. 6.22d). In well-separated flows, there is a streamwise fluctuation of the separation point (determined from skin-friction measurements) over approximately two thirds of the boundary-layer thickness (Fig. 6.23). As previously noted, in hypersonic flow, it is relatively difficult to separate a turbulent-boundary layer in a compression corner or with an incident shock. This occurs for turning angles below 14 degrees in supersonic flows; however, at hypersonic speeds, the flow remains attached on compression surfaces with turning angles as high as 30 degrees or with shock-generator angles of 15 degrees. In fact, the exact determination of incipient separation in these flows is more difficult because of their unsteady nature as illustrated in Fig. 6.23, in which the streamwise position is shown to fluctuate. For this reason, computations based on time-averaged formulations are unlikely to prove satisfactory. Further discussion about the unsteady aspects of SBLIs can be found in chapter 9.

It is surprising that the conditions to induce incipient separation and the salient characteristics of wedge- and shock-induced hypersonic flows can be correlated reasonably well in terms of the free-stream Mach number (M), the overall pressure rise ratio ($p_{inc} - p_0$)/p_0, and the skin friction coefficient (C_f) to the wall immediately upstream of the interaction. This is illustrated in Fig. 6.24, in which the conditions to promote incipient separation in wedge- and shock-induced interaction regions is shown to correlate well over a wide range of conditions with a combination of these variables.

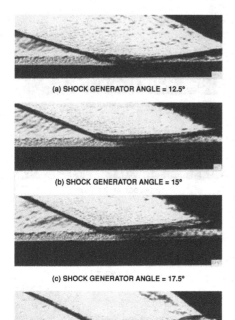

(a) SHOCK GENERATOR ANGLE = 12.5°

(b) SHOCK GENERATOR ANGLE = 15°

(c) SHOCK GENERATOR ANGLE = 17.5°

Figure 6.22. Turbulent separation induced by incident shock at Mach 8.

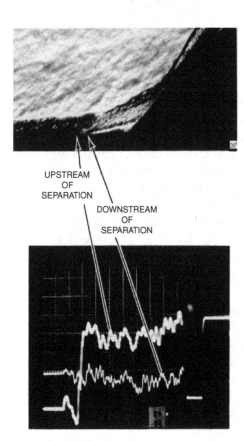

UPSTREAM OF SEPARATION

DOWNSTREAM OF SEPARATION

Figure 6.23. Flow separation and unsteady characteristics of the heat transfer recorded during the run time of a short-duration shock tunnel.

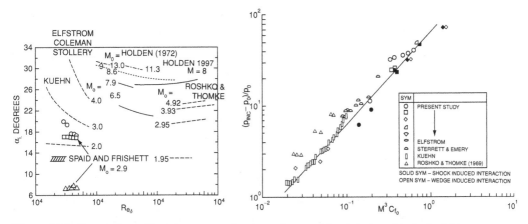

Figure 6.24. Correlation of incipient separation in wedge- and shock-induced turbulent flows.

Also, for well-separated flows, the plateau pressure can be correlated easily, as shown in Fig. 6.25. Most important, it is possible to correlate the maximum heating levels generated in the reattachment region – relative to the undisturbed upstream value – in terms of a simple power-law relationship with the pressure rise across the interaction region $q_{max}/q_0 = (p_{max}/p_0)^{0.85}$ (Fig. 6.26).

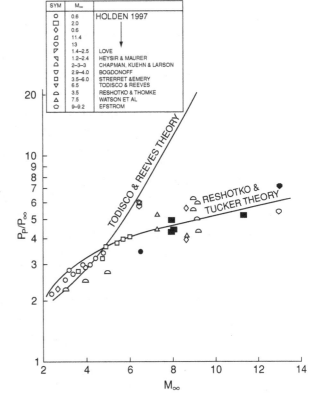

Figure 6.25. Correlation of plateau pressure in separated turbulent flows.

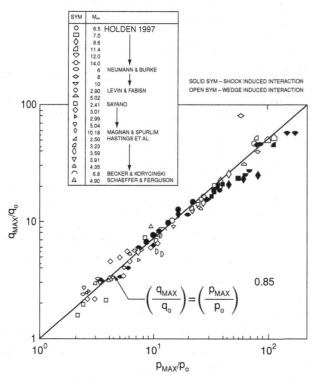

Figure 6.26. Correlation of peak heating in regions of SBLI.

6.3.3 Navier-Stokes Prediction of Shock Wave–Turbulent Boundary-Layer Interaction in Hypersonic Flow

More than a decade ago, Holden [27]concluded that there were no reliable techniques for predicting the size of shock-induced turbulent separated regions or the detailed distribution of heat transfer and skin friction to the walls bounding these regions. This conclusion still remains the case. For hypersonic flows, it is necessary to incorporate the influence of compressibility, so called low-Reynolds-number effects, and shock–turbulence-interaction effects into the turbulence models being used. For situations in which the flow is in the lower-Reynolds-number range, it is important to capture the large-scale instabilities observed. Currently, LES and DNS approaches are applicable to low-Reynolds-number turbulent flows, but it will require significant increases in available computational power to perform calculations of these flows at Reynolds numbers within a range that has practical relevance in hypersonic applications (i.e., $>10^7$). The rapid changes in the characteristics of the turbulence brought about by the deeply embedded shock waves at the separation and reattachment points, as well as the transition between attached and free-shear layers, pose extreme difficulties for modeling the turbulent structure at the higher Reynolds numbers.

Few solutions have been obtained for shock-induced separated flows in the hypersonic-flow regime. Horstman's [28] calculations for a Mach 11 shock-induced separated flow illustrate the basic problems encountered in describing the distributions of heat transfer and pressure through the interaction region (Fig. 6.27). The

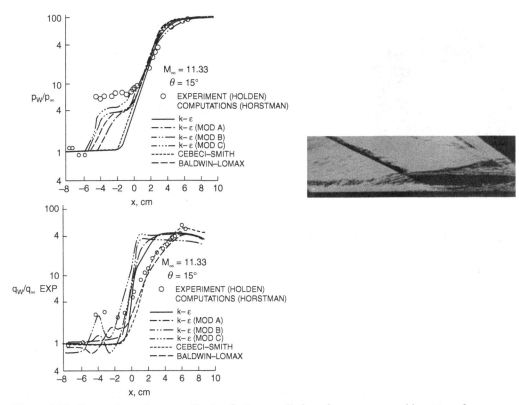

Figure 6.27. Comparisons between Navier-Stokes predictions for pressure and heat transfer with a range of turbulence models and measurements in a Mach 11 well-separated region of SBLI.

various turbulence models used in the calculations produce different answers for the separation length and the pressure rise in the separation process. Although most schemes are capable of predicting the peak heating and pressure at the end of the recompression process, they are unable to predict the distribution of heating during separation and in the separated region. It is difficult to blame this mismatch solely on the use of the time-averaged Navier-Stokes equations rather than their unsteady counterpart. However, the gross modeling of the turbulence in a manner that is unable to reflect the changes in its physical characteristics through these regions is certainly in question. There is a clear need for combined numerical and experimental studies to provide detailed flowfield and surface information with which to construct accurate models for development of turbulence in these flows in both the supersonic- and hypersonic-flow regime.

6.3.4 SBLI in Turbulent Hypersonic Flow on Axisymmetric Configurations: Comparison Between Measurements and Computations

Results from tests in which the SBLI occurs in a fully developed turbulent-boundary layer are sparse because of the experimental difficulties in obtaining adequate Reynolds numbers. Few hypersonic facilities are capable of accommodating suitably large models and, at the same time, operating at sufficiently high density to

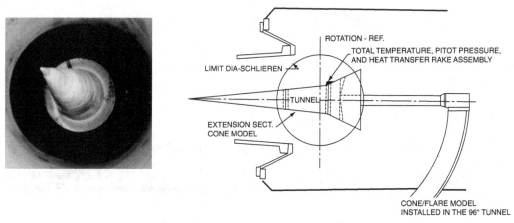

Figure 6.28. Large-cone–flare configuration used in studies of high Reynolds number shock wave–turbulent boundary layer interaction [27]. M_∞ 10.97 to 13.1; Re/m 1.12×10^6 to 1.39×10^6. Flare cone angles 36 and 42 degrees.

ensure that the flow is not transitional. One wind-tunnel experiment in which the necessary conditions were met was conducted by Holden [29] on a large cone–cylinder-flare configuration. This was an integral part of a program to investigate the flow on axisymmetric bodies aimed at providing boundary-layer transition data and measurements in regions of shock wave–turbulent boundary-layer interaction for code-validation studies in the high-Mach-number/high-Reynolds-number range. The model, shown in Fig. 6.28a,b, was tested at a free-stream Mach number of 11. This size was chosen so that transition occurred well upstream of the interaction region (i.e., more than one thousand boundary-layer thicknesses ahead) to eliminate concerns that the unsteadiness generated during the transition process would significantly influence the unsteady characteristics of the separated region. The mechanisms that cause this in turbulent separated flow have long been a concern of researchers; in particular, Dolling [30] investigated such effects in detail. In many studies of hypersonic flow, unsteadiness in the separated region was observed. However, it was not possible to rule out upstream transition as a trigger for these instabilities because there were extensive regions of boundary-layer transition upstream of the separated interaction region. In a more recent study conducted at a Reynolds numbers of 10^8 and a Mach number of 4.5, it was observed that the separated region ahead of a spherical dome was completely stable such that the propagation of optical beams through the shear layer ahead of the dome showed minimal distortion [31]. Measurements in the separated regions over the large-cone configuration also exhibited few large-scale instabilities, although it was observed that shock waves penetrating the boundary layer were distorted as a result of the interaction with boundary-layer turbulence. It appears from these observations that hypersonic-turbulent shock–boundary-layer interactions do not inherently exhibit large-scale unsteadiness but rather that they are susceptible to upstream disturbances. In the hypersonic-regime transition, the establishment of an equilibrium-turbulent boundary layer is a lengthy process and the residual disturbances are likely to

Figure 6.29. Installation photograph of full-scale HIFIRE configuration in LENS I shock tunnel.

perturb the SBLI if it is not located well downstream, thereby leading to apparent instabilities.

In a further phase of the combined study referred to at the beginning of Section 6.3.4, a slightly blunted cone followed by a cylindrical section and then a flare was selected for the initial ground-test studies of the HIFIRE vehicle configuration, in which a SBLI occurs at the flare junction (Fig. 6.29). High-frequency pressure and platinum thin-film instrumentation were used to define the characteristics of the transitional flow over the cone and cone–flare junction. These tests illustrate the problems encountered in properly predicting turbulent interactions.

Figure 6.30a,b shows the surface-pressure and heat-transfer measurements over two configurations of this model compared with code results. For the results presented here, the fore-cone half angle is 6 degrees and the flare angles are 36 and 42 degrees, respectively. For the computations, the selection of the compressibility model can significantly influence the heating level, even to the conical surface

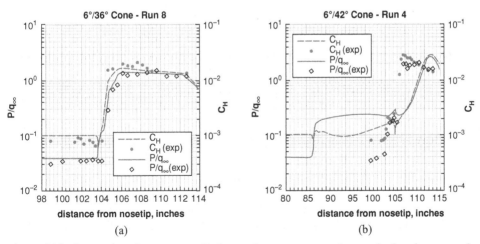

Figure 6.30. Comparison between prediction and measurements in attached and separated flows over the model shown in Figure 6.29: (a) pressure and heat transfer distributions in attached flows over 6°/36° cone–flare model; (b) pressure and heat transfer distributions in attached flows over 6°/42° cone–flare model.

(a) Schlieren photograph of flow over cylinder/37° flare configuration for M= 7.16 and Re₁/m = 0.98E+6.

(b) Gaitonde pre-test prediction using AVUS, unstructured NS-Solver

Figure 6.31. Comparison between measured and predicted flow pattern over cylinder–37-degree flare configuration: (a) Schlieren photograph of flow over cylinder–37-degree flare configuration for M = 7.16 and Re₁/m = 0.98E + 6; (b) Gaitonde pretest prediction using AVUS, unstructured NS-Solver.

upstream of the interaction. In these computations, the Shear Stress Transport (SST)[1] turbulence model was used. Figure 6.30a shows that the flow over the 36-degree configuration remains attached and the predictions of pressure and heat transfer are in good agreement with the experiment. However, for the 42-degree flare angle, shown in Fig. 6.30b, this model significantly overpredicts the length of the separated region at the cone–flare junction, although the pressure and heating levels downstream of the interaction are in reasonable agreement with the measurements.

The disagreement between the predicted and measured interaction physical scale and heating level is typical of many earlier comparisons. Figure 6.31a,b is a schlieren photograph of the 37-degree flare configuration compared with a more recent prediction made by Gaitonde et al. [32] (see also [23]) using an Air Vehicles Unstructured Solver (AVUS) parallel implicit unstructured Navier-Stokes solver. For this case, the RANS prediction gave a well-defined interaction region that was significantly smaller than found experimentally. In reality, as shown in the photograph, shear-layer reattachment occurred right at the base of the flare and thus did not provide a region of well-defined attached flow. Therefore, additional experiments were conducted with a 33-degree flare attached to the cone–cylinder configuration, the objective of which was to obtain a smaller separated region where the reattachment process was completed on the face of the flare (Fig. 6.32). Results from Navier-Stokes computations, made using the Data Parallel Line Relaxation (DPLR) method for the SST and the Spalart-Allmaras turbulence models including a compressibility correction, are shown in Fig. 6.32a,b. These two models provide different results and it is clear that significant efforts must be directed toward developing rational turbulence models for the separation, recirculation, and reattachment regions of the flow before a legitimate claim can be made for a successful turbulence model to describe these flows.

[1] The Shear Stress Transport (SST) model (see F. R. Menter, Zonal two-equation k-w turbulence models for aerodynamic flows. AIAA Paper 93–2906 [1993]) functions by solving a turbulence–frequency-based model (k-ω) at the wall and (k-ε) model in the bulk flow. A blending function ensures a smooth transition between the two models.

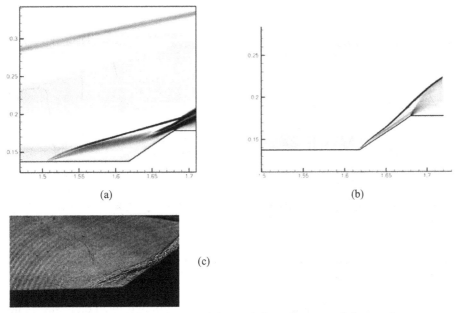

Figure 6.32. Comparison between predictions of size of separated flow and measurements on HIFIRE Flare: (a) DPLR solution using SST Model; (b) DPLR solution using Spalart-Allmaras model; (c) Schlieren photograph of separated region over 33-degree flare.

6.3.5 Swept and Skewed SBLIs in Turbulent Supersonic and Hypersonic Flows

The axial corner flow or swept-shock interaction has been one of the principal configurations selected for the investigation of three-dimensional regions of SBLI. In a frequently cited example of this flow category, a swept shock wave, generated by a wedge or fin mounted perpendicularly to a flat plate, impinges more or less normally on the flat-plate boundary layer (Fig. 6.33). Major studies in this area at Mach 3 with adiabatic-wall conditions were conducted by a Princeton University group headed by Bogdonoff and including Settles et al. [33], Dolling and Bogdonoff [34], Dolling and Murphy [35], and Dolling [30]. Measurements also were obtained by Stalker [36], Stanbrook [37], McCabe [38], Peake and Rainbird [39], and Cousteix and Houdeville [40] over a range of other conditions. Although incipient separation is relatively easy to define for two-dimensional turbulent interactions, this concept generated considerable controversy in three-dimensional flows. Whereas McCabe [38] suggested that separation should be defined on the basis of converging streamlines, Stanbrook [37] and others used criteria based on inflection points in the pressure distribution. The occurrence of separation in supersonic flows was correlated in simple terms by Korkegi [41] (see Fig. 6.34). He found that in low-Mach-number flow, the deflection angle Θ_{wi} for incipient separation varies as the inverse of the upstream Mach number (i.e., $\Theta_{wi} = 0.3/M_o$ radians), whereas for $2 < M < 3.4$, he suggested that p_i/p is independent of the Mach number. Goldberg's [42] and Holden's [43] measurements at Mach 6 and 11, respectively, do not agree with the Korkegi correlation. As noted previously, at low Mach numbers (i.e., $M = 2 \rightarrow 4$) and for adiabatic surfaces, a large body of data exists on the mean characteristics of

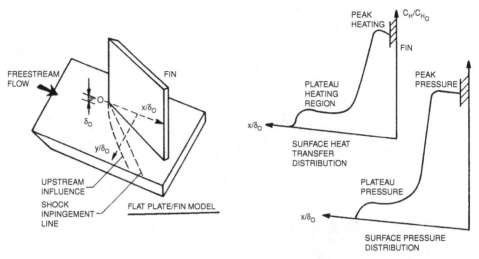

Figure 6.33. Typical surface heating and surface pressure distributions through swept shock–turbulent boundary layer interaction.

swept-shock interactions. Strangely, this body of three-dimensional data was found to be in better overall agreement with the solutions of Knight and Degrez [26], Horstman [29], Shang et al. [4, 5], and Settles and Horstman [46] to the Navier-Stokes equations than the relatively less complex two-dimensional flow separation over a flat plate/wedge. These results are not as sensitive to the turbulence model and suggest that the gross features of the flows are controlled principally by inviscid effects. Measurements of the peak-pressure ratio through the interaction and the plateau-pressure rise are in better agreement with calculations based on an inviscid-flow model in the two-dimensional theory of Reshotko and Tucker [47] than the correlations of Scuderi [48] (Fig. 6.34). It was found that, as in the studies of two-dimensional separated-interaction regions, peak heating can be related to the overall pressure rise by a simple power-law relationship, as shown in Figs. 6.35 and

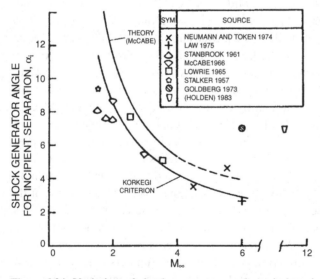

Figure 6.34. Variation of shock generator angle to induce incipient separation with Mach number [40].

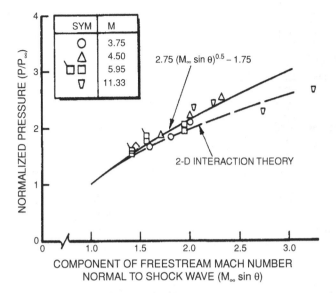

Figure 6.35. Correlation of plateau pressure measurement from swept-shock interaction studies [40].

6.36. Fig. 6.36 shows that the maximum pressure rise through the interaction region can be calculated with good accuracy from inviscid-flow relationships.

Another approach to exploring flow separation in regions of three-dimensional SBLI is to begin with a two-dimensional or axisymmetric interaction and sweep this interaction (or, for the axisymmetric case, introduce angle of attack) to progressively establish a crossflow in the interaction region. Experimental studies of this type were conducted by Settles et al. [33], who studied the interaction over swept and unswept flat-plate configurations over an adiabatic wind-tunnel wall in Mach 3 airflow. Considerable effort was expended in this study to determine the Reynolds-number scaling and the length from the upstream tip of the wedge for the flow to

Figure 6.36. Correlation of maximum pressures recorded in swept-shock interaction regions [40].

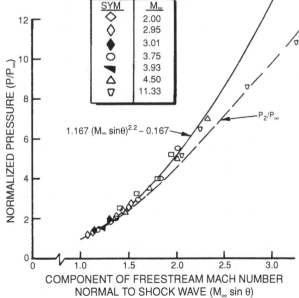

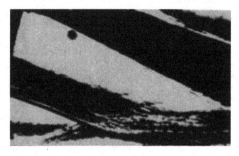

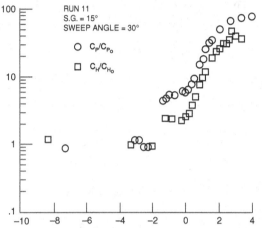

Figure 6.37. Effect of skewed shock on size and properties of interaction region.

become quasi-two-dimensional. However, the effect of changing the overall span-wise scale of the model on the scale of the interaction was not examined explicitly. The measurements of surface and pitot pressure through the interaction regions were in good agreement with solutions to the Navier-Stokes equations obtained by Horstman and Hung [49]; however, key features of the flow were poorly predicted. It is known that agreement with pressure data is not the most definitive test. Holden [43] conducted studies of crossflow effects on the size and properties of the inter-action region induced by a swept-oblique-shock incident on a turbulent-boundary layer over a flat plate at Mach 11 and $Re = 30 \times 10^6$. Experiments were conducted for two strengths of incident shock: the first (i.e., $\Theta_{SG} = 12.5$ degrees) was designed to produce a separated condition close to incipient separation and the second (i.e., $\Theta_{SG} = 15$ degrees) to generate a well-separated flow. Distributions of heat transfer and pressure and a schlieren photograph of the 30-degree swept condition are shown in Fig. 6.37. The plateau in these pressure and heat-transfer distributions provides evidence that the stronger incident shock produces a separated-flow region that, from this plot, can be inferred to extend a length of 2 inches (5 cm). This is cor-roborated by the existence of a shock wave seen in the photograph that emanates from the separation point. The measurements made of the heat-transfer and pres-sure distributions beneath the well-separated flow induced by both the 12.5- and 15-degree shock generators swept at angles of 0 and 30 degrees indicated that the induced crossflow has little effect on the size and characteristics of the interaction regions. This correlation is shown in Fig. 6.38. If there is a perceptible effect, it is a decrease in the length of the separated region with increased crossflow, a trend

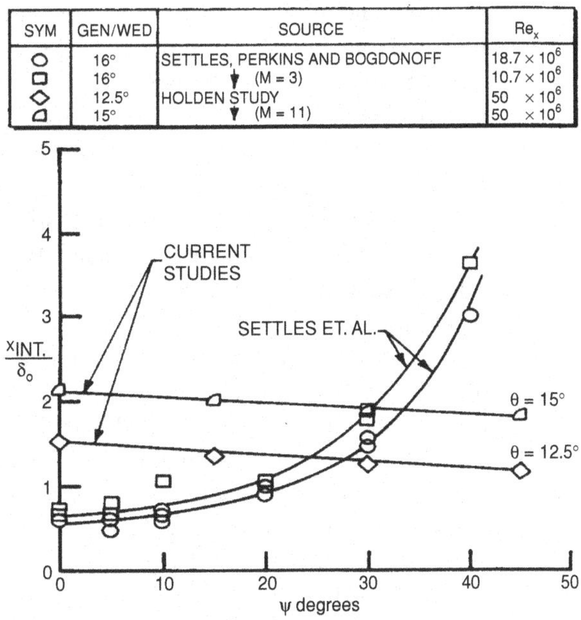

SYM	GEN/WED	SOURCE	Re_x
○	16°	SETTLES, PERKINS AND BOGDONOFF	18.7×10^6
□	16°	▼ (M = 3)	10.7×10^6
◇	12.5°	HOLDEN STUDY	50×10^6
◿	15°	▼ (M = 11)	50×10^6

Figure 6.38. Variation of interaction length with shock skew.

that is opposite to that observed by Settles et al. [33] in the Mach 3 flow of a wall-mounted swept wedge. It is also of interest that as in two-dimensional flows, the maximum heat transfer generated in the recompression regions of both swept and skewed interaction can be correlated with the standard power-law equation shown in Fig. 6.39.

6.3.6 Shock-Wave Interaction in Transitional Flows Over Axisymmetric/Indented Nose Shapes

In the late 1950s and early 1960s, there was interest in using laminar separated flows to obtain favorable heat-transfer characteristics over the nose tips of reentry

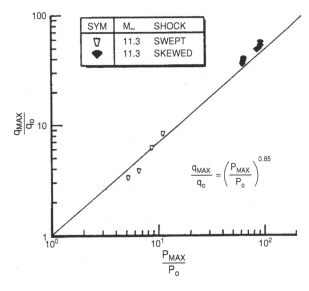

Figure 6.39. Correlation of peak-heating rates in skewed- and swept-shock interaction regions.

$$\frac{q_{MAX}}{q_o} = \left(\frac{P_{MAX}}{P_o}\right)^{0.85}$$

vehicles. Theoretical studies by Chapman [50] and experimental studies by Crawford [51] and Bogdonoff and Vas [52] suggested that by generating laminar separated flows over spiked bodies or over cavities elsewhere on the vehicles, it would be possible to reduce significantly the aerothermal loads at hypersonic speeds. Using a spike ahead of a blunt body also would significantly reduce drag. Crawford found that the total heat transfer to a spiked-hemispheric, fully capped cylinder was reduced below the basic hemispherical heating value for all spike lengths as long as the flow remained laminar. Bogdonoff and Vas investigated the separated flow promoted by a spiked flat-ended cylinder; their results indicated that the addition of the spike caused a 50 percent reduction in the total heat transfer to the front face of the cylinder for fully laminar flow. Although the measurements made in the Bogdonoff experimental studies provided total-heating information, they did not provide information about detailed distribution in the reattachment regions, which is key to understanding the major heating loads in regions of separated flows. Later, Holden [53] investigated the flow over a series of spiked, spherical, flat-ended, and conical nose tips to provide detailed information on heating rates in reattachment regions of the shear layer. Because of the large heat-transfer rates generated in the reattachment region in the transitional shear layer, integrated heating loads for all spiked lengths exceeded the total heating to an unspiked cylinder. Due to the ease with which laminar shear layers become transitional and because of the nonlinear effects associated with the extreme yaw sensitivity of the spiked-body flows, nose configurations incorporating them were deemed impractical for hypersonic vehicles.

Although interest in the separated flows over spiked bodies diminished, separated-flow and reattachment phenomena associated with these flows "resurfaced" when it was observed that indented nose shapes with imbedded shock interaction and separated regions were developed in transitional flows over ablating nose tips on ballistic missiles during reentry. These indentations and asymmetric nose shapes were of specific interest because they resulted in large destabilizing moments and side forces. Direct evidence to support the formation of the indented nose shapes was delivered from the unique postflight recovery of ablated nose shapes in the Nosetip Re-entry Vehicle (NRV) and Re-entry Transition Experiment (RTE) studies.

Photographs of the nose-tip replicas are in Fig. 6.40a,b with models that were used to examine the heating loads and determine the destabilizing forces developed as a result of these configurations [54]. Figure 6.40b shows a large-scale version of the NRV nose tip, which was highly instrumented with heat-transfer and pressure gauges to determine the heating and pressure loads on the nose tip at duplicated flight conditions. A schlieren photograph of the flow indicating regions of imbedded separated flow and shock-shock interaction and reported by English [55] is in Fig. 6.41. Typical heating patterns from these studies indicated heating loads in the reattachment region that exceed those at the stagnation point by a factor of more than two. Additional measurements [56] were made on a series of "ideal" ablated-nose shapes; Fig. 6.42 is a schlieren photograph of the flow of such configurations.

It is significant that during the studies of the flow over indented nose shapes, some of the first numerical solutions of the Navier-Stokes equations were obtained. Widhopf and Victoria [57] obtained solutions to the unsteady Navier-Stokes

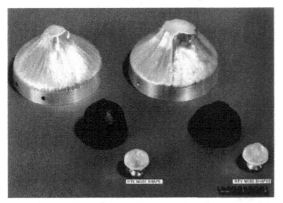

Figure 6.40. Replicas of the NRV and RTE nose tips recovered from flight tests and the installation of NRV model in 96-inch shock tunnel.

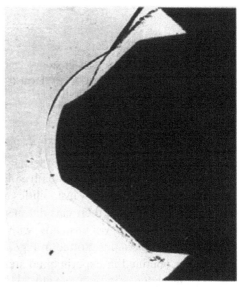

Figure 6.41. A Schlieren picture of the flow over the MRV nose tip.

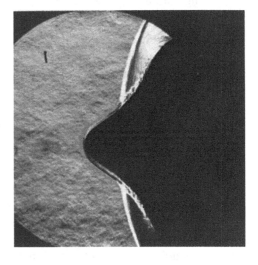

Figure 6.42. Shock interaction pattern over a rough "ideal" ablated nose shape.

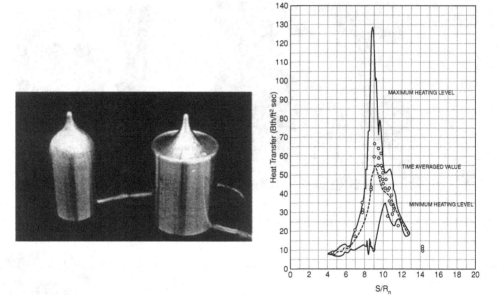

Figure 6.43. First Navier-stokes computation of a separated shock interacting flow over indented nose shape. *Right:* Heat-transfer comparison between theory and experiment for the Widhopf 1 configuration (1-10-10).

equations for laminar flows over the two idealized nose shapes shown in Fig. 6.43, designated as the Widhopf 1 and 2 shapes. These pioneering numerical computations also obtained unsteady solutions to the flow over the Widhopf 2 configuration, predicting flow instabilities not dissimilar to those observed earlier over spiked bodies. In fact, their calculations also suggested that the flow over the Widhopf 1 configuration was unsteady, with heating levels varying between the maximum and minimum values plotted in Fig. 6.43c. Also shown are measurements of heat transfer obtained in experimental studies by Holden [58], which did not indicate gross unsteadiness in the flowfield. The surface and flowfield measurements over this configuration demonstrated that regions of SBLI and shock-shock interaction were generated, resulting in large heating rates in the recompression region near the shoulder of the nose tip.

6.4 Characteristics of Regions of Shock-Shock Boundary-Layer Interaction

6.4.1 Introduction

The heating and pressure loads generated by regions of shock-shock boundary-layer interaction are among the largest that can be imposed on thermal-protection systems of vehicles traveling at hypersonic speeds. Typical heating loads can exceed more than thirty times the undisturbed stagnation-heating value and, in many cases, the flows are transitional, unsteady, and therefore extremely difficult to predict. As previously noted, shock-shock heating was responsible for catastrophic failures, including the loss of a National Aeronautics and Space Administration (NASA) scramjet engine when the supporting pylon was destroyed by shock-impingement

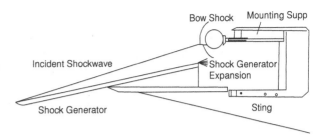

Figure 6.44. Schematic of shock generator model.

heating (see Fig. 6.2b) in an X-15 program [59]. Edney [60] classified a series of flow situations in terms of the relative position of the incident shock on the bow shock over a blunt body (see Section 2.2.3). The Type IV interaction (see Chapter 2 and Fig. 6.9a,b) generates a jet-like flow that impinges on the surface, forming a local stagnation region that can result in severe heating rates even if the flow remains laminar. However, for Type III interactions, in which a free shear layer is formed, transition in this layer can lead to even higher heating levels. Because this occurs at relatively low Reynolds numbers and therefore at high altitudes, in most practical situations, transitional shear layers are formed for both Types III and IV interactions, making the heating loads developed in these regions large and extremely difficult to predict correctly.

6.4.2 Shock-Shock Heating in Laminar, Transitional, and Turbulent Interactions

Because of the sensitivity of these flows to shear-layer transition, obtaining measurements for fully laminar shock-shock interaction regions is difficult and requires the ability to test at very low Reynolds numbers. The tests must be conducted with the Reynolds number falling between 10^3 and 10^4 based on the leading-edge diameter, which leads to flow conditions spanning the rarefied and continuum regimes. Experimental studies of shock-interaction heating typically use a test configuration like that shown in Fig. 6.44, in which the incident shock is generated by an angled flat plate. A photograph of a wind-tunnel model used for these studies is in Fig. 6.45.

Using different cylinder diameters, Holden et al. [61] conducted a series of tests in which they were able to obtain Reynolds numbers ranging from 500 to 800,000 based on cylinder diameter. Results are shown in Fig. 6.46 for the fully laminar cases. Heating-enhancement factors above the stagnation value are on the order of 10. For

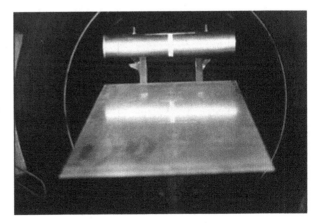

Figure 6.45. Photographs of models in the 48-inch shock tunnel.

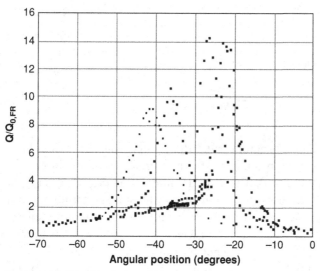

Figure 6.46. Distributions of heat transfer in fully laminar regions of shock-shock interactions.

flows with shear-layer Reynolds numbers higher than 5,000, transition occurs within this layer and, under these conditions, heating-enhancement factors of 15 and higher occur. Further increasing of the Reynolds number moves the transition point in the shear layer closer to the shock-interaction point, resulting in heating-enhancement factors for Type IV interactions higher than 20.

Similar variations in the peak heating with shock-impingement point are seen for transitional and fully turbulent shear layers [62]. The variation of the highest heating rate with Reynolds number for these flows is shown in Fig. 6.47. In the

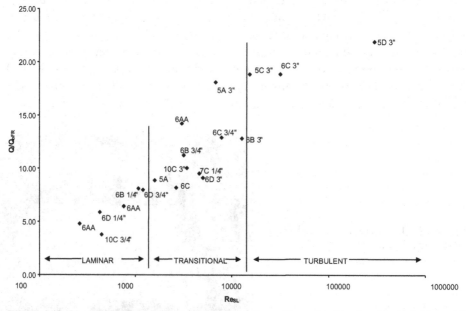

Figure 6.47. Variation of peak heating with shear layer Reynolds Number for both Types III and IV interactions.

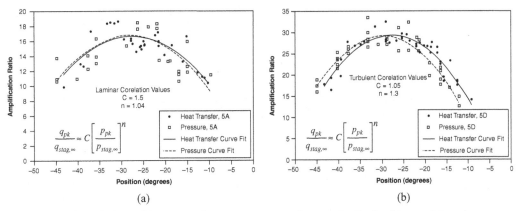

Figure 6.48. Correlation of heat-transfer measurements for nominally laminar and transitional and turbulent flows [61]: (a) nominally laminar; (b) transitional and turbulent.

former case, the shear-layer structure is such that the peak heating occurs for a Type III rather than a Type IV interaction, as is the case for a fully turbulent interaction. Because directly predicting transitional and turbulent heating using the Navier-Stokes code is particularly difficult for these flows, the most tractable approach may be to obtain peak-pressure estimates using Navier-Stokes codes (which is relatively simple with acceptable precision compared to obtaining the heat transfer) and then calculate a peak heating by using correlations such as those shown in Fig. 6.48a,b.

6.4.3 Comparison Between Measurements in Laminar Flows and Navier-Stokes and DSMC Predictions

As a part of code-validation activities supported by the ESA and later by NATO's Research Technology Organization (RTO), measurements obtained at ONERA [17] and at CUBRC were compared with computations for flows involving shock-shock interaction. Figure 6.49 is a typical example of calculations of pressure distributions obtained in studies conducted at ONERA using the short shock generator–cylinder model. Because of the combination of incident shock and expansion fan produced by this shock generator, it is a more complex flow than the configuration

Figure 6.49. Comparison between pressure distribution in shock-interaction region from ONERA studies and Navier-Stokes and DSMC calculations.

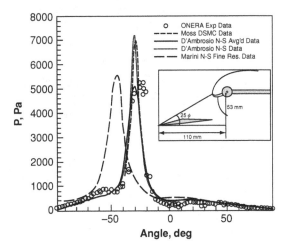

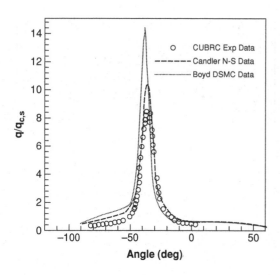

Figure 6.50. Comparison between heat-transfer measurements on cylinder in regions of SBLI and DSMC and Navier-Stokes computations by Boyd and Candler.

shown in the previous section. For these tests, the peak pressure and heating levels are significantly overpredicted by the numerical solutions. However, the pressure distribution is generally in good agreement.

Shown in Fig. 6.50 are comparisons between Navier-Stokes and DSMC solutions obtained by experienced computationalists and the measurements obtained in the CUBRC wind tunnels for a fully laminar interaction region. Typically, the calculations significantly overpredict the peak-heating and pressure levels, although general distributions are well characterized. Exactly how DSMC or Navier-Stokes predictions should be used to obtain peak-heating rates in laminar flows of this type remains to be evaluated.

6.5 SBLI Over Film- and Transpiration-Cooled Surfaces

6.5.1 Introduction

"Back face" vehicle-cooling methods, in which heat is extracted from beneath the skin, are frequently incapable of handling the large and dynamic heating loads developed in shock-interaction regions. Film- and transpiration-cooling are two alternative techniques that were proposed for hypersonic vehicles; the advantage is that they also reduce the skin friction to the walls. Although the relative merits must be evaluated against the complexities associated with the mechanical design of such systems, a more important consideration is their effectiveness in the severe environment generated by unsteady shock interactions. Film-cooling techniques using low-molecular-weight, high-specific-heat gases were used successfully to reduce the aerothermal loads on the optical windows of missile-seeker heads. Film-cooling is particularly attractive because of the flexibility of this technique for a range of free-stream conditions; however, recent studies show that the gaseous layers used for film-cooling may be separated readily and dispersed by incident shocks. Transpiration-cooling techniques also were shown to be highly successful in reducing heat-transfer and skin-friction levels on nose tips and the conical frustra of

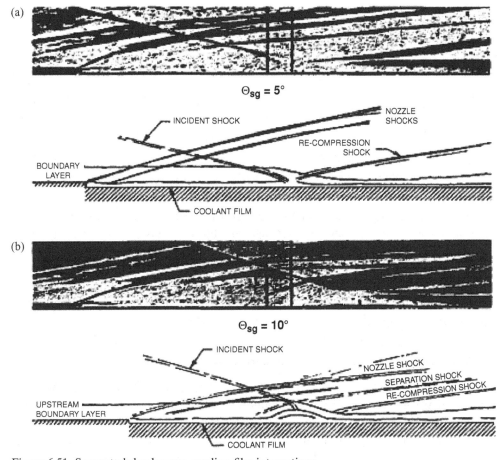

Figure 6.51. Separated shock-wave–cooling-film interactions.

hypersonic reentry vehicles. They also were effective in the combusting flows downstream of injectors in scramjet engines.

6.5.2 Shock Interaction with Film-Cooled Surfaces

Figure 6.51a,b illustrates major characteristics of the viscous-inviscid features of shock-wave–film-cooling interactions [63]. A critical aspect of these flows is whether the coolant layer can withstand the disturbance caused by the shock wave and remain intact throughout the interaction region. Figure 6.51a shows a case in which the flow remains attached and the coolant layer was not dispersed. In the separated region shown in Fig. 6.51b, the coolant layer was dispersed rapidly in the separation and reattachment regions, resulting in heating levels downstream of the incident shock that were not reduced by film-cooling.

Typical measurements of heat-transfer characteristics of a shock-coolant-layer interaction are shown in Figs. 6.52 and 6.53 for shock-generator angles Θ_{sg} of 5.5 and 8 degrees, respectively. For the 5.5-degree generator, the small separated region formed in the matched blowing case was swept away when the blowing rate (λ) was doubled. As illustrated in Fig. 6.52, the coolant remained intact and caused a

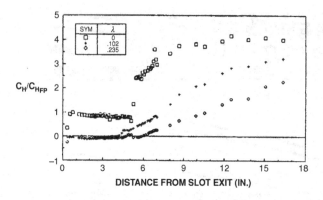

Figure 6.52. Heat-Transfer distribution in regions of incident-shock/wall-jet interaction ($\Theta_{sg} = 5.5°$, slot height = 3 mm [0.120 inch]).

reduction in peak-heating. For the configuration shown in Fig. 6.53, in which the flow was separated by the stronger incident shock, the coolant film was dispersed and the heating levels returned rapidly to those in the absence of the film. These studies suggest that in regions of strong shock interaction, the film-cooling technique is not robust and significantly less effective than transpiration-cooling (discussed in the following section). Predicting whether the film remains intact when impacted by the shock is a task that cannot be accomplished reliably.

6.5.3 Shock Interaction with Transpiration-Cooled Surfaces

Transpiration-cooling systems are particularly attractive in protecting the walls of scramjet engines in which large aerothermal loads can develop in the combustor in regions of shock wave–turbulent boundary-layer interaction. At the same time, they can reduce significantly the skin friction to the walls, thereby reducing a major component of engine drag. Experimental studies [64] examined shock interaction utilizing a transpiration-cooled model with the experimental configuration shown in Fig. 6.54. In these studies, measurements were made with nitrogen and helium coolants for a range of blowing rates and incident-shock strengths. Figure 6.55 is a typical distribution of heat transfer and pressure in regions of shock–coolant-layer interaction for a 5.3-degree shock generator and helium coolant. It is observed that introducing a helium coolant at the rate of 2 percent is required to reduce the heating level downstream of shock impingement to less than the level upstream of the shock on the smooth plate.

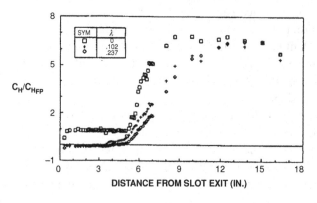

Figure 6.53. Heat-transfer distribution in regions of incident shock/wall-jet interaction ($\Theta_{sg} = 8.0°$, slot height = 3 mm [0.120 inch]).

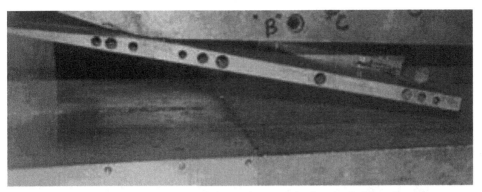

Figure 6.54. Shock generator supported above the transpiration-cooled flat plate.

From these experiments, a simple correlation relationship was developed (Fig. 6.56) based on reduction of the heat-transfer coefficient and a modified blowing parameter $\dot{m}/(\rho_s U_s Ch_s)(C_{p,inj}/C_{p,fs})^{0.7}(M_{fs}/M_{inj})^{0.5}$ determined from the local inviscid conditions downstream of the shock denoted by subscripts. By using this correlation with simple calculations to determine the local inviscid-flow conditions, it is possible to provide good estimates for the levels of mass addition required to control the peak-heating levels downstream of shock impingement. These correlations are more reliable than the computations for turbulent flow.

6.5.4 Shock-Shock Interaction on Transpiration-Cooled Leading Edges

Although transpiration cooling was shown to be effective in protecting flat surfaces in SBLI regions, using a similar technique to protect leading-edge surfaces from heating generated by shock-shock interaction proved less successful, as demonstrated by Holden et al. [65]. Transpiration-cooled spherical nose tips were tested to evaluate the effectiveness of this technique in the presence of Types III and IV shock-shock–interaction heating using the model shown in Fig. 6.57. The holes through which the coolant was passed are clearly visible. Measurements of the heating distribution on a smooth model without blowing (Fig. 6.58) are compared with those for a high blowing rate ($\lambda = 0.2$) in Fig. 6.59. It is clear from comparing these

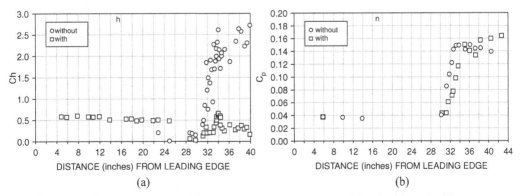

Figure 6.55. (a) Heat-transfer and (b) pressure measurements at Mach 6 with and without helium transpiration cooling for shock interaction from $5.3°$ shock generator [62].

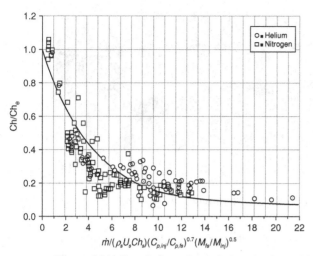

Figure 6.56. Correlation of heating reduction ratio with modified blowing parameter for shock-generator angles of 5°, 7.5°, and 10° and both nitrogen and helium coolants [61].

sets of measurements that transpiration-cooling had little or no effect on peak-heating or heat-transfer distribution in the shock-interaction regions (Fig. 6.60). Measurements of this type that were taken for a series of shock locations confirm that neither the magnitude nor the shape of the peak-heating region was altered significantly by transpiration-cooling.

6.6 Real-Gas Effects on Viscous Interactions Phenomena

6.6.1 Introduction

During the major part of the hypersonic-flight regime, the dissociation thresholds of oxygen and nitrogen are exceeded (Fig. 6.61) and the air in the shock and boundary layers enveloping a vehicle are in thermal and chemical nonequilibrium. At high altitudes, the flow also is likely to be ionized. The changes that these real-gas

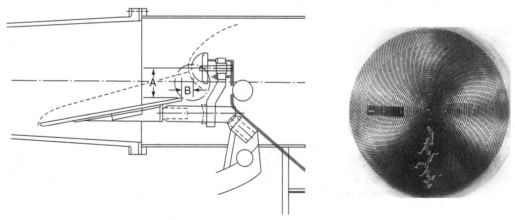

Figure 6.57. Schematic diagram of experimental configuration in CUBRC's 48-inch (1.22 m) hypersonic shock tunnel.

Figure 6.58. Heat-transfer distribution in shock-shock-interaction regions induced by a 10-degree shock generator over a transpiration-cooled hemisphere without blowing at Mach 12.

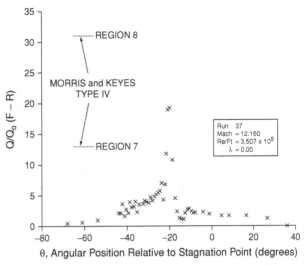

Figure 6.59. Heat-transfer distribution in shock-shock-interaction regions induced by a 10-degree shock generator over a transpiration-cooled hemisphere with $\lambda = 0.20$ at Mach 12.

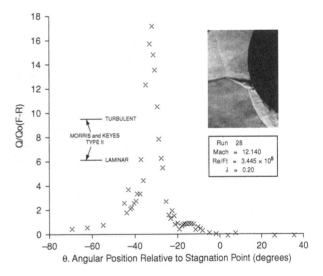

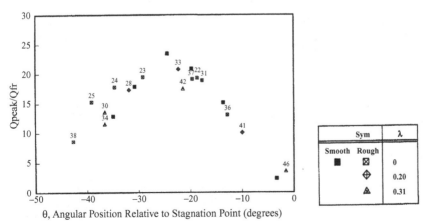

Figure 6.60. Variations of peak heating with angular position of the interaction region for various blowing parameters λ at Mach 12, $Re/m = 1.07 \times 10^6$.

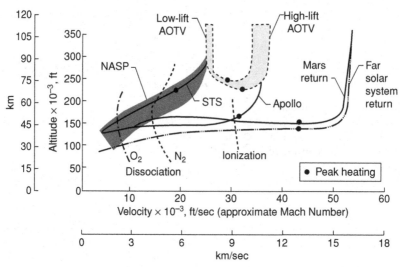

Figure 6.61. Velocity altitude trajectories for past and current reentry systems.

and chemical-reaction phenomena introduce significantly affect the flow structure, including the wake., For example, determining how they influence the stability of a vehicle or what the heating loads are in the presence of surface catalysis remain key problems for aerospace designers. The difficulties are exacerbated for ablating rough heat shields. As illustrated in Fig. 6.62, the influence of chemistry on control-surface performance in which SBLIs play a critical role remain of key importance; however, for velocities above 4.25 km/s (14,000 ft/s), direct validation of the models used to predict it and the effect on transition, vehicle stability, and blackout phenomena remains unavailable.

Real-gas effects are acknowledged as the principal reason for the control-surface anomalies experienced during the flight of STS-1 (Figs. 6.63 and 6.64).

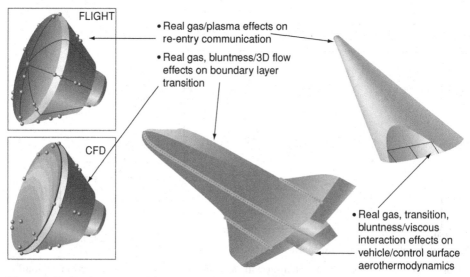

Figure 6.62. Real-gas, transition, and turbulent boundary-layer effects on reentry vehicle performance.

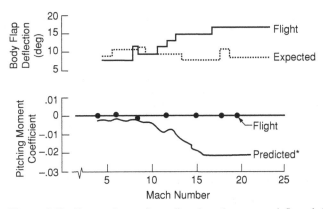

Figure 6.63. Comparison of predicted and measured flap deflection and pitching-moment coefficient from STS-1 flight.

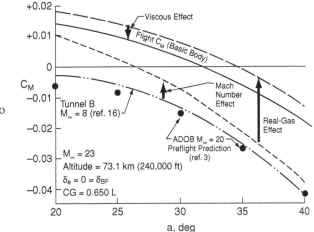

Figure 6.64. Postulated real gas effect to explain pitch trim anomaly.

Nevertheless, with the exception of tests in the Mach 5 CF_4[2] wind tunnel at NASA and preliminary studies conducted on the Hallis configuration in European high-enthalpy facilities, there are no definitive experimental measurements to validate the calculations on which the design was based. The exact scale of the influence of real-gas effects and chemistry on SBLI phenomena relative to flap controls has yet to be determined (Fig. 6.65).

Section 6.3 concludes that in the absence of real-gas effects, reliable solutions to the Navier-Stokes equations can be obtained for laminar separated flows and those involving SBLIs. Using methods such as those employed by Candler and Gnoffo [21], accurate predictions are possible for the nitrogen flow over the hollow-cylinder–flares, compression surfaces, and complex regions of shock-shock and SBLIs with imbedded separated flows over double-cone configurations. However, as shown in Fig. 6.66, this is not the case for flows when the nitrogen is replaced by air for velocities of 4.25 km/s (14,000 ft/s) and above, in which the fluid in the shock layer and separated regions are in a state of chemical nonequilibrium. The measured separated region is significantly larger than that predicted by the codes.

[2] The test medium in this facility is tetrafluoromethane. The density ratios across shock waves in this gas are heightened and, consequently, flows in the hypersonic regime with the Mach numbers roughly twice that in the test flow can be simulated in some measure.

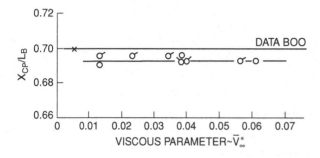

Figure 6.65. Shock-tunnel measurements indicating viscous interaction effects on shuttle stability.

The experimental heat transfer and pressure levels in the reattachment region also differ significantly from computations made by Candler [66] (see Fig. 6.66). A similar result was obtained independently by Gnoffo [67] with the Langley Aerothermodynamic Upwind Relaxation Algorithm (LAURA) code for the double-cone body at 4.25 km/s.

This inability to predict flows involving SBLIs with Navier-Stokes codes is in contrast to the experience for other situations in which real-gas effects have an important role. For example, in America and Europe, studies examined their

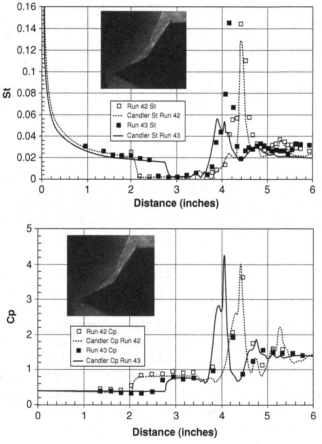

Figure 6.66. Comparisons between experiment and prediction by Candler for flows of air and nitrogen for the double-cone body at 4.25 km/s.

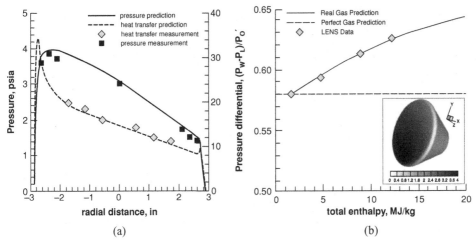

Figure 6.67. Comparisons between prediction and measurements on the Apollo capsule demonstrating reduced vehicle stability resulting from changes in the pressure distribution with real-gas effects: (a) comparison between prediction and measurement; (b) computation of flowfield.

influence on the flow over "Apollo-like" capsule configurations and, specifically, on their role in decreasing the stability. It was concluded that good agreement can be achieved between measurement and computation. In a further series of tests using the LENS I facility in the air with the Apollo capsule at angle of attack, in which the decrease in stability resulting from real-gas effects was investigated, measurements were taken of heat transfer and pressure on the front face of the model for total enthalpies of 2, 5, 10, and 13 MJ/kg. One set of measurements at the lower-enthalpy condition and predictions using the DPLR code are compared in Fig. 6.67a. Both the pressure and heat-transfer measurements are shown in excellent agreement with numerical predictions. In Fig. 6.67b, comparisons are shown between theory and prediction plotted in terms of the difference between the pressure on the windward and leeward edges of the capsule, nondimensionalized by the pitot pressure plotted against total enthalpy. Plotted in this format (which was used originally by ESA to evaluate flight data on an Apollo-shaped capsule), good agreement between measurements and predictions for the real-gas solution is shown, which differs significantly from the prediction based on perfect-gas assumptions. Further studies [68] of real-gas effects on capsule flows in air, nitrogen, and carbon dioxide with moderate favorable pressure gradients suggest that the routines incorporated to model flowfield chemistry in the current codes are accurate up to total enthalpies of 13 MJ/kg, as long as the steep gradients experienced in SBLI flows or rapidly expanding flows (e.g., converging-diverging nozzles) are not present – at least, for moderately dense flows.

6.6.2 Studies of Real-Gas Effects on Aerothermal Characteristics of Control Surfaces on a U.S. Space Shuttle Configuration

Little experimental information is available on the influence of chemistry and other real-gas effects on SBLIs. One exception is associated with the U.S. space shuttle

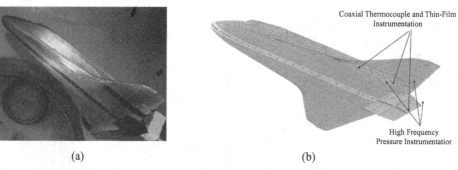

Figure 6.68. Models and instrumentation for studies of real-gas effects on control-surface performance and boundary-layer transition: (a) photograph of shuttle configuration in LENS I tunnel for real-gas studies; (b) instrumentation locations for high-enthalpy real-gas studies.

body flap. Because measurements of the aerodynamic performance of this vehicle during its first reentry in the earth's atmosphere were significantly different than the preflight predictions – which were based almost entirely on ground-test data in low-enthalpy flows – the exact reasons for the discrepancy remain controversial. Because during a critical phase of reentry the flap was deflected fully to maintain vehicle stability and to an angle significantly larger than had been predicted as necessary, the discrepancy became known as the Shuttle Flap Anomaly. In compiling ground-test data leading to the preflight predictions, engineers did not include measurements in high-Mach-number, low-Reynolds-number flows made in shock tunnels that indicated that viscous-interaction effects significantly could reduce flap force and the pressures generated on the adjacent surfaces. Soon after the flight, the anomaly apparently was explained in simple terms of a reduction in surface pressure to the aft-windward expansion surfaces of the shuttle resulting from real-gas effects [69]. This explanation later was reinforced with more detailed numerical solutions to the Navier-Stokes equations [70]. However, because the models of nonequilibrium-flow chemistry have yet to be validated at high-temperature, low-density conditions experienced by the shuttle when this stability problem occurred, there remain serious questions about the soundness of the modeling; hence, the exact reason for the Shuttle Flap Anomaly is still uncertain.

To further elucidate this problem, a 1.8 percent model of the space shuttle was tested in the CUBRC LENS I tunnel, as shown in Fig. 6.68. The replica is similar to that used in studies conducted in the United States and Europe to investigate the shuttle's aerothermal characteristics, with particular emphasis on control-surface performance.

Figure 6.69 shows correlations of the measurements made in these tests ahead of and over the shuttle flaps in air and nitrogen at unit Reynolds numbers of 4,000 and 6,000. These measurements demonstrate that at the end of the flap, the datasets collapse to a level consistent for laminar reattachment. At the 10 MJ/kg test condition, the tests in air result in significant oxygen dissociation in the shock layer approaching the flap, whereas those in nitrogen at the same enthalpy produce a shock layer in which nonequilibrium effects are small. Thus, by examining the two sets of measurements shown in the figure, real-gas effects are shown to decrease the size of the interaction region and increase the length of flap over which the flow is attached, and

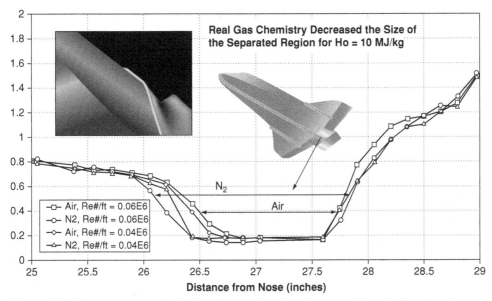

Figure 6.69. Measurements ahead of and over the flap demonstrating significant real-gas effects on flap effectiveness.

larger pressures are generated. The size of the separated-interaction region also is influenced by the Reynolds number, as shown in Fig. 6.69. The fact that the interaction increases in size with Reynolds number is another strong indication that the separated-interaction region remained laminar.

The effects of real-gas chemistry over the surfaces adjacent to a flap also were investigated by conducting tests at a fixed Reynolds number while varying the total enthalpy of the flow from 3 to 10 MJ/kg to change the level of oxygen dissociation

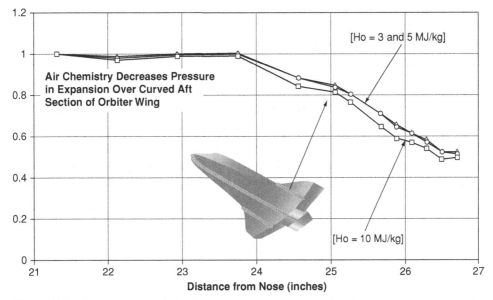

Figure 6.70. Measurements on shuttle wing demonstrating a reduction in pressure as a result of real-gas effects.

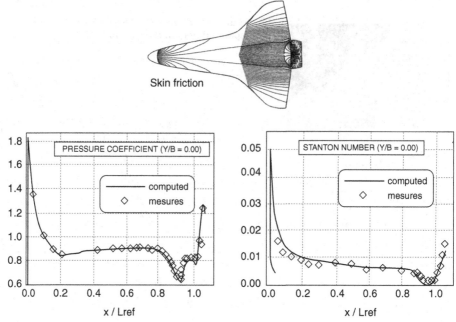

Figure 6.71. Comparison between Navier-Stokes predictions and pressure and heat-transfer measurements on Hallis shuttle configuration.

in the shock layer over the wing. The measurements obtained in this series of runs are in Fig. 6.70, nondimensionalized by a reference pressure ahead of the curved section of the wing to demonstrate that nonequilibrium-air chemistry acts to reduce the relative pressure in the expansion process on the curved aft section of the wing. Although pressure differences are relatively small, they act over a large area at the trailing edge of the wing, thereby resulting in a significant decrease in the pitching moment. Thus, measurements in these studies indicate that contrary to popular belief, the flap is relatively more effective as a result of nonequilibrium effects. However, as anticipated, the major contributor to the reduced stability of the shuttle relative to measurements made in low-enthalpy flows results from a decrease in the pitching moment caused by reduced pressures on the rearward part of the curved wing.

Since the studies surrounding the investigation of the shuttle stability problem were undertaken, little has been done to experimentally validate the numerical predictions used to explain these phenomena. However, more recent measurements taken on the Hallis shuttle configuration in Europe were in general agreement with numerical predictions (Fig. 6.71). This is in contrast to the unsatisfactory outcome shown in Fig. 6.66 and, although it provides encouragement, it is clear that further research is needed to develop and validate the computational models that reproduce the chemistry within the codes.

6.7 Concluding Remarks

In the past fifty years, major advances in computer power, coupled with the development of reliable Navier-Stokes and DSMC solvers, effectively eliminated the need to model viscous-inviscid interaction phenomena in terms of boundary-layer theory. Consequently, the emphasis on devising numerical techniques to predict accurately the characteristics of complex interacting flows now is refocused toward the development and validation of models of physical phenomena such as surface and flowfield chemistry (see Chapter 7); wall- and shear-layer transition; turbulence development in the shear layer; and wall-bounded flows in the presence of strong pressure gradients, shocks, flow chemistry, and combustion. This chapter explained that *fully laminar* SBLI flows – in the absence of real-gas and chemistry effects – can be predicted reliably with currently available computational methods. Whereas the flow can be expected to remain laminar during the high-altitude parts of the trajectories of practical vehicles, chemical reactions will occur and have a significant effect on flow structure. However, as yet, reliable methods to predict such flows are not available; hence, further studies are needed to develop and validate the models to calculate complex interacting flows above velocities of 3.5 km/s (11,500 ft/s). In these circumstances, computational predictions cannot be relied on to determine the critical details of these interactions in high-enthalpy flows without careful experimental confirmation of the results.

In hypersonic flows, boundary- and shear-layer transition is of major concern, particularly within regions of shock-shock or SBLI where it can occur within shear layers almost at the low-density limits of continuum flow. Predicting the characteristics of transitional regions of SBLIs is of major importance at these Mach flows; yet, their modeling remains in its infancy. In a similar way, predicting separated regions of shock wave–turbulent boundary-layer interaction in hypersonic flow is complicated by issues associated with the influence of thermodynamic and chemical nonequilibrium linked to boundary-layer transition, the effects of low Reynolds number and compressibility, shock–turbulence interaction, and the potential for macro and gross instabilities in separated flows. These flows are so complex that further major advances in this area clearly will be necessary to perform careful and extensive experimentation to provide datasets with surface and flowfield mean and fluctuation measurements integrated with computations using either RANS or (if possible) LES and DNS methods.

Obtaining fundamental measurements to examine and develop models of flowfields in which combustion and turbulence occur within the interaction (e.g., scramjet combustor) is also of significant practical importance. The ability to predict the high aerothermal loads developed in the shock-interaction regions – which requires accurate models of flowfield chemistry, boundary-layer transition, and turbulent–shock interaction in the presence of air and combustion chemistry – is another priority area. Again, there must be a strong coupling between the numerical computations and the design of appropriate experiments and analysis of the data. Finally, there remain significant gaps in our computational abilities to describe interactive flows with active cooling systems such as ablators and film- and transpiration-cooled surfaces. Prediction of SBLI regions in the presence of the combined effects of surface roughness and blowing and shock-interaction effects over actively cooled surfaces remains difficult. These significant problems require the development of models of transition and turbulence specific to these flowfields.

REFERENCES

[1] M. S. Holden. Study of flow separation in regions of shock wave–boundary layer interaction in hypersonic flow. *AIAA Paper 78–1169* (1978).

[2] M. J. Lighthill. On boundary layers and upstream influence. Part II: Supersonic flows without separation. *Proceedings of the Royal Society, London. Series A*, 217 (1953), 1131, 217, and 478.

[3] K. Oswatitsch and K. Weighardt. Theoretical analysis of stationary flows and boundary layers at high speed. *German Wartime Report* (1941). Translated as *NACA TM* 1189.

[4] G. E. Gadd. A theoretical investigation of laminar separation in supersonic flow. *Journal Aeronautical Sciences*, 24 (1957), 10, 759–71.

[5] H. S. Glick. Modified Crocco-Lees mixing theory for supersonic separated and reattaching flows. *Journal Aeronautical Sciences*, 29 (1962), 10, 1238–44.

[6] M. Honda. A theoretical investigation of the interaction between shock waves and boundary layers. *Journal Aeronautical and Space Science*, 25 (1958), 11, pp. 667–8; see also Tokyo University, Japan. *Report Institute High Speed Mechanics*, 8 (1957), pp. 109–30.

[7] L. Lees and B. L. Reeves. Supersonic separated and reattaching laminar flows: I. General theory and application to adiabatic boundary layer–shock wave interactions. *AIAA Journal*, 2 (1964), 1907–20.

[8] D. R. Chapman, D. M. Kuehn, and H. G. Larson. The investigation of separated flows in supersonic and subsonic streams with emphasis on the effects of transition. *NACA Report 1356* (1958).

[9] J. N. Nielsen, F. K. Goodwin, and G. D. Kuhn. Review of the method of integral relations applied to viscous interaction problems including separation. *NEAR Paper No. 7*, (presented at Symposium on Viscous Interaction Phenomena in Supersonic and Hypersonic Flow, May 7–8, 1969).

[10] J. E. Lewis, T. Kubota, and L. Lees. Experimental investigation of supersonic laminar, two-dimensional boundary layer separation in compression corner with and without cooling. *AIAA Paper 67–191* (1968).

[11] M. S. Holden. Theoretical and experimental studies of separated flows induced by shock wave-boundary layer interaction. *AGARD Specialists Meeting (Separated Flows, Brussels, Belgium, 10–13 May 1966)*; see also *AGARD Conference Proceedings No. 4, Part I, Separated Flows*, pp. 147–80 (1966).

[12] M. S. Holden. Theoretical and experimental studies of the shock wave–boundary layer interaction on curved compression surfaces. ARL Symposium on *Viscous Interaction Phenomena in Supersonic and Hypersonic Flow*, WPAFB, OH, USA (May 7–8, 1969).

[13] M. S. Holden. Shock wave–turbulent boundary layer interaction in hypersonic flow. *AIAA Paper 72–74* (1972).

[14] R..W. MacCormack and B. S. Baldwin. A numerical method for solving the Navier-Stokes equations with application to shock-boundary layer interactions. *AIAA Paper 75–1* (1975).

[15] C. M. Hung and R. W. MacCormack. Numerical solutions of supersonic and hypersonic laminar flows over a two-dimensional compression corner. *AIAA Paper 75–2* (1975).

[16] D. H. Rudy, J. L. Thomas, A. Kumar, P. A. Gnoffo, and S. R. Chakravarthy. A validation study of four Navier-Stokes codes for high-speed flows. *AIAA Paper 89–1838* (1989).

[17] T. Pot, B. Chanetz, M. Lefebvre, and P. Bouchardy. Fundamental study of shock-shock interference in low-density flow: Flowfield measurements by DLCARS. In *Rarefied Gas Dynamics* (Toulouse: Cépadués-Éditions, 1998), Part II-545.

[18] *Proceedings of the Aerothermodynamics Workshop Reentry Aerothermodynamics and Ground-to-Flight Extrapolation: Contributions to Hyperboloid Flare Test Cases, and Hallis in F4 Conditions Computation of case* 1.D. Eds. S. Menne and G. Hartmann, ESTEC, Noordwijk, the Netherlands (European Space Agency, 1996).

[19] M. S. Holden and T. P. Wadhams, Code validation study of laminar shock-boundary layer and shock-shock interactions in hypersonic flow, part a: experimental measurements, *AIAA Paper 2001-1031*, (2001).

[20] T. P. Wadhams and M. S. Holden. Summary of experimental studies for code validation in the LENS facility and comparisons with recent Navier-Stokes and DSMC solutions for two- and three-dimensional separated regions in hypervelocity flows. *AIAA Paper 2004–0917* (2004).

[21] M. S. Holden, T. P. Wadhams, J. K. Harvey, and G. V. Candler. Comparisons between DSMC and Navier-Stokes solutions on measurements in regions of laminar shock wave–boundary layer interactions in hypersonic flows. *AIAA Paper 2002–0435* (2003).

[22] G. A. Bird. *Molecular Gas Dynamics and the Direct Simulation of Gas Flows*, (Oxford: Clarendon Press, 1994).

[23] J. E. Greene. Interactions between shock waves and turbulent boundary layers. *Progress in Aerospace Science*, 11 (1970), 235–340.

[24] J. L. Stollery. Laminar and turbulent boundary layer separation at supersonic and hypersonic speeds. *AGARD CP-169* (1975).

[25] J. Délery and J. Marvin. Shock wave boundary layer interactions. *AGARDograph* 280 (1986).

[26] D. D. Knight and G. Degrez. Shock wave boundary layer interactions in high Mach number flows: A Critical survey of current numerical prediction capabilities. *AGARD Advisory Report 315*, 2 (1998).

[27] M. S. Holden. Studies of the mean and unsteady structure of turbulent boundary layer separation in hypersonic flow. *AIAA Paper 91–1778* (1991).

[28] C. C. Horstman. Prediction of hypersonic shock wave–turbulent boundary layer interaction flows. *AIAA Paper 87–1367* (1987).

[29] M. S. Holden, Studies of the mean and unsteady structure of turbulent boundary layer separation in hypersonic flow, *AIAA 91-1778* (1991).

[30] D. S. Dolling. Effects of Mach number in sharp fin-induced shock wave turbulent boundary layer interaction. *AIAA Paper 84–0095* (1984).

[31] M. S. Holden and T. P. Wadhams. A database of aerothermal measurements in hypersonic flows in "building-block" experiments for CFD validation. *AIAA Paper 2003–1137* (2003).

[32] D. V. Gaitonde, P. W. Canupp, and M. S. Holden. Evaluation of an upwind-biased method in a laminar hypersonic viscous-inviscid interaction. *AIAA Paper 2001–2859* (2001).

[33] G. S. Settles, T. J. Fitzpatrick, and S. M. Bogdonoff. A detailed study of attached and separated compression corner flowfields in high Reynolds number supersonic flow. *AIAA Paper 78–1167* (1978).

[34] D. S. Dolling and S. M. Bogdonoff. An experimental investigation of the unsteady behavior of blunt fin-induced shock wave turbulent boundary layer interactions. *AIAA Paper 81–1287* (1981).

[35] D. S. Dolling and M. Murphy. Wall-pressure fluctuations in a supersonic separated compression ramp flowfield. *AIAA Paper 82–0986*, (1982).

[36] R. J. Stalker. The pressure rise at shock-induced turbulent boundary layer separation in three-dimensional supersonic flow. *Journal Aeronautical Science*, 24 (1958), 547.

[37] A. Stanbrook. An experimental study of the glancing interaction between a shock wave and a turbulent boundary layer. *ARC CP 555* (1961).

[38] A. McCabe. The three-dimensional interaction of a shock wave with a turbulent boundary layer. *Aeronautical Quarterly*, 17 (1966), 3, 231–52.

[39] D. J. Peake and W. J. Rainbird. The three-dimensional separation of a turbulent boundary layer by a skewed shock wave and its control by the use of tangential air injection. *AGARD CP-168* (1975).

[40] J. A.Cousteix and R. Houdeville. Thickening and separation of a turbulent boundary-layer interacting with an oblique shock. *Recherche Aerospatiale*, 1 (1976), 1–11.

[41] R. H. Korkegi. A simple correlation for incipient turbulent boundary-layer separation due to a skewed shock wave. *AIAA Journal*, 11 (1973), 11, 1578–9.

[42] T. J. Goldberg. Three-dimensional separation for interaction of shock waves with turbulent boundary layers. *AIAA Journal*, 11 (1973), 11, 1573–5.

[43] M. S. Holden. Experimental studies of quasi-two-dimensional and three-dimensional viscous interaction regions induced by skewed-shock and swept-shock boundary layer interaction. *AIAA Paper 84–1677* (1984).

[44] J. S. Shang, W. L. Hankey, and J. S. Petty. Three-dimensional supersonic interacting turbulent flow along a corner. *AIAA Paper 78–1210* (1978); see also *AIAA Journal*, 17 (1979), 7, 706–13.

[45] J. S. Shang, W. L. Hankey Jr., and C. H. Law. Numerical simulation of a shock wave–turbulent boundary layer interaction. *AIAA Journal* 14 (1976), 10, 1451–7.

[46] G. S. Settles and C. C. Horstman. Flowfield scaling of a swept compression corner interaction: A comparison of experiment and computation. *AIAA Paper 84–0096* (1984).

[47] E. Reshotko and M. Tucker. Effects of discontinuity on turbulent boundary layer thickness parameters with applications to shock-induced separation. *NACA TN 3435* (1955).

[48] L. F. Scuderi. Expressions for predicting 3d shock wave-turbulent boundary layer interaction pressures and heating rates. *AIAA Paper 78–162* (1978).

[49] C. C. Horstman and C. M. Hung. Computations of three-dimensional turbulent separated flows at supersonic speeds. *AIAA Paper 79–2* (1979).

[50] D. R. Chapman. A theoretical analysis of heat transfer in regions of separated flow. *NACA TN 3792* (1956).

[51] D. H. Crawford. Investigation of the flow over a spiked-nose hemisphere-cylinder at a Mach number of 6.8. *NASA TN D-118* (1959).

[52] S. M. Bogdonoff and I. E. Vas. Preliminary investigations of spiked bodies at hypersonic speeds. *Journal Aeronautics and Space Sciences*, 26 (1959), 2, 65–74.

[53] M. S. Holden. Experimental studies of separated flows at hypersonic speeds. Part I: Separated flows over axisymmetric spiked bodies. *AIAA Journal*, 4 (1966), 4, 591–9.

[54] M. S. Holden. Accurate vehicle experimental dynamics program: Studies of aerothermodynamic phenomena influencing the performance of hypersonic reentry vehicles. *Calspan Rept.* AB-6072-A-2, SAMSO TR-79-47 (1979).

[55] E. A. English. Nose-tip recovery vehicle postflight development report. *Sandia Laboratories Report* SAND75–8059 (1976).

[56] M. S. Holden. A review of aerothermal problems associated with hypersonic flight. *AIAA Paper 086–0267* (1986).

[57] G. F. Widhopf and K. J. Victoria. Numerical solutions of the unsteady Navier-Stokes equations for the oscillatory flow over a concave body. *Proceedings of the 4th International Conference on Numerical Methods*, Boulder, CO, June 1974.

[58] M. S. Holden. Studies of the heat transfer and flow characteristics of rough and smooth indented nose shapes. Part I: Steady flows. *AIAA Paper 086–0384* (1986).

[59] M. S. Holden. Shock-shock boundary layer interaction. From AGARD-R-764, Special course on the three-dimensional supersonic-hypersonic flows including separation (1989).

[60] B. Edney. Anomalous heat-transfer and pressure distributions on blunt bodies at hypersonic speeds in the presence of an impinging shock. *FFA Report* 115 (Aeronautical Research Institute of Sweden, 1968).

[61] M. S. Holden, S. Sweet, J. Kolly, and G. Smolinski. A review of the aerothermal characteristics of laminar, transitional, and turbulent shock-shock interaction regions in hypersonic flows. *AIAA Paper 98–0899* (1998).

[62] M. S. Holden, J. M. Kolly, G. J. Smolinski, S. J. Sweet, J. Moselle, and R. J. Nowak. Studies of shock-shock interaction in regions of laminar, transitional and turbulent hypersonic flows. *Final Report for NASA Grant NAG-1–1339* (1998).

[63] M. S. Holden, R. J. Nowak, G. C. Olsen, and K. M. Rodriguez. Experimental studies of shock wave–wall jet interaction in hypersonic flow. *AIAA Paper 90–0607* (1990).

[64] M. S. Holden and S. J. Sweet. Studies of transpiration cooling with shock interaction in hypersonic flow. *AIAA Paper 94–2475* (1994).

[65] M. S. Holden, K. M. Rodriguez, and R. J. Nowak. Studies of shock-shock interaction on smooth and transpiration-cooled hemispherical nose tips in hypersonic flow. *AIAA Paper 91–1765* (1991).

[66] I. Nompelis, G. Candler, M. Holden, and M. MacLean. Investigation of hypersonic double-cone flow experiments at high enthalpy in the LENS facility. *AIAA Paper 2007–203* (2007).

[67] P. Gnoffo. Private communication, 2006.

[68] M. S. Holden, T. P. Wadhams, M. MacLean, E. Mundy, and R. Parker. Experimental studies in LENS I and LENS X to evaluate real-gas effects on hypervelocity vehicle performance. *AIAA Paper 2007–204*, (2007).

[69] J. R. Maus, B. J. Griffith, K. Y. Szema, and J. T. Best. Hypersonic Mach number and real-gas effects on space shuttle orbiter aerodynamics. *AIAA Paper 83–0343* (1983).

[70] P. Gnoffo. CFD validation studies for hypersonic flow prediction. *AIAA Paper 2001–1025* (2001).

[71] M. Marini and A. Schettino, MSTP Workshop 1996 – Reentry aerothermodynamics and ground-to-flight extrapolation contributions to hyperboloid flare test cases I.c and III.b, CIRA, Capua, Italy (1996) *Proceedings of the Aerothermodynamic Workshop held at Estec, Noordwijk, Netherlands, 1996.*

Numerical Simulation of Hypersonic Shock Wave–Boundary-Layer Interactions

Graham V. Candler

7.1 Introduction

Hypersonic flows are synonymous with high-Mach number flows and therefore are characterized by very strong shock waves. Every hypersonic vehicle has a bow shock wave in front of it, which bounds the flow around the vehicle. On the windward side of a vehicle, the bow shock usually is aligned closely with the vehicle surface, and the distance between the surface and the shock wave is usually small relative to the characteristic dimension of the vehicle. Thus, this shock-layer region is usually quite thin. Hypersonic vehicles tend to fly at high altitudes so that convective heating levels can be managed. Thus, the characteristic Reynolds numbers tend to be low and boundary layers are usually thick. In addition, shear heating in hypersonic boundary layers increases the temperature and viscosity, which also increases the thickness. The low Reynolds number and the relative stability of hypersonic boundary layers mean that many practical hypersonic flows are laminar or transitional. If the flow is turbulent, it is often only marginally turbulent. Therefore, hypersonic flows are particularly susceptible to shock wave–boundary-layer interactions (SBLIs).

The design of a hypersonic vehicle – whether a planetary-entry capsule, a high-lift reentry vehicle such as the Space Shuttle, or a scramjet-powered aircraft – is dominated by aerodynamic heating. Blunt bodies such as planetary-entry capsules are protected from large heat fluxes by the thick boundary layer that insulates the vehicle from the high-temperature stagnation-region flow. Anything that interrupts this insulating layer (e.g., an impinging shock wave) can be catastrophic to a vehicle. Shock interactions on wing leading edges and other vehicle components can be devastating to a vehicle, particularly at a high Mach number in which heat-transfer rates are especially severe. SBLIs can cause flow separation and change the effectiveness of control surfaces on hypersonic vehicles. The flow inside the isolator of a scramjet engine is dominated by interactions between shock waves and turbulent-boundary layers, and these interactions are responsible for raising the pressure and preventing engine unstart due to fluctuations in engine-operating characteristics. Thus, SBLIs can either destroy a hypersonic vehicle or be responsible for its effective operation.

The development of future hypersonic vehicles depends on high-fidelity simulation tools for their design and optimization. Current computational fluid dynamics

(CFD) methods can predict many aspects of practical hypersonic flows. However, the accurate simulation of SBLI flows remains a severe challenge for the best available CFD methods. This chapter discusses the physics of hypersonic SBLI flows, with emphasis on how the physics affects the numerical simulation of the flows. Three general types of SBLI flows are considered: compression-corner interactions, leading-edge shock interactions, and interactions on double-cone geometry at high-enthalpy-flow conditions.

Many examples in the literature show that hypersonic SBLI flows are difficult to simulate accurately. Even fully laminar flows have a huge range of length scales and often are unsteady and three-dimensional. At higher Reynolds numbers, shear layers become transitional and reattachment is from transitional to turbulent. Turbulent SBLI flows are notoriously difficult to predict because of unsteady interactions and the inability of Reynolds-averaged Navier-Stokes (RANS) turbulence models to represent these flows. Shock interactions on leading edges tend to be inherently unsteady for strong interactions. All of these issues result in extreme sensitivity to the grid quality, design, and alignment with the shock waves and other strong gradients. Simulations are also very sensitive to the numerical methods, and these flows provide a stringent test of CFD methods.

Experiments on hypersonic SBLI flows also are difficult for several reasons. Extreme instrument density is required to resolve strong peaks. Many flows are unsteady, and care must be taken to minimize the free-stream fluctuations that may drive unsteadiness and cause premature transition in shear layers. The flows are sensitive to any nonideal behavior in the wind-tunnel flow; therefore, the tunnel-flow state must be characterized with added care. Shock generators tend to contribute to fluctuations, especially at a high Mach number because the shock angles are very oblique, which tends to amplify disturbances.

7.2 Hypersonic SBLI Physics

This section discusses key features of several types of hypersonic SBLI flows. It emphasizes how hypersonic flows are different than their low-Mach-number counterparts. Because it is usually a critical design element in hypersonic vehicle design, particular attention is given to how the heat-transfer rate is affected by the shock interaction. In this section, these issues are illustrated with relatively simple two-dimensional interaction flows. Because of this simplicity, it is possible to perform highly resolved numerical simulations of these flows. The simulations are used to describe the key features of hypersonic SLBI flows. These simulations may not capture all of the critical physics of the flows, particularly when they become transitional or turbulent. However, the simulations are useful for understanding many important aspects of these complex flows. We are careful to note the limitations of the simulations that are presented.

7.2.1 Shock Wave–Laminar Boundary-Layer Interactions at High Mach Number

The simplest type of SBLI that can occur is an oblique shock wave impinging on a laminar-boundary layer. This is a classical problem and has been studied extensively;

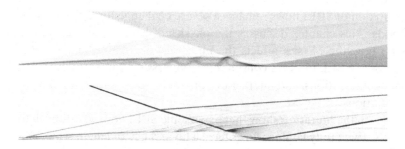

Figure 7.1. Temperature (*top*) and synthetic schlieren (*bottom*) in the flowfield of a Mach 6 boundary layer and a 10-degree wedge-shock generator.

it is discussed at length in other chapters of this book. At hypersonic conditions, several key features of these flows are noted, including the following:

- Under most conditions, the shock wave that impinges on a boundary layer is very oblique. This is a result of operating at hypersonic conditions in which shock layers are thin and any shocks that are generated are oblique.
- The quadratic variation of postshock pressure with Mach number results in a very strong pressure gradient at the impingement location.
- The large adverse pressure gradient usually produces a large separation zone, which is bounded by a long shear layer. The shear layer is likely to transition to turbulence well before the wall-boundary layer becomes turbulent. Once the shear layer becomes turbulent, the flow is usually unsteady, and the impingement-point heating may be transitional to turbulent.
- The large separation zone results in a strong separation shock wave that further strengthens the incoming shock wave. This increases the adverse pressure gradient and causes feedback between the separation zone and the strength of the interaction.
- The strong surface-normal pressure gradient due to the shock impingement causes a huge compression of the boundary layer, resulting in extreme localized heat-transfer rates.
- The size of the separation zone is affected by the surface-temperature condition and by the level of chemical reaction in the flow. In general, lower wall temperatures and increased chemical reactions tend to reduce the size of the separation zone.

Figure 7.1 is a computational simulation of a shock wave generated by a 10-degree wedge impinging on a flat-plate laminar-boundary layer at an edge Mach number of 6 and a unit Reynolds number of 10^6/m. The shock impinges at a distance of 1.5 meters (4.9 feet) from the plate leading edge, making $Re_x = 1.5 \times 10^6$ at that point. The interaction is visualized by contours of constant temperature and with a synthetic schlieren image (constructed as contours of constant density-gradient magnitude). The figure shows a classic SBLI, with the impinging shock reflecting from the surface. The shock impingement causes the boundary layer to separate, producing a separation shock that passes through the impinging shock and merges with the reflected shock. For this laminar boundary layer, the separation

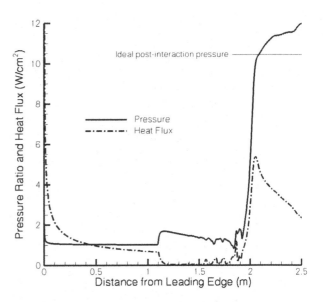

Figure 7.2. Surface pressure and heat flux for the shock-laminar boundary-layer interaction shown in Fig. 7.1; the ideal shock impingement point is at 1.5 m from the leading edge.

runs far upstream. In reality, the long shear layer at the edge of the separation zone likely undergoes transition to turbulent flow, which would reduce the separation-zone length. The CFD simulation cannot represent this process, and a fully laminar flow is assumed. The CFD also produces large-scale vortical structures in the separation zone, which generate waves that can be seen in both plots. These vortices are transient and therefore produce unsteadiness in the interaction. Turbulent SBLIs of this type are also known to be highly unsteady and driven by the same type of unsteadiness in the upstream separation zone.

Figrue 7.2 plots the computed pressure and heat flux on the surface. Notice the dramatic increase in the pressure downstream of the shock; the ideal inviscid postinteraction pressure ratio is 10.43. Here, the increased pressure is caused by the additional flow-turning due to the boundary layer and the complex system of shock waves generated by the interaction. The compression of the boundary layer in the region of shock impingement results in a large increase in the heat flux at the interaction point, followed by a gradual reduction as the boundary layer expands downstream of the interaction.

7.2.2 Hypersonic Compression-Corner Flows

A related class of shock interactions occurs on compression corners, on control surfaces, and at locations where the body surface changes angle. In Figs. 7.3 through 7.6, the Mach number and synthetic schlieren images for high-Mach-number flows over a 15-degree ramp are plotted. The incoming boundary layer is laminar and the Reynolds number at the corner is 10^6; a cold wall is assumed and a perfect-gas model for air is used. The compression corner causes the boundary layer to separate, producing a separation shock, which passes through the primary shock wave on the compression corner. The main effect of the SBLI is the compression of the boundary layer and a resulting dramatic increase in the heat flux. As the free-stream Mach number increases, the separation decreases slightly, and the wave angles decrease.

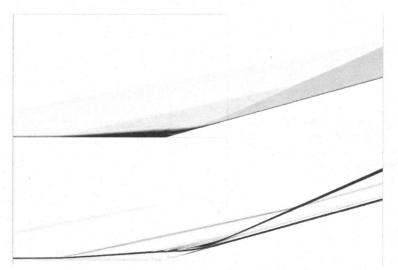

Figure 7.3. Mach number (*top*) and schlieren (*bottom*) in the flowfield of a 15-degree compression corner at Mach 6 and unit Reynolds number of 10^6 per meter.

At the Mach 12 condition (Fig. 7.6), the separation shock is strong enough to produce a significant secondary SBLI, which further raises the pressure and changes how the boundary layer adjusts to the pressure gradient.

As in the simple Mach 6 boundary-layer interaction shown in Fig. 7.1, the separation zones predicted in these simulations are excessively large because the flow is assumed to be laminar. However, a hypersonic transitional or turbulent compression-corner flow would have the same basic structure as shown in the figure.

Figrues 7.7 through 7.9 show the surface-pressure and heat-transfer-rate coefficients for 10-, 15-, and 20-degree compression-corner flows at Mach numbers from 6 to 12. Interesting conclusions can be drawn from these figures. For the 10-degree corner, the nondimensional pressure increase is largest for Mach 6, decreasing with

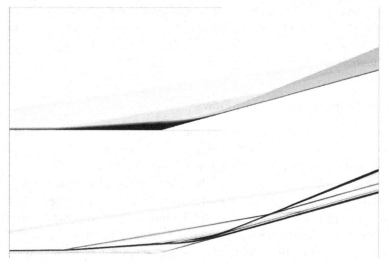

Figure 7.4. Mach number (*top*) and schlieren (*bottom*) in the flowfield of a 15-degree compression corner at Mach 8 and unit Reynolds number of 10^6 per meter.

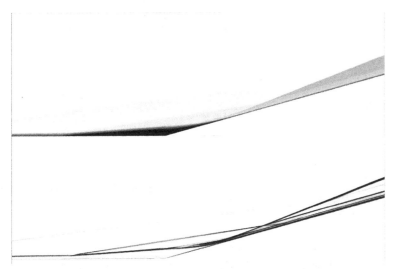

Figure 7.5. Mach number (*top*) and schlieren (*bottom*) in the flowfield of a 15-degree compression corner at Mach 10 and unit Reynolds number of 10^6 per meter.

Mach number. However, for the 15- and 20-degree turning angles, the Mach 12 flow has the largest pressure rise. This is a result of the separation-zone shock wave interacting with the primary shock, causing an additional pressure rise. This transmitted shock then reflects between the surface and the primary shock, which produces the pressure variations shown in Figs. 7.8 and 7.9.

In all cases, the pressure increase on the wedge is significantly larger than predicted by inviscid theory. Newtonian aerodynamics provide pressure coefficients of 0.060, 0.134, and 0.234 for the 10-, 15-, and 20-degree cases, respectively. The larger computed pressure increase is a result of the additional displacement of flow due to the boundary layer and, as in the flat-plate interaction, the more complex system of shock waves that the interaction causes. Far downstream from the corner, the

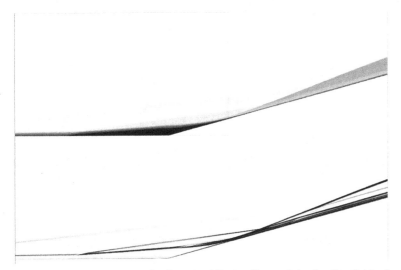

Figure 7.6. Mach number (*top*) and schlieren (*bottom*) in the flowfield of a 15-degree compression corner at Mach 12 and unit Reynolds number of 10^6 per meter.

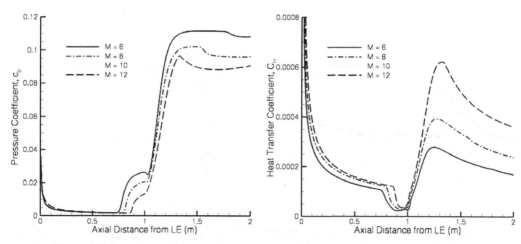

Figure 7.7. Pressure and heat-transfer coefficients on a 10-degree compression corner for laminar flow at a unit Reynolds number of 10^6 per meter.

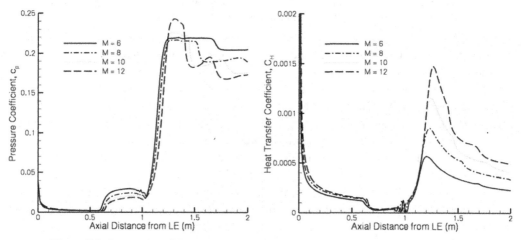

Figure 7.8. Pressure and heat-transfer coefficients on a 15-degree compression corner for laminar flow at a unit Reynolds number of 10^6 per meter.

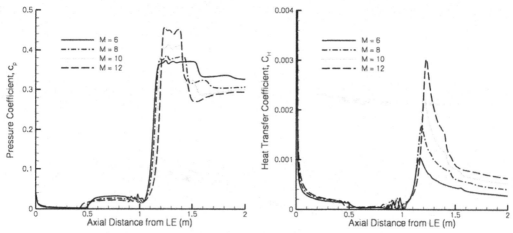

Figure 7.9. Pressure and heat-transfer coefficients on a 20-degree compression corner for laminar flow at a unit Reynolds number of 10^6 per meter.

pressure approaches the lower theoretical values. The pressure coefficient scales approximately with the corner angle squared, as predicted by inviscid theory.

The separation zone increases with Mach number for a given turning angle and with corner angle for a given Mach number. These increases are expected because the dimensional pressure rise increases with both Mach number and corner angle, resulting in a stronger adverse pressure gradient and a stronger separation.

The heat-transfer coefficient shows a simple variation with Mach number and turning angle; in all cases, it increases as the strength of the interaction increases. The largest heat-transfer rate occurs in the region of strongest boundary-layer compression, where the foot of the corner shock is located. The peak value has a strong Mach number and turning-angle scaling, particularly because the heat-transfer coefficient already has an M^3 scaling.

7.2.3 Hypersonic Shock-Shock Interactions

Edney's classification [1, 2] of shock-shock interactions is valid in the hypersonic regime. The most damaging interactions for hypersonic vehicles are the Types IV and V interactions that occur when a shock wave impinges on a leading edge. This could occur on an engine-cowl leading edge during off-design operation or on a wing leading edge on a lifting hypersonic vehicle. Vehicle designers should be careful to prevent or minimize these types of interactions because of the extreme increases in heat flux that can occur. In the hypersonic regime, the key aspects of these shock-shock interactions are as follows:

- The strong Types IV and V interactions result in massive increases in heat-transfer rates on blunt bodies. The interaction disturbs the insulating layer on the surface that normally protects the surface from high heat flux.
- Shock-shock and shock–leading-edge interactions are characterized by an extreme range of length scales. As a result, they are difficult to measure in experimental facilities. Even tiny thin-film heat-transfer gauges may not resolve the peak value of the heat-transfer rate.
- Type IV interactions appear to be unsteady in experiments. The location of peak-heating varies; it may be caused by inherent flow instability or amplification of free-stream disturbances by the oblique shock generator.
- At high enthalpy, finite-rate chemical reactions can affect the strength of the interactions. References [3, 4, 5] show that dissociation caused by the shock interaction changes the shock strengths, typically reducing the strength of the shock interaction and thereby reducing the peak-heat-transfer rates.

Consider a typical shock interaction on a cylindrical leading edge. A good example is the Run 43 conditions in the studies of Holden et al. [6, 7, 8]. That work considered Mach 13.94 flow over a cylinder of 1.5-inch (3.81 cm) radius, with a shock wave impinging on the surface generated by a 10-degree wedge shock generator. The free-stream conditions are at a sufficiently low Reynolds number (i.e., $Re = 42,600/m$) that the flow in the interaction remains fully laminar. Figure 7.10 is a series of simulations in which the location of the impinging shock is systematically changed, giving rise to a range of interactions. The grid used for these simulations has 1,200 elements around the half cylinder and 800 elements in the

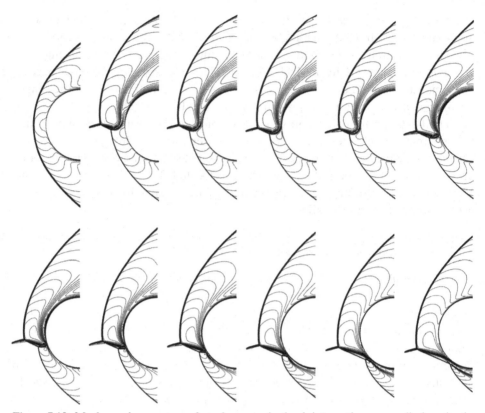

Figure 7.10. Mach number contours for a hypersonic shock interaction on a cylinder; shock-impingement locations vary for each image.

surface-normal direction, which provides sufficient resolution to capture the flow-field. Similar results were obtained by D'Ambrosio [9] on less-refined grids.

In Fig. 7.10, the impinging shock wave drastically disturbs the bow shock, resulting in a highly oblique shock wave emitted from the shock triple point. Depending on the location of the impinging shock, a jet of gas either flows over the top of the cylinder, impinges on the cylinder, or flows underneath the cylinder. The most damaging condition occurs when the jet impinges directly on the surface, resulting in a normal shock wave that compresses the thermal boundary layer on the surface. This is visualized more clearly in Fig. 7.11, which plots several flow variables in the vicinity of two of these interactions. (The first condition corresponds to the second figure from the left in the second row of Fig. 7.10; the second condition corresponds to the fifth image in the second row.) The first example is a Type IV interaction, in which the flow just below the shock triple point goes through a series of oblique shock waves before terminating with a normal shock near the surface. This shock is visualized most clearly in the synthetic schlieren image. As a result, the jet-gas entropy does not increase as much as if it traveled through a single normal shock wave. Thus, stagnation-pressure loss is reduced and postshock pressure is significantly larger; this is what causes the large increase in pressure due to shock impingement. More important, the large pressure behind the normal shock in the jet drastically compresses the thermal boundary layer on the surface of the cylinder. This is particularly evident in the temperature plot in Fig. 7.11, which shows that the thermal boundary layer

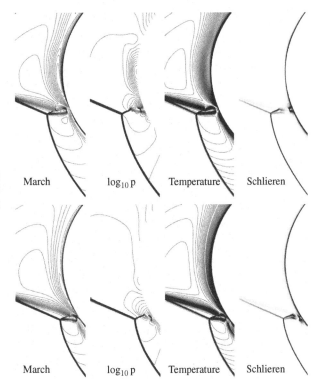

Figure 7.11. Contours of flow variables in the vicinity of Type IV (*top*) and Type V (*bottom*) shock interactions on a cylinder.

is compressed so much that it is essentially invisible at the plot scale. Naturally, this produces a huge increase in the heat flux at the interaction location. The second image shows a slightly different condition, in which the jet impinges obliquely on the cylinder surface. It is interesting that this condition produces a higher-peak heat flux than the condition with the surface-normal jet.

Figure 7.12 plots the variation of surface pressure and heat flux for the cases discussed herein. The normalization is with the stagnation-point pressure and heat

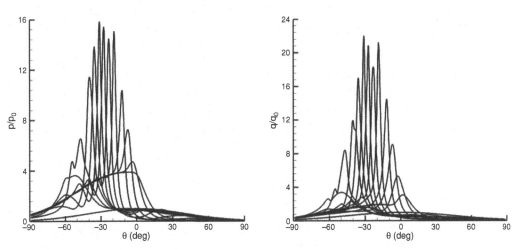

Figure 7.12. Surface pressure and heat flux for the shock-interaction flow shown in Fig. 7.10; each line corresponds to a different shock-impingement location. Normalization is with respect to the undisturbed stagnation-point pressure and heat flux.

flux for the undisturbed case. The shock interaction gradually strengthens as the impingement point is lowered (i.e., moved toward negative zero). First there is a strong interaction (shown at the top of Fig. 7.11), then a reduction in maximum pressure and heat flux, until the strongest interaction occurs (shown at the bottom of Fig. 7.11). This worst case gives a peak pressure of about sixteen times the undisturbed stagnation-point pressure and twenty-two times the stagnation-point heat flux. The peaks are narrow, with a width similar to the length scale of the terminal shock in the impinging jet. The spreading of the peak is due mostly to the finite thickness of the thermal boundary layer on the cylinder; as such, it is expected that higher-Reynolds-number conditions have even narrower pressure and heat-flux peaks. Clearly, these types of interactions must be designed against to avoid thermal-protection failure.

Comparisons with experimental data are not presented here, but interested readers are referred to the work of D'Ambrosio for details [9]. In general, the comparisons are not especially good; the computational simulations overpredict the peak pressure and heat-transfer rates by about 30 percent. D'Ambrosio makes a careful study of how the finite dimension of the pressure and heat-flux gauges and the varying surface temperatures affect the comparisons. Accounting for these nonideal effects improves the agreement and illustrates the difficulty of conducting shock-interaction experiments. The additional complication of flowfield unsteadiness was not addressed in D'Ambrosio's study, which would be difficult because the amplitude and frequency of the possible flowfield variations are not known.

7.3 Numerical Methods for Hypersonic Shock–Boundary-Layer Interaction Flows

The numerical simulation of hypersonic shock–boundary-layer interactions and shock-shock interactions is challenging for even the most advanced numerical methods. The primary reasons for this difficulty are illustrated in previous figures. The flows are characterized by a wide range of length and time scales; shock interactions create shock triple points that are difficult to resolve; separation zones produce complicated flows; and shock impingement drastically reduces the boundary-layer length scales.

The simulations presented here were performed on very fine grids (typically, one million grid elements) for relatively simple two-dimensional flows. This grid resolution is somewhat excessive; clearly, however, if these flows were three-dimensional, the grid requirements would rapidly become onerous. As important, the numerical methods used must be of high quality with low levels of numerical dissipation. Also, the method must be implicit to allow large time steps to be taken during the integration to a steady-state flowfield.

Thus, the critical features of a numerical method for hypersonic-shock interactions flows are as follows:

- The method must have a low level of numerical dissipation in the inviscid numerical flux function. The method must be at least second-order accurate, and the slope limiters and other means for controlling the solution must be chosen with care.

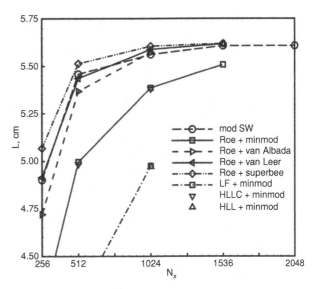

Figure 7.13. Computed separation length as a function of the number of axial-direction grid points for a double-cone flow at Mach 12; various widely used upwind methods are shown with a range of slope limiters for the numerical fluxes [23].

- The grid used must be of high quality with smoothly varying grid spacing and high resolution in the boundary layer and shock triple-point vicinity. The separation and reattachment points must be well resolved in both the streamwise and wall-normal directions so that the separation point is predicted correctly. Careful grid-convergence studies must be conducted to show that the grid resolution is sufficient.
- Simulations must be run for many flow times to produce a steady-state result. The separation zone takes a long time to establish, and care must be taken to verify that the solution has reached a steady state. If the flow is separated, the separation-zone size changes very slowly near the end of convergence; it must be monitored as a convergence metric. In addition, experimental data must be kept at arm's length to reduce the possibility of running the CFD solution until the separation zone matches that of the experiment. In some cases, there is a good reason why the simulated separation may be larger than the experiment (e.g., unmodeled physics in the simulation).
- Care must be taken when using local time-stepping methods or approximate implicit methods for the time integration. Due to the extreme range of length scales in the grid, these approximations in the time domain may cause aphysical solutions to be produced with CFD.
- Turbulent shock–boundary-layer interactions are notoriously difficult to simulate, and several turbulence models with appropriate realizability constraints should be used to compare the sensitivity of the predictions to these modeling choices.
- The separation-zone size and peak heating rates are strong functions of the numerical method and grid resolution. It is important to fully investigate this sensitivity before drawing conclusions about the accuracy of a CFD result.

The effect of numerical dissipation on the prediction of shock interactions cannot be understated. Figure 7.13 plots the computed size of the separation zone on a 25- to 55-degree double-cone geometry (this flowfield is discussed in detail in the next section) for various popular numerical methods and slope limiters. The exact details of the methods are not particularly important for this discussion; instead, there can be extreme variation of the separation-zone size depending on the numerical method used. Also, dissipative methods (i.e., those that tend to smooth gradients over large numbers of grid cells) may require massive numbers of grid points to obtain the correct asymptotic solution. It is interesting that this study shows that the choice of slope limiter used to control numerical errors near strong gradients can have a major effect on the accuracy of a method for predicting SBLIs. Therefore, it is mandatory that careful studies are conducted to verify the accuracy of CFD simulations of shock-interaction flows.

An additional numerical issue concerns the use of local time-stepping methods in the simulation of separated flows. It is not widely recognized in the literature that local time-stepping methods can cause artificial periodic shedding of vortices and other spurious numerical artifacts. Gnoffo [10] clearly illustrates this effect for a concave geometry and for the separated flow on the double-cone SBLI flow. Furthermore, he shows that using a time-accurate method eliminates this behavior and allows CFD to obtain the steady-state experimentally observed solution. Therefore, great care must be taken in how the equations are integrated in the temporal domain.

The numerical simulation of hypersonic shock interactions with turbulent flows remains an outstanding challenge to the CFD community. It is only recently that low-Reynolds-number turbulent shock interactions have been simulated with Direct Numerical Simulations (DNS) and Large-Eddy Simulations (LES) [11, 12, 13]. For realistic Reynolds numbers, we still must rely on RANS simulations for interactions with turbulent boundary layers. All RANS models use turbulent kinetic energy (TKE) production terms that depend on the local strain rate; in shock waves, the production becomes extremely large, often reaching aphysical values. Furthermore, it can be shown [14, 15] that the unsteady interaction of turbulence with shock waves actually suppresses TKE production. This effect is not represented in widely used RANS models. MacLean et al. [16] show the importance of this effect in comparisons of Mach 7.16 turbulent flow over a 33-degree flare. Figure 7.14 plots the experimental schlieren image of the turbulent shock interaction and results of two simulations. In the first simulation using the baseline Spalart-Allmaras (SA) one-equation RANS model [17], no separation is predicted. Then, using the strain-adaptive linear SA model of Rung et al. [18], the separation zone is in much better agreement with the experiment. This model modification limits the production of TKE in the high strain-rate regions of the flow. This approach improves the pressure distribution in the region of the separation; unfortunately, it does not significantly improve the heat-flux prediction [16]. Clearly, there is great need for improved turbulence modeling for these flows.

A related problem occurs in the isolator of scramjet engines. The isolator is usually a constant-area section of the propulsion system located between the exit plane of the inlet and the combustor section. The purpose of the isolator is to separate fluctuations in the inlet and combustor conditions so as to reduce the potential for engine unstart. Also, the isolator is responsible for providing additional pressure rise upstream of the combustor. The boundary layers on the isolator surface are usually turbulent (because of either natural transition or they have been tripped

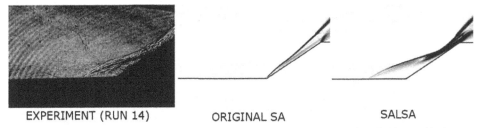

Figure 7.14. Comparison of predicted and measured separation-zone flowfield on a 33-degree flare at Mach 7.16, showing effect of the RANS turbulence model [16].

to turbulence); thus, shock interactions in the isolator occur with thick turbulent-boundary layers. The source of the shock waves is either from the imperfect canceling of shock waves by the inlet, due to off-design operation of the inlet, from the inlet cowl, or from the varying displacement of the turbulent-boundary layer itself. In any case, high-Mach-number isolator flows are characterized by shock wave–turbulent boundary-layer interactions, and these interactions are responsible for effective operation of the isolator. The resulting shock train can absorb pressure increases that occur in the combustor without allowing the pressure to travel through the isolator boundary layer, potentially resulting in a catastrophic unstart of the inlet.

Little is known about how to predict the isolator flowfield. The classic experimental work of Waltrup and Billig [19] with associated empirical expressions for the pressure rise in cylindrical ducts is probably state-of-the-art in this area. This problem is particularly difficult to simulate with CFD because of the importance of the turbulence model in turbulent interactions. As discussed previously, a small adjustment in the turbulence model can change a compression-corner flow from attached to separated. This dependence is magnified in an isolator flow because of the numerous shock interactions that can occur in the shock train. Any error resulting from missing physics in the turbulence model is amplified every time a shock wave interacts with the wall boundary layer.

7.4 Example: Double-Cone Flow for CFD Code Validation

Another interesting shock wave–laminar boundary-layer interaction occurs on the double-cone geometry used for code-validation studies [20, 21, 22]. Figure 7.15 plots a schematic of this flowfield for an approximate Mach 12 free-stream condition. Notice the attached shock wave that originates at the first cone tip, the detached shock wave formed by the second cone, and the resulting shock triple point. The transmitted shock impinges on the second-cone surface, which separates the flow and produces a large localized increase in the pressure and heat-transfer rate. This pressure rise causes the flow to separate and also produces a supersonic underexpanded jet that flows downstream near the second-cone surface. The size of the separation zone depends strongly on the location and strength of the shock impingement. This flowfield is sensitive to the wind-tunnel conditions, the physical models used in the CFD code, and the quality of numerical methods used to predict the flow [23, 20].

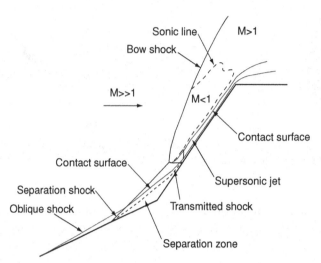

Figure 7.15. Schematic of the double-cone flow.

The double-cone flow was used for code-validation studies related to a NATO Research and Technology Organization (RTO) Working Group Study; as a result, several research groups simulated this flowfield. The experiments were performed at well-characterized hypersonic conditions in the CUBRC, Inc. (formerly the Calspan–University at Buffalo Research Center) Large Energy National Shock Tunnel (LENS) [24, 25]. These experiments used a large model with many surface-mounted heat-transfer-rate and pressure transducers. Nitrogen was used for the test gas to minimize the effects of chemical reactions, and the experiments were conducted at low pressure to ensure laminar-boundary layers and shear layers. In general, the comparisons between simulation and experiment were good [26]; however, there were several important differences. It is interesting that the simulations performed with high-quality numerical methods on the finest grids slightly over-predicted the size of the separation zone, and all simulations predicted excessive heating in the attached region prior to separation. This is shown in Fig. 7.16, which presents typical results for two double-cone cases. The error in heat-transfer rate on the first cone is as much as 20 percent, which is particularly puzzling because the pressure is accurately predicted in this region. Many attempts were made to

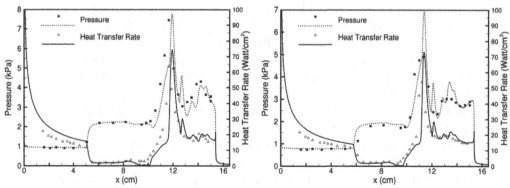

Figure 7.16. Comparison of Navier-Stokes predictions of surface quantities with experimental data for Run 28 (*left*) and Run 35 (*right*).

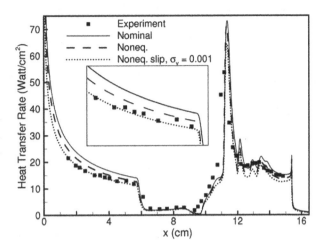

Figure 7.17. Comparison of Navier-Stokes predictions of surface-heat-transfer rate with experimental data for Run 35; various physical models for the free-stream and gas-surface interaction were used.

explain this difference by running CFD cases with finite nose-tip bluntness, model misalignment, and uncertainties in reaction rates. None of these effects explained the differences shown in Fig. 7.16.

The specification of the free-stream conditions in a hypersonic shock tunnel can be difficult because these facilities can be subject to nonideal effects in the nozzle – namely, a reflected shock wave is used to heat and compress the test gas, which results in vibrational excitation and chemical reaction. It is then rapidly expanded through the nozzle, and the thermochemical state may not fully de-excite during expansion. As a result, the gas flowing over the model may be in a nonideal thermochemical state. During design of these experiments, we were aware of the potential for this problem and chose nitrogen as the test gas to minimize chemical reactions. We also kept the enthalpy of the tests relatively low (i.e., $h_0 < 4$ MJ/kg) to further reduce this effect. As a result, there is essentially no chemical reaction of the test gas prior to expansion through the nozzle. Nitrogen vibrational modes relax very slowly; for these test conditions, this results in elevated vibrational energy in the wind-tunnel test section. A vibrational finite-rate simulation of the nozzle flow shows that the vibrational-energy modes are frozen near the throat temperature (i.e., $T_v = 2,560$ K). This has two major effects: (1) the kinetic-energy flux is reduced by about 10 percent; and (2) because nitrogen vibrational-energy modes are inefficient at accommodating to most metallic surfaces, they do not transfer their energy to the model. These two effects reduce the heat flux by about 20 percent and significantly improve the comparison between CFD and experiment (Fig. 7.17). More details about the double-cone code validation study are available in [20].

Consider the effects of high-enthalpy-flow conditions on the double-cone problem – specifically, how finite-rate chemical reactions affect shock interactions in this flowfield. As a follow-on to the code-validation study discussed previously, a series of high-enthalpy nitrogen and air experiments was performed on the double-cone geometry [24]. Results of this study are summarized in Fig. 7.18 in the form of heat-transfer rates to the double-cone for three air cases [25, 27, 28]. These runs were at approximately the same free-stream Reynolds number and Mach number, and the total enthalpy was increased from 4.5 to 15.2 MJ/kg (at 4.5 MJ/kg, $Re = 3.1 \times 10^5$/m per meter; at 10.4 MJ/kg, $Re = 2.9 \times 10^5$/m and at 15.2 MJ/kg, $Re = 2.3 \times 10^5$/m).

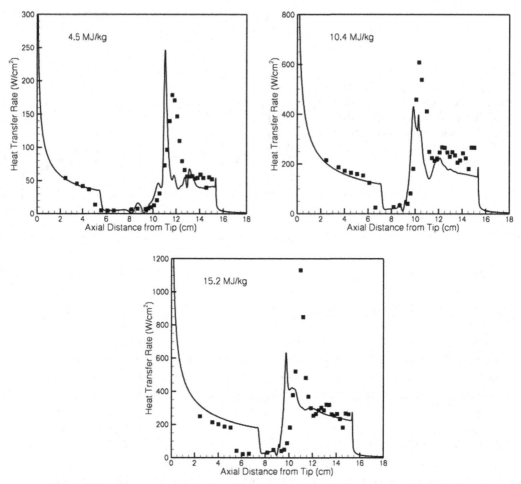

Figure 7.18. Measured and predicted heat flux to the double-cone model for high-enthalpy air conditions (4.5, 10.4, and 15.2 MJ/kg total enthalpy).

First, we focus on the computed results shown with the lines. As the free-stream enthalpy is increased, the simulations predict that the separation zone decreases in size (visualized by the extent of low heat-flux upstream of the $x = 10$ cm point). This is because the increasing energy of the flow increases the level of chemical reaction in the flow; in particular, there is an increased level of reaction in the separation zone. In air, the reactions are endothermic, which has the effect of decreasing the temperature and molecular weight of the mixture (i.e., diatomic molecules (O_2 in particular) dissociate to form atoms).

Why does the increasing level of reaction decrease the size of the separation zone? For a normal shock wave in air, the main effect of chemical reactions is a decrease in the postshock temperature and an increase in the postshock density relative to a perfect-gas shock wave. Second, the molecular weight of the mixture decreases. The combination of these changes results in approximately the same postshock pressure because the pressure is set by the momentum-flux balance across the shock wave. The result of this change in the postshock conditions is to decrease the shock standoff distance on blunt bodies. (The shock standoff distance is inversely proportional to the density rise across the bow shock wave.) Thus, the same effect occurs in the separated-flow region: increasing chemical reaction causes a larger density increase in the separation zone and a smaller separation zone. This effect is predicted by CFD simulations of the high-enthalpy air flows shown in Fig. 7.18. The CFD predicts a reduction in the heat flux caused by the chemical reactions; this occurs because the smaller separation zone produces more oblique shock waves and a weaker shock interaction between the separation shock and the bow shock on the second cone. Thus, the higher-enthalpy interactions are less clearly defined and result in a lower peak-heat-transfer rate (when nondimensionalized by the velocity cubed or by $h_o^{3/2}$).

Consider the experimental data points shown in Fig. 7.18. At the lowest enthalpy condition, the comparison between simulation and experiment is good but not perfect. CFD predicts a smaller separation zone, which results in a stronger and more upstream interaction. For this case, the heat flux to the first cone is well predicted, indicating that the nozzle energy flux is well characterized for this condition. At the higher-enthalpy conditions, the agreement worsens significantly. At 10.4 MJ/kg, the heat flux is underpredicted and the separation zone is much smaller than measured. At the highest enthalpy, the discrepancy is even worse, with the measured separation zone twice as large as predicted by CFD. Furthermore, it is interesting that the experimental trend is different than the CFD results: the smallest measured separation zone is at the intermediate enthalpy (10.4 MJ/kg), whereas CFD predicts a decreasing separation-zone size. The discrepancy in predicted heating on the cone forebody for the two higher-enthalpy cases indicates that the freestream conditions used for these simulations are likely wrong [28].

The discrepancy between CFD and the experimental measurements currently is not understood. However, there is no reason to believe that there is a fundamental problem with the simulations, especially because the lower-enthalpy double-cone flows in nitrogen were accurately predicted (once all the relevant physics were modeled). Clearly, there is more to be learned about simulating these flows and, most important, how to model expansion of the high-enthalpy flow in a reflected shock-tunnel test facility. More details about modeling uncertainties and attempts to understand the differences discussed here are in [27] and [28].

In related work by Holden, Wadhams, and MacLean [25], the control-surface effectiveness problem on the Space Shuttle Orbiter is examined in the LENS facility. This work revisited the problem identified during the first reentry of the Orbiter during STS-1 [29, 30], in which the predicted aerodynamic performance was significantly different than encountered during flight. Maus et al. [31] showed that the differences were caused primarily by Mach number and real-gas effects. Weilmeunster et al. [32] used more sophisticated CFD methods to show that real-gas effects decrease the size of the separation zone on the body flap, changing the pressure distribution and pitching moment. The new experiments of Holden et al. [25] further confirm this finding, showing that at the same Reynolds number and total enthalpy, the separation zone in air is smaller than in nonreacting nitrogen. This effect increases with increasing enthalpy, as expected. In addition to changes in the SBLI on the body flap, air chemistry (primarily oxygen dissociation) decreases the pressure in the expansion over the curved aft section of the Orbiter wing, further changing the pitching moment.

From this discussion, several conclusions are drawn about high-enthalpy effects on hypersonic shock interactions. First, because the separation zone is a slowly moving region, the gas tends to reach thermal and chemical equilibrium in the separation zone. This high level of chemical reaction absorbs energy into the internal- and chemical-energy modes, which affects the gas dynamics of the flow through changes in temperature and gas constants. This changes the size of the separation zone, just as the shock standoff distance is reduced by real-gas effects. It is interesting that the experimental data shown in Fig. 7.18 do not correspond to the expected trend; the CFD results follow this trend. Of course, there are other subtle effects that are difficult to classify: the chemical reactions change the viscosity of the flow; the reduced temperature increases the density, which affects the rate of entrainment across the shear layer; and the shock angles are changed due to the real-gas effects. These processes also affect the size of the separation zone and the strength of the interaction between the separation shock and the detached shock on the second cone. This changes the peak-heating level on the second cone. Again, CFD predicts a reduction in the interaction heating rate due to the reduced separation-zone size.

7.5 Conclusions

Hypersonic shock wave–boundary-layer interactions have many features in common with interactions at transonic and supersonic Mach numbers. In all conditions, the interactions result in complex flowfields characterized by a huge range of length scales. However, hypersonic interactions have features that distinguish them from their lower-Mach-number counterparts. Most important, hypersonic interactions can produce huge increases in the surface pressure and heat flux because they can interrupt the thick insulating boundary layer on hypersonic vehicles, allowing the high-energy external flow to directly attack the surface. Thus, to a large extent, strong shock interactions (e.g., Edney Types IV and V interactions) must be prevented at all costs.

Another interesting aspect of hypersonic interactions is that the impinging shock wave tends to be at a highly oblique angle to the boundary layer. However, even a shallow hypersonic-shock wave can produce a large adverse-pressure gradient, with resulting flow separation. For a laminar-boundary layer, the separation zone tends to run far upstream, producing a long shear layer at the edge of the separation zone. This free shear layer is unsteady and prone to transition, except at small Reynolds numbers. Thus, laminar separations tend to become turbulent in the shear layer, resulting in transitional or turbulent heating levels at reattachment. Current conventional CFD methods cannot capture the complex transitional-flow physics.

Hypersonic-shock interactions on compression corners can be computed accurately if the flow remains laminar, the free-stream conditions are well characterized, and there is little or no chemical reaction in the flowfield. As shown for the hypersonic double-cone code-validation studies, CFD can predict this rather narrow class of flow if the CFD method is not excessively dissipative and the grid is quite large. These flows are difficult to simulate because any error in the physical modeling or solution of the governing equations is amplified by the separation zone. A too-small separation zone results in a weaker interaction and smaller pressure rise, which reduces the adverse-pressure gradient and further reduces the separation-zone size.

Thus, there is a feedback loop that accentuates error in the numerical simulation or in the specification of free-stream conditions. Endothermic chemical reactions and other internal-energy excitation processes decrease the temperature rise and increase the density rise across shock waves. In general, this results in thinner shock layers and, as a result, smaller separation zones. However, it is difficult to perform experiments at relevant high-enthalpy conditions such that the flow remains laminar and the free-stream conditions are well understood. Current experimental data suffer from a degree of uncertainty, and CFD cannot replicate these flows.

Shock wave–turbulent boundary-layer interactions at hypersonic conditions remain a major difficulty for numerical simulations. Most RANS turbulence models are built on results from the incompressible literature. Simple compressibility corrections are added to account for nonuniform densities in compressible flows. It is rather optimistic (and unrealistic) to hope that such corrections can account for the huge density variations that occur in hypersonic flows. In practice, RANS models may be tuned to obtain reasonable agreement with experimental data for a narrow class of interactions, but a truly predictive capability currently does not exist. Shock interactions with turbulent boundary layers are particularly important in scramjet isolators because a shock train undergoes a series of boundary-layer interactions. Therefore, turbulence-modeling errors accumulate and the pressure rise through the isolator is poorly predicted with current CFD methods.

Acknowledgments

Most of the work presented herein was supported by the Air Force Office of Scientific Research (AFOSR) under Grants FA9550–04-1-0114 and FA9550–04-1-0341. The author acknowledges Dr. John Schmisseur for his support of the code-validation studies over an extended period. The views and conclusions in this chapter are those of the author and should not be interpreted as necessarily representing the official policies or endorsements – either expressed or implied – of the AFOSR or the U.S. Government.

REFERENCES

[1] B. E. Edney. Anomalous heat transfer and pressure distributions on blunt bodies at hypersonic speeds in the presence of an impinging shock. *FFA Report* 115, Aeronautical Research Institute of Sweden, Stockholm (1968).

[2] B. E. Edney. Effects of shock impingement on the heat transfer around blunt bodies. *AIAA Journal*, 6 (1968), 1, 15–21.

[3] S. R. Sanderson. Shock wave interaction in hypervelocity flow. Ph.D. thesis, California Institute of Technology, USA (1995).

[4] S. R. Sanderson, H. G. Hornung, and B. Sturtevant. Aspects of planar, oblique and interacting shock waves in an ideal dissociating gas. *Physics of Fluids*, 15 (2003), 5, 1638.

[5] S. R. Sanderson, H. G. Hornung, and B. Sturtevant. The influence of non-equilibrium dissociation on the flow produced by shock impingement on a blunt body. *Journal Fluids Mechanics*, 516 (2004), 1.

[6] M. S. Holden, J. Moselle, and S. Martin. A database for aerothermal measurements in hypersonic flow for CFD validation, *AIAA Paper* 1996–4597 (1996).

[7] M. S. Holden. Real gas effects on regions of viscous-inviscid interaction in hypervelocity flows. *AIAA Paper* 1997–2056 (1997).

[8] M. S. Holden, S. Sweet, J. Kolly, and G. Smolinksi. A review of aerothermal characteristics of laminar, transitional, and turbulent shock/shock interaction regions in hypersonic flows. *AIAA Paper* 1998–0899 (1998).

[9] D. D'Ambrosio. Numerical prediction of laminar shock/shock interactions in hypersonic flow. *AIAA Paper* 2002–0582 (2002).

[10] P. A. Gnoffo. "On the numerical convergence to steady state of hypersonic flows over bodies with concavities." In *West-East High-Speed Flow Fields 2002*, eds. D. E. Zeitoun, J. Periaux, A. Desideri, and M. Marini (Barcelona, Spain: CIMNE, 2002).

[11] M. S. Loginov, N. A. Adams, and A. A. Zheltovodov. Large-eddy simulation of shock-wave/turbulent boundary layer interaction. *Journal of Fluids Mechanics*, 565 (2006), 135.

[12] M. W. Wu and. M. P. Martin. Direct numerical simulation of supersonic turbulent boundary layer over a compression ramp. *AIAA Journal*, 45 (2007), 4, 879–89.

[13] M. W. Wu and M. P. Martin. Analysis of shock motion in shockwave and turbulent boundary layer interaction using direct numerical simulation data. *Journal of Fluid Mechanics*, 594 (2008), 71.

[14] K. Sinha, K. Mahesh, and G. V. Candler. Modeling shock unsteadiness in shock/turbulence interaction. *Physics of Fluids*, 15 (2003), 8, 2290.

[15] K. Sinha, K. Mahesh, and G. V. Candler. Modeling the effect of shock unsteadiness in shock/turbulent boundary layer interactions. *AIAA Journal*, 43 (2005), 3, 586–94.

[16] M. MacLean, T. Wadhams, M. S. Holden, and H. Johnson. A computational analysis of ground test studies of the HIFIRE-1 transition experiment. *AIAA Paper* 2008–0641 (2008).

[17] P. R. Spalart and S. R. Allmaras. A one-equation turbulence model for aerodynamic flows. *AIAA Paper* 1992–0439 (1992).

[18] T. Rung, U. Bunge, M. Schatz, and F. Theile. Restatement of the Spalart-Allmaras eddy-viscosity model in strain-adaptive formulation. *AIAA Journal*, 41 (2003), 7, 1396–9.

[19] P. J. Waltrup and F. S. Billig. Structure of shock waves in cylindrical ducts. *AIAA Journal* 11 (1973), 10, 1404–8.

[20] D. V. Gaitonde, P. W. Canupp, and M. S. Holden. Heat transfer predictions in a laminar hypersonic viscous/inviscid interaction. *Journal of Thermophysics and Heat Transfer*, 16 (2002), 4, 481.

[21] M. S. Holden, T. P. Wadhams, J. K. Harvey, and G. V. Candler. Experiments and DSMC and Navier-Stokes computations for hypersonic shock boundary layer interactions. *AIAA Paper* 2003–1131 (2003).

[22] I. Nompelis, G. V. Candler, and M. S. Holden. Effect of vibrational nonequilibrium on hypersonic double-cone experiments. *AIAA Journal*, 41 (2003), 11, 2162–9.

[23] M.-C. Druguet, G. V. Candler, and I. Nompelis. Effect of numerics on Navier-Stokes computations of hypersonic double-cone flows. *AIAA Journal*, 43 (2005), 3, 616–23.

[24] M. S. Holden, T. P. Wadhams, and M. MacLean. Experimental studies to examine viscous/inviscid interactions and flow chemistry effects of hypersonic vehicle performance. *AIAA Paper* 2005–4694 (2005).

[25] M. S. Holden, T. P. Wadhams, and M. MacLean. Experimental studies to examine viscous/inviscid interactions and flow chemistry effects of hypersonic vehicle performance. *AIAA Paper* 2005–4694 (2005).

[26] J. K. Harvey, M. S. Holden, and G. V. Candler. Validation of DSMC/Navier-Stokes computations for laminar shock wave/boundary layer interactions, Part 3. *AIAA Paper* 2003–3643 (2003).

[27] I. Nompelis, G. V. Candler, M. MacLean, T. P. Wadhams, and M. S. Holden. Numerical investigation of high enthalpy chemistry on hypersonic double-cone experiments. *AIAA Paper* 2005–0584 (2005).

[28] I. Nompelis, G. V. Candler, M. MacLean, and M. S. Holden. Investigation of hypersonic double-cone flow experiments at high enthalpy in the LENS facility. *AIAA Paper* 2007-0203 (2007).

[29] J. M. Underwood and D. R. Cooke. A preliminary correlation of the Orbiter stability and control aerodynamics from the first two Space Shuttle Flights STS-1 & 2 with preflight predictions. *AIAA Paper* 1982–564 (1982).

[30] J. C. Young, L. F. Perez, P. O. Romere, and D. B. Kanipe. Space Shuttle entry aerodynamic comparisons of Flight 1 with preflight predictions. *AIAA Paper* 1981–2476 (1981).

[31] J. R. Maus, B. J. Griffith, K. Y. Szema, and J. T. Best. Hypersonic Mach number and real-gas effects on Space Shuttle Orbiter aerodynamics. *Journal of Spacecraft and Rockets*, 21 (1984), 2, 136.

[32] K. J. Weilmeunster, P. A. Gnoffo, and F. A. Greene. Navier-Stokes simulations of Orbiter aerodynamic characteristics including pitch trim and body flap. *Journal of Spacecraft and Rockets*, 31 (1994), 3, 255.

8 Shock Wave–Boundary-Layer Interactions Occurring in Hypersonic Flows in the Upper Atmosphere

John K. Harvey

8.1 Introduction

Many hypersonic vehicles are designed to follow trajectories that extend well into the upper atmosphere where the density is extremely low. Despite this, aerodynamic heating is still a critical issue because of the very high flight velocity. The U.S. Space Shuttle Orbiter, for instance, experienced peak heating at a height of about 74 km even though ambient density at that altitude is not much more than one millionth of sea-level density. Shock wave–boundary-layer interactions (SBLIs) that occur within these flows are nearly always sites of intense localized heating; thus, it is essential to predict the level correctly to avoid vehicle structural failure or incurring unnecessary weight penalties by carrying excessive thermal protection.

Along vehicle trajectories in the upper atmosphere, with increasing altitude, the ambient density drops more rapidly than the velocity rises. For this reason, the Reynolds numbers drop to relatively low values and most (if not all) of the flow is likely to be laminar. An SBLI is then more likely to involve separation; nevertheless, predicting the flow should be relatively straightforward because uncertainties associated with deciding the location of transition and choosing an appropriate turbulence model are avoided. However, it is observed that above 50 to 60 km, discrepancies begin to appear between the measured flowfields and flow solutions obtained using Navier-Stokes methods. Initially, as the altitude increases, these differences are small and can be avoided by using modified surface-boundary conditions to account for velocity slip and temperature jump. However, at greater altitudes where the density drops farther, the discrepancies are more significant and cannot be eliminated using this approach. This genuine departure from the Navier-Stokes predictions is attributable to so-called 'rarefaction effects' and it is due to the underlying foundations on which these equations are based being no longer strictly applicable. Rarefaction leads to fundamental changes in the way the flowfield is structured and hence it can significantly influence the way that SBLIs develop. Generally, rarefaction reduces the severity of the heat transfer; however, in other respects, it can be detrimental – for example, by reducing the effectiveness of control surfaces. Hence, it is important to quantify accurately the impact.

Before reviewing specific examples that illustrate how rarefaction influences SBLIs, this chapter outlines general characteristics of rarefied flows and discusses a few methods available to predict them. Readers familiar with this background material may proceed to Section 8.4, in which implications for SBLI flows are considered.

8.2 Prediction of Rarefied Flows

8.2.1 Classical Kinetic Theory for Dilute Gases

At a fundamental level, all gases consist of particles – whether atoms, molecules, or a mixture of both. In conventional continuum aerodynamics, this particulate nature of a fluid is ignored, the justification being that the physical scale of the flowfield – as defined by a representative dimension L, such as the body length or diameter[1] – is much larger than the molecular mean-free path λ. The Knudsen number (Kn) is defined as follows:

$$Kn = \lambda/L \qquad (8.1)$$

and it is the key parameter that determines the degree of rarefaction. If it exceeds 10, an insignificant number of particle collisions occur within the gas near the body and the flow is then said to be collisionless, or free molecular (FM). These flows are relatively easy to analyze and appropriate methods to solve them are well documented [1]. The solutions are useful because they define a limiting behavior of rarefied flows at the low-density extreme. At the other extreme, the continuum limit corresponds strictly to vanishingly small values of Kn but, for hypersonic flows, it safely can be assumed to apply if the value based on the free-stream mean-free path is less than roughly 10^{-3}. For vehicles flying in the upper atmosphere,[2] this value is exceeded frequently; hence, it is necessary to acknowledge that the fluid is not a continuum and therefore rarefaction effects must be considered.

Even if the density is very low, it is not possible to solve these flows deterministically by following the motion of each particle because of the exceedingly large numbers involved. Fortunately, alternatives solution methods are available; the principal options come within what is termed the classical Kinetic Theory of Gases and the associated numerical methods. Vincenti and Kruger [2], Chapman and Cowling [3], and Cercignani [4] provide excellent explanations of this subject and their works are recommended for further reading. Analytic approaches to kinetic theory, for the most part, have been unsuccessful as a means for solving practical aerodynamic problems, and certainly nothing nearly as complex as a flow involving an SBLI has been tackled successfully. Nevertheless, the importance of this theory should not be underestimated because it provides the foundation for understanding the physics of rarefied flows and is the basis on which successful numerical particle-simulation methods have been devised. Of these methods, the powerful Direct Simulation Monte Carlo (DSMC) method devised by Bird [1] is the most successful tool for predicting practical rarefied flows. For readers unfamiliar with the

---

[1] In some situations (e.g., when studying the structure of shock waves), a more appropriate length scale to use is one based on local gradients within the flow, such as $\rho/(d\rho/ds)$, where ρ is the density and s is the scalar distance.

[2] The mean free path is about 0.005 m at 81 km altitude in the earth's atmosphere.

Kinetic Theory of Gases and the DSMC method, the relevant aspects are outlined in Appendix A.

Most of the SBLI discussion in this chapter is based on DSMC computations, although experimental results also are presented. However, there is a dearth of reliable measured data because it is inherently difficult to perform wind-tunnel experiments on rarefied flows. Signal levels from instruments are low, the flows are sensitive to probe interference, and unique measurement errors arise that are attributable to rarefaction effects. Furthermore, providing a suitable environment in which to conduct the tests is a challenging task and it is difficult to obtain properly characterized and gradient-free wind-tunnel flows. Conventional flow-visualization techniques such as schlieren and shadowgraph are usually insufficiently sensitive; however, the low density allows alternative nonintrusive optical diagnostic methods such as the widely used electron-beam fluorescence technique to be used. This technique readily yields information on flow density, and the translational, vibrational, and rotational temperatures also can be inferred using spectrometric methods. This technique was used to obtain measurements shown in Appendix Figs. A.1 and A.3. Conventional methods of measuring heat transfer (e.g., thin-film resistance gauges) are generally satisfactory except for problems in addressing low signal levels. Static-pressure measurements are more problematic and sizeable errors arise if the conventional technique of using small holes (or "tappings") in the model surface is used.[3] As Kn is increased, these errors become larger and more difficult to quantify because they are dependent on the unknown velocity distribution of the gas particles in the flow just above the tapping. The error can be avoided by using pressure transducers with sensing diaphragms set flush to the surface of the body. Small instruments with adequate sensitivity are not widely available but were used in the experiment discussed in Section 8.4.

8.3 Characteristics of Rarefied Flows

When the density of a flow is reduced, molecular collisions occur less frequently. Hence, internal-energy exchange, dissociation, recombination, chemical reactions, and ionization proceed more slowly because these processes are a consequence of particle collisions. For this reason, rarefied flows are characterized by heightened levels of molecular and thermodynamic nonequilibrium compared with their continuum counterparts. This is manifested in various ways within the thermodynamics of the flow. For example, distributions associated with the thermal part of the particles' translational velocity and their internal energy may depart from their equilibrium values (i.e., f_o and f_{ε_i}, respectively). There may be an imbalance between the amounts of energy within these modes; thus, the concept of equi-partition of energy does not apply in parts of the flow. Most reactions that take place in air at elevated temperatures depend on the level of vibrational excitation as well as

---

[3] This is because low-density flows exhibit molecular nonequilibrium and the velocity-distribution function of the gas molecules approaching the sensing hole from just above the surface (the incident distribution) differ from that of molecules moving in the opposite direction from within the tapping. The latter will have accommodated to the body temperature and the distribution function of this flux of particles will correspond to this. For there to be no net flow in the plane of the surface, the flux of particles in each direction must be equal. However, because their velocity-distribution functions differ, a pressure gradient will be generated around the entrance to the pressure tapping. Hence, the incorrect pressure is recorded.

the translational energy. If the balance of energy between these modes varies, the reaction probabilities will change, which will bear directly on the species concentrations within the gas. The reduced collision frequency also can lead to the population of the quantum levels within any specific mode being perturbed, which can cause further deviation in the reaction rates from those of continuum (i.e., Arrhenius) chemistry. For these reasons, determining thermodynamic changes and chemical reactions within rarefied flows is difficult and only with the advent of powerful particle-simulation methods, that even extend to dealing with the chemical reactions at a quantum physics level, has any significant progress been made in realistically solving these problems.

8.3.1 Structural Changes that Occur in Rarefied Flows

In addition to the markedly higher degrees of molecular nonequilibrium in rarefied flows, unique and far-reaching changes to the flow structure are observed that cannot be explained by continuum theory. These changes impact the way that SBLIs develop. To explain these phenomena, we digress to examine two simple hypersonic flows that illustrate a number of features of rarefied flow that bear directly on the formation of SBLIs in low-density flows.

In the classical continuum supersonic viscous flow over a flat plate at zero angle of attack, an oblique shock wave is generated by the displacement due to growth of the boundary layer that begins at the leading edge. The interaction between the shock and the boundary layer is described as either strong or weak, depending on the degree of their mutual coupling. This flow is analyzed, for example, by Mikhailov et al. [5] and to first order the pressure along the plate for a strong interaction is proportional to the following viscous-interaction parameter:

$$\overline{X} = M_\infty^3 \sqrt{C_\infty} / \sqrt{Re_{x,\infty}}. \tag{8.2}$$

C_∞ comes from the approximate expression $\mu/\mu_\infty = C_\infty T/T_\infty$ relating the viscosity to the temperature, M_∞ is the incident Mach number, and $Re_{x,\infty}$ is the Reynolds number based on the distance x from the leading edge. Equation 8.2, which is based on continuum arguments, implies that the pressure, which is proportional to $x^{-1/2}$, tends toward an unphysical infinite value at the leading edge. Of course, in reality, a different flow has to be established near the leading edge, the characteristics of which can be understood only by acknowledging the particulate nature of the fluid. In the schematic depiction of the hypersonic flat-plate leading edge flow shown in Fig. 8.1, several regions are indicated and each exhibits a different physical structure. They are best characterized by considering the magnitude of Kn based on the free-stream conditions and the distance from the leading edge. On the far right of Fig. 8.1, there is a region of continuum flow with strong viscous interaction. Upstream of this, Kn rises and anomalies in the surface boundary conditions know as velocity slip and temperature jump become evident. These two effects have an impact on the structure of SBLIs in the rarefied regime by changing the characteristics of the boundary layer approaching the interaction; they can, for example, delay separation. These effects are considered in more detail later in this chapter.

In the center of Fig. 8.1, the boundary layer and the shock wave are shown to have converged; from this point upstream, they are said to have 'merged'. This reduces the shock wave's strength by modifying the internal structure such that it

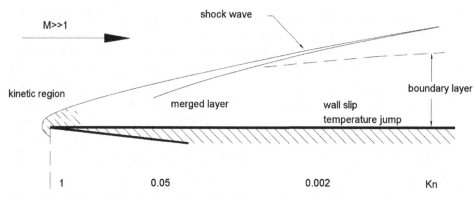

Figure 8.1. Schematic depiction of a rarefied hypersonic flat-plate flow.

ceases to satisfy the Rankine-Hugoniot conditions even though the inclination virtually is unchanged. Merging occurs in other situations, and the existence of shock waves that do not satisfy the Rankine-Hugoniot conditions is a unique feature that distinguishes rarefied flows from continuum low-Reynolds-number flows. The merging and weakening of the shock wave reduces the pressure and heat transfer on the flat plate nearby below the levels predicted by strong-interaction theory. This first occurs when Kn is roughly 0.005; upstream of this point, kinetic-theory or particle-simulation methods are the only viable ways to analyze the flow correctly. Merging and the consequential weakening of the shock waves are major contributors leading to the reduced intensity of SBLIs in the rarefied flow regime.

Close to the leading edge, Kn increases toward unity. The flow then can be best visualized as a cloud of gas molecules surrounding the tip of the plate that experiences relatively few collisions to change their velocity. Within this so-called 'kinetic region', the shock wave completely loses its identity.

The changes in the flow character toward the leading edge create a significant departure in the pressure and heat-transfer distributions from the continuum values. Fig. 8.2a shows a typical pressure distribution obtained using the DSMC method compared with results from the strong-interaction theory. The data are plotted against the inverse Kn; this corresponds to the physical streamwise dimension. The heat-transfer distribution to a flat plate for a higher-Mach-number example is plotted in Fig. 8.2b and shows more detail of the flow near the leading edge. Both the pressure and heat transfer rise to peaks near the tip and then drop within the kinetic region when $1/Kn$ is less than about 10.

The second rarefied flow example exhibits merging but it also is a reminder that at low densities, shock waves have significant thickness and therefore occupy a finite space within the flowfield. They typically have a thickness of ten to twenty times the incident mean-free path; therefore they can no longer be regarded as discontinuities. The flow considered here is of a flat-ended circular cylinder with the axis aligned to the approaching free stream. Figrue 8.3 shows three sets of DSMC computations[4] in which the degree of rarefaction is increased from near

---

[4] The validity of the computed results shown here is supported by evidence from similar computations for the same flow shown in Fig. 3.A in the Appendix of this chapter, where predictions compare favorably with experimental data.

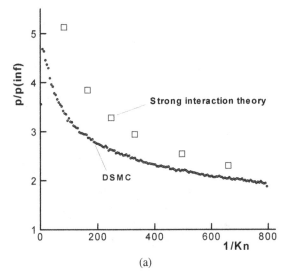

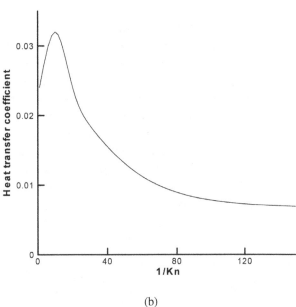

Figure 8.2. (a) Pressure distribution on a flat plate at $M = 9.55$ in nitrogen for $T_{wall}/T_\infty = 1.51$. (b) Heat-transfer coefficient distribution on a flat plate at $M = 20.6$ in nitrogen for $T_{wall}/T_\infty = 14.4$.

continuum toward FM flow. The figure shows the density variation along the stagnation streamline as a fraction of the free-stream value, where x is the coordinate in the free-stream direction measured from the front face of the cylinder and n is the particle number density. Kn is based on the cylinder's radius R, which is 0.01m.[5]

For the densest case for which $Kn = 0.0083$, the shock wave – indicated by a sharp rise in density – forms at around $x/R = -0.45$. Although this Kn places the flow close to the continuum limit, the finite thickness and internal structure of the shock wave are evident. Downstream of the shock wave, a plateau in density indicates that this wave is separated from the viscous flow on the front face of the cylinder. Apart

---

[5] Other flow conditions in this example are $T_\infty = 40K$, $n_\infty = 3.0 \times 10^{21}/m^3$, $T_\infty/T_{wall} = 0.135$, n is the particle number density, and the gas is nitrogen.

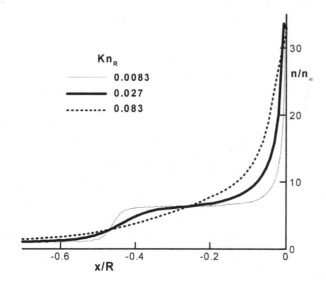

Figure 8.3. Comparison of the density profiles along the stagnation streamline for a blunt-ended cylinder for different Knudsen numbers at Mach 23.

from the shock-wave thickening, the flow structure otherwise appears to be similar to the continuum counterpart for a cold-walled body.

At the intermediate density, for which $Kn = 0.027$, a true plateau downstream of the shock wave no longer can be identified. This is due to the shock wave merging with the viscous flow on the front face of the cylinder, and it is not possible to state exactly where the shock wave ends. Nevertheless, it is clear from the structure that the shock wave is considerably thicker than in the first case. In the final example, Kn is increased to 0.083, which is only ten times larger than in the first flow. However, it is apparent that over this relatively small Kn range, the structure of the flow has altered completely. There is no trace of a shock wave; rather, a smooth monotonic rise in density occurs as the flow approaches the surface. If the density is reduced further, much the same pattern persists, but the thickness of the perturbed layer on the front face of the body decreases with rising Kn. This new flow structure can be understood by first visualizing the equivalent "near FM" flow in which the incident stream of particles will not experience a significant number of collisions until they almost reach the body's face, where they congregate in a dense layer on the surface before flowing radially outward. A similar process is evident in the lowest density example (i.e., $Kn = 0.083$), although the flow cannot be classed as truly FM because collisions occur within the gas well ahead of the body. Nevertheless, for this flow, the pressure and heat-transfer coefficients on the front face approach closely the FM values, which are higher than the corresponding continuum values. For SBLIs in rarefied conditions, at least the primary shock wave over the body can take on similar characteristics to those noted herein, thus affecting the interaction. Again, attention is drawn to the rapidity in which the changes in the flow structure occurred: once within the rarefied regime, they occur within a one-order-of-magnitude increase in Kn.

Throughout most continuum hypersonic flows, the vibrational temperature is frozen, whereas the rotational component, which can be changed by relatively fewer molecular collisions, usually follows closely the variations in the translational value. However, in rarefied flows, this is not generally the case; in much of the flow the

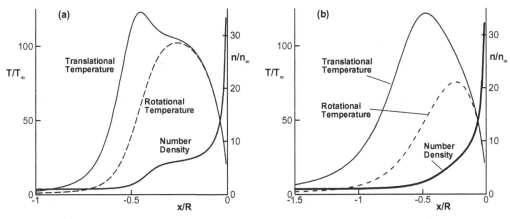

Figure 8.4. DSMC-computed profiles of number density and temperature along the axis of a blunt-ended cylinder: (a) $Kn = 0.027$; (b) $Kn = 0.083$ with the density and temperature normalized with respect to the free-stream values.

rotational component also lags behind changes in the translational value as a consequence of the reduced collision frequency. It is most noticeable in shock waves in which the changes are rapid. This is illustrated clearly in the blunt-ended cylinder flow. The two plots shown in Fig. 8.4 correspond to the intermediate- and lowest-density examples in Fig. 8.3; it is shown in both cases that these two temperature components come into equilibrium only near the surface, where the flow is denser and not subjected to rapid changes. In the $Kn = 0.027$ flow, this occurs at about $x = -0.25$ but nearer the surface in the more rarefied example. The vibrational temperature, which requires many more collisions to affect a change, is completely frozen in both examples except for very close to the surface, and hence not shown. Also note that the translational temperature rises well upstream of any change in the density, which is a consequence of a small number of particles being reflected from the denser flow within the shock layer and penetrating upstream into the low-density-approaching flow. The high relative velocity between the two sets of particles leads to intense collisions that elevate the translational temperature. This effect is also evident in the results shown in Appendix Fig. A.4a for a blunted cone, and can lead to chemical changes in the gas upstream of the main body of the shock wave in high-enthalpy flows.

8.4 Examples of SBLIs in Rarefied Hypersonic Flows

8.4.1 Introduction

The previous section concluded that as Kn increases, rarefied flows experience structural changes that distinguish them from their continuum counterparts. These changes eventually lead to the transition to FM flow as density falls. They are due in part to an increase in the relative importance and extent of viscous phenomena that can be thought of as a low-Reynolds-number effect. However, there are other phenomena – such as velocity slip, temperature jump, shock-wave merging, and weakening – that are strictly rarefied-flow effects, which (when they occur) lead to fundamental physical changes in overall flow structure. Also notice that a degree of molecular nonequilibrium is generally present, which increases in significance as *Kn* rises. This is responsible for changes in the thermodynamic and chemical composition of the flows.

Two configurations that exhibit SBLIs are considered in detail. These examples illustrate how the flow structure is changed when the density is reduced to levels experienced during flight in the upper atmosphere. Because this occurs only at high speeds, the discussion is restricted to hypersonic Mach numbers.

8.4.2 SBLIs on a Hollow-Cylinder–Flare Body

The first example is flow over a hollow-cylinder–flare body. This is an idealization of the type of flow that occurs over the rear of a cylindrical missile with a flared tail, and it is the axisymmetric equivalent of a flat plate terminated by a ramp or flap. This flow was the subject of an extensive study instigated by Working Group #10 of the NATO Research and Technology Organization (RTO). It was selected as a suitable subject for a computational fluid dynamics (CFD) code-validation exercise that was an integral part of a wider-ranging investigation into hypersonic flows that exhibit regions of complex viscous-inviscid interaction. With U.S. Air Force Office of Scientific Research (AFOSR) support, several carefully conducted experimental studies and numerical investigations using both Navier-Stokes and DSMC codes were undertaken [6]. The flare angle chosen for the RTO validation exercise was 30 degrees, which ensured that separation occurred for the range of Mach numbers and Reynolds numbers included.

Although the shape of the hollow-cylinder–flare body is simple, it creates a complex flow involving shock-shock interactions as well as SBLIs. This flow proved to be a good subject for the critical evaluation of CFD codes, and the challenge caught the imagination of an international group of distinguished researchers who participated in the study by providing experimental data and computed results. The lowest-density test conditions included in this exercise provided only a moderate degree of rarefaction. For this reason, additional computed results are presented here that extend the range toward the FM limit. An axisymmetric shape was chosen to avoid the experimental uncertainties that arise for supposedly two-dimensional configurations that, in reality, always exhibit three-dimensional anomalies. Most of the measurements were made in shock tunnels at CUBRC, Inc (formerly Calspan-University of Buffalo Research Center) in New York State, and a full set of tabulated results is in the CUBDAT database [7] and the Holden and Wadhams paper [6]. These experiments were conducted at nominal Mach numbers of 9.5, 11.4, and 15.7 for a range of Reynolds numbers that in every instance ensured fully laminar flow. All of the flows considered were in nitrogen.[6] Further experiments were performed in France in the ONERA R5Ch wind tunnel [8, 9].

Main features of the flow are illustrated in Fig. 8.5. The leading edge of the cylinder is aerodynamically sharp and chamfered on the inner side to ensure that the flow through the core remains supersonic. A weak bow shock wave is formed on the outer surface of the cylinder due to the boundary-layer growth. The flow

---

[6] The outer radius (*R*) of the cylinder was 32.5 mm (1.25 inches) and the length (*L*) before the flare was 101.7 mm (4.0 inches).

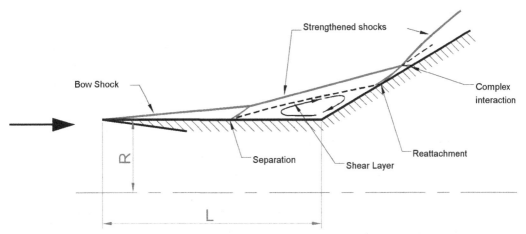

Figure 8.5. Schematic of the hollow-cylinder–flare body flow under continuum conditions. (The axis is shown as the dashed line.)

then separates on approaching the corner (if the flare angle is sufficiently large) and subsequently reattaches on the flare. A secondary shock originates at the separation point, which then impinges on the bow shock wave, thereby intensifying and steepening it. This secondary shock compresses the flow between the primary shock and the separation bubble; for a free-stream Mach number of 12 and a 30-degree flare, the Mach number is typically reduced to about 4.5 after the second shock. The size of the recirculation region varies considerably with the Reynolds number, Mach number, and flare angle; however, similar flow structures nevertheless are produced unless separation is suppressed. If the ambient density is reduced, velocity slip and temperature jump occur on the cylinder, which delay separation, thereby reducing the size of the bubble.

Reattachment occurs on the flare if it is long enough. The resultant flow deflection produces a succession of compression waves that quickly coalesce into a third shock wave that impinges on the bow shock wave, causing it to be deflected farther away from the body and significantly strengthened. The form of the resultant shock-shock interaction depends on the flare angle and flow conditions. For small angles, two shock waves and a shear layer are formed; however, if the interaction is more intense, two triple points occur separated by a short planar shock wave. In either case, a shock wave from the interaction is directed toward the body where it impinges on the flare-boundary layer. By this time, the fluid near the surface has passed through four shock waves or sets of compression waves; thus, the density and temperature will have risen sharply. This process thins the boundary layer on the flare and creates localized and intense peaks in the pressure and heat-transfer distributions downstream of reattachment. In flight applications, this intense heating can seriously endanger the structure of a vehicle.

Figures 8.6 and 8.7 are examples of the computed and measured distributions of pressure coefficient (C_p) and Stanton number (St) for the lowest-density conditions included in the RTO study. For this, Kn based on L equals 0.00087; hence, this flow is at the dense extreme of the rarefied flow regime. The measured results are taken from the CUBRC experiment [7] and the DSMC results were computed by Markelov et al. [10] using code they developed. The two coefficients are plotted

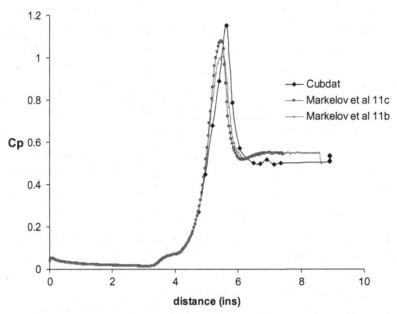

Figure 8.6. Pressure-coefficient (Cp) distribution for the hollow-cylinder–flare body: $M = 12.4$; $Kn = 0.00087$; $n_\infty = 1.197E22$; $T_\infty = 95.6$ K; $T_{\text{wall}} = 93$K; nitrogen. Experimental results are from CUBDAT [7] and the DSMC data are from Markelov et al. [10]. The cone intersection is at 4.0 inches (101.6 mm).

against the distance s along the external surface of the body. In Fig. 8.6, a small rise in pressure is shown at about $s = 3.5$ inches (90 mm: $x/L = 0.874$), which indicates separation. In the flow upstream, strong viscous interaction occurs that causes the rise in pressure near the leading edge. After separation, a small plateau in pressure is observed; however, once on the flare (i.e., $s > 4$ inches $= 101.6$ mm), the pressure

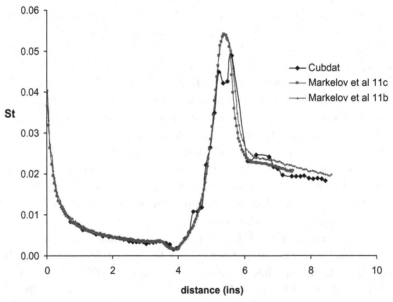

Figure 8.7. The corresponding heat-transfer distribution for the same hollow-cylinder–flare flow shown in Figure 7.6.

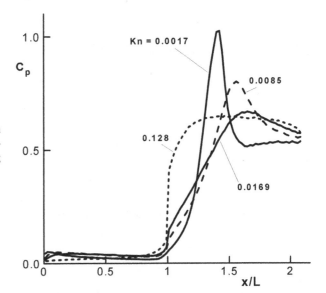

Figure 8.8. The effect of varying Kn on the pressure distribution on the hollow-cylinder–flare. $M = 12.4$; $T_\infty = 95.6$ K; $T_{wall} = 293$ K; nitrogen.

and heat transfer both rise increasingly rapidly toward the intense peaks located downstream of reattachment (for reasons explained previously). From these plots, it is not possible to discern precisely where reattachment occurs. Downstream of the peaks, the pressure and heat transfer quickly recover toward the laminar values for a 30-degree cone, thereby confirming that transition to turbulent flow had not occurred in the experiments.

An impressive correlation between the computed and measured data is evident especially for the heat transfer. Not shown but equally good results for this flow were obtained using Bird's DSMC code [11] by Moss and Bird [12]. This is a complex flow involving separation, reattachment, and shock-shock interactions and SBLIs, and the agreement between experiment and computation is a remarkable outcome. Achieving the correct solution is crucially dependent on accurately predicting the positions of separation and reattachment, something that has proven historically to be a struggle to achieve consistently well using CFD.

Great care had to be exercised in performing the experiments for this study because the absolute levels of pressure and heat flux were very low. Potential errors due to rarefaction effects were avoided by using flush-mounted pressure transducers. That being said, it is believed that the two heat-transfer measurements in the region of the peak of the curve in Fig. 8.7, where there is some discrepancy, are anomalous due to experimental error.

Figure 8.8 shows the effect on the computed pressure distributions of progressively increasing Kn, in this case using Bird's DSMC code. The increase was achieved by reducing the density while retaining the other flow parameters constant. Thus, the Reynolds number also drops. The densest case is for $Kn = 0.0017$, which is twice the value in the previous example shown in Figs. 8.6 and 8.7. This change has the effect of moving the separation point downstream, almost to the beginning of the flare. The increase in Kn thickens the shock waves and there also is an increase in the level of velocity slip on the cylinder. The intense peak in the pressure distribution is still present, although the magnitude is reduced slightly. However, the peak is at

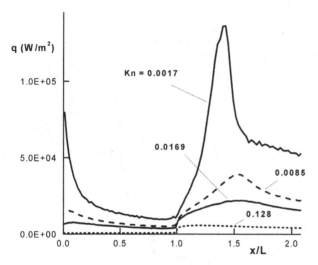

Figure 8.9. The effect of varying Knudsen number on the heat flux to the hollow-cylinder–flare. $L = 0.102$ m; $M = 12.4$; $T_\infty = 95.6$ K; $T_{wall} = 293$ K; nitrogen.

almost the same position as it was for the lower Kn flow, despite the separation bubble being much smaller and with reattachment occurring closer to the corner. The peak in the pressure is linked to the reflection of the reattachment shock toward the body from the shock-shock interaction; therefore, it is expected that the peak would have moved upstream as the size of the separation bubble is reduced. (Even when separation is completely suppressed, a compression wave is formed at the junction with the cone that interacts with the primary shock.) However, the increase in Kn creates a thickening of the shock waves and results in the flow taking on a generally more viscous and benign character, which has the effect of reducing the sharpness of the gradients within the flow especially in the primary shock because the density upstream of it is the lowest in the flow. With increasing Kn, this "smearing" of flow structures attenuates the peaks and moves them downstream, eventually leading to a complete merging of the shocks with the viscous flow on the body. Thus, two factors influence the position of the peak; for the increase of Kn to 0.0017, they roughly balance one another. With further increases in Kn, the peak moves downstream and quickly decays. For the most rarefied example shown in the figure, for which Kn is 0.128, there is no discernable peak in the pressure distribution. Instead, the flow begins to resemble closely the FM pattern for which there would be a constant-pressure coefficient on the flare equal to 0.650 solely dependent on the inclination of the surface. What is striking about the results shown in Fig. 8.8 is how quickly the flow structure changes from the continuum pattern to one resembling FM flow. The transition occurs principally in the Kn range from 10^{-3} to 10^{-2}.

Figure 8.9 shows how the absolute value of the aerodynamic heating for this flow varies with increasing rarefaction. The overall levels drop due to the reduction in density, but it is also clear that the sharpness of the peak is diminished, which echoes the trend seen in the pressure. As expected, the pressure and heat-transfer peaks coincide at the same point on the flare for each flow condition. The heat-flux data are replotted in Fig. 8.10 in terms of the Stanton number; it is shown here that as the density is reduced – that is, as Kn is increased and the Reynolds number falls – the overall levels of this coefficient rise. This follows the Reynolds-number trend observed in laminar-continuum flows. From this plot, we can observe more clearly

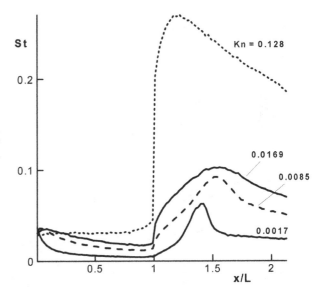

Figure 8.10. The effect of rarefaction on the corresponding heat-transfer coefficient distribution on the hollow-cylinder–flare.

how the peak moves back along the flare as Kn increases from 0.0017 to 0.0169. In the most rarefied case, for which $Kn = 0.128$, there is no longer a discernable down-stream peak produced by the reflected shock wave, which is now very diffuse and impinging on the flare. Although this flow appears from the pressure distribution to have taken on the character of a FM flow, it is not as evident in the heat-transfer dis-tribution; the FM value on the flare would have been constant and equal to 0.28 but instead it is shown falling away from this value with increasing x. Thus, the computed heat-transfer distribution still retains some of the continuum character, illustrating the complex nature of these transitional flows within this Kn range.

8.4.3 Velocity-Slip and Temperature-Jump Effects

Values of the velocity – expressed as a fraction of that of the free stream – are plotted in Fig. 8.11 for the computational cells adjacent to the surface of the hollow-cylinder body. The curves are plotted for varying degrees of rarefaction, and it is immediately apparent that if the density is low, the magnitude of what is effectively the slip velocity u_{slip} can be substantial. This departure from the usually assumed "no-slip" surface boundary condition reduces the retardation of fluid in lower parts of the boundary layer. This delays separation; hence, this phenomenon is significant when an SBLI occurs.

The velocity-slip and temperature-jump effects are usually the first rarefied-flow phenomena to influence SBLIs when the density is reduced. Their importance is increased if the inclination of the surface ahead of the interaction is small with respect to the free stream as the pressure level and hence also the density within the boundary layer is then low. Neither of these two effects can be understood prop-erly without acknowledging the particulate nature of the flow, to which – because of their importance – we now digress in detail.

Consider the flow within a boundary layer that is formed on a plane surface (Fig. 8.12). The velocity is $u = f(y)$ in the x direction and y is the coordinate normal

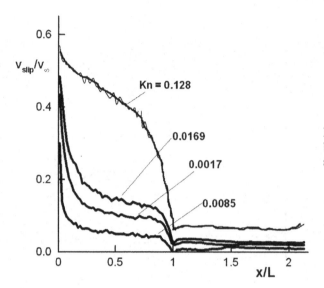

Figure 8.11. Slip velocities in the boundary layer along the biconic surface.

to the surface. We propose that close to the surface but for y a little greater than the local mean-free path λ, the velocity gradient is assumed to be constant such that $du/dy = k$, which depends on the shear level in the boundary layer above the region being considered. We concentrate on the region up to one mean free path from the surface. The flow within it consists of particles that have a mean velocity plus a random component due to their thermal motion. Therefore, some move toward the surface (i.e., the incident stream); others that have collided with the surface move away from it into the flow (i.e., the reflected stream). Consider now an incident particle that is moving toward the surface as a consequence of its random thermal motion. Once it is within a mean-free path of the wall, it does not (on average) experience a further collision to change the velocity before reaching the surface. Thus, it carries with it the momentum acquired from the last collision within the gas; it arrives at the wall with a positive finite velocity in the x direction, the magnitude of which depends on the values of k and λ. Each particle that reaches the surface

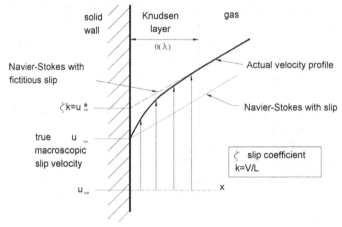

Figure 8.12. A schematic impression of the velocity profile close to the surface in a boundary layer on a plane wall.

is returned to the flow after adjusting to the wall conditions. For "practical engineering surfaces," it is justifiable to assume that these particles are returned fully accommodated and diffusely re-emitted into the flow. This is equivalent to them being returned as if they were coming from an imaginary stationary gas beneath the surface at the wall temperature and pressure.[7] Thus, the re-emitted particles have zero-mean tangential velocity. The flow velocity at the wall is obtained by combining the properties of the incident and reflected streams, and there is thus a resultant finite velocity in the x direction at the wall, referred to as the slip velocity. This is in contravention to the usual no-slip boundary condition assumed in continuum-fluid mechanics. The region in which this happens is approximately one mean-free-path thick and is referred to as a Knudsen layer. Velocity slip is accompanied by a corresponding temperature-jump effect, in which a difference occurs between the temperature of the wall and that of fluid adjacent to it. The velocity-slip and temperature-jump values depend on the degree of rarefaction and the level of shear and temperature gradient in the boundary layer, respectively. They always have the effect of reducing the shear stress and usually the heat transfer.

These effects are present in all boundary layers but, in most circumstances, the mean-free path is so small that the scale renders them completely insignificant. They generally can be ignored and the no-slip condition can be applied legitimately if Kn is less than 10^{-4}. Above this value, continuum-flow calculations should be adapted by modifying the wall boundary conditions, which can be accomplished using semi-empirical expressions to determine the velocity-slip and temperature-jump effects. This procedure is acceptable for a small Kn; however, as the value increases, the approximation becomes less reliable and the real influence can be accounted for only by analyzing the flow using kinetic theory or particle simulation – at least, for the flow near the surface. Although the changes in the boundary-layer profiles due to velocity slip and temperature jump will be experienced mostly near a leading edge or tip of a body where Kn is highest, they can materially affect the development of the subsequent boundary-layer flow, thereby influencing SBLIs that occur downstream.

The velocity-slip effect has been the subject of numerous kinetic-theory studies that properly consider any particle collisions that occur within Knudsen layer. It has proved to be a difficult problem to solve; however, a successful result using a linearized approximation that disregards temperature gradients was provided by Cercignani [4]. He gives the velocity profile near the surface as follows:

$$u(y) = k \left[x + \zeta - \frac{\pi^{-1/2}\lambda}{2} I \left(\frac{y}{\lambda} \right) \right] \tag{8.3}$$

where ζ is a so-called slip coefficient equal to 1.01615λ and k is (as before) the gradient of mean velocity in the boundary layer just beyond Knudsen layer. The function $I(y/\lambda)$ is plotted by Cercignani and rises close to the surface, but it is virtually zero outside Knudsen layer. A velocity profile corresponding to this expression is illustrated in Fig. 8.12. Although the layer is only about one mean-free-path thick, some collisions between particles do occur that lead to the curvature of the profile. Other

---

[7] The re-emitted particles therefore have velocities corresponding to an equilibrium distribution for a stationary gas $f_0(c, T_{wall})$, for $v > 0$ where v is the outward y component of velocity and T_{wall} is the wall temperature.

attempts were made to refine the kinetic-theory analysis; however, for the most part, they failed to provide results that agree well with measured data. Recent numerical solutions using the DSMC method, however, confirm Cercignani's results [13]. From the figure, it is evident that the true slip velocity at the surface is not the most appropriate value to use as the extrapolated boundary condition for Navier-Stokes calculations of the flow beyond Knudsen layer. A higher value, indicated as u^* in the figure, is more suitable. When appropriate, this value can be incorporated into continuum calculations as the wall boundary condition using an expression such as the following:

$$u^* = \sigma_p \frac{\mu}{p} \left(\frac{2\hat{k}T}{m} \right)^{1/2} \frac{\partial u}{\partial y}\bigg|_e \tag{8.4}$$

where u^* is the effective tangential velocity of the gas at the solid surface, μ is the gas viscosity, p and T are the local pressure and temperature (respectively), m is the molecular mass of the gas, and $\hat{k}$ is the Boltzmann constant. The degree of velocity slip is proportional to the shear stress evaluated at the outer edge (i.e., subscript e) of Knudsen layer. The constant σ_p is an alternative velocity-slip coefficient first introduced by Maxwell in 1879 [32], but his value was inaccurate. For practical application of this theory, the extensive reviews by Sharipov et al. [14,15] can be consulted for appropriate suggestions for the coefficients covering various situations including gas mixtures. Corresponding expressions for the temperature jump also are discussed by these authors.

The velocity profile in Fig. 8.12 is consistent with the viscosity, effectively decreasing in Knudsen layer as the wall is approached because there is no mechanism for the shear stress to vary greatly within the short distance. A DSMC solution by Torczynski et al. [16] confirmed this and demonstrated that the effective viscosity and thermal conductivity fall sharply by as much as 50 percent very close to the wall when compared with the Chapman-Enskog [33] (i.e., the continuum) values. This again underlines the care with which the Navier-Stokes methods should be used for flows exhibiting rarefaction effects because these changes in the transport coefficients cannot be incorporated in any way.

The kinetic-theory solution from which equation 8.3 is derived assumes that the gas just outside Knudsen layer is in local molecular equilibrium. This normally would be a valid assumption in continuum flows; however, for low-density hypersonic situations, some degree of nonequilibrium is likely even in the lower strata of the boundary layer. This can occur in situations where it otherwise would be legitimate to use Navier-Stokes methods with slip-boundary conditions. For example, it could arise due to reactions taking place in the flow. In developing the velocity-slip theory, it also is normally assumed that the gas re-emitted from the surface is in full equilibrium at the wall temperature. This may be an unsound assumption for two reasons. First, the accommodation of the internal-energy modes on surfaces is acknowledged to be frequently incomplete; for example, an accommodation coefficient of only 0.35 was predicted [17] for the rotational mode for nitrogen in certain circumstances (e.g., a value of 1.0 indicates full accommodation). Furthermore, it is conjectured that the vibrational mode is even less likely to be fully accommodated to wall conditions. Second, surfaces are important in chemically reacting flows: If they are not fully catalytic, the re-emitted gas will not be in full chemical equilibrium.

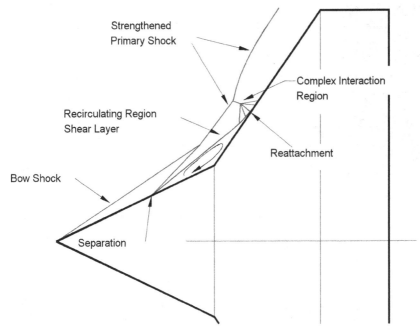

Figure 8.13. Schematic of the flow over the biconic body showing flow topology.

The degree to which these factors influence the velocity-slip and temperature-jump phenomena has not been investigated and cannot be quantified easily.

8.4.4 SBLIs Occurring on a Sharp Biconic Body

The second example considered to illustrate the effects of rarefaction on the physical structure of SBLIs is the flow over a double cone, otherwise referred to as a biconic body. This flow has much in common with the hollow-cylinder–flare-body flow, but it demonstrates what happens when the interaction is more intense due to the greater deflections being imposed on the flow. This configuration was included in the RTO code-validation exercise [6, 18].

Figure 8.13 is a schematic diagram of the main features of the continuum hypersonic flow for this body. Separation occurs upstream of the junction between the two cones, producing a recirculating region roughly centered on this point. For the validation exercise, the angles of the fore and aft cones were selected as 25 and 55 degrees, respectively, which ensured that the flow was steady and separated. As with the hollow-cylinder–flare body, the extent of the recirculation region varies with changes in Knudsen, Reynolds, and Mach numbers. Figure 8.14 shows the velocity contours and streamlines for this flow from a DSMC simulation for the region near the intersection of the two cones at a *Kn* of 0.0024. This corresponds to test case No. 7 in the CUBDAT dataset, which was performed at $M = 10$ using nitrogen for the test gas. Other results discussed in this section are restricted to this gas as well to avoid the complication of any chemical reactions. The shape of the primary shock wave can be inferred from Fig. 8.14 as the edge of the dark outer region.

As in the hollow-cylinder–flare flow, a secondary shock wave is generated just ahead of the separation point and this wave impinges on the bow shock wave,

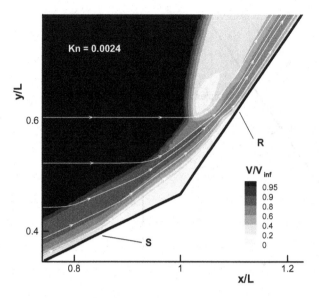

Figure 8.14. DSMC computation showing the velocity and stream lines near the corner of the biconic body for $Kn = 0.0024$; CUBDAT test case No. 7; $M = 10$; nitrogen.

thereby strengthening and deflecting it away from the body. The path of the separation shock can be inferred from the kinks in the streamline and the change in slope of the primary shock at about $x/L = 0.95$. A complex interaction, similar to that observed on the hollow-cylinder–flare body, occurs near reattachment; however, in comparison, a more intense flow is established here due to the greater flow deflections imposed by this body. From the shape of the streamline defining the edge of the separation bubble, it is apparent that immediately upstream of reattachment point (R), the shear layer is concave. This produces a set of focusing compression waves that radiate away from the surface and interact with the bow shock wave to produce the shock-shock interaction. These waves usually (but not always) focus to form a shock wave before interacting with the primary shock. The interaction is similar to those categorized by Edney [19] for flows in which an oblique shock wave impinges on the bow shock wave ahead of a blunt leading edge. Flows analogous to his Type III and, more important, Type IV interactions are possible on the biconic body. The structure of an Edney interaction for a similar degree of rarefaction is illustrated in Fig. 8.15a, in which DSMC results for an oblique shock impinging on a circular body obtained by Moss et al. [20] are shown. Mach-number contours are plotted from which the shock-wave pattern can be identified clearly. Two triple points (TP) are produced by the interacting shock waves between which a near-planar oblique shock wave forms. Downstream, a supersonic jet forms that directs fluid toward the surface and creates intense localized peaks of heat transfer and pressure on the leading edge. Examples of this occurring in hypersonic flows are discussed in Chapter 7. The effect of reducing the density on a Edney type IV interaction similar to that computed by Moss et al. is illustrated in Fig. 8.15b. As in previously considered examples, increasing the degree of rarefaction reduces the intensity of the interaction. The thickening of the shock waves smears the flow structure and the two triple points that create the high-velocity jet in Fig. 8.5a cannot be identified. The intense peaks in heat transfer and pressure on the surface correspondingly are greatly attenuated. The biconic body does not exhibit exactly the same flow structure as the Edney interaction; nevertheless, a very-high-velocity

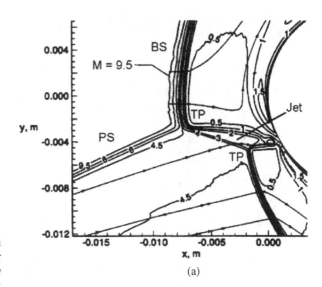

(a)

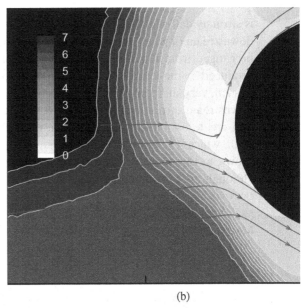

(b)

Figure 8.15. (a) Computation of an Edney-type interaction. Mach number contours. $Kn_{\infty,D} = 0.0067$. Key: BS, bow shock wave; PS, incident primary shock wave; TP, triple point. (Moss et al. [20]). (b) Mach-number contours for a shock impinging in a similar place on a cylinder under more rarefied conditions. $Kn_{\infty,D} = 0.049$; $M = 7.71$.

jet is produced that is clearly discerned in Fig. 8.14 as the dark strip running close to the rear-cone surface. The velocity within this jet reaches a value of about 75 percent of the incident-flow value. Instead of being directed toward the surface as in the Edney interaction, this jet travels considerable distance parallel to the cone. If we trace back the streamlines bounding this jet, it is evident that the fluid within it previously passed through several oblique shock waves: the primary shock, the separation shock, and then the shock or compression waves emanating from the reattachment region. As in a multishock supersonic intake, high pressures can be achieved this way because the process tends toward an isentropic compression; indeed, Fig. 8.16, which shows the pressure contours, confirms this. A pressure peak is generated immediately downstream of the shock-shock interaction that is more than 650 times the free-stream value. The primary shock immediately upstream of this region

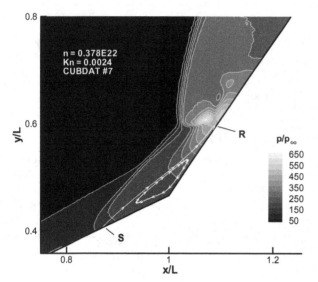

Figure 8.16. Contour of normalized pressure (p/p_∞) in the region of the corner of the biconic body. CUBDAT test case No. 7, $Kn = 0.0024$. $M = 10$, $T_{wall} = 293$ K; $T_\infty = 94$ K; nitrogen.

is strongly deflected and it can be seen that it becomes essentially a normal shock downstream of which the flow is subsonic. The extent of the low-velocity region can be gauged from Fig. 8.14. The strong compression combined with the high shear produced by the jet causes a thinning of the boundary layer on the aft cone and, for this Kn, very sharp and intense peaks in the pressure and heat-transfer distributions downstream of reattachment (Figs. 8.17 and 8.18). Disturbances from the interaction create oscillations in the pressure distributions along the cone after the first peak, which – to a lesser extent – also are evident in the heat transfer. The curves shown in this example correspond to Case No. 7 in the CUBDAT dataset and, although the experimental points are not shown, there is good agreement between the two.

Computations also were made for this body for three higher Kn at the same Mach number and flow temperature. With increasing rarefaction, separation is

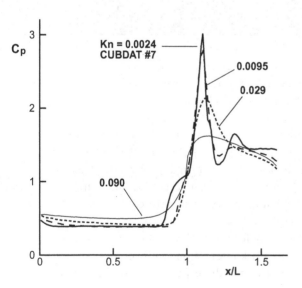

Figure 8.17. The pressure distribution on a sharp biconic body for varying degrees of rarefaction. $M = 10$, $T_{wall} = 293$ K; $T_\infty = 94$ K; nitrogen.

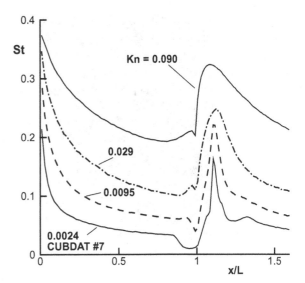

Figure 8.18. Heat-transfer coefficient distribution for a sharp biconic body for varying degree of rarefaction. $M = 10$, $T_{wall} = 293$ K; $T_\infty = 94$ K; nitrogen.

delayed due to the reduction in Reynolds number and the influence of increased velocity slip on the fore cone (although the latter is less significant than on the hollow-cylinder–flare body because the density levels on the surface are higher). Examination of the streamline patterns confirms that up to $Kn = 0.029$, separation still occurs, but then it is almost at the cone intersection; by $Kn = 0.0899$, it is suppressed. Also, with increasing Kn, the intensity of the SBLIs diminishes and, as in the hollow-cylinder–flare flow, thickening of the shock waves, merging, velocity slip, and the increased viscous nature of the flow all contribute to this process. Peaks in the surface pressure and absolute heat-transfer distributions decay with rising Kn and pressure oscillations are suppressed.

Figure 8.19 shows the velocity contours and streamlines near the corner for the $Kn = 0.029$ flow. The interaction is more dispersed than in the previous example and the primary shock wave, far from being a discontinuity, shows a structure similar

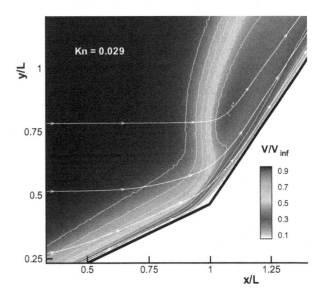

Figure 8.19. A DSMC solution for the flow near the corner for $Kn = 0.029$ showing streamlines and velocity.

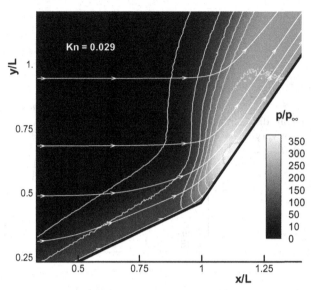

Figure 8.20. Contour of normalized pressure in the region of the corner of the biconic body. $Kn = 0.029$; $M = 10$, $T_{wall} = 293$ K; $T_\infty = 94$ K; nitrogen.

to the compression wave seen ahead of the blunt-ended cylinder at similar values of Kn (see Figs. 8.3 and 8.4). Elsewhere, a general smearing of the flow structure occurred, which leads to an attenuation of the peaks in the surface fluxes and the intense shock-shock interaction does not occur. An extra, darker band of contours was added, centered at approximately 30 percent of the free-stream velocity. This indicates where the boundary layer is; from its position above the surface of the aft cone, it can be inferred that the high-velocity jet generated by the shock-shock interaction process, observed in Fig. 8.14 for the lower Kn, is no longer present. Consequently, the peak in the surface pressure that occurs downstream of reattachment is attenuated because the flow structure is more benign. Figure 8.17 shows that the maximum value drops by a factor of almost 2 when Kn changes from 0.0024 to 0.0899. The effect of increasing Kn on the heat transfer is shown in Fig. 8.18. As in the hollow-cylinder–flare body, overall levels of the coefficients increase due to the falling Reynolds number, but the peaks in the actual heat flux that occur on the aft cone are weakened and broaden with increasing rarefaction (for reasons previously explained). However, because the pressure levels are higher on the cone than in the corresponding cylinder–flare flows, even for the most rarefied case (i.e., $Kn = 0.09$), the profiles do not approximate to the constant FM values on either cone.

Figure 8.20 shows contours of p/p_∞ for $Kn = 0.029$. These contours can be compared directly to Fig. 8.14, in which Kn is one tenth the value. Again, it is apparent that all of the intense compression regions were eliminated. The maximum p/p_∞ levels in this flow are about half of those in the denser example. Nevertheless, the compression of the flow at the corner is still strong enough to require the primary shock to be deflected into a near-normal wave upstream of the reattachment region.

Figure 8.21, the final figure in the sequence, indicates the levels of molecular nonequilibrium that can be expected in this type of flow. Because this flow is pure

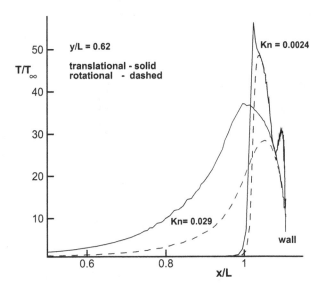

Figures 8.21. Profiles of translational and rotational temperature through the reattachment region along the line $y/L = 0.62$.

nitrogen and the vibrational mode is frozen, non-equilibrium can only be manifested as a difference between the translational and rotational temperatures. Values of these are plotted along the line at $y/L = 0.62$, which is a little farther outboard from the axis of the body than the point where the reattachment shock intersects the primary shock. The line goes through the low-velocity patch shown in Fig. 8.19. Curves are shown for Kn equal to 0.0024 and 0.029. The contrast is obvious: For the denser flow, the sharp rise at $x/L = 1$ indicates the compact nature of the shock wave in which the rotational temperature lags minimally behind the translational value. The flow comes quickly into equilibrium after this and remains so up to the surface. With only one order-of-magnitude increase in Kn, the shock changes completely in character and becomes very diffuse. The curve now closely resembles profiles shown for the blunt-ended cylinder in Fig. 8.4b. Throughout, the rotational temperature lags far below the translational value and only comes into equilibrium in the denser flow close to the surface.

8.4.5 Flows Involving Chemical Reactions

So far, all of the examples considered are for flows of nitrogen in which the only real-gas effects are associated with the exchange of internal energy. Practical hypersonic flow take places in air or other planetary atmospheres in which dissociation and chemical reactions are expected to occur. For this reason, we now examine how rarefaction influences SBLI flows that involve these phenomena as well. The intension is not to provide a comprehensive picture of the influence of reactions on SBLIs but rather to illustrate the interplay between rarefaction and chemical activity. We again use the example of the flow over the 25–55-degree biconic body to illustrate, using as the fluid idealized air – that is, a mixture of 80 percent nitrogen and 20 percent oxygen. Conclusions are drawn from computed results using the DSMC method because suitable experimental data in the rarefied regime are not available. The calculations include the full range of possible reactions but exclude those involving ionization; the surface of the body is assumed to be fully catalytic for nitrogen and

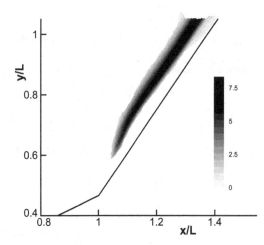

Figure 8.22. Contours of NO fraction (as %) downstream of the cone intersection at *Kn* of 0.0015. Stagnation enthalpy: 10 MJ/kg. The solid line depicts the cone surface.

oxygen recombination. Because the DSMC method is based on particle collisions, it is possible to model the reactions in a sophisticated way and account for the influence of enhanced levels of molecular nonequilibrium present in these flows on the reaction probabilities. The calculations cover a range of *Kn* similar to those considered in previous examples for this configuration. However, the incident velocity and temperature were increased so that the stagnation enthalpy is sufficient to promote chemical activity while retaining approximately the same incident Mach number.

An overarching effect of increased rarefaction is a reduction in the collision frequency. Because all chemical reactions are a direct outcome of molecular encounters, we can anticipate that this will slow these processes and effectively reduce reaction rates. At the same time, shock waves become thicker and less intense; the way in which the reactions are initiated through the rise in temperature that they impart is therefore more benign, especially in the shock-shock–interaction region. The combination of these two effects results in the level of chemical activity for a given stagnation enthalpy becoming progressively less with increased rarefaction and the area in which the reactions take place more spread out. As the density is reduced, the point where products of the reactions first appear occurs farther downstream. The lowering of the collision frequency also affects any subsequent recombination of the reaction products, which requires three body collisions to initiate within the flow.

The effect of increasing *Kn* on the flow chemistry is illustrated graphically for the biconic body in Figs. 8.22 and 8.23. In these figures, contours of the fraction of nitrous oxide (NO) in the region near the junction between the two cones are plotted for a 10-MJ/kg flow of air for *Kn* = 0.0015 and 0.048, respectively. (Refer to the pressure-contour plots in Figs. 8.16 and 8.20, although for nitrogen and not at exactly the right *Kn* values, which provide a good indication of the topology of the shock-shock interaction for these flows.) In the denser flow in Fig. 8.22, NO appears almost immediately downstream of the shock-shock interaction. This species is an effective marker of the chemical activity within the flow, and it is a consequence of either exchange reactions following oxygen or nitrogen dissociation or the direct recombination of these atomic species. The dissociation of nitrogen and oxygen are strongly endoergic processes that have the effect of reducing peak temperatures that

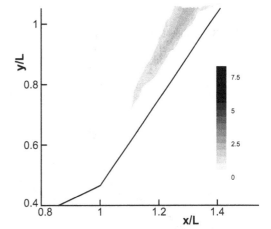

Figure 8.23. Corresponding contours of NO fraction (as %) for *Kn* of 0.048 for the same stagnation enthalpy.

would have occurred in the absence of reactions. However, as the fluid approaches the surface, the temperature falls and reverse reactions occur that return the heat to the flow or the surface. The latter is fully catalytic and complete recombination of nitrogen and oxygen would occur at the wall. However, it appears that virtually all of this recombination occurred before the fluid reaches the body.[8] Figure 8.22, in which $Kn = 0.0015$, shows that the NO generated almost immediately downstream of the shock-shock interaction persists in the flow between the outer shock and the surface of the cone above the jet. In this flow, dissociation of NO is relatively slow and the concentrations remain roughly the same along the length of the rear cone.

Figure 8.23 shows the effect on the NO concentration of lowering the density while maintaining the same stagnation enthalpy. *Kn* for this flow is 0.048. The same contour levels were used as those in the previous figure, and it is immediately evident from the lower concentrations that the formation of NO is both delayed and attenuated. There is now a physical separation between the shock-shock interaction (which is more spread out and in which dissociation is initiated) and the place where the reactants first appear in the flow. This is a direct consequence of the reduced collision frequency.

The effects of changing the stagnation enthalpy on the surface-flow properties for the reacting flow are illustrated in Figs. 8.24 through 8.28. First, we consider the distributions of pressure and heat transfer near the reattachment region for a flow at $Kn = 0.0015$. This is close to the continuum limit, although some influence of rarefaction can be expected. Figure 8.24 shows the pressure coefficient for three different enthalpies; as expected, increasing the energy of the incident flow raises the peak-pressure level. It also delays separation, reducing the extent of the recirculation region; consequently, the peak moves closer to the junction between the two cones at $x/L = 1$. A result for a nonreacting nitrogen flow (i.e., $H_0 = 2.1$ MJ/kg) also is shown, and it is evident from comparisons with the similar 3-MJ/kg air flow that changing from inert nitrogen to a reacting gas results in changes in the flow structure. Most notably, the size of the recirculating region increases with the separation point moving forward, causing the peak in C_p to move rearward.

---

[8] Results obtained with surface recombination turned off show virtually identical heat-transfer distribution in the vicinity of the peak, confirming that this occurs before the fluid reaches the surface.

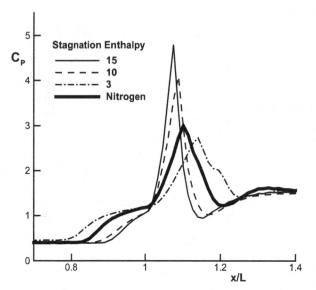

Figure 8.24. Pressure distribution in the reattachment region for $Kn = 0.0015$ in air.

Reducing the density by an order of magnitude profoundly affects these distributions, as shown in Fig. 8.25. There is now an almost total insensitivity to the enthalpy level except for the location of separation, which is delayed with increased enthalpy. This follows the trend shown in the denser-flow examples. At this level of rarefaction, increasing the enthalpy has the effect of slightly reducing the peaks in pressure for the 10- and 15-MJ/kg flows compared with the value at the lowest enthalpy level.

Figures 8.26 and 8.27 show the corresponding heat-transfer distributions for the same enthalpies and Kn as for the pressure, and previous trends are repeated. For the $Kn = 0.0015$ cases, intense heating occurs just downstream of reattachment. Although the peak for the 15-MJ case is very high, it is only about three times greater than that for the 3-MJ flow. This is because of the absorption and convection of energy away from the body through endoergic reactions. Nevertheless, this peak is a serious threat to the integrity of the structure of a hypersonic vehicle, which underlines the importance of predicting the flows as precisely as possible. Reducing the density to raise Kn by an order of magnitude to 0.015 has a similar effect on heat transfer as on pressure. Again, there is a much-reduced sensitivity to

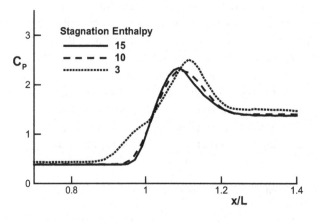

Figure 8.25. Pressure distribution in the reattachment region for $Kn = 0.015$.

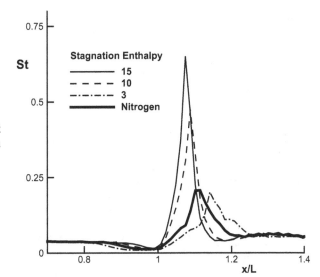

Figure 8.26. Heat-transfer coefficient distribution in the reattachment region for $Kn = 0.0015$ in air.

the stagnation-enthalpy level; however, for the heat transfer, there is an approximate 35 percent increase in the peak value when increasing the enthalpy from 3 to 15 MJ/kg. Recall that the pressure was almost completely unaffected by this change. The direct effects of chemical activity now are largely displaced downstream from the SBLI region due to the reduced collision frequency and the more benign shock-shock interaction.

The influence of rarefaction on the heat transfer for the reacting flow is summarized in Fig. 8.28, in which data are replotted covering two orders-of-magnitude variation of Kn. Increasing this is equivalent to reducing the Reynolds number; hence, the overall levels of the Stanton number rise – a trend that is most obvious on the fore cone ($x/L < 1$). Results for $Kn \approx 0.14$ are added, which is a condition approaching FM flow. The pressure for this case was not shown and is almost constant on each cone and very close to the FM values, which are determined simply by the surface inclination. However, this is not the case for the heat transfer and there is still a noticeable peak near the corner with a value in excess of the FM on the aft cone, echoing the trend in the nonreacting flow (see Fig. 8.18).

8.5 Concluding Remarks

This chapter examines how rarefaction influences the physical characteristics of hypersonic SBLIs. From the study of two sample flows that include SBLIs, we

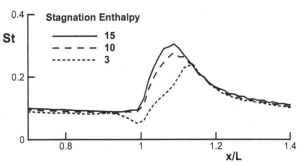

Figure 8.27. Heat-transfer distribution in the reattachment region for $Kn = 0.015$.

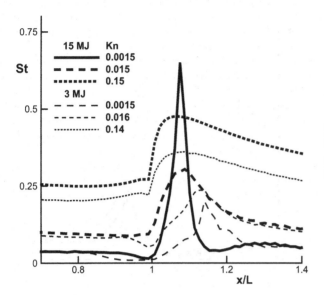

Figure 8.28. Heat-transfer distribution in the reattachment region for a range of Knudsen numbers for the biconic body in air.

observed a steady evolution in the structure from continuum toward FM as *Kn* was increased. A relatively small increase in the scale of *Kn* is needed to effect this change in the transition region. Superficially, the initial manifestations of this process (e.g., the delay or suppression of separation) can be viewed simply as low-Reynolds-number effects. However, as the density reduces further, it becomes increasingly apparent that phenomena unique to the rarefied regime also are important in determining structural changes that occur in these flows. From an aerospace designer's perspective, an important factor is the considerable attenuation of the sharp peaks in the surface pressure and heat transfer that dominate the reattachment region for the continuum flows. We attempt to show when and why this happens and observe that it occurs well before the flow can be seen as approaching the FM-flow limit. It is more marked in the less-intense hollow-cylinder–flare-body example than on the biconic body, in which the compression by the stronger-bow shock wave delays the transition towards FM flow to a higher *Kn* value.

Several phenomena are unique to rarefied flows that influence the development of SBLIs as *Kn* is increased. Velocity slip and temperature jump at the wall boundary and shock-wave thickening and attenuation due to merging with the adjacent viscous layer each have a significant impact. At the same time, there is an overarching effect due to the decrease in the collision frequency that results from rarefaction. This increases the levels of molecular and chemical nonequilibrium throughout the flow. Practical hypersonic flows that include SBLIs, for the most part, are high-enthalpy flows in which there is significant chemical activity and real-gas effects. The reduction in the collision frequency delays the onset of chemical reactions until farther downstream from where the temperature rise occurs that initiates these processes. This is particularly significant for those triggered by the rapid rise in temperature in shock-shock interaction and reattachment zones of the SBLI. For the more rarefied flows, this has the effect of greatly reducing the sensitivity of the surface-pressure and heat-transfer coefficient distributions to increases in enthalpy. This is not a result that would be predicted accurately by continuum CFD methods, and it underlines the value of particle-simulation schemes in these circumstances (e.g., the DSMC method).

The conclusions in this chapter are drawn primarily from computations, presupposing that they truly represent actual flows. Although there are attendant risks, validation studies indicate that the DSMC results are accurate – at least, for the nonreacting flows. Experimental data are not available to assess the accuracy of the code for the chemically reacting flows.

Appendix A: Kinetic Theory and the DSMC Method

The basis of kinetic theory is the determination – either explicitly or implicitly – of the molecular-velocity-distribution function $f = f(c, x; t)$ for a given set of boundary conditions. This function defines the probability of any particle having a specific value of the velocity c ($= u, v, w$) at a position x ($= x, y, z$) and at time t. As explained in [1] through [4], all of the macroscopic properties of a flow can be derived from f. If the gas is multispecies or has internal-energy modes, more complex forms of the distribution functions are required.

A dilute gas is one for which the collisions between particles solely involve two partners. For this, the governing equation from which f can be solved is the Boltzmann equation [3]. This is a notoriously difficult integro-differential equation that is virtually impossible to solve directly, even for the simplest of flows; except for a few isolated and relatively trivial examples, all of the available solutions were obtained using approximate methods. The most successful approaches were based on the Chapman-Enskog expansion of the velocity-distribution function, which takes the form $f_0(1 + \varepsilon)$, where f_0 is the local equilibrium velocity distribution[9] and ε is a small quantity. Thus, these solutions are restricted strictly to flows in which there is only a small perturbation from local molecular equilibrium. For high-altitude hypersonic flows, this is frequently not the case; thus, this method of solution is not strictly appropriate. For further information on this topic, readers are referred to [2, 3, 4].

The usual form of four transport coefficients for gases – namely, the viscosity, thermal conductivity, mass diffusivity, and thermal diffusivity – is evaluated in kinetic theory using the Chapman-Enskog approximation. The use of any of these coefficients is thus strictly legitimate when there is only a small departure from local molecular equilibrium (see [2], chapter X, sec. 7). This is generally true for continuum flows but if Kn is above 10^{-3}, for example, there may be regions

---

[9] The velocity distribution for a stationary gas that has relaxed to a state of equilibrium within an isothermal container is given by the Maxwellian function[2]:

$$f_0 = (\beta^3/\pi^{3/2}) \exp(-\beta^2 c'^2) \qquad [8.A.1]$$

where c' is the random thermal velocity of the gas molecules and

$$\beta = (2RT)^{-1/2} = \{m/(2\hat{k}T)\}^{1/2}. \qquad [8.A.2]$$

R is the gas constant, T is the translational temperature, m is the molecular mass, and $\hat{k}$ is the Boltzmann constant ($= 1.38066 \times 10^{-23}$ J/K). For a polyatomic gas that has rotational and vibrational internal-energy components, there is a corresponding equilibrium distribution for these modes given by:

$$f_{\varepsilon_i} = \varepsilon^{\zeta/2-1} \exp\left(-\varepsilon_i \hat{k} T_i\right) \qquad [8.A.3]$$

where ε is the internal energy associated with each mode and ζ is the corresponding number of internal degrees of freedom associated with this.

where this requirement is not met due to the reduction in collision frequency. In these circumstances, the use of these coefficients is unsound and explains one reason why the Navier-Stokes equations are unable to predict satisfactorily rarefied flows.

A.1 Particle-Simulation Methods

A group of numerical methods was developed that explicitly bypass solving the velocity-distribution function. Instead, the methods simulate the flow using particles that mimic the motion and collisions of the real-gas atoms and molecules. Even in low-density flows, the number of particles is prodigiously large, and it is not possible to track every one. Instead, a smaller number of simulators are used to represent statistically the entire fluid. As the simulators move through the computational domain, they are required to reproduce the properties of the real-gas atoms and molecules and to model (with acceptable realism) the inter-particle collisions and encounters with solid surfaces that they experience. For hypersonic flows in which high temperatures occur, the collision models also must replicate internal-energy exchange, dissociation, recombination, and chemical reactions in the flow and on the surfaces. For higher-enthalpy flows, ionization also can occur, which may require special schemes to address the long-range Coulomb forces that arise between charged particles if there is significant charge separation.

These numerical methods avoid the explicit evaluation of $f(c, x; t)$, but the information relating to the spatial and temporal distributions of the particles' velocity components, composition, internal energies, and so on is held in a statistical sense within the population of simulators as they pass through the flow domain. Within this group of methods, the DSMC method is the most successful for aerodynamic studies and it has proven to be an exceptionally powerful and versatile prediction tool. Alternative particle-simulation methods exist – for example, the Particle in Cell codes much favored by physicists. These methods are successful for certain applications, but they are better suited to situations in which charged particles and electromagnetic fields occur, and they do not perform nearly as well as the DSMC method for gas-dynamic problems in which complex flow topologies and body shapes must be addressed. For these problems, the DSMC is unrivaled, offering greater computational flexibility and efficiency. It is no exaggeration that it has completely revolutionized rarefied-gas dynamics and brought within grasp practical and accurate numerical solutions for many complex flow problems.

A.2 The DSMC Method

The DSMC method was formulated originally in 1970 by Professor Graeme Bird [21], who developed it from an intuitive standpoint relying heavily on physical reasoning. It was twenty-five years before a formal mathematical proof was published [22] showing that fully resolved computations produce the same solutions as the Boltzmann equation. For this reason, experimental validation was emphasized to confirm the accuracy of the method [23]. Figure A.1 shows results for one comparison that demonstrates impressive agreement between computed and measured results. The plots are for a nitrogen flow over a 70-degree spherically blunted

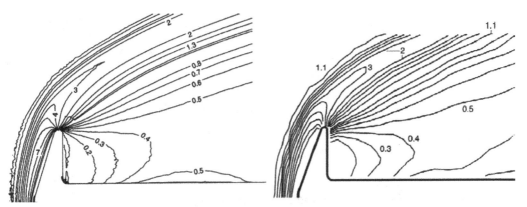

Figure A.1. DSMC computed (*left*) and measured (*right*) density contours for a 70° spherically blunted at zero angle of attack. $M = 20$; $Kn = 0.0045$; nitrogen.

cone – typical of a planetary-probe reentry capsule [24] – and they show calculated density contours compared with measurements taken in the ONERA R5Ch wind tunnel using the nonintrusive electron-beam fluorescence technique. The same contour intervals are used in both images but, for clarity, values above seven times the free-stream value were omitted in plotting the measured results.

These results reveal the structure of the flowfield, thereby providing a more sensitive assessment of the code's ability to capture the detailed physics of the flow than would be possible, for example, from surface-pressure or heat-transfer information. The agreement between the two is impressive and it is clear that most of the important features of the flow were captured precisely. Although not initially appreciated, it soon became evident that Bird's DSMC method was potentially extremely versatile and, in time, successful ways were devised to adapt it to include dissociation, chemical reactions, ionization, and high degrees of molecular nonequilibrium. Several sophisticated codes currently exist that are capable of addressing multispecies reacting flows and geometrically complex shapes. Only an outline of the DSMC method is presented here because a full description is provided in [21], in which Bird comprehensively and practically explains his own code and the underlying theory. Further explanation, more within the context of kinetic theory, is in the work of Cercignani [4].

DSMC computations advance in real time and track the progress of an array of simulator particles using time-steps Δt that must be smaller than the local mean-collision time. The simulators have three components of velocity even if the flow that is being computed is not three-dimensional. A defining and innovative feature of Bird's scheme is that he decouples the movement of the particles from their collisions by alternately simulating the two processes, depicted schematically in Fig. A.2. The Boltzmann equation [3] for a dilute gas is expressed as follows:

$$\frac{\partial f}{\partial t} + v \cdot \nabla_x f = \frac{1}{\varepsilon} Q(f) \qquad (8.A.4)$$

where v is the velocity and Q is the collision operator that is a complex function to determine how f changes as a consequence of intermolecular collisions. Over the small time increment $[n\Delta t, (n+1)\Delta t]$, first the left-hand side of the equation

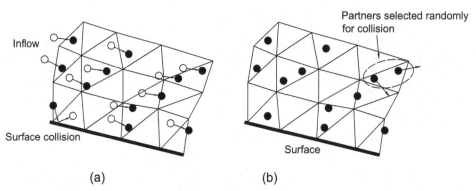

Figure A.2. A schematic representation of the two steps in a two-dimensional DSMC computation: (a) the move phase, and (b) the collision phase.

(i.e., $\partial f/\partial t + n\nabla_x f = 0$) is replicated by allowing the free motion of all of the simulators using the following approximation:

$$x_i((n+1)\Delta t) = x_i(n\Delta t) + v_i(n\Delta t)\Delta t v_i \quad v_i((n+1)\Delta t) = v_i(n\Delta t). \qquad (8.A.5)$$

This is depicted in Fig. A.2(a). If a simulator encounters a solid surface during this phase, it is returned to the flow with new velocity components that are determined using an assumed surface-scattering kernel. During this phase, new simulator particles are introduced at inflow boundaries that are selected randomly from distributions commensurate with the state and velocity of the fluid entering the domain. The number entering at each time-step is set by the ratio chosen between the density of the simulators and that of the real flow.

In the next time-step (see Fig. A.2(b)), the right-hand side of the Boltzmann equation, $\partial f/\partial t = Q(f)$, is simulated. An appropriate number of "collision partners" are selected from within a small localized partition Δx of the spatial domain. These "cells" must be smaller than the local mean-free path. If the selection of partners within a cell is random, $Q(f)$ is effectively replaced by the spatially smoothed operator $Q^{\Delta x}(f)$. Alternatively, nearest-neighbor collision pairs can be identified to improve on this smoothing. Efficient numerical schemes were devised to implement this method and they have provided significant improvements in the accuracy of the results, especially for denser flows. After each simulated collision, new properties of the two particles are determined using an appropriate representation of the interaction. Only ten to twenty simulators are required in each cell at any time. The alternate steps are repeated many times and the flow properties are determined from information collected about the population of simulators as they pass through small sampling regions of the flow. Bird's time-step–splitting procedure was an inspired move that greatly increased the numerical efficiency of the DSMC method over alternative particle-simulation methods, such as molecular-dynamics computations.[10] Although Fig. A.2 depicts a two-dimensional flow, the method can be extended readily to three dimensions.

The computations advance in real time from an impulsive start; for problems for which the boundary conditions are such that a steady-state solution exists, an

---

[10] The DSMC method is N dependent, whereas the molecular-dynamics methods are N^2, where N is the number of particles in the simulation.

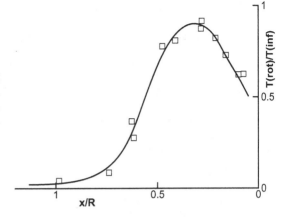

Figure A.3. The rotational temperature along the stagnation stream line of a blunt-ended cylinder. $M = 20.6$; $T_\infty = 52K$; $Kn = 0.02$; nitrogen. DSMC solution; □ experimental results.

initial transient settling period must be allowed before sampling begins. Typically, well-resolved computations are performed using on the order of 10^7 simulators. Hence, the particle-collision routines must be called many times during a computation, which limits the level of sophistication that can be tolerated. Usually, phenomenological models are used in which the postcollision properties are determined by randomly sampling from appropriate predetermined distributions. The number of collisions allowed during each time-step can be determined in different ways, but the no-time-counter method is commonly used [21].

The behavior of molecules during the simulated collisions depends on the choice of the intermolecular-force potential. Models used range in complexity from simple hard spheres and inverse-power representations to the Morse and Lennard-Jones potentials [25], which include the realistic long-range attraction between molecules as well as the short-range repulsion. It is accepted that the latter two models impose too high a computational overhead for their use to be justifiable in an engineering context. Simpler alternatives that proved to be very good yet remain computationally efficient were devised; the most commonly used is the Koura and Matsumoto [26] Variable Soft Sphere model. For polyatomic particles, the transfer of energy to and from the internal modes also must be considered. The most popular is a stochastic model devised by Borgnakke and Larsen [27]. It was proven successful in describing the macroscopic distribution of energy in the gas even though the treatment of individual collisions is not rigorous. Uncertainties about the real distributions that prevail led them to adopt an implicit local equilibrium assumption, insofar as the postcollision properties are sampled from a varying but equilibrium distribution for each collision. Empirical evidence indicates that this model works well for nonreacting flows;[11] however, the validity of using it for situations in which chemical reactions occur is not obvious because the local equilibrium assumption for the products is not always appropriate. This has not been properly tested.

An example of results showing the success of the Borgnakke and Larsen model for a low-density nonreacting flow is shown in Fig. A.3. The body is a blunt-ended circular cylinder with the axis aligned to the incident stream, which in this case is

---

[11] This is probably because in these interactions, all the available energy states tend to be populated without any specific preference. This leads to rapid energy equilibration within just a few collisions. However, this is not generally the case for collisions in which chemical reactions occur.

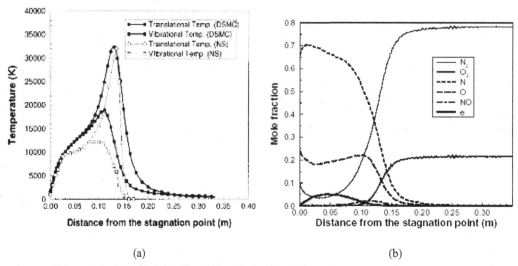

(a)   (b)

Figure A.4. (a) DSMC [30] and Navier-Stokes [31] results showing the temperature components along the stagnation streamline for a spherically blunted 70° cone in real air; (b) Species
concentrations along the stagnation streamline. (Note that the flow is from right to left.)

a Mach 25 nitrogen flow. Results for this flow also were cited in Figs. 8.3 and 8.4.
Figure A.3 shows the rotational temperature profile along the stagnation line. The
computed result is compared with experimental data obtained using the electron-
beam fluorescence technique. From the close agreement, it is clear that the transfer
of energy between the translational and rotational internal mode was reproduced
with good precision using this model.

The requirement for all of the computational cells to be smaller than the local
mean-free path while also retaining approximately the same number of particles
per cell limits the density for which DSMC computations can be made with the
computer resources currently available. Nevertheless, it is possible to predict flows
that overlap the range in which the Navier-Stokes methods also are valid. This has
allowed informative validation exercises to be undertaken, in which results from the
two methods were compared with one another and with experimental data. In one
study [28] that included flows involving SBLIs, it was concluded that the DSMC
method is capable of accurately predicting flows in the low-density end of the continuum regime.

A feature that sets the DSMC technique apart from the continuum-CFD methods is the level of sophistication that can be incorporated in modeling dissociation,
chemical reactions, and ionization. In a complex way, these processes are dependent on energy components of individual particles involved in each collision. The
continuum codes generally use global Arrhenius equilibrium-reaction rates combined with a simplistic dependence on the translational and vibrational temperatures (e.g., Park's model [29]) to compute changes due to chemical processes.
Particle-simulation codes offer the possibility of using more realistic modeling of
the reactions, including those occurring on the surface by considering the energy
states of individual particles before each collision. This is especially significant for
flows in which there is a substantial level of molecular nonequilibrium in the gas.

Sample results [30] that demonstrate the use of this method when reactions
occur are presented in Fig. A.4. The gas is air and the full spectrum of possible

chemical reactions and ionization is modeled using sophisticated methodology. The computation is for the stagnation region of a spherically blunted aerobrake vehicle reentering the earth's atmosphere at an 83-km altitude at a speed of 9.848 km/s. *Kn* based on the nose radius of the vehicle is 0.005. In Fig. A.4a, the translational and vibrational temperature components along the stagnation streamline are compared with the Navier-Stokes predictions of Greendyke et al. [31]. In the figure, the flow is from right to left, with the stagnation point at the origin. Significant difference between the two sets of results is evident, especially for the vibrational temperature that, for nearly all of the flow, is considerably higher in the DSMC results than predicted by the Navier-Stokes (NS) method. There is good reason to believe that the DSMC results are likely to be nearer the truth. Most of the chemical reactions that occur in air are strongly influenced by the vibrational energy; hence, these discrepancies are significant. Figure A.4b shows the mole fractions of species as predicted by the DSMC code. It shows that the precursor rise in temperature ahead of the main body of the shock initiates dissociation almost twice as far upstream of the body as the position of the shock wave predicted by the Navier-Stokes code. The latter is located at about $x = 0.15$ m; upstream from here, no reactions can be predicted using the continuum method.

REFERENCES

[1] G. A. Bird. *Molecular Gas Dynamics and the Direct Simulation of Gas Flows* (Oxford: Clarendon Press, 1994).

[2] W. G. Vincenti and C. H. Kruger. *Introduction to Physical Gas Dynamics* (New York: Wiley, 1967).

[3] S. Chapman and T. G. Cowling. *The Mathematical Theory of Non-Uniform Gases*, 3rd edition (London: Cambridge University Press, 1970).

[4] C. Cercignani. *Rarefied Gas Dynamics* (London: Cambridge University Press, 2000).

[5] V. V. Mikhailov, V. YA. Neiland, and V. V. Sychev. The theory of viscous hypersonic flow. *Annual Review in Fluid Mechanics*, 3 (1971), 371–96.

[6] M. S. Holden and T. P. Wadhams. Code validation studies of laminar shock–boundary layer and shock-shock interaction in hypersonic flow. Part A: Experimental measurements. *AIAA Paper* 2001-1031 (2001).

[7] M. S Holden and T. P. Wadhams. A database of aerothermal measurements in hypersonic flows in "building-block" experiments for CFD validation. *AIAA Paper* 2003-1137 (2003).

[8] B. Chanetz, R. Benay, J.-M. Bousquet, R. Bur, T. Pot, F. Grasso, and J. Moss. Experimental and numerical study of the laminar separation in hypersonic flow. *Aerospace Science and Technology*, 2 (1998), 3, 205–18.

[9] B. Chanetz and M.-C. Coet. Shock wave–boundary layer interaction analyzed in the R5Ch wind tunnel. *Aerospace Research*, 5 (1993), 43–56.

[10] G. N. Markelov, M. S. Ivanov, S. F. Gimelshein, and D. A. Levin. Statistical simulation of near-continuum flows with separation. *Rarefied Gas Dynamics: 23rd International Symposium*, AIP Conference Proceedings, 663 (2003), 457–64.

[11] Available online at www.aeromech.usyd.edu.au/dsmc_gab/.

[12] J. N. Moss and G. A. Bird. DSMC simulation of hypersonic flow with shock interactions and validation with experiments. *AIAA Paper* 2004–2585 (2004).

[13] C. Cercignani. "Knudsen layer theory and experiment." In *Recent Developments in Theoretical and Experimental Fluid Mechanics*, eds. U. Müller, K.G. Rösner, and B. Schmidt (Berlin: Springer-Verlag, 1979), pp. 187–95.

[14] F. Sharipov and V. Seleznev. Data on internal rarefied gas flows. *Journal of Physical Chemistry Ref. Data*, 27 (1998), 657.

[15] F. Sharipov and D. Kalempa. Velocity slip and temperature jump coefficients for gaseous mixtures. I: Velocity slip coefficients. *Physics Fluids*, 15 (2003), 6, 1800–6; see also Velocity slip and temperature jump coefficients for gaseous mixtures. II: Thermal slip coefficients. *Physics of Fluids* 16 (2004), 3, 759–64.

[16] J. R. Torczynski, M. A. Gallis, and D. J. Rader. DSMC simulations of Fourier and Couette flow: Chapman–Enskog behaviour and departure therefrom. *Rarefied Gas Dynamics* (AIP Conference Proceedings #762, 2005), pp. 620–5.

[17] H. Takeuchi, K. Yamamoto, and T. Hyakutake. Behaviour of reflected molecules of a diatomic gas at a solid surface. *Rarefied Gas Dynamics* (AIP Conference Proceedings #762, 2005), pp. 987–92.

[18] M. S. Holden, T. P. Wadhams, J. K. Harvey, and G. V. Candler. Comparisons between measurements in regions of laminar shock wave boundary layer interaction in hypersonic flows with Navier-Stokes and DSMC solutions. Technologies for Propelled Hypersonic Flight, *RTO-TR-AVT-007-V3*, pp. 4-1–56 (Research and Technology Organisation [NATO], 2006).

[19] B. Edney. Anomalous heat transfer and pressure distributions on blunt bodies at hypersonic speeds in the presence of an impinging shock. *Aeronautical Research Institute of Sweden, FFA* Report 115, Stockholm (1968).

[20] J. N. Moss, T. Pot, B. Chanetz, and M. Lefebvre. DSMC simulation of shock-shock interactions: Emphasis on Type IV interactions. *Proceedings of the 22nd International Symposium on Shock Waves*, London (1999).

[21] G. A. Bird. Direct simulation of the Boltzmann equation. *Physics of Fluids*, 13 (1970), 11, 2676–82.

[22] W. Wagner. Monte Carlo methods and numerical solutions. *Rarefied Gas Dynamics*, ed. M Capitelli (AIP Conference Proceedings, 762 2005), pp. 459–66.

[23] J. K. Harvey and M. A. Gallis. Review of code validation studies in high-speed low-density flows. *Journal of Spacecraft and Rockets*, 37 (2000), 1, 8–20.

[24] F. Coron, J. K. Harvey, and H. Legge. "Synthesis of the rarefied flow test cases." In *Hypersonic Flows for Reentry Problems Part III*, eds. R. Abgrall, J-A. Desideri, R. Glowinski, M. Mallet, and J. Periaux (Berlin: Springer Verlag, 1992).

[25] M. N. Macrossan. Diatomic collision models used in the direct simulation Monte Carlo method applied to rarefied hypersonic flows. Ph.D. Thesis, University of London (1983).

[26] K. Koura and H. Matsumoto. Variable soft sphere molecular model for inverse-power-law or Lennard Jones Potential. *Physics of Fluids* A, 3 (1991), 10, 2459–65.

[27] C. Borgnakke and P. S. Larsen. Statistical collision model for Monte Carlo simulation of polyatomic gas mixtures. *Journal of Computational Physics*, 18 (1975), 405–20.

[28] J. K. Harvey, M. S. Holden, and T. P. Wadhams. Code validation studies of laminar shock–boundary layer and shock-shock interaction in hypersonic flow. Part B: Comparison with Navier-Stokes and DSMC solutions. *AIAA Paper* 2001–1031, (2001).

[29] C. Park. *Nonequilibrium Hypersonic Aerothermodynamics* (New York: Wiley, 1990), p. 114.

[30] M. A. Gallis and J. K. Harvey. The modelling of chemical reactions and thermochemical non-equilibrium in particle simulation computations. *Physics of Fluids*, 10 (1998), 6, 1344–58.

[31] R. B. Greendyke, P. A. Gnoffo, and R. W. Lawrence. Calculated electron number density profiles for the aero-assisted flight experiment. *Journal of Spacecraft and Rockets*, 29 (1992), 621.

[32] J. C. Maxwell. On stresses arising in rarefied gases from inequalities in temperature. Phil. Trans. Roy. Soc. 170 (1879), 231–56.

[33] S. Chapman and T. G. Cowling. *The Mathematical Theory of Non-Uniform Gases* (London: Cambridge University Press, 1940).

9 Shock-Wave Unsteadiness in Turbulent Shock Boundary-Layer Interactions

P. Dupont, J. F. Debiève, and J. P. Dussauge

9.1 Introduction

If the shock wave associated with a shock wave–boundary-layer interaction (SBLI) is intense enough to cause separation, flow unsteadiness appears to be the almost-inevitable outcome.[1] This often leads to strong flow oscillations that are experienced far downstream of the interaction and can be so severe in some instances as to inflict damage on an airframe or an engine. This is generally referred to as "breathing" or, simply, "unsteadiness" because it involves very low frequencies, typically at least two orders of magnitude below the energetic eddies in the incoming boundary layer. The existence of these oscillations raises two questions: "What is their cause?" and "Is there a general way in which they can be understood?"

There are several distinct types of SBLIs, depending on the geometry and whether the flow separates, and it is possible that these create fundamentally different types of unsteadiness. An interpretation was proposed by Dussauge [1] and Dussauge and Piponniau [2] using the diagram reproduced in Fig. 9.1. The organization of the diagram requires comment: In the upper branch, unseparated flows are depicted; those that separate are restricted to the lower branch. In both cases, the shock wave divides the flow into two half spaces: the upstream and the downstream layers. Hence, the shock wave can be considered an interface between the two conditions and its position and motion vary accordingly. With these various elements in mind, the shock motion can be analyzed from the perspective of the upstream and downstream conditions. The discussion in this chapter is a commentary about flow organization and other phenomena related to the two branches of the diagram.

9.2 The Upper Branch: Unseparated Flows

An overview is provided for the development of the mean and turbulent fields, as constitutive parts of the shock motion. Even if no separation occurs, the incoming turbulence is strongly distorted on passing through the shock wave. This

---

[1] This observation does not include laminar flows, and instances of very high Reynolds-number hypersonic ramp-induced separated flows in which no associated unsteadiness is detectable have been reported.

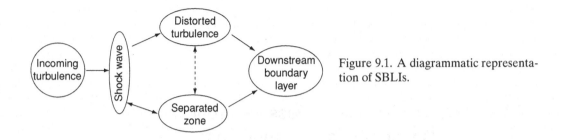

Figure 9.1. A diagrammatic representation of SBLIs.

creates new initial conditions for the downstream flow and, as the perturbed boundary layer relaxes, it evolves into a new equilibrium state. It is necessary to consider the distortion of the boundary layer as it passes through the shock if separation occurs because this is known to be sensitive to the turbulence level. For both separated and unseparated flows, the change in characteristics of the upstream turbulence can be described adequately by local theories (i.e., those that consider only what happens at a single point).

In general, shock waves are not unstable. Nevertheless, they may have a particular frequency response that depends on the shape of the shock and the flow around it (Culick and Rogers [3]; Robinet and Casalis [4]). In particular cases (e.g., the transonic experiments of Sajben and Kroutil [5]), shock waves were shown to be frequency-selective; however, in general, the transfer function is not known and depends on the flow. If shock waves are in any way frequency-selective, the apparent overall trend is that they behave rather like low-pass filters; therefore, they are expected to be more sensitive to low frequencies and to reject the higher ones. This behavior is shown in experiments in which, away from the boundaries, the shocks seem not to move despite the foot being imbedded in a turbulent boundary layer. Similarly, in numerical simulations of shocks in which the foot is subjected to unsteadiness by being located within a turbulent boundary layer, perturbations are not found in the outer flow. This is because they are damped as they propagate outward along the oblique shock in the direction of the tangential velocity (see, e.g., Garnier [6] and Wu and Martin [7]).

The action of shocks on the mean flow and turbulent field is summarized as follows: For mean flow, the overall effect of the shock is to alter the average velocity distribution of the incoming boundary layer so as to produce profiles more linear or less energetic near the wall. Turbulence subjected to a shock wave is distorted and generally amplified in passing through it; the anisotropy is modified in this process. If the distortion occurs fast enough, the nonlinear effects can be neglected and the evolution of the turbulence can be described by a (linear) theory of rapid distortion. There are several versions of such theory and a common feature is that they all consider a linearized set of equations. The differences result from the degree of complexity of the formulations. The simpler formulations assume that the shock is steady but, even so, they enable the level of the downstream turbulence to be determined with considerable precision (see, e.g., Ribner and Tucker [8]; Debiève, Gouin, and Gaviglio [9]; Jacquin, Cambon, and Blin [10]). Other approaches (e.g., Ribner [11]) account for adaptation of the shock geometry to the incoming turbulence and, hence, provide an idea of the shock motion. Such theories are constructed from a weakly deformed shock and produce compatibility conditions for the upstream and downstream flows. Moreover, they can represent the conversion in passing through the shock between the thermodynamic and kinematic fields. For example, a shock interacting with an entropy spot can produce vorticity by a mechanism similar to baroclinic generation of turbulence or, in a related situation, noise is radiated from eddies after they are subjected to a shock wave. These aspects, examined in Ribner's work, also are described by Kovasznay's mode theory (Kovasznay [12]; Chu and Kovasznay [13]). Numerical studies, which also examined these aspects by performing simulations of the linearized or full Navier-Stokes equations, show reasonable or even good agreement between linear theory and direct simulations for weak shocks (among others, see Anyiwo and Bushnell [14]; Lee et al. [15]; Hannapel and Friedrich [16]; and Garnier et al. [17]).

In cases in which no separation occurs, or in shock–turbulence interactions far from any walls, experimental and numerical results indicate that the shock motion is directly related to the incoming turbulence. Debiève and Lacharme [18] showed in their experiments on a shock–homogeneous-turbulence interaction that the scale of the shock corrugation is on the order of the integral scale of incoming turbulence. In their direct simulations, Lee et al. [15] also found similar results. Moreover, their computations show that the amplitude of the corrugations depends on the level of the upstream turbulence, as is expected. Therefore, for weak SBLIs, the incoming turbulence is the main source of shock unsteadiness, which corresponds to fluctuations at much higher frequencies than in the separated cases (see Section 9.3).

If we now consider SBLI situations in which there is no feedback from the far downstream flow, behavior resembling the shock–homogeneous-turbulence interaction is observed. In this case, the flow downstream of the shock wave does not impose particular conditions, and it allows the shock motion to be specified by the incoming turbulence. The turbulence is amplified through the shock, but no turbulent structures with a particular dynamic are formed just downstream of it. This corresponds, for example, to compressions caused by turning, as studied by Poggie and Smits [19]. Their flow consisted of a free-shear layer developed over a cavity, which reattached on an inclined flat plate. The fluid could flow freely downstream. Their measurements show clearly that the resulting mean-pressure gradient and the spectra of pressure fluctuations scaled with the size of the incoming coherent turbulent structures that developed in the shear layer.

Finally, distorted turbulence is convected downstream and contributes to the formation of a nonequilibrium boundary layer after the interaction. If long-distance coupling can exist, this distorted turbulence will possibly indirectly influence the shock motion. This effect, however, is mentioned only for completeness; if the layer separates, the downstream perturbations are much more powerful and create the possibility of efficient upstream coupling in transonic conditions.

To summarize, the main finding for the nonseparated case – in the absence of acoustic coupling – is that the shock unsteadiness depends primarily on the perturbations produced by incoming turbulence. For the separated cases, the influence of upstream perturbations still exists, but the conditions imposed by separation are more likely to become predominant.

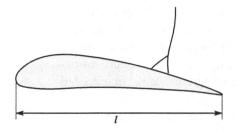

Figure 9.2. An airfoil in transonic conditions.

9.3 The Lower Branch: Separated Flows

9.3.1 Introduction

This section considers situations in which the shock wave is strong enough to cause the boundary layer to separate. The diagram in Fig. 9.1 shows that the lower branch is constructed with double-ended arrows, which indicate that the downstream flow potentially can control the motion of the shock. Therefore, couplings between the downstream layer and the shock wave may exist.

9.3.2 Separated Flows with Far Downstream Influence

This section considers a specific type of interaction that is characterized by the couplings mentioned previously, extending over distances much greater than the length scale of the incoming flow. Such a situation can arise if the flow downstream of the shock is mostly subsonic, such as in a plane–shock interaction in a channel or in an SBLI on a profile (i.e., airfoil) in the transonic regime. This occurs when the shock motion depends on the turbulent flow far downstream of the interaction. It is probably also the case in transonic buffeting for which, according to classical interpretations, there is an acoustic feedback between the flow at the trailing edge of the profile and shock wave (Lee [20]). More recent interpretations (Crouch et al. [21]) suggest that this also could be linked to global-flow instability properties. A classical example is given by the flow around a biconvex airfoil in which discrete frequencies are produced. A consequence of this type of far-field influence is that it is possible to generate shock motion by imposing conditions a long way downstream. This is achieved classically in wind-tunnel experiments in which a downstream rotating cam is used to produce shock motion in the test section, as in experiments by Galli et al. [22] and Bruce and Babinsky [23]. For wind-tunnel experiments in the transonic regime, another consequence is that if there are sufficiently strong fluctuations in the nozzle-diffuser flow, they may contribute significantly to the movement of a shock wave upstream in the test section. These flows represent a particular class of flows, which generally produce very low frequencies.

For an example, consider the case of buffeting on a profile in transonic flow with acoustic coupling between the trailing edge and the shock wave (Fig. 9.2): ℓ is the space scale of the problem that is, for example, on the order of the chord of the profile. If index 2 refers to conditions downstream of the shock, the period of the feedback loop is approximated by $T \approx \ell/(a_2 - u_2)$ and the frequency by $f_{buf} \approx (a_2 - u_2)/\ell$. Notice that with these approximations, f_{buf} is independent of the upstream-boundary-layer scales and depends only on the characteristic size of the

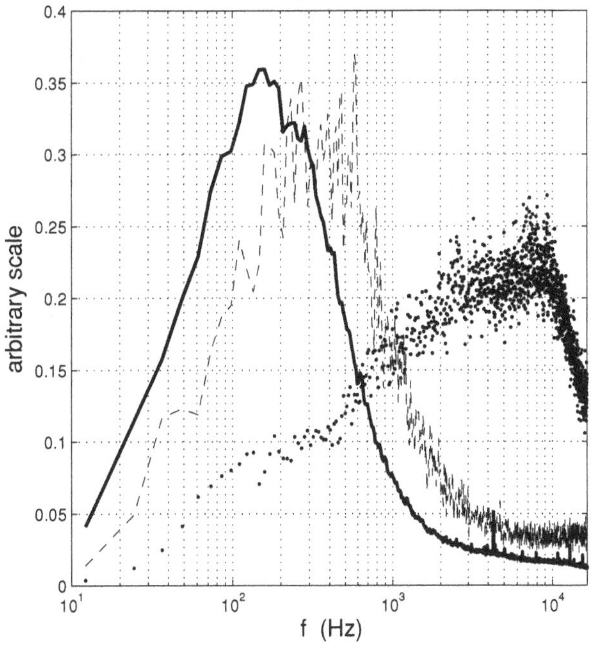

Figure 9.3. Spectra at the foot of the unsteady shock and in the external flow; oblique-shock reflection. Solid line: wall pressure, flow deviation 9.5°; dashed line: wall pressure, flow deviation 8°; dots: hot-wire anemometer, external flow, flow deviation 5.5°.

airfoil. This frequency is compared with other characteristic scales observed with SBLIs in Section 9.4.

9.3.3 Separated Flows without Far-Downstream Influence

9.3.3.1 General Organization

We now consider the cases in which separation can occur but only over a finite distance. This is typical of compression ramps, shock-wave reflections, blunt fins, and overexpanded nozzles in which separation is followed by a reattachment. The latter is referred to in nozzle studies as the Restricted Shock Separation (RSS) configuration. The separation can be either incipient or well developed. The latter occurs if an entire specific zone experiences reverse velocity during periods long enough to produce an average separated bubble. The cases in which isolated and intermittent spots contain fluid with negative velocity but produce no average separation belong to the incipient-separation class. These two cases of incipient and well-developed separation can produce different shock dynamics, illustrated in Fig. 9.3 in which results from Piponniau et al. [24] are shown. In their experiment, an oblique-shock reflection on a flat plate was studied at Mach number $M = 2.3$ for various shock intensities. For sufficiently strong shock waves (flow deviations of 9.5 and 8 degrees, in this case), the boundary layer separates and the reflected shock is unsteady. For a weakest deviation of 5.5 degrees, mean separation is not detectable but intermittent spots of separated fluid are produced. Figure 9.3 shows the wall pressure spectra for the 9.5- and 8-degree cases. For the 5.5-degree deviation, the flow was investigated using hot-wire anemometry with the probe located at the mean position of the unsteady shock. The authors verified that these two types of measurements provide the same information; therefore, it makes sense to compare the frequency contents of the signals derived using the different techniques.

In this example, the dominant frequency of the shock unsteadiness is defined as the maximum of the premultiplied power spectrum $fE(f)$, where f is the frequency and $E(f)$ is the power-spectral density normalized to 1.0. Monochromatic fluctuations are not observed for the separated cases (i.e., 9.5 and 8 degrees), but a characteristic low-frequency range (around 200–500 Hz) is obtained in both situations. These low frequencies are associated with large-amplitude shock motions.

In contrast, for the incipient configuration, no such "bump" of frequency is present: The apparent peak at approximately 9 KHz is a consequence of setting the frequency cutoff of the anemometer around this value, which causes the spectrum to drop above it. Nevertheless, it is evident that low-frequency content – relative to the upstream energetic scales – is still present but not dominant. This suggests that the three-dimensional motions of the shock are limited and correspond mainly to small corrugations, producing higher frequencies. These small, unsteady ripples were observed previously in numerical simulation (Garnier, Sagaut, and Deville [17]; Touber and Sandham [25]; Wu and Martin [7]). In cases in which no separation occurs, the small corrugations of the shock are linked to the incoming turbulence (Debiève and Lacharme [18]; Garnier [6]; Lee et al. [15]), leading again to shock motion at higher frequencies.

Therefore, it may be speculated that these two frequency contents are superimposed in the case of incipient separation. Most of the time, the flow remains attached, which would lead to high-frequency motions mainly related to upstream influences. When separated regions occur – which would be randomly distributed in time and space – a dynamic corresponding to the separated case could develop with large shock motions over a longer time scale; this would contribute to the low-frequency content observed in Fig. 9.3. This seems to correspond to the Particle Image Velocimetry (PIV) observations of Souverein et al. [26] in an incipient-separation case at Mach 1.7.

Finally, fluid leaving the separated region merges with the turbulence coming from upstream (which was distorted on passing through the shock system) to form a new boundary layer downstream of the interaction. When an obstacle or a large separation zone is present, the downstream conditions often predominate in controlling the shock dynamics. This was shown unambiguously by the experiments of Dolling and Smith [29] for the interaction produced by a cylinder normal to a plate. They found that the dominant frequency of the shock unsteadiness varies like the inverse of the cylinder diameter. Similar results were found in interactions produced by oblique-shock reflections.

A more comprehensive view can be obtained from the compilation of results for different types of interactions proposed by Dussauge et al. [28], which is recalled in Fig. 9.4. In this figure, the dominant frequency f of the shock unsteadiness, recorded at the foot, is normalized by the length of the interaction and the external velocity downstream of the leading shock to form the following Strouhal number:

$$S_L = \frac{fL}{U_\infty} \tag{9.1}$$

The length of interaction L is defined consistently, depending on the particular geometrical situation, and is the distance between the mean position of the separation shock and:

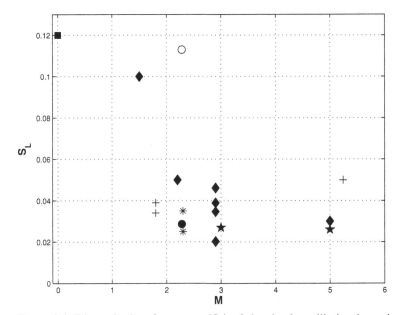

Figure 9.4. Dimensionless frequency (S_L) of the shock oscillation in various configuration: (■) subsonic separation, (Kiya & Sasaki [27]; (♦) compression ramp cases; (*) IUSTI reflection cases; (+) overexpanded nozzle (RSS, (⋆) blunt fin (Touber & Sandham [25]); (○) estimated superstructures upstream influence for the 8° IUSTI case. Adapted from Dussauge et al. [28].

- the extrapolation to the wall of the incident shock, for reflection cases
- the corner or, if known, the location of reattachment for compression ramps
- the beginning of the obstacle, for blunt fins

The collapse of the data versus Mach number for the different flows (i.e., shock reflections, compression ramps, blunt bodies, and channel flows) is not excellent; the scatter is about 20 percent. Moreover, no particular trend with Mach number is observed, except for the compression-ramp flow of Thomas et al. [30]. Higher frequencies were found than in other data; however, because the Mach number was 1.5, transonic effects may have been present. This collapse, even if only partial, is surprising because the dominant frequency reasonably could be expected to depend on the flow geometry. It suggests that the different flow cases share common features for the origin of the unsteadiness. However, the scatter indicates that some details are not well represented. Nevertheless, it is evident that most of the experiments have a Strouhal number around 0.03–0.04. As for well-developed interactions, L is much larger than the boundary-layer thickness δ, and it appears unambiguously that the unsteadiness is several orders of magnitude below the frequencies U_∞/δ produced by the energetic eddies of the incoming layer of typical size δ.

The amplitude of the shock oscillations is now considered. As shown by Dussauge et al. [28] and recalled by Dussauge and Piponniau [2], in many SBLIs, the shock can move at a low frequency over a longitudinal distance L_{ex}, which is on the order of a significant fraction of the interaction length (typically, $L_{ex}/L \approx 0.3$). Different types of interactions – compression ramps, incident-shock waves, and blunt fins – produce similar results (Fig. 9.5).

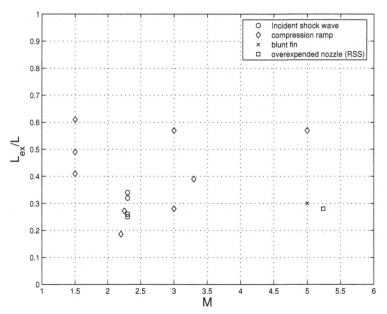

Figure 9.5. Amplitude of the oscillation of the separation shock. From Dussauge et al. [28].

A shock velocity U_s can be derived from this excursion length and from the following frequency scale:

$$U_s / U_\infty = 2 L_{ex} f / U_\infty \approx S_L / 2 \qquad (9.2)$$

Because $S_L \ll 1$, this implies that $U_s \ll U_\infty$. In supersonic conditions, this suggests that the shock wave behaves quasistatically, the additional velocity U_s being insufficient to produce a significant added dynamic effect (i.e., the shock intensity can be regarded as independent of velocity). This is in contrast to transonic situations in which the shock intensity depends generally on the velocity (see, e.g., Bruce and Babinsky [23]).

Other properties of these interactions can be recalled. First, we examine features shared by different flows at the beginning of separation. In this region, high levels of velocity fluctuations are observed with the maxima in the turbulence intensity being located away from the wall, as shown in Fig. 9.6 for the shock-reflection case.

Smits and Dussauge [31] assessed the amplification of turbulence levels through the shock using the "rapid-distortion approximation" for the case of a 24-degree compression-ramp flow, assuming that the dilatation effects are predominant as well as by using Debiève's formulation (Debiève et al. [9]). (Incidentally, this is an appropriate way to assess the levels of turbulence in the 8-degree shock-reflection case illustrated in Fig. 9.6.) Smits and Dussauge concluded that the resulting levels were not consistent with linear-amplification mechanisms. Instead, they observed that they are consistent with the level of turbulence in mixing layers, $u'/\Delta U \sim 10$ percent, for example. The spatial organization of the flow indicated in Fig. 9.6 is similar to what is observed in subsonic separated flows (Kiya and Sasaki [27]; Cherry et al. [32]). It is related to the development of a mixing layer starting at the separation line: From this point, the mean velocity profiles are strongly deformed with

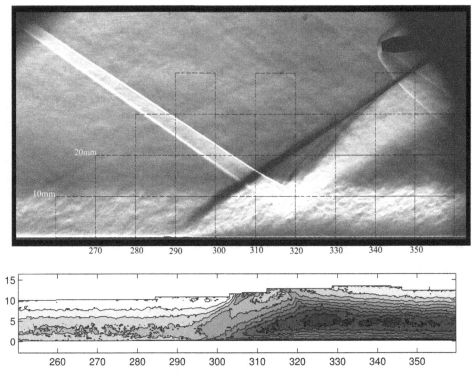

Figure 9.6. A schlieren image of an oblique-shock reflection interaction at $M = 2.3$; flow deviation $8°$ below which, to the same scale, is a plot of isocontours of the RMS values of vertical velocity fluctuations.

inflectional shapes, as observed in plane mixing layers. The high level of turbulence is related to the development of large coherent vortical structures of the Kelvin-Helmholtz type [56]. In the second part of the separated region, subsonic studies showed that these large vortices are shed into the reattached boundary layer and that they can be observed over long distances downstream of reattachment.

The same type of organization is found in shock-induced separated flows. PIV measurements (Dupont et al. [33]) for the oblique-shock reflection at $M = 2.3$ (see Fig. 9.6) revealed the development of large vortical structures in the first part of the interaction. In the second part of the interaction, these large eddies are shed downstream, as observed in subsonic separated flows. They contribute to high-turbulence intensities well away from the wall in the downstream boundary layer and persist over several interaction lengths. This is shown in Fig. 9.7, in which the vorticity detector proposed by Graftieaux et al. [34] was applied to the PIV measurements. The development of a mixing layer is detected along with vortex shedding into the reattached layer.

Therefore, to summarize different characteristics of the interactions, it is possible to list the following five distinct types of fluctuations that may have different frequency content and levels:

1. *Turbulence of the upstream boundary layer.* This involves mainly high frequencies ($f \sim U_\infty/\delta$), although very large scales (or superstructures) on the order of 30δ also were noted (Adrian et al. [35] and Ganapathisubramani et al. [36])

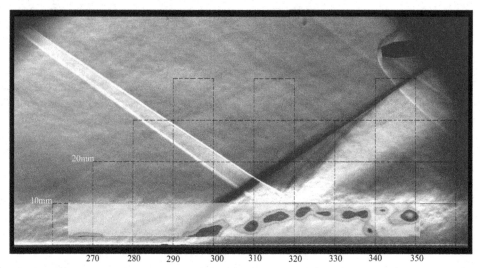

Figure 9.7. A contour plot of vorticity obtained from PIV measurements in the region near the wall superimposed on a schlieren image of an oblique-shock reflection, $M = 2.3$; flow deviation $8°$.

with corresponding lower frequency. However, these superstructures do not contribute significantly to the energy of the signal.

2. *Fluctuations resulting from the (linear) amplification of the incoming turbulence caused by passing through the shock wave.* This mechanism acts on turbulence in all conditions, whether nonseparated or separated. The associated frequency range is the same as in the upstream layer, and rapid-distortion theories can estimate the amplification.

3. *Fluctuations related to the development of large eddies in the mixing layer that develops in the separated region.* The resulting fluctuations are much larger than the levels determined by rapid-distortion theory. The associated frequencies depend on the longitudinal position.

4. *Large persistent fluctuations observed in the downstream flow, well away from the wall (typically, $y/\delta \sim 0.5$).* Near the wall, the relaxation is very fast with a time scale given by $t \sim (\partial U/\partial y)^{-1}$; however, farther away, large maxima of turbulence are observed that persist over several lengths of the interaction (see Fig. 9.6). These correspond to the transport and diffusion of the large coherent structures shed from the interaction (see Fig. 9.7).

5. *Fluctuations related to the shock motions, which create an intermittent signal. This is not usual turbulence and can be effectively described by an intermittent boxcar signal.* The maximum value of rms fluctuations is observed at the median position of the unsteady shock wave and is equal to $\Delta q/2$, where Δq is the local step across the shock of the considered quantity (e.g., pressure or velocity) (Dupont et al. [37]). The frequencies related to these intermittent fluctuations are low and constitute the core of the problem of interaction unsteadiness.

9.3.3.2 Separated Flows: Frequency Content

This section demonstrates that the frequency content of shock interactions may depend strongly on size and geometry. Predicting such frequencies thus requires knowledge of the mean structure of the interactions. In turn, this depends on knowing the correct separation-point location, which is difficult to ascertain precisely. Examples are provided in Chapters 4 and 5 of this book as well as in several published experimental compilations. The results often are flow-sensitive, for example, whereas in nozzle flows at large Reynolds numbers, the onset of separation seems to be relatively independent of this number (see, e.g., Zukoski [38] and Reijasse [39]), and it appears that for compression-ramp and oblique-impinging shock flows, the onset of separation at moderate Reynolds numbers is sensitive to it. To date, the numerical prediction of separation and reattachment remains a challenging task for many turbulence models. Although the precise extent and magnitude of the reverse flow is generally difficult to estimate, when the interaction is strong enough to cause the flow to separate, general observations can be made.

Drawing on classical considerations (e.g., Chapman's [57] free-interaction theory), one observation concerns the intensity of the leading shock formed upstream of the separated region. As noted by Délery and Marvin [40] and by Haddad [41], the strength of this shock does not depend on the particular downstream conditions – that is, the ramp angle in compression-ramp flows or the incident-shock angle in oblique-shock reflections. It also is observed that wall pressure starts to increase upstream of the mean separation-shock location. In the past, this initial rise was referred to as the "upstream influence" but, as noted by Dolling [42], it is likely due to the intermittency caused by the shock traveling backward and forward. This does not appear to depend on the specific conditions creating the interaction but rather is much more general; it is tempting to conclude that this is a universal feature that can be applied to explain the dominant frequency of the shock beating.

The frequency content of the different types of fluctuations is now discussed. First, we consider the influence of the incoming turbulence in more detail. As shown in Fig. 9.1, it is evident that upstream turbulence can influence the shock motion. As noted previously, there is evidence of very long structures within boundary layers (see Adrian et al. [35] and Ganapathisubramani et al. [43] for subsonic boundary layers; see Ganapathisubramani et al. [43] for supersonic flows). It is possible to determine the frequency scale generated by these superstructures in the upstream flow and check whether it is consistent with the observed Strouhal numbers. Such structures are supposed to be 30δ long, with a convection speed of $0.75U_\infty$. The resulting frequency is therefore $f = \frac{0.75U_\infty}{30\delta}$ and the corresponding Strouhal number will be $S_L = 0.025\frac{L}{\delta}$. It turns out that such structures can generate fluctuations with the right Strouhal number for interactions so that $L/\delta \sim 1$. For compression ramps and oblique-shock reflections, such low ratios generally correspond to nonseparated interactions or incipient separations but not to cases in which a separated zone is well developed. Interactions like the oblique-shock reflections studied by Dupont et al. [45 and 46] have ratios L/δ about 5 to 7 for Strouhal numbers about 0.03–0.04. For separated compression ramps with M > 2, typical values of 0.02–0.05 are obtained (Dussauge et al. [28]).

Further insight was obtained from tomographic visualizations of the oblique-shock reflection by Piponniau et al. [47]. These authors recorded photographs of this interaction for different values of the flow deviation. For small values of the deflection angle (typically, 4 or 5 degrees), the flows were not separated and the reflected shock did not move much. However, on increasing the angle, the amplitude of the

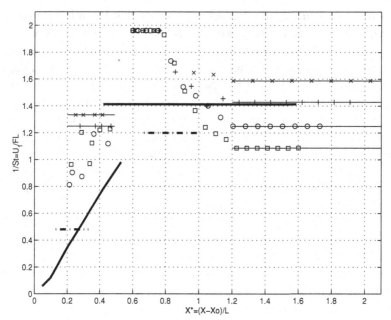

Figure 9.8. Longitudinal evolution of the inverse Strouhal numbers in subsonic and supersonic separations, (▬) subsonic separated flow Cherry et al. [32]; (·▬·▬·) $M = 1.5$ compression-ramp flow Thomas et al. [30]; oblique-shock reflection $M = 2.3$ (×) $\theta = 7°$, (+) $\theta = 8°$, (○) $\theta = 8.8°$, and (□) $\theta = 9.5°$.

shock oscillations rose correspondingly, particularly when separation occurred. The maximum oscillations occurred for their highest angle of deviation, which corresponds to the longest separation length. Moreover, it was found that the frequencies decreased on increasing the interaction length (i.e., the shock intensity). The conclusion from this simple experiment is that for the same incoming boundary layer – and, therefore, for the same incoming large-eddy structure – the shock unsteadiness can vary considerably. There are only two possible explanations: (1) an increase in the excitation of the shock provided by a stronger separation; or (2) a modification of the transfer function of the shock, which might depend on its intensity. This latter effect, however, is not believed to be dominant. Note that as previously stated, for most separated flows, the mean wall pressure at the reflected shock – when normalized by upstream value – is independent of the total flow deviation. Therefore, the low frequencies produced by the very large structures cannot explain the unsteadiness observed in the shock reflection and generally do not answer the question of the origin of the shock-wave unsteadiness in separated flows. Thus, another source of shock unsteadiness must be found. In cases of shock-induced separation, because the shock is considered an interface providing a link between the upstream and downstream conditions, the only remaining possibility is to consider the separated region downstream of the leading shock wave.

First, the dynamics of the large eddies developed in the separated bubble are discussed in detail. The dominant frequency of these large coherent structures can be derived from unsteady wall-pressure measurements (see Section 9.3.3.1). They are summarized in Fig. 9.8 from Dupont et al. [46] and compared with results in the subsonic regime. The inverse of the Strouhal number $S_L = fL/U_\infty$ is plotted versus the longitudinal coordinate $X^* = (X - X_0)/L$, where X_0 is the location of the separation shock and L is the interaction length. The intermediate case of a 14-degree compression-ramp flow at $M = 1.5$ also is included (Thomas et al. [30]). The different regions are easily identifiable, as follows:

- $0 < X^* < 0.5$ corresponds to the region of development of the mixing layer
- $0.5 < X^* < 0.8$ corresponds to the shedding of large structures
- $0.8 < X^* < 1.2$: For the shock-reflection case, an increase in the frequency associated with the structures shed downstream is observed in this region where the flow is turned down to return parallel to the wall. This zone depends explicitly on the geometry of the interaction and does not appear in compression-ramp flows or the subsonic case.
- $X^* > 1.2$: The frequency is associated with eddies that persist in the reattached boundary layer over large distances after the shedding process, whatever the configuration.

Figure 9.8 clearly shows that the longitudinal evolution of the Strouhal number associated with large structures is globally similar for the different cases. Nevertheless, regarding the actual numbers, significant differences are observed between supersonic and subsonic or transonic cases: For the $M = 2.3$ shock-reflection case, the dimensionless shedding frequency is reduced by about 40 percent compared with the incompressible value. Because mixing layers are known to be particularly sensitive to compressibility effects, an attempt to relate this behavior to the characteristic Mach number of the flow was made (Dupont et al. [46]), leading to the following results. The Strouhal number can be expressed as follows:

$$S_L = \frac{fL}{U_\infty} = \frac{U_c}{U_\infty} \frac{L}{\delta_\omega} \frac{f \delta_\omega}{U_c} = \frac{U_c}{U_\infty} \frac{L}{\delta' x} S_\delta \approx \frac{U_c}{U_\infty} \frac{S_\delta}{\delta'} X^{*-1} \tag{9.3}$$

where $S_\delta = f\delta_\omega / U_c$, $\delta_\omega = \delta' x$, δ' is the spreading rate of the mixing layer, and U_c is the convection velocity of the large structures. Equation 9.3, written at the location of vortex shedding (i.e., $x \sim L/2$), provides the following:

$$S_{L, \text{shed}} = 2 \frac{U_c}{U_\infty} \frac{S_\delta}{\delta'} \tag{9.4}$$

Moreover, the spreading rate of the canonical plane mixing layer can be expressed as follows (Papamoschou and Roshko [48]):

$$\delta' = \delta'_{\text{ref}} \frac{\Delta U}{U_c} \Phi(M_c) \tag{9.5}$$

where ΔU is the velocity difference across the mixing layer and δ'_{ref} is the spreading rate of the subsonic mixing layer, with constant density and $U = 0$ on the low velocity side. Based on the convection velocity of the large coherent scales U_c and the external velocities U_1 and U_2 (Papamoschou and Roshko [48]), $\Phi(M_c)$ is a decreasing function of the convective Mach number. The function $\Phi(M_c)$ expresses the drastic reduction of the mixing-layer spreading rate and, consequently, of the entrainment process when the convective Mach number increases.

In incompressible flows, the convection velocity can be estimated from the following equations:

$$\frac{U_c}{U_\infty} = \frac{1 + r\sqrt{s}}{1 + \sqrt{s}} \quad \text{or} \quad \frac{\Delta U}{U_c} = \frac{(1-r)\,(1+\sqrt{s})}{1 + r\sqrt{s}} \tag{9.6}$$

in which $r = U_2/U_1$ and $s = \rho_2/\rho_1$. These expressions are derived using classical isentropic relationships (Papamoschou and Roshko [48]) for both incompressible and compressible mixing layers. The dimensionless convection velocity thus does not depend on Mach number, only on velocity and density ratios. The generality of these relations is still a matter of debate, and this point is addressed later in this chapter. Relation 9.5 can be evaluated for incompressible flows – that is, for $M_c <$ 0.3 and, in this case, $\Phi(M_c) \approx 1$. Considering standard subsonic values such as $S_\delta \approx$ 0.22 and $\delta'_{\text{ref}} \approx 0.16$, and bearing in mind that the maximum velocity in the reverse flow often remains relatively low, it was found that for constant-density flows, $\Delta U \approx U_\infty$ and $U_c \approx U_\infty/2$. Finally, a value $S_{L,\text{shed}} \approx 0.7$ is obtained that is in very good agreement with the measurements of Cherry et al. [32] (see Fig. 9.8).

When we consider compressible cases, equation 9.4 – together with the isentropic approximation 9.6 – implies that the Strouhal number related to the shedding process should vary mainly as the ratio S_δ/δ'. Referring to experimental work or linear-stability analysis, it is found that this ratio does not vary much with the convective Mach number and remains close to 1 (Blumen [49]; Muscat [50]). The consequence is that the Strouhal number for the shedding frequency depends mainly on the velocity and density ratios. This conflicts with results shown in Fig. 9.8, in which differences in the values of the shedding Strouhal number are observed for different Mach numbers. However, several experimental studies (Papamoschou [51]; Barre, Dupont, and Dussauge [52]) show that when the convective Mach number is larger than 0.5, large departures from the isentropic relation are observed. Such behavior generally is found when the mixing layer develops in the vicinity of a wall (Tam and Hu [53]; Greenough et al. [54]), as is the case here. For this reason, direct measurements of the convection velocity of the larger-scale eddies were taken for the shock-reflection case at Mach number 2.3 to assess the validity of relation 9.6. From multi-point, unsteady wall-pressure measurements, a value of $U_c/U_\infty \approx 0.3$ was obtained. This differs significantly from the value obtained using the isentropic relation 9.6 ($U_c/U_\infty \approx 0.45$) – by about 40 percent – and is consistent with the experimental results reported in Fig. 9.8. If, conversely, we consider the $M = 1.5$ compression-ramp case of Thomas et al. [30], which is referred to in the same figure, much smaller convective Mach numbers are expected (i.e., typically, around 0.5). In this case, the validity of the isentropic relationships is more likely and values for the shedding Strouhal number similar to the subsonic values are expected, which is confirmed by the experiments (see Fig. 9.8).

Finally, returning to relation 9.3, an expression for the wavelength $\lambda_{\text{shed}} = U_c/f_{\text{shed}}$ associated with the frequency f_{shed}, (i.e., the distance between two consecutive eddies) is found, as follows:

$$S_{L,\text{shed}} = \frac{f_{\text{shed}}\,L}{U_\infty} = \frac{U_c}{U_\infty}\frac{L}{\lambda_{\text{shed}}} \quad \text{thus} \quad \frac{\lambda_{\text{shed}}}{L} = S_{L,\text{shed}}^{-1}\frac{U_c}{U_\infty} = \frac{1}{2}\frac{\delta'}{S_\delta} \approx \frac{1}{2} \tag{9.7}$$

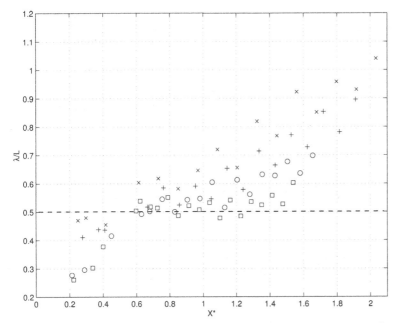

Figure 9.9. Average distance between large eddies for several shock-reflection interactions at $M = 2.3$.

Therefore, because the variation of the shedding Strouhal number is related to the ratio U_c/U_∞, relation 9.7 indicates that typical wavelengths of about $L/2$ could be expected, whatever the Mach number. This value is indicated by a dotted line in Fig. 9.9, which agrees well with measurements taken in the rear half of the separation region (i.e., $0.5 < X^* < 1.0$).

From these considerations, the dominant frequencies in the separated region are indicated. These flows have many features in common with their subsonic counterparts; in particular, the recirculating zones do not generate a single frequency but rather a range of frequencies that evolve with longitudinal distance. Experiments with subsonic flows revealed that very-low-frequency perturbations are present in the separated bubble. However, significant differences can be expected when the speed is high enough to generate convective Mach numbers larger than 0.5. Fig. 9.8 shows that the associated Strouhal numbers are much larger than the shock Strouhal numbers (see Fig. 9.4). Their frequency is an order of magnitude lower than the unsteadiness associated with the large coherent structures formed in the mixing layer. Such low frequencies are associated with the flapping of the mixing layer or with the bubble "breathing," with a typical Strouhal number of 0.12.

Erengil and Dolling [55] in an $M = 5$ compression-ramp experiment were probably the first to remark that the subsonic low-frequency flapping ($S_L = 0.12$) could not be compared directly to the low-frequency shock motions ($S_L = 0.03$). Systematic compilations performed by Dussauge et al. [28] (see Fig. 9.4) confirmed that for any geometry, similar conclusions can be drawn. For shock-induced separations, if the external Mach number is larger than two, the shock-motion frequency is at least four times smaller than their subsonic counterpart. Moreover, the same reduced frequencies are observed for both the low-frequency bubble "breathing" and the

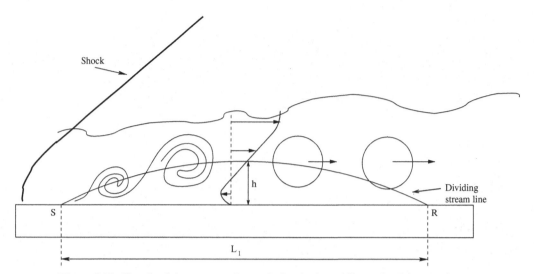

Figure 9.10. Sketch of the separated zone in impinging oblique-shock interactions.

shock motions ($S_L \approx 0.03$). Although not the most energetic band, these low frequencies were found in the wall-pressure spectra in the separated zone (Dupont et al. [33]). Furthermore, the coherence between wall-pressure fluctuations at the foot of the shock and in the recirculation region are high, typically 0.8 (Dupont et al. [46]). Subsequently, Piponniau et al. [24] took conditional measurements of the size of the separated bubble and found that it is strongly intermittent, with a few events during which intense backward flow penetrates the separation pocket. Their measurements showed clearly that these events generate large vertical-amplitude oscillations of the mixing layer, which are linked to significantly large longitudinal excursions of the shock. In their analysis of the origin of the unsteadiness, Piponniau et al. reasoned that the large shock pulsations are closely related to the flapping of the mixing layer formed at the edge of the separated bubble, as illustrated in Fig. 9.10. Therefore, they proposed an explanation based on considerations of air entrainment by this mixing layer.

Their objective was to find the parametric dependence of the shock-motion frequency rather than a complete theoretical description. Piponniau et al. suggested that the low-frequency flapping of the shear layer must be related to an unbalanced mass budget in the separated region. The basic mechanism is illustrated in Fig. 9.10. Large vortical structures develop in the initial part of the interaction ($X^* < 0.5$), entraining mass from the reverse flow. Then, in the second part of the interaction, the structures are shed into the downstream layer, causing a mass defect in the separated region. Considering these elements, they evaluated the amount of mass contained in the bubble and the rate of mass entrainment. The ratio of these two quantities provides a time scale, which represents the interval necessary to drain a significant amount of mass from the separated zone. The inverse of this time provides an order-of-magnitude estimate of the frequency scale at which new air is injected into the recirculating zone. Piponniau et al. assumed that the dependence of the spreading rate of the mixing layer on the density and velocity ratios and on the convective Mach number is the same as in canonical mixing layers. It appears that a length scale introduced naturally by this model is the height h of the separated bubble. The analysis provides a Strouhal number $S_h = fh/U_\infty$ based on this height of the following form:

$$S_h = \Phi(M_c)\, g(r, s) \tag{9.8}$$

in which $\Phi(M_c)$ is again the normalized spreading rate of the mixing layer. The function g is from the contributions of density and velocity ratios (i.e., r and s, respectively) to the spreading rate and the rate of mass entrainment. It is a weak function of r and s and varies minimally for the usual values found in boundary-layer separations over adiabatic plates. An expression was proposed in which it is assumed that the velocity profiles in the mixing layer follow an error function, as follows:

$$g(r, s) = \frac{\delta_{\text{ref}}}{2} \frac{(1 - r)(1 + \sqrt{s})}{1 + r\sqrt{s}} \left\{ (1 - r)\, C + \frac{r}{2} \right\}, \quad \text{with } C \approx 0.14 \tag{9.9}$$

at low speed, with constant density, and with $r = 0$, $g(0, 1) \approx 0.02$.

Equation 9.8 implies that in separations, the dominant frequency varies like $1/h$: If h is small, there is little mass involved, so that it is rapidly drained by the mixing layers; this produces high frequencies. In practice, it is difficult to determine h from experiments, and most measurements involve the length L of the separated flow or the interaction. From equation 9.8, it is straightforward to derive a Strouhal number based on the following length:

$$S_L = \Phi(M_c)\, g(r, s)\, \frac{L}{h} \tag{9.9}$$

from which it appears that S_L is proportional to the aspect ratio L/h of the separated bubble. If the aspect ratio of the bubble is taken as constant, this implies that the Strouhal number varies with the convective Mach number like the spreading rate of the mixing layer. This hypothesis is supported by the existing data shown in Fig. 9.11.

In most of the interactions considered in this chapter, the convective Mach number of the large structures in the mixing layer is close to 1, which corresponds to a value of $\Phi(M_c)$ of about 0.2. This implies that the aspect ratio of the separation bubbles is about 6, which is consistent with what is known about their geometry. Therefore, it appears that this simple model provides a more general representation of the unsteadiness. Of course, this scheme is limited to two-dimensional situations in which a reattachment point exists – for example, in the so-called RSS in nozzle flows. Moreover, it can be inferred from the previous analysis that S_L depends on the geometry of separated zones and thus will have different values if the shape of the separation bubble varies considerably. The model, however, indicates the leading elements for analyzing other situations and how to control them.

9.4 Conclusions: A Tentative Classification of Unsteadiness and Related Frequencies

It is now possible to summarize the main characteristics of the natural unsteadiness present in SBLIs. An interesting property, shown herein, is that the frequency

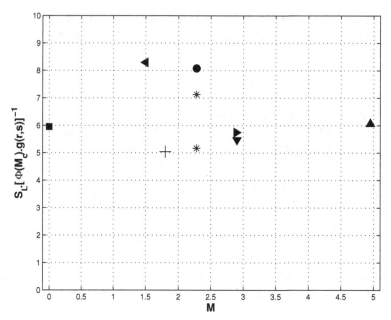

Figure 9.11. Strouhal number corrected for compressibility effects, according to Piponniau et al. [24].

range of the unsteadiness depends on the nature of the interaction. Three typical situations considered in previous sections are used for illustration: the transonic interaction with acoustic coupling, the supersonic-separated case, and the supersonic-nonseparated case. The transonic case is exemplified by a case of acoustic coupling between the trailing edge and the shock wave on a wing profile. As noted in Section 9.3.2, this leads to the following frequency:

$$f_{buf} \approx (a_2 - u_2)/\ell \qquad (9.10)$$

where ℓ is on the order of the chord of the profile.

In the supersonic-separated case, the frequency is derived from equation 9.8, as follows:

$$f_{ss} \approx \frac{U_\infty}{h} \Phi(M_c) \, g(r, s) \qquad (9.11)$$

and corresponds to requirements imposed by mass conservation for the bubble.

To compare frequencies in these three situations, it is instructive to normalize them by the frequency f_{ec} generated by the passing of the energetic eddies in the upstream turbulence. The eddies have a typical size δ and are convected at a speed on the order of U_∞. Thus, the frequency is $f_{ec} = U_\infty/\delta$, and the results are summarized in Table 9.1. In the supersonic nonseparated case, the main source of excitation is considered to be the incoming turbulence; hence, for this case, the normalized frequency is on the order of 1.

The normalized frequency for the transonic interaction f_{buf}/f_{ec} was derived in the following way. First:

$$\frac{f_{buf}}{f_{ec}} = \frac{a_2 - u_2}{\ell} \frac{\delta}{u_\infty} = \frac{\delta}{\ell} \left(\frac{1}{M_2} - 1\right) \frac{u_2}{U_\infty} \qquad (9.12)$$

Table 9.1. *Comparison of unsteadiness frequency for various types of interactions*

Flow	Phenomenon	Frequency	Normalized frequency	Order of magnitude
Transonic interaction	Acoustic coupling	$(a_2 - u_2)/\ell$	$\dfrac{\delta}{\ell}(M_\infty - 1)$	$<10^{-2}$
Separated, supersonic	Mass conservation	$\dfrac{U_\infty}{h}\,\Phi(M_c)\,g(r,\,s)$	$\dfrac{\delta}{h}\,\Phi(M_c)\,g(r,\,s)$	$<10^{-1}$
Nonseparated	Eddy convection	U_∞/δ	1	1

assuming that the shock on an airfoil can be approximated by a normal shock, accepting the following approximation:

$$M_\infty\,M_2 \approx 1 \qquad (9.13)$$

and noting that $u_2/U_\infty < 1$, the following approximation is obtained:

$$\frac{f_{buf}}{f_{ec}} < \frac{\delta}{\ell}\,(M_\infty - 1) \qquad (9.14)$$

The ratio δ/ℓ is on the order of the spreading rate of the boundary layer (i.e., typically 10^{-2}). For transonic profiles, a typical Mach-number value upstream of the compression shock is 1.4, for example; thus, $(M_\infty - 1)$ is less than 1 and $u_2/U_\infty \approx 0.6$, so that the frequency ratio is much less than 10^{-2}.

In many supersonic interactions with separation, $\delta/h \sim 1$ or $\delta/h < 1$, for which typically (according to Piponniau et al. [24]) $\Phi(M_c) \sim 0.2$ and $g(r,\,s) \sim 0.02$. Therefore, the estimate in Table 9.1 is probably rather conservative.

The observations from these representative situations can be applied to other flow cases. For example, the shock-pumping in front of the supersonic air intake of the Pitot type is probably governed by acoustic coupling; therefore, the estimate of buffeting frequency can be applied to this flow.

The incipient-separation case is likely to be more difficult because of the complex nature of the physics. There is no mean separation but instead many intermittent spots of separation; further physical analysis is required in this case. The spectrum shown in Fig. 9.3 suggests that for this particular case, the frequency range matches the scales of the incoming turbulence. There is evidence that a low-frequency content is present. Only simple cases are considered herein and other parameters (e.g., heat and mass transfer, Reynolds number, laminar-turbulent transition, and three-dimensional geometry) may influence the unsteadiness. The reviewed examples clearly indicate the trends observed in simple physical situations; however, this should not obscure the fact that in practical situations, many factors influencing the unsteadiness can be present at the same time. There is no unique source of unsteadiness, and shocks respond to all influences. However, apart from provoking boundary-layer separation, the shocks seem to have a passive role in the unsteadiness. This analysis suggests that it is the recirculating zone that imposes fluctuations on the flow.

Probably the more spectacular result is that in most cases, the interactions are capable of producing low-frequency perturbations from the high-frequency convective-flow elements, such as the incoming turbulence or Kelvin-Helmholtz–like structures. These perturbations have frequencies that often are two or three orders of magnitude below other flow components. Their origin is localized to a limited region of the flow, but their effect can be experienced in the entire field downstream.

REFERENCES

[1] J. P. Dussauge. Compressible turbulence in interactions of supersonic flows. *Proceedings of the Conference TI* 2006 (Berlin: Springer Verlag [in press]).

[2] J. P. Dussauge and S. Piponniau. Shock-boundary layer interactions: Possible sources of unsteadiness. *Journal of Fluids and Structures*, 24 (2008), 8, 1166–75.

[3] F. E. C. Culick and T. Rogers. The response of normal shocks in diffusers. *AIAA Journal*, 21 (1983), 10, 1382–90.

[4] J. C. Robinet and G. Casalis. Shock oscillations in a diffuser modelled by a selective noise amplification. *AIAA Journal*, 37 (1999), 4, 1–8.

[5] M. Sajben and J. C. Kroutil. Effect of initial boundary layer thickness on transonic diffuser flow. *AIAA Journal*, 19 (1981), 11, 1386–93.

[6] E. Garnier. *Simulation des grandes échelles en régime transsonique*. Thèse de Doctorat, Univ. Paris XI Orsay, Paris, France (2000).

[7] M. Wu and M. P. Martin. Analysis of shock motion in shockwave and turbulent boundary layer interaction using direct numerical simulation data. *Journal of Fluid Mechanics*, 594 (2008), 71–83.

[8] H. S. Ribner and M. Tucker. Spectrum of turbulence in a contracting stream. *NACA TN* 2606 (1952).

[9] J. F. Debiève, H. Gouin, and J. Gaviglio. Evolution of the Reynolds stress tensor in a shock-turbulence interaction. *Indian Journal of Technology*, 20 (1982), 90–7.

[10] L. Jacquin, C. Cambon, and E. Blin. Turbulence amplification by a shock wave and rapid distortion theory. *Physics of Fluids A*, 5 (1993), 2539–50.

[11] H. S. Ribner. Convection of a pattern of vorticity through a shock wave. *NACA TN* 2864 (1953).

[12] L. S. G. Kovasznay. Turbulence in supersonic flow. *Journal of Aeronautical Sciences*, 20 (1953), 657–74.

[13] B. T. Chu and L. S. G. Kovasznay. Nonlinear interactions in a viscous heat-conducting compressible gas. *Journal of Fluid Mechanics*, 3 (1958), 494–514.

[14] J. C. Anyiwo and D. M. Bushnell. Turbulence amplification in shock wave boundary layer interactions. *AIAA Journal*, 20 (1982), 893–9.

[15] S. Lee, S. K. Lele, and P. Moin. Direct numerical simulation of isotropic turbulence interacting with a weak shock wave. *Journal of Fluid Mechanics*, 251 (1993), 533–62.

[16] R. Hannapel and R. Friedrich. Direct numerical simulation of a Mach 2 shock interacting with isotropic turbulence. *Applied Science Research*, 54 (1995), 205–21.

[17] E. Garnier, P. Sagaut, and M. Deville. Large eddy simulation of shock-homogeneous turbulence interaction. *Computer and Fluids*, 31 (2002), 2, 245–68.

[18] J. F. Debiève and J. P. Lacharme. "A shock wave–free turbulence interaction." In *Turbulent Shear Layer–Shock Wave Interactions*, ed. J. Délery (Berlin: Springer Verlag, 1985).

[19] J. Poggie and A. J. Smits. Shock unsteadiness in a reattaching shear layer. *Journal of Fluid Mechanics*, 429 (2001), 155–85.

[20] B. H. K. Lee. Self-sustained shock oscillations on airfoils at transonic speeds. *Progress in Aerospace Sciences*, 37 (2001), 147–96.

[21] J. D. Crouch, A. Garbaruk, and D. Magidov. Predicting the onset of flow unsteadiness based on global instability. *Journal of Computational Physics*, 224 (2007), 2, 924–40.

[22] A. Galli, B. Corbel, and R. Bur. Control of forced shock wave oscillations and separated boundary later interaction. *Aerospace Science and Technology*, 9 (2005), 8, 653–60.

[23] P. J. K. Bruce and H. Babinsky. Unsteady shock-wave dynamics. *Journal of Fluid Mechanics*, 603 (2008), 463–73.

[24] S. Piponniau, J. P. Dussauge, J. F. Debiève, and P. Dupont. A simple model for low frequency unsteadiness in shock-induced separation. *Journal of Fluid Mechanics*, 629 (2009), 87–108.

[25] E. Touber and N. Sandham. Oblique shock impinging on a turbulent boundary layer: Low frequency mechanisms. *AIAA Paper* 2008–4170 (2008).

[26] L. J. Souverein, B. W. van Oudheusden, F. Scarano, and P. Dupont. Unsteadiness characterization in a shock wave turbulent boundary layer interaction through dual PIV. *AIAA Paper* 2008–4169 (2008).

[27] M. Kiya and K. Sasaki. Structure of a turbulent separation bubble. *Journal of Fluid Mechanics*, 137 (1983), 83–113.

[28] J. P. Dussauge, P. Dupont, and J. F. Debiève. Unsteadiness in shock wave boundary layer interactions with separation. *Aerospace Science and Technology*, 10 (2006), 85–91.

[29] D. S. Dolling and D. R. Smith. Unsteady shock-induced separation in Mach 5 cylinder interactions. *AIAA Journal*, 27 (1989), 12, 1598–706.

[30] F. Thomas, C. Putman, and H. Chu. On the mechanism of unsteady shock oscillation in shock wave–turbulent boundary layer interaction. *Experiments in Fluids*, 18 (1994), 69–81.

[31] A. J. Smits and J. P. Dussauge. *Turbulent shear layers in supersonic flows*. (New York: Springer Verlag, 2006).

[32] N. J. Cherry, R. Hillier, and M. E. M Latour. Unsteady measurements in a separated and reattaching flow. *Journal of Fluid Mechanics*, 144 (1984), 14–46.

[33] P. Dupont, S. Piponniau, A. Sidorenko, and J. F Debiève. Investigation by Particle Image Velocimetry measurements of oblique shock reflection with separation. *AIAA Journal*, 46 (2008), 6, 1365–70.

[34] L. Graftieaux, M. Michard, and N. Grosjean. Combining OIV, POD, and vortex identification algorithms for the study of unsteady turbulent swirling flows. *Measurement Science and Technology*, 12 (2001), 1422–9.

[35] R. J. Adrian, C. D. Meinhart, and C. D. Tomkins. Vortex organization in the outer region of the turbulent boundary layer. *Journal of Fluid Mechanics*, 422 (2000), 1–53.

[36] B. Ganapathisubramani, N. T. Clemens, and D. S. Dolling. Large-scale motions in a supersonic turbulent boundary layer. *Journal of Fluid Mechanics*, 556 (2006), 271–82.

[37] P. Dupont, C. Haddad, and J. F. Debiève. Space and time organization in a shock-induced separated boundary layer. *Journal of Fluid Mechanics*, 559 (2006), 255–77.

[38] E. E. Zukoski. Turbulent boundary layer separation in front of a forward-facing step. *AIAA Journal*, 5 (1967), 10, 1746–53.

[39] P. Reijasse. *Aérodynamique des tuyères propulsives en sur-détente: Décollement libre et charges latérales en régime stabilise*. Thèse de Doctorat, Univ. Paris VI, Paris, France (2005).

[40] J. Délery and J. G. Marvin. Shock wave–boundary layer interactions. *AGARDograph* 280, NATO, Neuilly sur Seine, France (1986).

[41] C. Haddad. *Instationnarités, mouvements d'onde de choc et tourbillons à grande échelle dans une interaction onde de choc–couche limite avec décollement*. Thèse de Doctorat, Université de Provence, Marseille, France (2005).

[42] D. S. Dolling. Fifty years of shock wave–boundary layer interactions: What next? *AIAA Journal*, 39 (2001), 8, 1517–31.

[43] B. Ganapathisubramani, E. K. Longmire, and I. Marusic. Characteristics of vortex packets in turbulent boundary layers. *Journal of Fluid Mechanics*, 478 (2003), 35–46.

[44] B. Ganapathisubramani, N. T. Clemens, and D. S. Dolling. Planar imaging measurements to study the effect of spanwise structure of upstream turbulent boundary layer on shock-induced separation. *AIAA Paper* 2006–324 (2006).

[45] P. Dupont, C. Haddad, J. P. Ardissone, and J. F. Debiève. Space and time organisation of a shock wave–turbulent boundary layer interaction. *Aerospace Science and Technology*, 9 (2005), 561–72.

[46] P. Dupont, C. Haddad, and J. F. Debiève. Space and time organization in a shock-induced separated boundary layer. *Journal of Fluid Mechanics*, 559 (2006), 255–77.

[47] S. Piponniau, P. Dupont, J. F. Debiève, and A. Sidorenko. Unpublished work, IUSTI (2007).

[48] D. Papamoschou and A. Roshko. The compressible turbulent shear layer: An experimental study. *Journal of Fluid Mechanics*, 197 (1988), 453–77.

[49] W. Blumen. Shear layer instability of an inviscid compressible fluid. *Journal of Fluid Mechanics*, 40 (1970), 769–81.

[50] P. Muscat. *Structures à grandes échelles dans une couche de mélange supersonique. Analyse de Fourier et analyse en ondelettes.* Thèse de Doctorat, Université de la Méditerranée (Aix-Marseille II), Marseille, France (1998).

[51] D. Papamoschou. Structure of the Compressible Turbulent Shear Layer. *AIAA Journal*, 29, 5 (1991).

[52] S. Barre, P. Dupont, and J. P. Dussauge. *Estimation de la vitesse de convection des structures turbulentes à grande échelle dans les couches de mélange supersonique. Aerospace Science and Technology*, 1 (1997), 4, 355–66.

[53] C. K. W. Tam and F. Q. Hu. The instability and acoustic wave modes of supersonic mixing layers inside a rectangular channel. *Journal of Fluid Mechanics*, 203 (1989), 51–76.

[54] J. A. Greenough, J. J. Riley, M. Soetrisno, and D. S. Eberhardt. The effect of walls on a compressible mixing layer. *AIAA Paper* 89–0372 (1989).

[55] M. E. Erengil and D. S. Dolling. Unsteady wave structure near separation in a Mach 5 compression ramp interaction. *AIAA Journal*, 29 (1991), 5, 728–35.

[56] Jie-Zhi Wu, Hui-Yang Ma, Ming-De Zhou, Vorticity and Vortex Dynamics (Berlin Heidelberg: Springer Verlag, 2006).

[57] D. Chapman, D. Huehn, H. Larson, Investigation of separated flows in supersonic and subsonic streams with emphasis on the effect of transition. NACA TM 3869, 1957.

Analytical Treatment of Shock Wave–Boundary-Layer Interactions

George V. Inger

10.1 Introduction

10.1.1 Motivation for Analytical Work in the Computer Age

Notwithstanding the success of powerful CFD codes in predicting complex aerodynamic flowfields, analytical methods continue to be a valuable tool in the study of viscous-inviscid interaction problems for the following reasons:

1. Such methods appreciably enhance physical insight by illuminating the underlying basic mechanisms and fine-scale features of the problem, including the attendant similitude properties [1]. An example in the case of shock wave–boundary-layer interaction (SBLI) is the fundamental explanation of the phenomena of upstream influence and free interaction provided by the pioneering triple-deck–theory studies of Lighthill [2], Stewartson and Williams [3], and Neiland [4].
2. Analysis provides an enhanced conceptual framework to guide both the design of related experimental studies and the correlation and interpretation of the resulting data. This was exemplified in a recent study of wall-roughness effects on shock-wave–turbulent boundary-layer interaction wherein a two-layered analytical theory revealed key features and appropriate scaling properties of the problem that were then used to design and evaluate a companion experimental program [5].
3. Analytical solutions can enhance substantially the efficiency and cost-reduction of large-scale numerical codes [6] by both providing accurate representation of otherwise difficult far-field boundary conditions and serving as an imbedded local element within a global computation to capture key smaller-scale physics. An example of the latter is the application of a small-perturbation theory of transonic normal shock–turbulent boundary-layer interaction in a global inviscid-boundary layer [7]; the resulting hybrid code provided more than 100-fold savings in design-related parametric-study costs.
4. A final noteworthy benefit is the occasional revelation of the deeper basic explanation for well-known empiricisms, such as the local pressure-distribution inflection-point criteria for incipient separation that are widely used by experimentalists.

These attributes argue strongly for the inclusion (whenever possible) of a vigorous analytical component to fluid-mechanics research in general and certainly to studies of SBLI in particular. A focused combination of analytical, computational, and experimental approaches wherein each is used to complement the other remains the most powerful strategy for the investigation of the complex physics involved.

10.1.2 Scope of the Present Survey

This chapter is characterized by the following five features. First, attention is restricted to steady and mainly two-dimensional high-Reynolds-number flows of an ideal gas. Second, only unseparated-flow conditions are considered; hence, issues such as reversed-flow–separation-bubble development and turbulence modeling are not addressed. Third, large-scale global interactions – such as those associated with leading-edge nose-bluntness–entropy-layer effects [8], global boundary-layer displacement-thickness growth [9], and abrupt algebraic branching disturbances [10] encountered in hypersonic flow – are excluded and the focus is exclusively on the smaller-scale localized events in the immediate vicinity of an SBLI zone. Fourth, the survey focuses primarily on the interests of engineers rather than mathematicians by providing a comprehensive assessment of the fundamental analytical work on both laminar and turbulent interaction and what it reveals about interactive physics. Consequently, the more esoteric aspects of the theory – such as certain intricacies of asymptotic-matching procedures and the ultrafine-scale resolution of shock-wave diffraction in the boundary layer – are always underemphasized to provide a general working understanding from an engineering perspective. Fifth, notwithstanding the existence of other analytical approaches, a triple-deck model of the interaction zone is adopted as providing the most general, overarching conceptual framework within which the distinctly different features of laminar-versus-turbulent interactions can be displayed in a unified way for Mach-number regimes ranging from transonic to hypersonic, including the effects of heat transfer.

10.1.3 Content

Section 10.2 is a qualitative overview of the interaction process, the primary triple-deck structure, and the essential physical distinctions between laminar and turbulent boundary-layer responses to either externally impinging or wall-compression-corner-generated shocks. Section 10.3 is a detailed, step-by-step analysis of either laminar or turbulent nonadiabatic-disturbance flow within the triple-deck layers and how they are matched, followed by a discussion of the conditions imposed by the various Mach-number regimes of the interactive-inviscid outer deck. Section 10.3 concludes by combining all of these elements into sets of final, canonical, nondimensional triple-deck formulations, along with determination of the corresponding interactive flow-scaling factors. To provide background for the "uninitiated," Section 10.3 offers more intermediate detail and explanation on the triple-deck theory than typically found in the literature.

Section 10.4 addresses application of the previously mentioned theory to laminar interactions involving supersonic flow past adiabatic walls, hypersonic nonadiabatic flows, and transonic flows. The mechanisms of free interaction, upstream influence, and incipient separation are particularly emphasized in this discussion. Also included is a brief survey of analytical work on simplified "quasi-two-dimensional" three-dimensional interactions, addressing the effects of sweepback and axisymmetric-body geometry. Section 10.5 is an analogous discussion of turbulent interactions in the various Mach-number regimes. Section 10.6 concludes with an overview of the various limitations of the triple-deck–theory approach encountered in practice.

10.2 Qualitative Features of SBLIs

10.2.1 High-Reynolds-Number Behavior: Laminar versus Turbulent

It is well known that significant differences exist between the physical properties of laminar and turbulent boundary layers and their dependence on the Reynolds number [11, 12]. In particular, a turbulent boundary-layer profile is much "fuller" than a laminar profile (see Fig. 2.12); hence (except very near the wall where $y/\delta_0 \leq .01$), it involves velocities that depart only slightly from the outer-edge value. This is described by the classical defect-form of the turbulent boundary-layer Law of the Wall/Law of the Wake velocity profile (see Section 2.3.1 and Fig. 2.13).

Because high Reynolds numbers imply very small C_{f_0} values on the order of 10^{-3}, they also imply that the friction-velocity ratio $\varepsilon_\tau \equiv U_\tau / U_{0e}$ is much less than unity and therefore a suitable small-perturbation parameter to characterize the large Reynolds-number limit for purposes of asymptotic analysis. For example, the nondimensional velocity defect $[U_{0e} - U_{0e}(y)]/U_{0e}$ is on the order of ε_τ, whereas the corresponding boundary-layer thickness is shown [13] to be of the order of $\delta_0/L \sim \varepsilon_\tau$, with the underlying laminar sublayer thickness of even smaller order $\varepsilon_\tau \exp[1/\varepsilon_\tau]$. These properties are different in both magnitude and Reynolds-number dependence from the purely laminar values $C_{f_0} \sim Re_\ell^{-1/2} \sim \varepsilon_L^4$ and $\delta_0/\ell \sim \varepsilon_L^4$ that involve the laminar-asymptotic small parameter $\varepsilon_L \equiv Re_\ell^{-1/8}$.

A useful auxiliary parameter that characterizes the shape of the velocity profile is the so-called incompressible shape factor H_i defined by the displacement-thickness/momentum-thickness ratio equation (see Chapter 2, section 2.3.1). Typical values for unseparated turbulent boundary layers range from 1.0 in the extremely large Reynolds-number limit $\varepsilon_T \to 0$ up to 1.3–1.6 at ordinary Reynolds numbers $10^5 \leq Re_L \leq 10^8$, as compared to the much larger value of 2.6 pertaining to laminar flat-plate boundary layers. When illustrated in terms of H_i (see Fig. 2.14), the much greater fullness of the turbulent boundary-layer profile relative to the laminar case is evident. It also is shown that even a small decrease in H_i implies a significant reduction in the velocity defect or "filling out" of a turbulent profile. Thus, it is not surprising that H_i was found to have a significant effect on turbulent-boundary interactions at practical Reynolds numbers [14].

The profiles shown in Fig. 2.12 display another important feature that governs the interactive response of the incoming boundary layer: the sonic height within the profile. Whereas this height occurs near the outer edge of a laminar boundary layer, it lies deep within a turbulent layer because of its much fuller velocity profile (i.e., typical values of y_{sonic}/δ_0 are less than 0.01; see Fig. 2.16). Because a shock wave exists only in the supersonic part of the boundary layer, the resolution of the

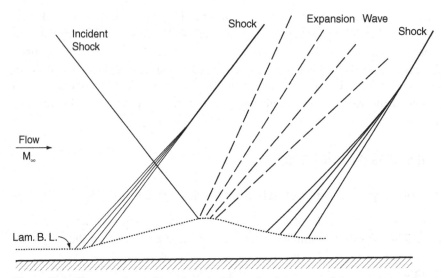

Figure 10.1. Flow configuration of an incident shock-generated interaction with a laminar boundary layer.

detailed shock diffraction in short-ranged turbulent boundary-layer interactions is a more important issue than in the case of well-spread-out laminar interactions (see the following section).

10.2.2 General Scenario of a Nonseparating SBLI

The presence of a boundary layer on a surface causes an incident-oblique shock wave to reflect as a complicated, spread-out wave system [15] rather than as a simple "inviscid" shock. Likewise, the incoming boundary layer causes the compressive disturbance field generated by a wall-mounted ramp to involve a dispersed wave system rather than a single outgoing corner shock. The features of both interactions have much in common [16], and each is dramatically affected by whether the boundary layer is laminar or turbulent due to their different streamwise scales, sonic heights, and shape factors.

10.2.2.1 Incident-Oblique Shock

The typical flow pattern observed when a weak oblique shock impinges on a laminar wall boundary layer (Fig. 10.1) consists of the following three main features:

1. A sizeable upstream-influence region $\ell_u \sim (10\text{–}20)\delta_0$ involving a streamwise-dispersed fan of weak outward-running compression waves ahead of the incident shock that appears to emanate from the outer part of the boundary layer
2. The generation of an outgoing expansion fan in the immediate vicinity of the shock impingement on the boundary-layer outer edge
3. A subsequent downstream region of dispersed, outgoing compression waves that ultimately overcome the preceding expansion and coalesce into a far-field final shock that appears to have been reflected from a point on the surface downstream of the incident-shock wall-impingement point

Figure 10.2. Flow configuration of an incident ramp-generated interaction with a laminar boundary layer.

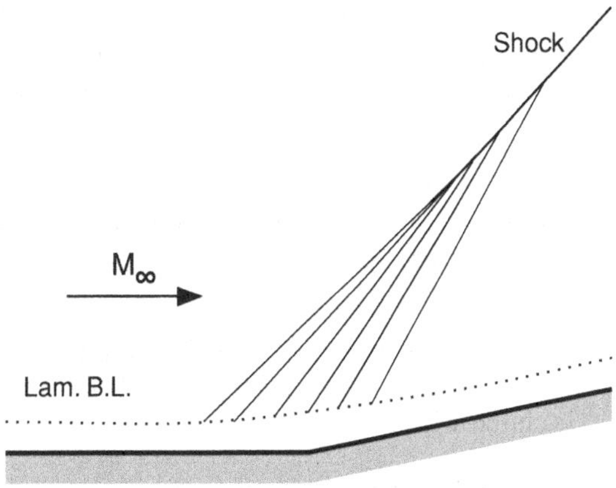

The prominent expansion feature (2), which effectively eliminates the external shock jump at the boundary-layer edge, is due to the previously mentioned fact that the sonic line in a laminar boundary-layer profile occurs near the outer edge, thereby causing the incident shock to locally reflect as if from a constant-pressure surface (i.e., as an expansion). The combined effect of all of these features from the standpoint of a theoretical model is that the wave system of a laminar interaction is dispersed so smoothly both streamwise and laterally that the resolution of the incident-shock–diffraction process across the thin, upper, supersonic region near the boundary-layer edge is of negligible importance.

Examination of the typical pattern of a nonseparating turbulent interaction, conversely, discloses a different physical situation in two respects (see Figs. 2.22 and 2.23): (1) the upstream-influence region is much shorter ($\ell_u \sim \delta_0$), the outgoing curved compression waves emanating from deep within the boundary layer; and (2) the incident shock penetrates far down into the boundary layer because of the deeply buried sonic line, with the resulting dispersed expansion reflections from this sonic line exiting the boundary layer to externally interact with the upstream-generated compression waves of (1). In contrast to the laminar case, these features – with the much tighter streamwise focus of the interaction – make the resolution of the shock diffraction within the boundary layer a more prominent aspect of the theoretical problem.

10.2.2.2 Compression Corner

The upstream disturbance pattern resulting from the encounter of a laminar-incoming boundary layer with a weak corner-generated shock is similar to the previously mentioned incident-shock case: a sizeable upstream-influence region of $\ell_u \gg \delta_0$ with outgoing compression waves emanating from the outer edge of the boundary layer (Fig. 10.2). Downstream of the corner, additional compression waves coalesce into an emergent oblique shock approaching the inviscid shock angle associated with the ramp deflection and appearing to originate upstream of the corner. As in the incident-shock case discussed previously, the overall wave system is thereby so gradually dispersed that the details of the emerging shock structure inside the thin supersonic zone near the boundary-layer edge are of negligible interest.

The upstream-interaction region for a turbulent boundary layer encountering a compression corner (see Fig. 2.29) is similar to that caused by an incident shock: a very short-ranged ($\ell_u \sim \delta_0$) upstream-influence zone with outgoing compression waves emanating from the deeply buried sonic line within the boundary layer. Downstream, the flow consists of an interacting wave system in the rotational flow above the local sonic line involving reflections from the curved corner shock that forms within the boundary layer. The resulting emergent external wave system coalesces into a single oblique shock that appears to have originated upstream of the corner. These events influence much of the boundary layer over such a short streamwise scale such that they dominate the determination of the pressure field right behind the corner and require a finely scaled ($\Delta x \ll \delta_0$) analytical treatment. Again, this behavior is in stark contrast to the more dispersed disturbance field associated with laminar flows.

10.2.3 Basic Structure of the Interaction Zone

10.2.3.1 Triple Deck: General Features

When a fully developed high-Reynolds-number boundary layer encounters an abrupt change in outer conditions (e.g., an impinging external shock) or in wall-boundary conditions (e.g., a compression corner), it was rigorously established by asymptotic analyses of the full Navier-Stokes equations for both laminar [3] and turbulent [17, 18] flow that the resulting local disturbance field astride the change organizes into a vertically layered structure, or "triple deck" (see Fig 2.18). These decks consist of the following:

1. An outer region of inviscid flow above the boundary layer, which contains the inviscid shock and interactive-wave systems
2. An intermediate deck of negligible shear-stress-perturbation rotational inviscid disturbance flow occupying the outer 90 percent or more of the incoming boundary-layer thickness
3. An inner shear-stress disturbance sublayer consisting of both laminar and turbulent eddy (i.e., Reynolds) stress contributions, which contains the interactive skin-friction perturbations (and, hence, any possible incipient separation) and the origin of the upstream influence of the interaction

The inviscid shock associated with the "forcing function" of the problem is impressed by the outer deck on the boundary layer. The middle deck couples this to the response of the inner deck but can modify the disturbance field to some extent, whereas the slow flow in the thin inner deck reacts strongly to the pressure-gradient disturbances imposed by these overlying decks. Qualitatively, this triple-deck structure also was established for nonasymptotic Reynolds numbers $Re_\ell \sim 10^5\text{--}10^8$ by Gadd et al. [19], Lighthill [2], and Honda [20] and is supported by a large body of experimental evidence and numerical studies with the full Navier-Stokes equations [21, 22]

Triple-deck theory consists of applying a systematic analysis of the disturbance flow in each deck (to the leading approximation in an appropriate basic small parameter) and then matching them to form a mutually interactive, self-consistent set of equations explicitly linked to the properly scaled forcing function involved. The solution to such a formulation then provides a description of important interaction properties such as the upstream influence associated with free-interaction behavior; the rapid streamwise variation of pressure, skin-friction, and heating across the interaction zone (including possible incipient separation); and the downstream "wake" behavior. Considering the significant differences between laminar and turbulent behavior, the general triple-deck concept forms the framework of this discussion.

Due to the features outlined in Section 10.2.1, an incoming turbulent boundary layer encountering a shock wave proves far more resistant to interactive perturbation than a laminar layer. Thus, as indicated by Lighthill's pioneering theoretical studies [2] that have since been corroborated by experiments [19, 20, 23, 24, 25, 26, 27] and asymptotic analyses [17, 18, 28, 29, 30, 31, 32, 33], a turbulent interaction zone exhibits (1) a nearly inviscid-like behavior (especially upstream) dominated by the defect portion of the velocity profile; (2) an extremely small vertical-velocity displacement effect of the inner-deck region compared with the large viscosity-dominated displacement effect of a laminar inner deck; (3) a much smaller upstream influence (and, indeed, overall interaction-zone length) that is only weakly affected by viscosity; and (4) a larger magnitude of the forcing function (e.g., incident shock strength) required to provoke local incipient separation. Thus, the detailed "content" of the triple-deck structure within turbulent boundary layers differs significantly from the laminar case. The typical vertical and streamwise scales of triple-deck structures for laminar and turbulent interactions revealed by analysis (see Section 10.3.6) are summarized in Fig. 10.3 to illustrate their small size and to further emphasize the significant difference between laminar and turbulent flow.

10.2.3.2 Further Local Subdivisions

The vertical triple-deck organization suffices to resolve all of the significant physics of a laminar SBLI due to the well-dispersed disturbance field (see Section 10.2.2). Conversely, because turbulent interactions present a more tightly focused problem involving resolution of the diffracted shock deep in the boundary layer, analysis of the interaction pressure and skin friction in the immediate neighborhood of the shock near the wall requires the study of events on a local streamwise scale even smaller than δ_0. Indeed, asymptotic analysis [28, 31, 34] indicates that such localized subscaling must be at least on the order of $\varepsilon_\tau^{1/2}$ times smaller than δ_0 and y_{sonic} in x and y, respectively (Fig. 10.4); in fact, this may require even finer nested subscales to entirely resolve certain singularities near the sonic line as $x \to 0^+$. Although the elucidation of such finely scaled features is fundamentally important, this discussion treats such matters only briefly and we refer readers to more details available in the cited literature.

10.3 Detailed Analytical Features of the Triple Deck

10.3.1 Middle Deck

10.3.1.1 General Aspects

The middle deck is the main deck of the triple-deck structure. It comprises negligible viscous and turbulent (i.e., Reynolds) stress disturbances – that is, the stresses being "frozen" along the streamlines at the undisturbed upstream values. This is

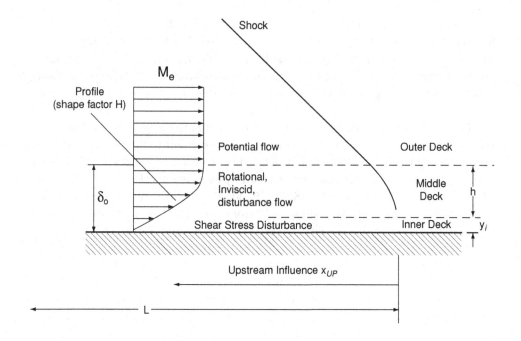

Asymptotic theory values
for supersonic flow

	LAMINAR	TURBULENT
h/L	δ_0/L	δ_0/L
x_{UP}/L	$(\delta_0/L)^{3/4}$	$(\delta_0/L)^{3/2}$
y_i/L	$(\delta_0/L)^{5/4}$	$(\delta_0/L)^2$
δ_0/L	$Re_L^{-1/2}$	Cf_0

Figure 10.3. The basic triple-deck structure of an interaction zone.

a good approximation for high-Reynolds-number unseparated flows because of the short extent of the shock-boundary interaction zone [35, 36, 37]. (If, however, separation occurs with the attendant lengthening of the zone and modification of the turbulent-eddy stress relationships, this approximation becomes questionable.)

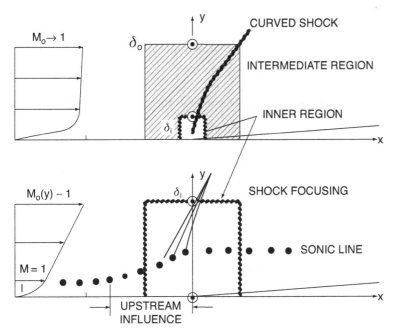

Figure 10.4. Further smaller-scale regions needed to resolve the local shock structure in a turbulent interaction. (Enlarged view of inner region shown in lower figure.)

Thus, the governing equations of the middle deck may be considered as those of a particle-isentropic rotational inviscid flow.

Imagine that the flow consists of disturbances proportional to an appropriate small parameter ε (which may be Reynolds-number dependent) about the undisturbed state of the incoming boundary layer (denoted by subscript zero) and so we write the following:

$$u = U_0(x, y) + \varepsilon u'(x, y) \tag{10.3.1}$$

$$v = v_0(x, y) + \varepsilon v'(x, y) \tag{10.3.2}$$

$$p = p_0(x) + \varepsilon p'(x, y) \tag{10.3.3}$$

$$\rho = \rho_0(x, y) + \varepsilon \rho'(x, y) \tag{10.3.4}$$

$$T = T_0(x, y) + \varepsilon T'(x, y) \tag{10.3.5}$$

$$H = H_0(y) + \varepsilon H'(x, y) \tag{10.3.6}$$

where $H \equiv C_p T + u^2$ is the total-enthalpy variable with $C_p = \gamma R/(\gamma - 1)$. Then, if we further neglect the streamwise variations of the basic flow over the short extent of the interaction zone and treat it as an isobaric parallel shear flow, for which $U_0 = U_0(y)$, $\rho_0 = \rho_0(y)$, $T_0 = T_0(y)$, $p_0 = \text{const.} = p_{0e}$ with $v_0 = 0$, then the general disturbance field is governed to leading order in ε by the following rotational inviscid perturbation flow equations:

$$\rho_0 \frac{\partial u'}{\partial x} + \frac{\partial \rho'}{\partial x} U_0 + \frac{\partial}{\partial y}(\rho_0 v') = 0 \quad <\text{continuity}> \tag{10.3.7}$$

$$U_0 \frac{\partial u'}{\partial x} + v' \frac{\partial U_0}{\partial y} \simeq -\frac{1}{\rho_0} \frac{\partial p'}{\partial x} \quad <\text{x-momentum}> \tag{10.3.8}$$

$$U_0 \frac{\partial v'}{\partial x} \simeq -\frac{1}{\rho_0} \frac{\partial p'}{\partial y} \quad <\text{normal momentum}> \tag{10.3.9}$$

$$p'/p_{0e} \simeq \frac{\rho'}{\rho_0} + \frac{T'}{T_0} \quad <\text{equation of state}> \tag{10.3.10}$$

$$U_0 \frac{\partial H'}{\partial x} + v' \frac{dH_0}{dy} \simeq 0 \quad <\text{steady adiabatic energy}> \tag{10.3.11}$$

where $H' = (\gamma RT')/(\gamma - 1) + U_0 u'$. An important immediate consequence of these equations can be realized by combining equations (10.3.7), (10.3.8), (10.3.10), and (10.3.11) to obtain the following vertical-velocity relationship:

$$\frac{\partial (v'/U_0)}{\partial y} \simeq \frac{(1 - M_0^2)}{\gamma M_0^2} \frac{\partial (p'/p_{0e})}{\partial x} \tag{10.3.12}$$

where $M_0^2(y) = \rho_0(y) U_0^2(y)/\gamma p_{0e}$ is the undisturbed Mach-number distribution across the incoming layer. Integrating equation (10.3.12) across the boundary layer from the effective edge $y = y_i$ of the inner deck to the edge of the incoming boundary layer $y = \delta_0$, and assuming $\partial p'/\partial y \simeq 0$, we obtain:

$$\frac{v'_e}{U_{0e}} = \left(\frac{v'}{U_0}\right)(y_i) + \frac{\partial p'/\partial x}{\gamma p_{0e}} \int_{y_i}^{\delta_0} \left[\frac{1 - M_0^2}{M_0^2}\right] dy \tag{10.3.13}$$

which describes the streamline slope change across the middle deck (the so-called streamline-divergence effect). Equation (10.3.13) is important because it connects the streamline deflection of the outer deck to the vertical displacement of the inner deck. At this point, we might further integrate the normal-momentum equation (10.3.9) to obtain the lateral pressure variation across the deck, as follows:

$$\frac{p'_e - p'_w}{\rho_e U_{0e}} \simeq \int_{y_i}^{\delta_0} \left(\frac{M_0}{M_{0e}}\right)^2 \frac{\partial (v'/U_0)}{\partial x} dy \tag{10.3.14}$$

which is seen as very small except where the streamwise (x) scale of the interaction is much less than the already-small lateral y-scale (δ_0). Thus, the influence of this variation compared to that of the rapid streamline pressure changes along the interaction can be neglected unless the finer details of the wave structure within the boundary layer are of interest [32]. Hereafter, we regard p' as a function of x only.

Equation (10.3.13) combined with the streamwise-momentum equation (10.3.8) – as well as the energy equation if the heat-transfer aspects are of interest – constitute the key relationships needed to link the bottom of the middle deck (i.e., the top of the inner deck) to the outer-deck disturbance flow. Further insight into the behavior near the body now requires a more detailed consideration of the velocity profile and the scaling properties of the flow, both of which depend strongly on whether the flow is laminar or turbulent.

10.3.1.2 Purely Laminar Flows

In the case of purely laminar flows in which the appropriate perturbation parameter is $\varepsilon_L = Re_L^{-1/8}$, it was firmly established [3, 14, 30, 38] that the entire length of the interaction zone is dominated by the vertical-displacement effect of the inner deck as represented by the first right-hand side term in equation (10.3.13), with the middle deck consisting of nearly parallel disturbance streamlines. Consequently, the appropriate formulation of the momentum and energy equations (10.3.8) and (10.3.11) near the body surface $y \to y_w$ is to substitute the value of $v_i'/U_0(y_i)$ given by equations (10.3.13) and obtain the following:

$$\lim_{y \to y_w} \left(\frac{\partial u'}{\partial x} \right) = - \left(\frac{v_e'}{U_{0e}} \right) \left[\frac{dU_0}{dy} \right]_{y \to y_w} + \frac{dp'/dx}{\gamma\, p_{0e}} \left\{ \left[\frac{dU_0}{dy} \right] I_m(y) - \frac{U_0(y)}{M_0^2(y)} \right\}_{y \to y_w}$$

(10.3.15)

$$\lim_{y \to y_w} \left(\frac{\partial H'}{\partial x} \right) = - \left(\frac{v_e'}{U_{0e}} \right) \left[\frac{dH_0}{dy} \right]_{y \to y_w} + \frac{dp'/dx}{\gamma\, p_{0e}} \left\{ I_m(y) \left[\frac{dH_0}{dy} \right] \right\}_{y \to y_w} \qquad (10.3.16)$$

where $\rho_0 U_0 = \gamma\, p_{0e} M_0^2 / U_0$ is used and where:

$$I_m(y) \equiv \int_y^{\delta_0} \left[(1 - M_0^2)/M_0^2 \right] dy \qquad (10.3.17)$$

is an important boundary-layer Mach-number-profile integral in interaction theory. With proper handling of the indicated limit $y \to y_w$ on the right side, equations (10.3.15) and (10.3.16) provide the outer boundary conditions on the inner-deck solution and therefore a link to the outer deck, as shown herein. In this connection, the dp'/dx terms in these equations, which represent the middle-deck streamline-divergence effect (see equation 10.3.13), turn out to be negligible except for strongly hypersonic external-flow conditions.

10.3.1.3 Turbulent Flows at Large Reynolds Numbers

At very high Reynolds numbers in which the appropriate perturbation parameter is $\varepsilon_T = U_\tau / U_{0e}$, the turbulent boundary layer becomes far less interactive vertically and – unlike the laminar case – yields dominant terms in the middle-deck formulation that are significantly different downstream of the incident shock compared to those upstream. Upstream, it has been shown [17, 18, 28–33] that the vertical-velocity disturbance produced by the inner deck is so small ($v_i'/U_{0e} \sim \varepsilon_T^2$, in fact) that it has a negligible role in the interaction. Thus, in the leading approximation for $\varepsilon_T \to 0$, equations (10.3.13) and (10.3.17) reduce to the following interaction equation:

$$\frac{v_e'}{U_{0e}} \simeq \frac{dp'/dx}{\gamma\, p_{0e}} I_m(y_i) \qquad (10.3.18)$$

from which the pressure disturbance $p'(x)$ can be found once the inner-deck height and a pressure deflection-angle relationship for the outer deck are specified (discussed herein). Furthermore, because $U_{0e} - U_0$, $\rho_{0e} - \rho_0$, dU_0/dy and dH_0/dy in turbulent flow are each proportional to ε_T in the defect region occupied by the middle deck [12], the leading inner approximations of equations (10.3.8) and (10.3.11) to first order in ε_T become:

$$\lim_{y \to y_w} \left(\frac{\partial u'}{\partial x} \right) \simeq -\frac{1}{(\rho_{0e} + \cdots)} \frac{dp'/dx}{(U_{0e} + \cdots)} \tag{10.3.19}$$

$$\lim_{y \to y_w} \left(\frac{\partial H'}{\partial x} \right) \simeq 0 \tag{10.3.20}$$

because the terms involving v_i' are negligibly small. These equations may be recognized as those governing adiabatic inviscid small-perturbation flows in a parallel stream.

The turbulent middle deck downstream of the shock is significantly different because the interaction there must be dominated by the imposed forcing function (e.g., an incident-shock pressure jump) that is necessarily included in the v_e'/U_{0e} term. Consequently, although the vertical-velocity effect is still small, it cannot be dismissed; rather, it must be retained in the momentum equation (10.3.8) by again using equations (10.3.15) and (10.3.16) but now with high-Reynolds-number turbulent values for the incoming boundary-layer properties. Thus, we have the inner conditions for $x > 0$ that:

$$\lim_{y \to y_w} \left(\frac{\partial u'}{\partial x} \right) = -\frac{dp'/dx}{\rho_{0e} U_{0e}} - \left(\frac{v_e'}{U_{0e}} \right) \lim_{y \to y_w} \left(\frac{dU_0}{dy} \right) + \frac{dp'/dx}{\gamma \, p_{0e}} \lim_{y \to y_w} \left(I_m \frac{dU_0}{dy} \right) \tag{10.3.21}$$

$$\lim_{y \to y_w} \left(\frac{\partial H'}{\partial x} \right) = -\left(\frac{v_e'}{U_{0e}} \right) \lim_{y \to y_w} \left(\frac{dH_0}{dy} \right) + \frac{dp'/dx}{\gamma \, p_{0e}} \lim_{y \to y_w} \left(I_m \frac{dH_0}{dy} \right) \tag{10.3.22}$$

where the first two terms on the right-hand side of equation (10.3.21) are dominant. This downstream situation significantly interacts with the inner deck, unlike the disturbance flow upstream. If the incoming boundary layer is nonadiabatic, so is this downstream interaction.

10.3.2 Inner Deck

10.3.2.1 General Aspects

The thin inner deck is dominated by the combined effects of pressure-gradient and total shear-stress (as well as heat-flux) disturbances under the influence of the no-slip condition along the wall. The flow within has a boundary-layer–like character in that lateral pressure-gradient and streamwise-diffusion effects are negligible. Whereas in purely laminar flow, the shear stress is due entirely to molecular viscosity, the composite Law of the Wall/Law of the Wake structure of an incoming turbulent boundary layer [11, 12] (see Fig. 2.13) dictates that its disturbance stress field is the sum of both a molecular- and an eddy-viscosity (i.e., Reynolds stress) components; thus, the inner deck is two-layered, involving laminar stresses in an extremely thin sublayer near the wall overlaid by a Reynolds stress (or "blending") layer.

In formulating the governing flow equations, we assume that the inner deck lies within the Law of the Wall region of the incoming turbulent boundary layer (which appreciably overlaps the inner part of the outer-defect region) and treat the turbulent Reynolds stress and heat transfer therein by the well-known Van Driest eddy-viscosity model appropriate for nonseparating flows [12]. Furthermore, we neglect the effect of streamwise variations of the undisturbed flow over the short extent of nonseparating interactions, consistent with the same approximation in the main-deck analysis.

To allow for the presence of a possible local wall-shape distortion $y_w(x)$ relative to the incoming upstream flow, such as a "bump" or compression corner, it is convenient to formulate the inner-deck equations in terms of the following shifted variables:

$$z \equiv y - y_w \tag{10.3.23}$$

$$w \equiv v - u \cdot (dy_w/dx) \tag{10.3.24}$$

which are the vertical distance relative to the distorted surface and the vertical velocity relative to that associated with an inviscid streamline along y_w, respectively. Application of the Prandtl Transposition Theorem [39] shows that the resulting continuity, momentum, and energy equations are invariant under such a transformation. In addition, it is convenient to absorb much of the compressibility effect in the problem by further introducing the following Howarth-Dorodnitzn transformations [40]:

$$\left. \begin{aligned} &U(X, Z) \equiv u(x, z) \\ &W(X, Z) \equiv U\frac{\partial Z}{\partial X} + \rho w(x, z)/\rho_{0w} \\ &dZ \equiv \rho dz/\rho_{0w} \\ &X = x \end{aligned} \right\} \tag{10.3.25}$$

which render the continuity equation into the following incompressible form:

$$\frac{\partial U}{\partial X} + \frac{\partial W}{\partial Z} = 0 \tag{10.3.26}$$

When the Chapman-Rubensin [40] approximation $\rho\mu = \text{const.} = \rho_{0w}\mu_{0w}$ is further applied to the laminar portion of the shear stress, the compressible momentum and energy equations resulting from the transformations (i.e., equations [10.3.23] and [10.3.25]) assume the following forms when expressed in terms of the perturbations $u' = u - U_0(y)$ and $H' = H - H_0(y)$:

$$(U_0 + u')\frac{\partial u'}{\partial X} + W'\left(\frac{dU_0}{dZ} + \frac{\partial u'}{\partial Z}\right) + \frac{1}{\rho_0}\frac{dp'}{dX}$$
$$= \frac{\partial}{\partial Z}\left[v_{0w}\frac{\partial u'}{\partial Z} + 2\left(\frac{\rho_0}{\rho_{0w}}\right)\frac{\mu_{T_0}}{\rho_{0w}}\left(\frac{\partial u'}{\partial Z}\right)\right] \tag{10.3.27}$$

$$(U_0 + u')\frac{\partial H'}{\partial X} + W'\left(\frac{dH_0}{dZ} + \frac{\partial H'}{\partial Z}\right)$$
$$\simeq \frac{\partial}{\partial Z}\left[\frac{v_{0w}}{P_R}\frac{\partial H'}{\partial Z} + \frac{\mu_{T_0}}{P_R}\frac{\rho_0}{\rho_{0w}}\frac{\mu_{T_0}}{\rho_{0w}}\left(\frac{\partial H'}{\partial Z} + \frac{dH_0}{dU_0}\frac{\partial u'}{\partial Z}\right)\right] \tag{10.3.28}$$

where the undisturbed flow was subtracted and the small effect of a term $(1 - P_R)/\partial/\partial y[\partial(u^2/2)/\partial y]$ was neglected on the right-hand side of equation (10.3.28). The undisturbed background flow is governed by the Law of the Wall relationships [12]:

$$\left[v_{0w} + \frac{\rho_0}{\rho_{0w}} \left(\frac{\mu_{T_0}}{\rho_{0w}} \right) \right] \frac{dU_0}{dZ} = \frac{\tau_{0w}}{\rho_{0w}} \equiv U_\tau^2 \tag{10.3.29}$$

$$\left[\frac{v_{0w}}{P_R} + \frac{1}{P_{RT}} \frac{\rho_0}{\rho_{0w}} \left(\frac{\mu_{T_0}}{\rho_{0w}} \right) \right] \frac{dH_0}{dZ} = \frac{-\dot{q}_{w0}}{\rho_{0w}} \tag{10.3.30}$$

whereas the perturbed- and basic-flow eddy viscosities, respectively, are:

$$\varepsilon_T' \simeq \left[\frac{\partial u'}{\partial z} \Big/ (dU_0/dZ) \right] \mu_{T_0} \tag{10.3.31a}$$

and

$$\frac{\mu_{T_0}}{\rho_{0w}} = \left[kD\frac{\rho_0}{\rho_{0w}} \left(\frac{z}{Z} \right) Z \right] \frac{dU_0}{dZ} \tag{10.3.31b}$$

with D, the Van Driest damping factor:

$$D \equiv 1 - \exp\left(-\mu_{T_0} Z/2v_{w0}\right) \tag{10.3.32}$$

These equations are solved subject to either the no-slip conditions that W', u', and H' vanish along the wall $Z = 0$ for all X or equivalent conditions outside the laminar sublayer (discussed herein), as well as outer boundary conditions on u' and H' as $Z \to \infty$ that enforce matching with the previously mentioned inner behavior of the overlying middle deck. In conjunction with the pressure gradient in momentum equations (10.3) through (10.27), near the surface, the variable-density coefficient $\rho_0^{-1} = RT_0/p_0 \simeq (T_0/T_w)\rho_{0w}^{-1}$ can be expressed as the following Taylor series expansion in the distance Z across the inner deck after noting that $U_{0_w} = 0$:

$$\frac{\rho_{0w}}{\rho_0} = \frac{\left[T_{0w} + \left(\frac{\partial T_0}{\partial Z} \right)_w Z + \cdots \right]}{T_{0w}} \simeq \left[1 + \frac{1}{H_{0w}} \left(\frac{\partial H_0}{\partial Z} \right)_w Z \right] \tag{10.3.33}$$

This relationship indicates that ρ_0 may be taken as a constant across the thin inner deck when the incoming boundary layer is adiabatic or when $(\partial H_0/\partial Z)_w Z/H_{0w}$ is negligible compared to unity.

Although we know at least the vertical scale of the middle deck ($y_m \sim \delta_0$), at this point we have no a priori knowledge of the scales of the inner deck and the disturbance flow. To determine them, the "hated" nondimensional variables $\hat{x} \equiv X/X_R$, $\hat{z} \equiv Z/Z_R$, $\hat{u} \equiv u/u_R$, $\hat{w} \equiv W'/W_R$, $\hat{\rho} \equiv \rho/\rho_R$, $\hat{p} \equiv p'/\rho_{0e}U_{0e}2\pi_R$ are introduced with $\hat{H} \equiv H/H_R$, where the R-subscripted reference parameters are scaling factors to be found. These variables are substituted into the previous governing equations and they require that the U_R, W_R, and so on be chosen such that the resulting nondimensional terms are all of equal order-unity. Thus, for example, the continuity equation (10.3.26) reduces to:

$$\partial\hat{u}/\partial\hat{x} + \partial\hat{w}/\partial\hat{z} = 0 \tag{10.3.34}$$

if we set:

$$W_R/U_R = Z_R/x_R \tag{10.3.35}$$

whatever the value of the individual parameters. Addressed next are the momentum and energy equations and a detailed turbulence model is introduced.

10.3.2.2 Laminar Flows

Here, u' and U_0 are expected to be of the same order near the surface; therefore, we retain the full nonlinear convective-acceleration effect. Then, introducing the previously mentioned scaling factors into equation (10.3.27) and requiring that the pressure gradient and shear terms each be of the same order-unity as the convective-acceleration terms, we obtain the following two scaling relationships:

$$\pi_R = (\rho_{0w}/\rho_e)\,(U_R/U_{0e})^2 \tag{10.3.36}$$

$$v_{0w}\,X_R = U_R\,Z_R^2 \tag{10.3.37}$$

and the simplified nondimensional laminar–turbulent inner-deck momentum equation:

$$(\hat{U}_0 + \hat{u}')\frac{\partial \hat{u}'}{\partial \hat{x}} + \hat{w}'\left(\frac{d\hat{U}_0}{d\hat{z}} + \frac{\partial \hat{u}'}{\partial \hat{z}}\right) + \frac{\rho_{0w}}{\rho_0}\frac{d\hat{p}}{d\hat{x}} = \frac{\partial}{\partial \hat{z}}\left(\frac{\partial \hat{u}'}{\partial \hat{z}}\right) \tag{10.3.38}$$

The corresponding treatment of energy equation (10.3.28) yields:

$$(\hat{U}_0 + \hat{u}')\frac{\partial \hat{H}'}{\partial \hat{x}} + \hat{w}'\left(\frac{d\hat{H}_0}{d\hat{z}} + \frac{\partial H'}{\partial \hat{z}}\right) = \frac{\partial}{\partial \hat{z}}\left(P_r^{-1}\frac{\partial \hat{H}'}{\partial \hat{x}}\right) \tag{10.3.39}$$

where $\hat{H}_0 \equiv H_0/H_R$ and the evaluation of $d\hat{H}_0/d\hat{U}_0$ (i.e., dividing equation [10.3.30] by equation [10.3.29]) indicates that it is rendered of order unity if H_R is chosen as follows:

$$H_R = (-\dot{q}_{w_0} U_R P_R)/\rho_{0w} u_{T_0}^2. \tag{10.3.40}$$

Equations (10.3.38) and (10.3.39) are to be solved subject to the wall no-slip conditions $u_w = 0$, $T = T_w$ along the body $\hat{z} = 0$ as well as the outer boundary-conditions provided by the appropriate inner-middle-deck matching relationships.

10.3.2.3 Turbulent Flows

At very high Reynolds numbers, the laminar sublayer of the incoming turbulent boundary layer becomes so thin (exponentially thin, in fact) that the inner deck may be regarded as fully turbulent virtually to the wall; it is then appropriate to use an asymptotic approach with $U_\tau/U_{0e} = \varepsilon_T$ as the small parameter. Furthermore, the streamwise velocity is regarded as a small perturbation of order ε_τ on U_{0e} within the defect part of the incoming boundary-layer region, so that $\rho u\,\partial u'/\partial x \simeq \rho_{0e} U_{0e}\partial u'/\partial x$ to the first order. Then, noting that $D = 1$ and $dU_0/dy \simeq kU_\tau/z$ for such fully turbulent flow, introducing scaled variables again and consistently taking $\rho_{0w} = \rho_{0e}$ in the pressure-gradient term, equation (10.3.27) yields the following nondimensional momentum equation governing the leading approximation to the fully turbulent inner-deck flow, where $\hat{z} \approx 0(1)$:

$$\frac{\partial \hat{u}'}{\partial \hat{x}} + \pi_R \frac{U_{0e}}{U_R} \frac{d\hat{p}}{d\hat{x}} \simeq \frac{2k^2 X_R U_R}{Z_R U_{0e}} \frac{\partial}{\partial \hat{z}} \left\{ [f_D \hat{z}]^2 \frac{d\hat{U}_0}{d\hat{z}} \frac{\partial \hat{u}'}{\partial \hat{z}} \right\} \tag{10.3.41a}$$

$$f_D \equiv (\rho_0/\rho_{0w})^{3/2} z/Z \tag{10.3.41b}$$

where a term $(\hat{w}/U_{0e}) dU_0/d\hat{z}$ was dropped compared with $\partial \hat{u}'/\partial \hat{x}$ because it was shown [28, 29, 30, 31, 32] to be of negligible second order. Likewise, equation (10.3.28) yields the following companion first-order energy equation:

$$\frac{\partial \hat{H}}{\partial \hat{x}} \cong \frac{k^2 X_R U_R}{P_{r_t} Z_R U_{0e}} \frac{\partial}{\partial \hat{z}} \left\{ [f_D \hat{z}]^2 \left[\frac{\partial \hat{H}'}{\partial \hat{z}} + \left(\frac{d\hat{H}_0}{d\hat{U}_0} \right) \frac{\partial \hat{u}'}{\partial \hat{z}} \right] \right\} \tag{10.3.42}$$

The undisturbed flow equations (10.3.29) through (10.3.32) with $D = 1$ and the laminar-viscosity term neglected indicate that the proper velocity scale in this asymptotic limit case is as follows:

$$U_R = U_\tau = \varepsilon_\tau U_{0e} \tag{10.3.43}$$

rendering the factors $d\hat{U}_0/d\hat{z}$ and $d\hat{H}_0/d\hat{U}_0$ of order unity as given by the relationships:

$$\frac{d\hat{U}_0}{d\hat{z}} = [k f_D \hat{z}]^{-1} \tag{10.3.44}$$

$$\frac{d\hat{H}_0}{d\hat{U}_0} = \frac{-\dot{q}_{w_0} P_{r_t} U_R}{\rho_{0w} u_{\tau_0}^2 H_R} \tag{10.3.45}$$

Inspection of equations (10.3.41) through (10.3.45) shows that the proper choice of asymptotic-scaling relationships is as follows:

$$\pi_R = \varepsilon_T \tag{10.3.46}$$

$$Z_R = k \varepsilon_T X_R \tag{10.3.47}$$

$$H_R = -\dot{q}_{w_0} P_{r_t} / \rho_{0w} \varepsilon_\tau U_{e0} \tag{10.3.48}$$

thereby giving $d\hat{H}_0/d\hat{U}_0 = 1$ and reducing equations (10.3.41a) and (10.3.42), respectively, to their ultimate simplified form:

$$\frac{\partial \hat{u}'}{\partial \hat{x}} + \frac{d\hat{p}}{d\hat{x}} \simeq 2 \frac{\partial}{\partial \hat{z}} \left\{ [f_D \hat{z}]^2 \frac{\partial \hat{u}'}{\partial \hat{z}} \right\} \tag{10.3.49}$$

$$\frac{\partial \hat{H}}{\partial \hat{x}} \simeq P_{r_t}^{-1} \frac{\partial}{\partial \hat{z}} \left\{ [f_D \hat{z}]^2 \left[\frac{\partial \hat{H}'}{\partial \hat{z}} + \frac{\partial \hat{u}'}{\partial \hat{z}} \right] \right\} \tag{10.3.50}$$

The streamwise scaling $X_R = \delta_0 \sim \varepsilon_T \ell$ demonstrated herein for turbulent interactions implies from equation (10.3.47) a small inner-deck thickness scale $Z_R \sim \varepsilon_T \delta_0 \sim \varepsilon_T^2 \ell$.

    Because these equations completely neglect the laminar sublayer, their solution cannot be carried all the way to the wall; the usual wall no-slip conditions therefore must be replaced by appropriate inner slip conditions at some nonzero "cutoff" height z_c that falls within the inner part of the velocity-defect region and yet contains the integrated effect of the true wall no-slip condition across the laminar sublayer. The classical logarithmic Law of the Wall for turbulent boundary layers, which appreciably overlaps the defect region (see Fig. 2.13), in fact provides a good engineering account of these features over a wide range of conditions, even in the presence of moderately adverse pressure gradients if significant separation does not occur. Thus, we may apply it locally along the interaction both at the height $z_c \simeq 60v_{0w}/U_\tau$ just outside the laminar sublayer at the bottom of the defect region, and along the undisturbed boundary-layer edge, where dU_0/dy and dH_0/dy vanish. As shown in Appendix 10.A, this procedure yields a pair of local perturbation relationships linking the desired slip velocity along z_c and the corresponding skin-friction disturbance to the overlying interactive disturbance field. After introducing scaled variables, they are as follows:

$$\hat{u}'(\hat{x},\, \hat{z}_c) = -(B_i/B_0)\,\hat{p}(\hat{x}) \tag{10.3.51}$$

$$\tau_w'(\hat{x}) \cong -2\tau_{0w}\,\hat{p}\,B_0^{-1} \tag{10.3.52}$$

where B_i and B_0 are Law of the Wall parameters for the incoming boundary layer defined in Appendix A. An analogous procedure applied to the Temperature Law of the Wall yields a corresponding pair of relationships for the local total-enthalpy slip and perturbation heat transfer, respectively (Appendix A):

$$\hat{H}'(\hat{x},\, z_c')/C_p T_R = -(B_t - k_t^{-1})\hat{p}B_0^{-1} - B_i\,(\dot{q}_w'/\dot{q}_{w0}) \tag{10.3.53}$$

$$(\dot{q}_w)'/\dot{q}_{w0} = -\lambda_{q1}\hat{p}B_0^{-1} + \lambda_{q2}\varepsilon_\tau U_{0e}^2/C_p T_R \tag{10.3.54}$$

where $T_R = \dot{q}_{w0}/P_{rr}C_P u_{\tau_0}$ and the coefficients B_t, k_t, λ_{q1} and λ_{q2} also are defined in Appendix A. These slip relationships apply to all $\hat{x}$ along the interaction zone, including downstream.

10.3.3 Middle-Inner-Deck Matching

There are now four relationships in eight unknown scaling parameters; therefore, additional information is required, some of which is obtained by enforcing the matching of the inner and middle decks.

10.3.3.1 Laminar Flows
The outer behavior required of the inner-deck solution is linked to events along the bottom of the middle deck (see Section 10.3.1) by applying Van Dyke's [41] heuristic first-order matching principle as follows: The outer limit of the inner solution equals the inner limit of the outer (i.e., middle-deck) solution when both are expressed in the same coordinate, here taken as Z. Writing the following:

$$u(X, Z \to 0) = u_0(Z \to 0) + u'(X, Z)$$
$$\simeq (dU_0/dZ)_w \cdot Z + u'(X, Z) \tag{10.3.55}$$

$$H(X, Z \to 0) = H_0(Z \to 0) + H'(x, Z)$$
$$\simeq H_{0w} + (dH_0/dZ)_w \cdot Z + H'(x, Z) \tag{10.3.56}$$

and substituting the values of u' and H' obtained by streamwise integration of the middle-deck relations, (equations (10.3.15) and (10.3.16)) the principle yields the following outer-boundary conditions on the inner-deck solution:

$$\lim_{\hat{z} \to \infty} [\hat{u}_{INNER}(\hat{x}, \hat{z})] \simeq \lim_{\hat{z} \to 0} [\hat{u}_{MIDDLE}(\hat{x}, \hat{z})]$$

$$= \frac{Z_R \left(\frac{dU_0}{dZ}\right)_w}{U_R} \left\{ \hat{z} - \frac{\int_{-\infty}^{\hat{x}} (v'_e / U_{0e}) \, d\hat{x}}{Z_R / X_R} + \frac{\pi_R M_e^2 \hat{p}}{Z_R} \lim_{z \to 0} \left[I_m(z) - \frac{z}{M_0^2(z)} \right] \right\} \quad (10.3.57)$$

and

$$\lim_{\hat{z} \to \infty} \left[\hat{H}(\hat{x}, \hat{z})_{INNER} - \hat{H}_{0w} \right] = \lim_{\hat{z} \to 0} \left[\hat{H}_{MIDDLE} - \hat{H}_{0w} \right]$$

$$= \frac{Z_R \left(\frac{dH_0}{dZ}\right)_w}{H_R} \left\{ \hat{z} - \frac{\int_{-\infty}^{\hat{x}} (v'_e / U_{0e}) \, d\hat{x}}{Z_R / X_R} + \frac{\pi_R M_e^2 \hat{p}}{Z_R} \lim_{z \to 0} I_m(z) \right\} \quad (10.3.58)$$

Inspection reveals that these relationships optimally simplify if we scale $v'_e \sim (Z_R / X_R) U_{0e}$ as expected and also require that:

$$U_R \equiv Z_R (dU_0 / dZ)_w \quad (10.3.59)$$

$$H_R = U_R (dH_0 / dU_0)_w \quad (10.3.60)$$

where the wall-gradient values pertain to a laminar compressible flat-plate boundary layer. Because in laminar flow $(dH_0/dU_0)_w = -q_{w0} P_r / \rho_w U_\tau^2$, equation (10.3.60) exactly agrees with equation (10.3.40) and therefore adds no new information. Using the limit values for the last terms in equations (10.3.57) and (10.3.58) given in Appendix 10.B, these outer boundary conditions reduce to:

$$\lim_{\hat{z} \to \infty} [\hat{u}_{INNER}(\hat{x}, \hat{z})] \simeq \hat{z} - X_R Z_R^{-1} \int_{-\infty}^{\hat{x}} (v'_e / U_{0e}) \, d\hat{x}$$

$$+ \pi_R M_e^2 \hat{p} Z_R^{-1} \lim_{z \to 0} I_m (z_i / \delta_0) \quad (10.3.61)$$

$$\lim_{\hat{z} \to \infty} \left[\hat{H}_{INNER}(\hat{x}, \hat{z}) \right] \simeq \lim_{\hat{z} \to \infty} [\hat{u}_{INNER}(\hat{x}, \hat{z})] \quad (10.3.62)$$

10.3.3.2 Turbulent Flows

We apply the matching principle to the middle-deck relationships discussed in Section 10.3.1.3, which therefore gives different upstream and downstream results. When $x < 0$, introduction of the scaled variables, equations (10.3.43) and (10.3.46), into equations (10.3.19) and (10.3.20) and integration with respect to x, yields the following outer boundary conditions on the inner-deck solution in the following leading asymptotic approximation:

$$\lim_{\hat{z} \to \infty} [u'_{INNER}(\hat{x} < 0, \hat{z})] \simeq -\hat{p} \quad (10.3.63)$$

$$\lim_{\hat{z} \to \infty} [\hat{H}'_{INNER}(\hat{x} < 0, \hat{z})] \simeq 0 \quad (10.3.64)$$

Consideration of equation (10.3.18) in light of equation (10.3.46) suggests that the appropriate choice of streamwise scale is $X_R = \delta_0$; this yields:

$$(v'_e / U_{0e}) / \varepsilon_T = I_m (Z_i / \delta_0) M_e^2 (d\hat{p} / d\hat{x}) \quad (10.3.65)$$

which indicates that we should scale $v'_e \sim \varepsilon_T U_{0e}$ as indeed verified later herein.

Turning to the more interactive downstream region $x \geq 0$, introduction of scaled variables into equations (10.3.21) and (10.3.22), and integration with respect to x yields the following boundary conditions on using $dU_0/dZ \simeq U_\tau/k\,Z$:

$$\lim_{\hat{z}\to\infty}[u'_{\text{INNER}}(\hat{x} > 0,\, \hat{z})] \simeq -\hat{p} - \frac{1}{\hat{z}}\int_{0^+}^{\hat{x}} \left(\frac{v'_e/U_{0e}}{k^2\varepsilon_T}\right)d\hat{x} + k^{-2}M_{0e}^2\frac{\hat{p}}{\hat{z}}I_m(z_i/\delta_0) \tag{10.3.66}$$

$$\lim_{\hat{z}\to\infty}[\hat{H}'_{\text{INNER}}(\hat{x} > 0,\, \hat{Z})] \simeq -\frac{dH_0/dU_0}{(H_R/U_\tau)}\left\{\frac{1}{\hat{z}}\int_{0^+}^{\hat{x}}\left(\frac{v'_e/U_{0e}}{k^2\varepsilon_T}\right)d\hat{x} + \frac{M_{0e}^2\hat{p}}{k^2\hat{z}}I_m(z_i/\delta_0)\right\} \tag{10.3.67}$$

where the lower limit on the streamwise v'_e integral is reckoned from $x = 0^+$ to be consistent with the neglected effect of this integral in the upstream region. Equation (10.3.66) indicates that a scaling of $v'_e \sim \varepsilon_\tau U_{0e}$ is also appropriate for the downstream region, whereas equation (10.3.67) shows that the appropriate enthalpy scaling for the asymptotic turbulent case is:

$$H_R = U_R\,(dH_0/dU_0) = \varepsilon_T\,(dH_0/dU_0)_w\,U_{0e} \tag{10.3.68}$$

which, using equations (10.3.29) and (10.3.30), agrees exactly with the previous result in equation (10.3.48). Furthermore, as discussed in Section 10.2.1, we now bypass a detailed examination of the finer scales of the shock structure near $x \to 0^+$ by limiting attention to events on the scale of $x \sim \delta_0$; the previously mentioned boundary conditions then simplify to the following:

$$\lim_{\hat{z}\to\infty}[u'_{\text{INNER}}(\hat{x} > 0,\, \hat{z})] \simeq -\hat{p} - \frac{1}{\hat{z}}\int_{0^+}^{\hat{x}}[(v'_e/U_{0e})/k^2\varepsilon_\tau]d\hat{x} + \varepsilon_T M_{0e}^2 f(\hat{z})\hat{p} \tag{10.3.69}$$

$$\lim_{\hat{z}\to\infty}[\hat{H}'_{\text{INNER}}(\hat{x} > 0,\, \hat{Z})] \simeq \lim_{\hat{z}\to\infty}[u'_{\text{INNER}}(\hat{x} > 0,\, \hat{z})] + \hat{p} \tag{10.3.70}$$

where, in obtaining the last term of equation (10.3–70), we used $Z_R/\delta_0 = k\varepsilon_T$ from equation (10.3.47) and where:

$$f(\hat{z}) \equiv (k\hat{z})^{-1}\int_{\hat{z}_i}^{k\varepsilon_T^{-1}}\left\{\left[1 - M_0^2(\hat{z})\right]/M_0^2(\hat{z})\right\}d\hat{z} \tag{10.3.71}$$

is a modified Mach-number distribution function presumed to be of order unity. The streamwise v'_e integral on the right-hand side of equation (10.3.69) is an important aspect of the downstream problem because it contains the forcing function that drives the overall interaction (see the next section). Regarding the last term $\sim M_0^2\hat{p}$, as in the laminar case, it is shown to be of negligible higher-order importance unless the Mach number is strongly hypersonic.

10.3.4 Inviscid-Pressure–Flow Deflection Relationships for the Outer Deck

Completion of the triple-deck formulation and determination of the remaining scaling relationships requires consideration of the outer-deck behavior, to the extent of providing a relationship between the pressure-disturbance field and the streamline deflection v'_e/U_{0e} along the bottom of the upper deck. Although the specific nature of this relationship depends on the prevailing external Mach-number regime, in general, the pressure-perturbation field $p'(x)$ consists of the pressure jump Δp_s due to the imposed inviscid shock (whether from an externally imposed wave or due to a wall-compression corner), as well as a local viscous displacement-induced interactive component p_{INTER} as follows [42, 43]:

$$p'(X) = J\Delta p_s + p'_{\text{INTER}} \tag{10.3.72}$$

where $J = 0$ for $X < 0$ and $J = 1$ for $X \geq 0$ introduces the step-function driving-pressure jump* imposed at $X = 0$ (see Fig. 2.24), whereas p'_{INTER} is proportional to the local interactive streamline deflection along the outer edge of the middle deck and vanishes both far upstream and far downstream of the interaction zone. Thus, an appropriate expression relating p'_{INTER} to v'_e/U_{0e} is needed, and it is here that the particular Mach-number regime of the outer flow primarily enters the problem.

When the external flow is supersonic and outside the transonic range ($M_e \geq 1.2$ or so), the tangent-wedge approximation [8, 9] as well as Van Dyke's combined Supersonic–Hypersonic Similarity Rule [44] are satisfactory for engineering purposes throughout the entire supersonic–hypersonic regime unless the far-field aspects of the interaction well above the boundary layer are of interest. This gives the following:

$$p'_{\text{INTER}}/\rho_{0e}U_{0e}^2 \simeq (v'_e/U_{0e})\beta^{-1}\left[\sqrt{1 + \kappa_H^2(v'_e/U_{0e})^2} + \kappa_H(v'_e/U_{0e})\right] \tag{10.3.73}$$

where $\kappa_H \equiv (\gamma + 1)\beta/4$. When $\kappa_H v'_e/U_{0e}$ is small compared to unity, equation (10.3.73) passes over to classical linearized supersonic theory, whereas large values yield the Newtonian-like limit of classical hypersonic small-disturbance theory. For these purposes, equation (10.3.73) conveniently inverts to the following:

$$v'_e/U_{0e} = \beta\left(p'_{\text{INTER}}/\rho_e U_{0e}^2\right)/\sqrt{1 + 2\kappa_H\beta\left(p'_{\text{INTER}}/\rho_e U_{0e}^2\right)} \tag{10.3.74}$$

which, via equation (10.3.72), relates the ubiquitous integral of v'_e/u_{0e} appearing herein to a comparable streamwise integral of the total interactive pressure $p'(x)$. Equation (10.3.73) applies to both the fully laminar and fully turbulent asymptotic inner-deck-flow models.

In the transonic range with purely supersonic flow throughout, equation (10.3.73) must be replaced by the combined transonic–supersonic pressure–deflection-angle relationship [45]:

$$\frac{p'_{\text{INTER}}}{\rho_e U_{0e}^2} \sim \frac{\beta^2}{(\gamma+1)M_{0e}^4}\left\{1 - \left[1 - \frac{3}{2}\frac{(\gamma+1)}{\beta^3}M_{0e}^4(v'_e/U_{0e})\right]^{2/3}\right\} \tag{10.3.75}$$

which is seen to pass over to the linearised-supersonic–theory result when $v'_e/\beta^3 U_{0e} \ll 1$, while giving the classical nonlinear transonic result $p'/p_{0e} \sim -[M_{0e}^2/(\gamma+1)]^{1/3}(v'e/U_{0e})^{2/3}$ as $\beta \to 0^+$.

The required inverted form is as follows:

$$\frac{v'_e}{U_{0e}} \simeq \frac{(2/3)\beta^3}{(\gamma+1)M_{0e}^4}\left\{1 - \left[1 - \frac{(\gamma+1)M_{0e}^4}{\beta^2}\left(\frac{p'_{\text{INTER}}}{\rho_e U_{0e}^2}\right)\right]^{3/2}\right\} \tag{10.3.76}$$

---

* For example, for a compression corner of small angle θ_w at moderately supersonic Mach numbers, $\Delta p_s \cong \rho_e U_e^2\theta_w/\beta$, whereas for a weak impinging oblique shock wave of incident-pressure-jump strength, Δp_i, $\Delta p_s \cong 2\Delta p_i$.

If instead the transonic problem involves a mixed supersonic–subsonic external flow astride a near-normal shock (see Fig. 2.34), as often is encountered in practice, equation (10.3.76) applies only in the upstream-supersonic region $X < 0$; downstream, where1 $M_e = M_2 < 1$, it must be replaced by the subsonic Cauchy-integral relationship:

$$\frac{v_e'}{U_{0e}} \cong \frac{\sqrt{1 - M_2^2}}{\pi \rho_e U_{0e}^2} \int_{0^+}^{\infty} \left[\frac{p'(\xi) - J \Delta p_s}{x - \xi} \right] d\xi \tag{10.3.77}$$

with $\Delta p_s / \rho_e U_e^2 = 2\beta^2 / (\gamma + 1) M_{0e}^2$ for a normal shock. Thus, the mixed-flow nature of this problem is reflected in a discontinuous $p'_{\text{INTER}}(v_e')$ relationship.

10.3.5 Combined Matching of All Decks

10.3.5.1 Laminar Flows
When the external flow lies in the supersonic–hypersonic regime, the boundary conditions can be simplified further, as follows. Expressing equation (10.3.74) in terms of the scaled pressure, the term $(X_R/Z_R) v_e'/U_{0e}$ appearing in the matching conditions becomes:

$$\frac{X_R}{Z_R} (v_e'/U_{0e}) = \frac{\beta \pi_R}{Z_R/X_R} \left[\frac{\hat{p}'_{\text{INTER}}}{\sqrt{1 + \chi_H \hat{p}'_{\text{INTER}}}} \right] \tag{10.3.78}$$

where $\chi_H \equiv 2\kappa_H \beta \pi_R$ is a hypersonic triple-deck interaction parameter [46] with $\chi_H \ll 1$ pertaining to the linearised supersonic limit. This expression optimally simplifies if we now take:

$$\beta \pi_R = Z_R/X_R \tag{10.3.79}$$

which then yields from equations (10.3.61), (10.3.62), and (10.3.72) the following final form of the outer matching conditions:

$$\lim_{\hat{z} \to \infty} [\hat{u}_{\text{INNER}} (\hat{x}, \hat{z})] = \hat{z} - \int_{-\infty}^{\hat{x}} \frac{[\hat{p} - J\Delta\tilde{p}] \, d\hat{x}}{[1 + \chi_H (\hat{p} - J\Delta\tilde{p})]^{1/2}}$$
$$+ \frac{\delta_0 \tilde{p} \, M_{0e}^2}{\beta X_R} \lim_{\hat{z} \to 0} [I_m (z_i / \delta_0)] \tag{10.3.80}$$

$$\lim_{\hat{z} \to \infty} \left[\hat{H}_{\text{INNER}} (\hat{x}, \hat{z}) \right]_A = \lim_{\hat{z} \to 0} [\hat{u}_{\text{INNER}} (\hat{x}, \hat{z})] \tag{10.3.81}$$

where $\Delta\tilde{p} \equiv \Delta p_s / \rho_e U_{0e}^2 \pi_R$ is the nondimensional scaled imposed pressure jump where $\hat{p}$ refers to the total pressure-perturbation field such that the term $\hat{p} - J\Delta\hat{p}_s$ vanishes far downstream. Anticipating that $\delta_0 / X_R \sim \varepsilon_L$ (see Section 10.3.6.1), the last term in equation (10.3.80) representing the streamline-divergence effect can be neglected in the first asymptotic approximation $\varepsilon_L \to 0$ unless the flow is hypersonic ($M_{e0}^2 \gg 1$).

An analogous procedure is carried out for the case of transonic flow but with different results. When the external flow is everywhere supersonic and equation

(10.3.76) with $J = 0$ applies, we now have:

$$\frac{X_R}{Z_R} \frac{v_e'}{U_{0e}} \simeq \frac{2\beta^3}{3(\gamma+1) M_{0e}^4} \frac{X_R}{Z_R} \left\{ 1 - \left[1 - (\gamma+1) M_{0e}^4 \beta^{-2} \pi_R \hat{p} \right]^{3/2} \right\} \tag{10.3.82}$$

Bodonyi and Kluwick [47] showed that this most conveniently is scaled by introducing the independent laminar-transonic interaction parameter:

$$\chi_T \equiv \beta^2 / (\gamma+1) M_{0e}^4 \pi_R \tag{10.3.83}$$

and choosing:

$$Z_R / X_R = \sqrt{(\gamma+1) M_{0e}^4 \pi_R^{3/2} \chi_T^{1/2}} = \beta \pi_R, \tag{10.3.84}$$

Equation (10.3.82) applied to equation (10.3.61) then yields the following form of the outer boundary condition:

$$\lim_{\hat{z} \to \infty} \left[\hat{u}_{\text{INNER}} (\hat{x}, \hat{z}) \right] = \hat{z} - \frac{2}{3} \chi_T \int_{-\infty}^{\hat{x}} \left\{ 1 - \left[1 - \hat{p} \chi_T^{-1} \right]^{3/2} \right\} d\bar{x} \tag{10.3.85}$$

where we dropped the resulting last term $(\delta_0/\chi_R)\,\hat{p}$ in equation (10.3.61) because it is of negligible higher order $(\varepsilon_L^{8/5})$ in transonic flow [47].

When mixed-transonic external flow with a near-normal shock is involved, this still applies to the upstream region (the linearized supersonic version usually is sufficiently accurate in practice), whereas downstream equation (10.3.77) is appropriate. After using equations (10.3.83) and (10.3.84), we then obtain that:

$$\frac{X_R}{Z_R} \frac{v_e}{U_{0e}} = \sqrt{\frac{1 - M_2^2}{M_e^2 - 1}} \frac{1}{\pi} \int_0^\infty \left[\frac{\hat{p}_e(x) - J\Delta\hat{p}}{\hat{x} - \hat{\xi}} \right] d\hat{\xi} \tag{10.3.86}$$

where equation (10.3.61), without the last term, yields the downstream boundary condition in the form of equation (10.3.85) with the right-hand side replaced by:

$$= \hat{z} - \frac{\sqrt{1 - M_2^2}}{\pi\beta} \int_0^{\hat{x}} \left\{ \int_0^\infty \left(\frac{\hat{p} - J\Delta\hat{p}s}{\hat{x} - \hat{\xi}} \right) d\hat{\xi} \right\} d\hat{x} \tag{10.3.87}$$

10.3.5.2 Turbulent Flows

In the supersonic–hypersonic flow regime, the term $v_e'/\varepsilon_T U_{0e}$ appearing in the boundary conditions takes the following nondimensional form after using equations (10.3.46) and (10.3.74):

$$v_e'/\varepsilon_T U_{0e} = \beta \hat{p}_{\text{INTER}}' / \sqrt{1 + \chi_{HA} \hat{p}_{\text{INTER}}'} \tag{10.3.88}$$

where $\chi_{HA} \equiv 2\beta\kappa_H \varepsilon_T$ is a turbulent hypersonic interaction parameter with $\chi_{HA} \ll 1$ pertaining to the linearized supersonic limit. Because we already established six of the eight scaling relationships needed by the asymptotic-turbulent model (and we are free to choose ρ_R and μ_R; see following discussion), no further scaling choices remain. Thus, equation (10.3.88) indicates that the Mach-number parameter β cannot be "scaled out" of this $\varepsilon_T \to 0$ problem. Thus, in the upstream free-interaction region $X < 0$ with $J = 0$, we have the boundary-condition equations (10.3.63) and

(10.3.64) with the pressure-governing relationship from equations (10.3.65) and (10.3.88) that:

$$\hat{p}/\sqrt{1 + \chi_{HA}\hat{p}} \cong I_m (z_i/\delta_0) M_{0e}^2 \beta^{-1} (d\hat{p}/d\hat{x}) \qquad (\hat{x} < 0) \qquad (10.3.89)$$

which describes for $\chi_{HA} = 0$ an exponential pressure decay in $\hat{x} < 0$ that depends explicitly on β and M_{e0}^2. Downstream, we have the β-dependent boundary conditions from equations (10.3.69) through (10.3.72) and (10.3.88) that:

$$\lim_{\hat{z} \to \infty} [\hat{u}'_{\text{INNER}} (\hat{x} \geq 0, \hat{z})] = -\hat{p} - \frac{\beta}{\hat{z}k^2} \int_{0^{+}}^{\hat{x}} \frac{(\hat{p} - J\Delta\tilde{p}) \, d\hat{x}}{\sqrt{1 + \chi_{HA}(\hat{p} - J\Delta\tilde{p})}} + M_{0e}^2 \hat{p} f(\hat{z})$$

$$(10.3.90)$$

$$\lim_{\hat{z} \to \infty} [\hat{H}'_{\text{INNER}}(\hat{x} \geq 0, \hat{z})] = \hat{p} + \lim_{\hat{z} \to 0} [\hat{u}_{\text{INNER}}(\hat{x}, \hat{z})] \qquad (10.3.91)$$

In the transonic regime with purely supersonic-inviscid flow, the $v'_e/\varepsilon_T U_{0e}$ term from equation (10.3.76) can be expressed as follows:

$$v'_e/\varepsilon_T U_{0e} = (2/3)\,\beta\chi_{T_i} \left\{ 1 - [1 - (\hat{p}/\chi_{T_i})]^{3/2} \right\} \qquad (10.3.92a)$$

where $\chi_{T_i} \equiv \beta^2/(\gamma + 1)\varepsilon_T M_{0e}^4$ is the turbulent counterpart of the laminar-transonic interaction parameter (10.3.83). When combined with equation (10.3.65), (10.3.92a) provides the upstream free-interaction pressure relationship:

$$(2/3)\,\beta\chi_{T_i} \left\{ 1 - [1 - (\hat{p}/\chi_{T_i})]^{2/3} \right\} = I_m M_{0e}^2 (d\hat{p}/d\hat{x}) \qquad (10.3.92b)$$

which as $\chi_{T_i} \to \infty$ reduces to the linearized-supersonic–theory result (10.3.89) when $\chi_{HA} = 0$. The corresponding upstream outer boundary condition remains that of equation (10.3.65). Downstream, where equation (10.3.92b) is inapplicable, we obtain from equations (10.3.69) and (10.3.76) the following boundary condition:

$$\lim_{\hat{z} \to \infty} [\hat{u}'_{\text{INNER}} (\hat{x}, \hat{z})] = -\hat{p} - (2\beta\chi_{T_i}/3\hat{z}k^2) \int_{-\infty}^{\hat{x}} \left\{ 1 - \left[1 - \left(\hat{p}\chi_{T_i}^{-1} \right) \right]^{3/2} \right\} d\hat{x}$$

$$(10.3.92c)$$

If the transonic flow is mixed with a subsonic downstream flow, the right-hand side of equation (10.3.92c) must be replaced by the expression:

$$-\frac{\sqrt{1 - M_2^2}}{\pi \hat{z}k^2 \beta} \int_0^{\hat{x}} \left[\int_0^{\infty} \left(\frac{\hat{p} - \Delta p_s}{\hat{x} - \hat{\xi}} \right) d\hat{\xi} \right] d\hat{x} \qquad (10.3.92d)$$

10.3.6 Summary of Scaling Properties and Final Canonical Forms of Triple-Deck Equations

The remaining undefined scaling parameters ρ_R and μ_R, in the absence of any further analytical constraints, may be chosen on the basis of convenience. For example, in view of the inner-deck proximity to the surface, we might base them on either the undisturbed upstream wall temperature or the so-called reference temperature that conveniently characterizes the upstream boundary-layer properties. We choose the latter in this chapter, based on Eckert's empirical result (further supported by fundamental compressible boundary-layer theory [48]) for air:

$$T_R/T_\infty \cong 0.50 + 0.039 M_\infty^2 + 0.05 T_w/T_\infty \tag{10.3.93}$$

10.3.6.1 Laminar Flows

Summarizing the results from the previous analysis in the case of supersonic–hypersonic external flow, we have the following set of six independent relationships in the six scaling parameters U_R, V_R, Z_R, X_R, π_R, and H_R:

$$V_R = (Z_R/X_R)\,U_R \tag{10.3.94}$$

$$\pi_R = (\rho_R/\rho_e)\,(U_R/U_e)^2 \tag{10.3.95}$$

$$v_w X_R = U_R Z_R^2 \tag{10.3.96}$$

$$U_R = Z_R (dU_0/dZ)_w \tag{10.3.97}$$

$$Z_R/X_R = \beta \pi_R \tag{10.3.98}$$

$$H_R = Z_R (dH_0/dZ)_w = U_R (dH_0/dU_0)_w \tag{10.3.99}$$

When we supply appropriate relationships for the two wall gradients $(dU_0/dZ)_w$ and $(dH_0/dZ)_w$, these relationships then can be solved to express the scalings in terms of the basic laminar triple-deck small parameter $\varepsilon_L \equiv Re_L^{-1/8}$ and the properties of the undisturbed incoming flow. Introducing the laminar compressible flat-plate boundary-layer values [49]:

$$(dU_0/dZ)_w = \tau_{w_0}/\mu_{w_0} = (U_{0e}\lambda/L)\,C_R^{-1/2}\,(T_e/T_R)\,\varepsilon_L^{-4} \tag{10.3.100}$$

$$(dH_0/dZ)_w = (H_{\text{WAD}} - H_{0w})\,C_R^{-1/2}\,P_R^{1/3}\,(\mu_e/\mu_R)\,\lambda\varepsilon_L^{-4}/L \tag{10.3.101}$$

with $\lambda = 0.332$, $\mu = \mu_R (T/T_R)^\omega$ and $C_R \equiv \mu_R T_e/\mu_e T_R$, equations (10.3.94) through (10.3.99) yield the following values:

$$X_R/L = C_x \varepsilon_L^3 \tag{10.3.102}$$

$$Z_R/L = C_z \varepsilon_L^5 \tag{10.3.103}$$

$$\pi_R = C_\pi \varepsilon_L^2 \tag{10.3.104}$$

$$U_R/U_{0e} = C_u \varepsilon_L \tag{10.3.105}$$

$$H_R = C_H \varepsilon_L \tag{10.3.106}$$

$$V_R/U_{0e} = C_v \varepsilon_L^3 \tag{10.3.107}$$

where the six Mach-number–wall-temperature-dependent coefficients C_x, C_y, C_z, C_u, C_h, and C_v are given in Appendix C. The hypersonic interaction parameter χ_H in equation (10.3.78) correspondingly takes the form:

$$2\chi_H = \beta^2\,(\gamma+1)\,\pi_R = (\gamma+1)\,C_R^{1/4}\,\sqrt{\lambda/\beta}\,(\beta\varepsilon_L)^2 \tag{10.3.108}$$

from which it is apparent that χ_H is actually only of second-order importance $(\sim \varepsilon_L^2)$ even in moderately strong hypersonic flows. Thus, as in the contemporary triple-deck literature, we neglect the effect on the previously mentioned boundary conditions and thereby appreciably simplify the theory. The other parameter of possible importance at high Mach number – the coefficient of $\hat{p}$ in the last term of equation (10.3.80) – takes the following form after applying the above scalings and the value $\delta_0/L = C_\delta \varepsilon_L^4$ with $C_\delta = 5.2 C_R^{1/2}(T_R/T_e)$:

$$\frac{\delta_0}{\beta X_R} M_e^2 I_m = \varepsilon_L S I_m \qquad (10.3.109)$$

where $S = C_\delta M_e^2 / C_x \beta$. Although formally of order ε_L and hence negligible at ordinary supersonic speeds, it is observed that this streamline-divergence term can be significant at hypersonic speeds [50] because then the coefficient $S \sim M_{0e}$ and hence it can be large, especially if the influence of heat transfer (which enters via the integral I_m; see Appendix B) is of interest [51]. Finally, when the values of equations (10.3.102) through (10.3.107) are applied to the heat-transfer factor on the right-hand side of equation (10.3.33), this factor becomes:

$$(Z_R/H_w)\,(d H_0/d Z)_w\,\hat{z} = \varepsilon_L C_M (H_{0w} - H_w)\,\hat{z}/H_w \qquad (10.3.110)$$

where $C_M \equiv P_R^{1/3} C_Z C_R^{1/2} \lambda(\mu_e/\mu_R)$. This term is thus a higher order ε_L effect across the inner deck where $\hat{z} \leq 0$ (1), unless abnormally large levels of surface cooling $[(H_{0w} - H_w)/H_w \leq \varepsilon_L^{-1}]$ are present in the incoming flow; thus, it may be neglected for practical purposes [46].

Turning to the case of purely supersonic-transonic external flow, there are five scaling relationships (i.e., equations [10.3.94] through [10.3.98] with equation [10.3.98] replaced by its transonic version $Z_R/X_R = \sqrt{\gamma + 1/M_{0e}^4}\,\pi_R^{3/2}$); combined again with equations (10.3.100) and (10.3.101), they lead to the following final scaling relationships involving different exponents on ε_L:

$$X_R/L = k_x \varepsilon_L^{12/5} \qquad (10.3.111)$$

$$Z_R/L = k_z \varepsilon_L^{24/5} \qquad (10.3.112)$$

$$\pi_R = k_\pi \varepsilon_L^{8/5} \qquad (10.3.113)$$

$$U_R/U_{0e} = k_u \varepsilon_L^{4/5} \qquad (10.3.114)$$

$$V_R/U_{0e} = k_v \varepsilon_L^{16/5} \qquad (10.3.115)$$

where the five coefficients k_x, k_z, k_π, k_u, and k_v are also given in Appendix C. The attendant laminar-interaction parameter χ_T (equation [10.3.83]) assumes the form:

$$\chi_T = \left[(M_{0e}^2 - 1)/(\gamma + 1)\, M_{0e}^4 k_\pi \right] \varepsilon^{-8/5} \qquad (10.3.116)$$

which clearly implies that this (prescribed) value is asymptotically large. The hypersonic terms involving χ_H and $\delta_0 M_e^2 \hat{p}/X_R$ are both of negligible higher order $\varepsilon_L^{8/5}$ under this scaling.

Applying these simplifications to the inner-deck equations (10.3.38) and (10.3.39) with $\rho_0 \simeq \rho_w$, they read as follows in terms of $\hat{u} = \hat{u}_0 + \hat{u}'$:

$$\hat{u}\frac{\partial \hat{u}}{\partial \hat{x}} + \hat{w}'\frac{\partial \hat{u}}{\partial \hat{z}} + \frac{d\hat{p}}{d\hat{x}} = \frac{\partial^2 \hat{u}}{\partial \hat{z}^2} \tag{10.3.117}$$

$$\hat{u}\frac{\partial \hat{H}}{\partial \hat{x}} + \hat{w}'\frac{\partial \hat{u}}{\partial \hat{z}} + \frac{d\hat{H}}{d\hat{y}} = P_r^{-1}\frac{\partial^2 \hat{H}}{\partial \hat{z}^2} \tag{10.3.118}$$

These equations are to be solved subject to the wall boundary conditions $\tilde{u}(\hat{x}, 0) = \hat{v}(\hat{x}, 0) = 0$ on a solid surface of known temperature and the following outer (i.e., middle–outer deck-matching) conditions from equations (10.3.80) and (10.3.81) or equations (10.3.85) and (10.3.87) with $\chi_H = 0$:

$$\lim_{\hat{z}\to\infty} \hat{u}(\hat{x}, \hat{z}) = \hat{z} - I_u(\hat{x}) \tag{10.3.119}$$

$$\lim_{\hat{z}\to\infty} \hat{H}[\hat{x}, \hat{z}] = \lim_{\hat{z}\to\infty} \hat{u}(\hat{x}, \hat{z}) \tag{10.3.120}$$

where:

$$I_u(\hat{x}) \equiv \int_{-\infty}^{\hat{x}} [\hat{p} - J\Delta\tilde{p}]\,d\hat{x} + \varepsilon_L S I_m \hat{p} \quad \langle Supersonic/Hypersonic\rangle \tag{10.3.121}$$

$$= \frac{2}{3}\chi_T \int_{-\infty}^{\hat{x}} \left\{1 - \left[1 - \tilde{p}\chi_T^{-1}\right]^{3/2}\right\} d\hat{x} \quad \left\langle \begin{array}{l} Transonic: \text{ at all } \hat{x} \text{ for } M_2 > 1 \\ \text{or at } \hat{x} < 0 \text{ only for } M_2 < 1 \end{array}\right\rangle \tag{10.3.122}$$

$$= \frac{\sqrt{1 - M_2^2}}{\beta}\frac{1}{\pi}\int_0^{\hat{x}}\left[\int_0^\infty \left(\frac{\hat{p} - \Delta\hat{p}}{\hat{x} - \hat{z}}\right) d\hat{\xi}\right] d\hat{x} \quad \left\langle \begin{array}{l} Mixed\ Transonic\ Flow, \\ M_2 < 1:\ \text{at } \hat{x} > 0 \end{array}\right\rangle \tag{10.3.123}$$

The energy-equation aspects are ignored in the transonic case, whereas in the moderate supersonic case, the streamline-divergence term $\varepsilon_L S I_m \hat{p}$ can be consistently neglected.

10.3.6.2 Turbulent Flows

In this model, the previous analysis established five relevant scaling relationships, independent of the external-flow regime. Summarizing, they are as follows:

$$\frac{\chi_R}{L} = \frac{\delta_0}{L} = \varepsilon_T K_\delta \tag{10.3.124}$$

$$\frac{Z_R}{L} = k\frac{X_R}{L}\varepsilon_T = k K_\delta \varepsilon_T^2 \tag{10.3.125}$$

$$\pi_R = \varepsilon_T \tag{10.3.126}$$

$$\frac{U_R}{u_{0e}} = \varepsilon_T \tag{10.3.127}$$

$$H_R = \varepsilon_T (dH_0/dU_0)_w\, U_{0e} = \varepsilon_T P_R^{0.6}\left[(H_{W,AD} - H_w)/(T_e/T_w)\right] \tag{10.3.128}$$

Because these relations directly involve the basic small-perturbation parameter $\varepsilon_T \equiv U_\tau/U_{0e} = \sqrt{(\rho_e/\rho_w)C_{f_0}/2}$, no further development is needed.

The turbulent hypersonic-interaction parameter χ_{HT} appropriate to this model thus becomes:

$$2\chi_{HT} = (\gamma + 1)\beta^2 \varepsilon_T \qquad (10.3.129)$$

which is seen to be $0(\varepsilon_T)$ and, hence, a higher-order effect in the leading approximation unless the flow is so strongly hypersonic that $\varepsilon_T M_\infty^2 \geq 0(1)$, which typically occurs at $M_e \geq 5$. Because fully turbulent boundary layers usually do not usually exist at such Mach numbers, in practice, we may ignore the influence of this parameter. The last term in equation (10.3.69) involving $\hat{p}$, which is also proportional to $\varepsilon_T M_\infty^2$, may likewise be neglected. Finally, the variable density coefficient of the pressure-gradient term in the inner-deck equations under the scalings (i.e., equations [10.3.124] to [10.3.128]) is of the following order:

$$\frac{\rho_{0w}}{\rho_0} - 1 \cong \left(\frac{H_{0e} - H_w}{H_w}\right) P_r^{1/3} \varepsilon_T \qquad (10.3.130)$$

and, hence, for $\hat{z} \sim 0(1)$ may be neglected unless very large surface cooling (such that $H_w \leq \varepsilon_T H_{0e}$) is present.

Implementing these simplifications with $\rho_0 \cong \rho_{0w}$ ($f_D = 1$), we thus obtain from equations (10.3.49) and (10.3.50) the following final form of the inner-deck equations in the leading very-high-Reynolds-number approximation:

$$\frac{\partial \hat{u}'}{\partial \hat{x}} + \frac{d\hat{p}}{d\hat{x}} = 2\frac{\partial}{\partial \hat{z}}\left(\hat{z}\frac{\partial \hat{u}'}{\partial \hat{z}}\right) \qquad (10.3.131)$$

$$\frac{\partial \hat{H}'}{\partial \hat{x}} \cong P_{r_t}^{-1}\frac{\partial}{\partial \hat{z}}\left(\hat{z}\left[\frac{\partial \hat{H}'}{\partial \hat{z}} + \frac{\partial \hat{u}'}{\partial \hat{z}}\right]\right) \qquad (10.3.132)$$

The inner boundary conditions for these diffusion-type equations are the slip relations, equations (10.3.51) and (10.3.52), applied at the effective wall location $\hat{z}_c \cong 60(T_w/T_e)^{\omega+1}/kK_\delta Re_L \varepsilon_T^3$ all along the interaction. The outer boundary conditions that derive from matching with the middle deck depend on the external-flow Mach-number regime and are discontinuous across $\hat{x} = 0$. Summarizing, we have for supersonic–hypersonic flow that upstream ($\hat{x} < 0$):

$$\lim_{\hat{z}\to\infty}[\hat{u}'(\hat{x},\hat{z})] = -\hat{p} \qquad (10.3.133)$$

$$\beta\hat{p} \cong I_m(z_i/\delta_0)M_{0e}^2\frac{d\hat{p}}{d\hat{x}} \qquad (10.3.134)$$

$$\lim_{\hat{z}\to\infty}\left[\hat{H}'(\hat{x},\hat{z})\right] = 0 \qquad (10.3.135)$$

whereas downstream ($\hat{x} \geq 0$), we have:

$$\lim_{\hat{z}\to\infty}[\hat{u}'(\hat{x},\hat{z})] = -\hat{p} - \frac{\beta}{k^2\hat{z}}\int_{0+}^{\hat{x}}(\hat{p} - J\Delta\hat{p})\,d\hat{x} \qquad (10.3.136)$$

$$\lim_{\hat{z}\to\infty}\left[\hat{H}'(\hat{x},\hat{z})\right] = \lim_{\hat{z}\to\infty}[\hat{u}'(\hat{x},\hat{z})] + \hat{p} \qquad (10.3.137)$$

where we reiterate that this $\hat{x} > 0$ formulation does not resolve the detailed shock-diffraction effects in the region $0 \leq \hat{x} \ll 1$. When the external flow is transonic with $M_e > 1$, the boundary conditions assuming adiabatic flow are instead those given by equation (10.3.92a).

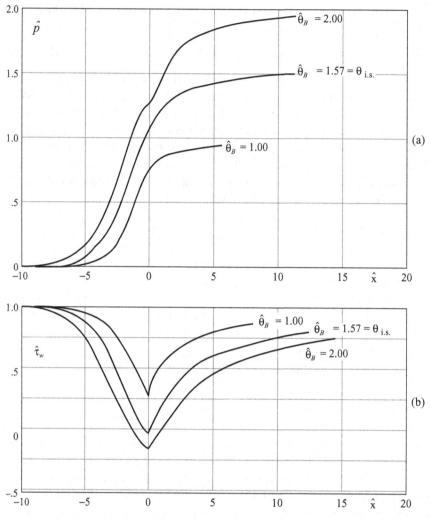

Figure 10.5. Typical features of am ramp-generated laminar interactions zone according to triple-deck theory for supersonic flow. (a) Nondimensional pressure distribution. (b) Nondimensional wall-shear-stress distribution.

10.4 Application to Laminar-Flow Interactions

10.4.1 Supersonic Adiabatic Flows

10.4.1.1 General Aspects

Analytical and numerical solutions of the canonical triple-deck equations (10.3.117) and (10.3.119) with the streamwise-divergence term $\varepsilon_L S I_m \hat{p}$ dropped were studied extensively for adiabatic flow; see, for example, reviews by Délery and Marvin [14] and Kluwick [33]. Representative results for the predicted pressure and skin-friction distributions along an unseparated ramp-provoked interaction are shown in Fig. 10.5. These curves illustrate the typical rapid upstream pressure rise and corresponding skin-friction decrease associated with free interaction (see the following discussion), followed by a slower further compression and skin-friction recovery downstream of the ramp. Careful comparisons of such results with experiments and the predictions of nonasymptotic interacting boundary-layer theory over a wide range of Reynolds numbers [52, 53] showed that the error involved in the first-order asymptotic approximation becomes significant at $Re_L \leq 5 \times 10^5$, primarily due to neglect of the streamline-divergence effect (Fig. 10.6). This situation is not improved by including second-order terms in the asymptotic theory [54].

Surface bump-generated interactions in supersonic flow also were studied [55]. They involve a compression on the front face followed by a downstream overexpansion and then ultimate recompression in the wake. Very small bumps much thinner than the incoming boundary layer permit a linearized treatment of the tripledeck equations that yields analytical solutions for both the upstream free-interaction zone and the algebraic-disturbance decay in the far downstream wake (see [55] for details).

10.4.1.2 Free Interaction and Upstream Influence

Free interaction, a term originally coined by Chapman et al. [56], designates the spontaneous inviscid-viscous interaction process that occurs upstream of any imposed disturbance (e.g., an impinging shock), whereby the local thickening of the boundary layer and the corresponding induced-pressure rise mutually reinforce to produce a departure from the incoming undisturbed flow independent of the nature or size of the disturbance. These branching or departure phenomena – which exist even in a nonseparating flow – were first identified by Oswatitsch and Wiegardt [57]. The eigensolutions subsequently were analyzed by Lighthill [2], Stewartson [3], and Neiland [4], who revealed their unique scaling properties that have since been corroborated experimentally.

Due to the presence of a subsonic region near the surface of the incoming boundary-layer profile, free interaction self-initiates in the small-disturbance region upstream of a compression corner or impinging shock. The initial stage of this process can be examined analytically by regarding the flow as a small perturbation on the incoming boundary layer $\hat{u} = \hat{z}$. Then, if we take $\hat{u} \simeq \hat{z} + \hat{u}'$ with $\hat{u}'$ and $\hat{w}'$ small, the inner-deck equations yield the following linearized problem:

$$\frac{\partial \hat{u}'}{\partial \hat{x}} + \frac{\partial \hat{w}'}{\partial \hat{z}} = 0 \tag{10.4.1}$$

$$\hat{z}\frac{\partial \hat{u}'}{\partial \hat{x}} + \hat{w}' + \frac{d\hat{p}}{d\hat{x}} \simeq \frac{\partial^2 u'}{\partial \hat{z}^2} \tag{10.4.2}$$

with $\hat{u}'(\hat{x}, 0) = \hat{w}'(\hat{x}, 0) = 0$. The solution of these equations is readily obtained by differentiating equation (10.4–2) with respect to $\hat{z}$ and using equation (10.4–1) to eliminate $\hat{w}'$ so as to obtain the following disturbance vorticity equation:

$$\hat{z}\frac{\partial}{\partial \hat{x}}\left(\frac{\partial \hat{u}'}{\partial \hat{z}}\right) = \frac{\partial^2}{\partial \hat{z}^2}\left(\frac{\partial \hat{u}'}{\partial \hat{z}}\right) \tag{10.4.3}$$

We seek a solution on $\bar{x} < 0$ that vanishes upstream and satisfies the outer-interactive boundary condition:

$$\hat{u}'(\hat{x} \to \infty) = -\int_{-\infty}^{\hat{x}} \hat{p}\, d\hat{x} \tag{10.4.4}$$

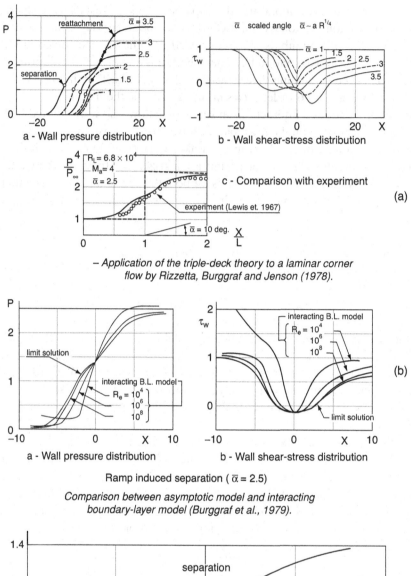

– Application of the triple-deck theory to a laminar corner flow by Rizzetta, Burggraf and Jenson (1978).

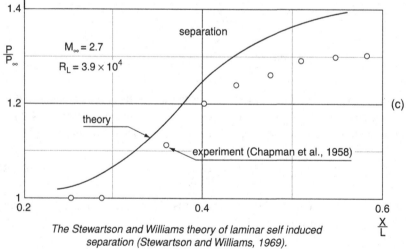

The Stewartson and Williams theory of laminar self induced separation (Stewartson and Williams, 1969).

Figure 10.6. Comparisons of laminar triple-deck results with experiment and interacting boundary-layer theory.

along with the requirement that the vorticity $\partial \hat{u}'/\partial \hat{z}$ and, hence, the disturbance shear stress vanish as $\hat{z} \to \infty$. Inspection suggests separation of variables with an x-wise exponential behavior, so we set:

$$\frac{\partial \hat{u}'}{\partial \hat{z}} (\hat{x}, \hat{z}) = a e^{\kappa \hat{x}} S(\hat{z}) \tag{10.4.5}$$

$$\hat{p}(\hat{x}) = a e^{\kappa \hat{x}} \tag{10.4.6}$$

where the nondimensional amplitude a is determined from downstream events and proportional to the forcing function of the problem, whereas κ is a universal upstream-influence factor to be found.

Then, substitution into equation (10.4.3) gives an Airy equation governing the vorticity profile $s(z)$:

$$zs = \frac{d^2 s}{dz^2} \quad (z \equiv \kappa^{1/3} \hat{z}) \tag{10.4.7}$$

whose solution that vanishes at $z \to \infty$ is:

$$s = C A_i(z) \tag{10.4.8}$$

where C is a constant. Next, we integrate equation (10.4.5) subject to the zero wall-slip condition to obtain:

$$\hat{u}'(x, z) = a e^{\kappa \hat{x}} (C/\kappa^{1/3}) \int_0^z A_i(z) \, dz \tag{10.4.9}$$

This result, along with equation (10.4.6), now can be applied to the matching requirement equation (10.4.4) and the so-called wall-compatibility condition that the momentum equation (10.4.2) be satisfied at $z = 0$. This yields the following two equations in C and κ:

$$C = -\kappa^{-2/3} \Big/ \left[\int_0^\infty A_i(z) \, dz \right] \tag{10.4.10}$$

and

$$C = -\kappa^{2/3} / [-A_i'(0)] \tag{10.4.11}$$

where we note that enforcing these relationships serves to couple interactively the outer and inner decks independently of the disturbance amplitude or cause. Eliminating C between equations (10.4.10) and (10.4.11) and applying the known Airy derivative function properties $A_i'(0) = -0.259$ and $\int_0^\infty A_i(z) \, dz = 1/3$, we find:

$$\kappa = (0.259 \times 3)^{3/4} = 0.827 \tag{10.4.12}$$

This upstream-influence factor, a universal eigenvalue for all interactions, is thus self-determined by the "free" interaction within the fluid independently of downstream events. The upstream pressure-disturbance field for large $x < 0$ thus consists of the exponential pressure rise ($a > 0$):

$$\hat{p}(\hat{x}) = a e^{\kappa \hat{x}} = \hat{p}(\hat{x}_0) \exp\left[\hat{x}_{\text{eff}}/\hat{\ell}u\right] \tag{10.4.13}$$

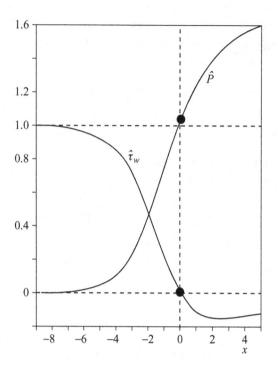

Figure 10.7. Typical properties of a numerical free-interaction solution of the nonlinear laminar triple-deck equations.

where $\hat{l}u \equiv \kappa^{-1} = 1.29$ is the characteristic, nondimensional, upstream-influence factor, whereas $\hat{p}(\hat{x}_0) \equiv a\exp(\kappa\hat{x}_0)$ along with the effective origin shift $\hat{x}_{\text{eff}} = \hat{x} - \hat{x}_0$ (see Fig. 2.24) are determined by downstream events (e.g., incident shock or compression corner) or by imposing a small arbitrary value of a. The corresponding upstream exponential drop in local skin friction as $x \to 0$ is given by $\tilde{\tau}_w' \simeq -1.209\hat{p}$, indicating local incipient separation ($\tau_w = 0$ or $\tilde{\tau}_w' \simeq -1$) at $\hat{p} \simeq 0.83$ (a more exact value of 1.03 is obtained from the full nonlinear inner-deck solution). An analogous behavior also occurs in turbulent flow but on a much shorter scale, as discussed in Section 10.5.1.

This free-interaction, small-perturbation field grows as the flow proceeds downstream and, if the applied perturbation is sufficiently large, evolves into the nonlinear behavior governed by the full triple-deck equations, including possible separation near $x = 0$. If not locally altered near the origin by a specific incident shock or body deflection but instead allowed to freely develop downstream, this behavior ultimately passes through separation and approaches a pressure plateau. An example of such a solution, generated by a small arbitrary "kick" Δa administered upstream, is shown in Fig. 10.7 (further discussion of "unbound" free-interaction solutions and their numerical techniques is in the literature [33, 58]). Indeed, experimental data confirm this free-interaction concept that the upstream flow physics – well up to and even beyond incipient separation – is locally self-determining independently of downstream events. Moreover, as shown herein, the concept yields valuable correlation laws governing the Mach- and Reynolds-number effects on both the length scale and magnitude of the interactive pressure field.

Inversion of the streamwise scaling relation (equation (10.3.102)) gives the following expression for the physical upstream-influence distance ℓu in supersonic laminar flow:

$$\frac{\ell u}{L} \simeq \frac{\hat{\ell} u}{\lambda^{5/4}} \cdot \frac{(T_R/T_e)^{3/2} (T_w/T_R)^{1+\frac{\omega}{2}}}{\beta^{3/4} Re_L^{3/8}} \tag{10.4.14a}$$

$$\frac{\ell u}{L} \simeq \frac{\hat{\ell} u}{\lambda^{5/4}} \underbrace{\left\{ \frac{(T_R/T_e)^{\frac{1}{2}[1-\frac{3}{4}(1-\omega)]}}{\beta^{1/4} Re_L^{1/8}} \right\}}_{A} \underbrace{\left\{ \frac{T_R/T_e}{[(M_\infty^2 - 1) Re_L]^{1/4}} \right\}}_{B} \tag{10.4.14b}$$

where $\lambda = 0.332$ for flat-plate boundary layers. Here, we have written the basic triple-deck result equation (10.4.14a) in the alternative factored form equation (10.4.14b) to show that only the factor B – which is based on Chapman's original, less sophisticated free-interaction analysis assuming an x-scale $\sim 0 (\delta_0)$: (see Section 2.7.1) – is needed to successfully correlate laminar-interaction data [59]. The factor A is evidently a correction for the fact that the interactive scale is slightly longer by a factor $0(1/\varepsilon_L)$. Conversely, because use of the factor B already agrees well with experiment, we may alternatively view factor A as representing the weakly Mach- and Reynolds-number-dependent error incurred when the asymptotic theory for $\varepsilon_L \to 0$ is applied at ordinary Reynolds numbers (Stewartson and Williams [3]). In any case, equation (10.4.14) predicts that ℓu decreases with increasing Mach number; noting that $\delta_0/L \sim Re_L^{-1/2}$, the equation also predicts that the ratio $\ell u/\delta_0$ increases slowly with increasing Reynolds number.

10.4.1.3 Wall-Pressure Distribution and Incipient Separation

Inversion of the pressure-scaling relationship equation (10.3.104) yields the following expression for the physical pressure-coefficient distribution:

$$(p - p_\infty)/\rho_e U_{0e}^2 = \lambda^{1/2} C_R^{1/4} \beta^{-1/2} Re_L^{-1/4} \cdot \hat{p}(\hat{x}) \tag{10.4.15}$$

where $\hat{p}(\hat{x})$ is the universal curve shown in Fig. 10.5. Equation (10.4.15) clearly suggests that either theoretical results or experimental data for laminar-interaction pressure distributions should fall on a single universal curve when plotted in the nondimensional form $(p - p_\infty) \beta^{1/2} Re_L^{1/4}/C_R^{1/4} \rho_e U_{0e}^2$ versus $\hat{x}$ when $\hat{x}$ is based on the factor B of equation (10.4.14b), as discussed previously. As shown in Section 2.7.1, this indeed is found to be the case over a wide range of Mach and Reynolds numbers (see Fig. 10.5). Thus, asymptotic theory confirms the similarity laws established by Chapman's free-interaction concept.

Numerical solutions of the full nonlinear triple-deck equations show that the wall shear vanishes in the nonlinear region just upstream of $\hat{x} = 0$ (Fig. 10.5) at a value $\hat{p}_{is} = 1.03$ (in the case of ramp-generated interactions, this corresponds to a scaled ramp-angle value of $\hat{\theta}_w = 1.57$). Thus, equation (10.4.15) predicts that the pressure coefficient at incipient-separation scales as:

$$(C_p)_{is} = 2.06 C_R^{1/4} (M_\infty^2 - 1)^{-1/4} Re_L^{-1/4} \tag{10.4.16}$$

a result supported by a large body of experimental data on laminar interactions over a wide range of supersonic Mach and Reynolds numbers [56, 59, 60, 61] (see Section 2.7.1).

10.4.1.4 Linearized Solutions

When the magnitude $\hat{h}$ (for example) of the particular forcing function of the problem, such as the incident-shock strength or body-deflection angle, is sufficiently weak to avoid separation, the triple-deck equations can be simplified further by expanding the flow properties in ascending powers of $\hat{h}$ [ie., $\hat{p}(\hat{x}) = \hat{h}p_1(\hat{x}) + \hat{h}^2 p_1(\hat{x})$, $\hat{u}(\hat{x}, \hat{z}) = \hat{z} + \hat{h}u_1(\hat{x}, \hat{z}) + \cdots \hat{w} + \hat{h}w_1(\hat{x}, \hat{z}) + \cdots$ etc.]. The leading terms in such a scheme, when applied to all $\hat{x}$ along the interaction zone, then constitute a linearized approximation to the entire interactive physics, which – because of the resulting simplified triple-deck equations (equations [10.4.1] and [10.4.2]) – can be treated analytically by the Fourier x-wise transform method. Upstream, the resulting solution yields the exponential free-interaction behavior described in Section 10.4.1.2 with the exact numerical value of the amplitude factor a ($\sim\hat{h}$), now depending on the specific forcing function (e.g., incident shock, compression corner, or wall bump). Due to the intricacies of the Fourier inversion process, the downstream behavior is far more complicated but ultimately leads to a fractional power law $\hat{x}$-decay of the pressure and skin-friction disturbances in the far wake (the exponent again depends on the specific problem). Typical details are provided in several papers that address various types of interactions [55, 62, 63].

These linearized solutions reveal much about the physics of an interaction problem up to incipient separation, including the scaling properties and dependence on the particular type of forcing function involved. Moreover, the linearized version of triple-deck theory serves as a valuable guide in the purely numerical solution of the fully nonlinear problem by providing an exact check solution in the small-amplitude limit and establishing the proper conditions (which necessarily approach a small-disturbance behavior) both upstream and in the downstream wake. Finally, these closed-form solutions are of interest in their own right – for example, in hydrodynamic-stability work [55, 62].

10.4.2 Hypersonic Nonadiabatic Flows

10.4.2.1 Streamline Divergence Effect

The influence of the term $\varepsilon_L I_m S \hat{p}$ in equation (10.3.121), representing the streamline divergence effect, was studied numerically for adiabatic-hypersonic interactions by Rizzetta et al. [50]. Their results (Fig. 10.8) show that this effect slightly reduces both the scaled upstream-influence distance $\hat{\ell}u$ and the scaled pressure $\hat{p}$ at incipient separation. The effect of wall-cooling on these results subsequently was examined [60] and shown to be attributable to the wall-temperature effect on the basic Mach-number profile integral I_m. For the case of a so-called supercritical incoming boundary layer with $M_{AV}^2 \gg 1$ and $I_m < 0$, cooling was found to further reduce $\hat{\ell}u$ while delaying the onset of separation. Indeed, it was found that a sufficiently large degree of wall-cooling $T_w \ll T_{w,AD}$ could eliminate completely both the upstream influence and incipient separation. These findings are in agreement with a numerical study of cooled compression-corner interactions based on interacting boundary-layer theory [64].

10.4.2.2 Upstream Influence

It is customary in the hypersonics community to express analytical results concerning viscous-inviscid interactions in the term of the hypersonic interaction

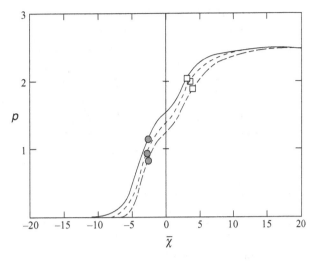

Figure 10.8. Streamline divergence effect on an adiabatic hypersonic-laminar interaction $\alpha = 2.5; -, \sigma = 0;$ $---, \sigma = 1; —, \sigma = 2; \bullet,$ separation point; $\square,$ re-attachment point.

parameter $\bar{\chi} \equiv M_e^3 (C_R / Re_L)^{1/2}$. Accordingly, the upstream-influence distance equation (10.4.14a) in hypersonic flows with $\beta \simeq M_e$ and $T_0 \simeq (\gamma - 1) M_e^2 T_e / 2$ can be expressed in the following form:

$$\frac{\ell u}{L} \simeq \frac{\hat{\ell} u}{\lambda^{5/4}} \left(\frac{\gamma - 1}{2} \right)^{3/2} \left(\frac{T_w}{T_0} \right)^{1 + \frac{\omega}{2}} C_i \, (\bar{\chi})^{3/4} \qquad (10.4.17)$$

where $C_i \equiv \{[(\gamma - 1)/2]^{1/4} M_e^{1/2} T_R / T_0\}^{1-\omega}$ is of order one (and exactly unity for $\omega = 1$). Because of the effect on $\hat{\ell} u$ combined with the (T_w / T_0) factor, equation (10.4.17) predicts that wall-cooling significantly reduces the physical upstream-influence distance, in agreement with available experimental evidence on hypersonic compressive interactions (see Fig. 2.59). Moreover, incorporation of the cooling effect on the nondimensional ($\hat{x}$) scalings was shown [65] to effectively correlate both cooled ($T_w \simeq 0.24 T_0$) and adiabatic-wall pressure data on a single curve (Fig. 10.9). The hypersonic effect on ℓu according to equation (10.4.17) scales directly as $(\bar{\chi})^{3/4}$; hence, it decreases significantly with increasing Mach number as $M_e^{-3/4}$ when $\omega = 1$.

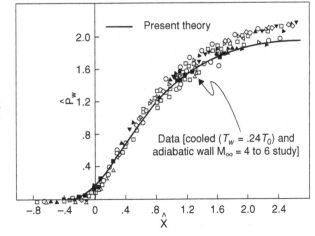

Figure 10.9. Correlation of wall-cooling effect experiments according to triple-deck concepts.

10.4.2.3 Incipient Separation

Based on an examination of triple-deck solutions over a wide range of conditions, as well as analysis of the associated vertical velocity field, Inger [46] found that incipient separation occurs at approximately the same nondimensional local deflection-angle value $\hat{\theta}_{is} \cong 1.57$ regardless of the streamwise location. This observation was developed into an incipient-separation criterion by noting that the scaled outer-disturbance streamline slope is necessarily proportional to $\hat{\theta}_{is}$. Using a tangent-wedge relationship and working back through the triple-deck scaling, the following physical incipient-separation pressure relationship was obtained:

$$\frac{p_{i.s.} - p_\infty}{\gamma p_\infty} \cong C_4 \bar{\chi}^{1/2} \left\{ \sqrt{1 + \left[\left(\frac{\gamma+1}{4} \right) C_4 \bar{\chi}^{1/2} \right]^2} + \left(\frac{\gamma+1}{4} \right) C_4 \bar{\chi}^{1/2} \right\} \qquad (10.4.18)$$

where c_4 is a constant of order unity. In the weak-interaction limit at moderate supersonic speeds where $\bar{\chi} \ll 1$, this equation predicts that:

$$(p_{i.s.} - p_\infty)/\gamma p_\infty \cong C_4 \bar{\chi} \qquad (10.4.19a)$$

or

$$(C_p)_{is} \sim (C_R/Re_L)^{1/4} \left(M_\infty^2 - 1 \right)^{-1/4} \qquad (10.4.19b)$$

in agreement with equation (10.4.16). Conversely, for strong hypersonic interactions with $\bar{\chi} \gg 1$, it predicts the separation pressure scaling:

$$(p_{i.s.} - p_\infty)/\gamma p_\infty \cong 0.5 \, C_4^2 \, (\gamma + 1) \, \bar{\chi} \qquad (10.4.20)$$

which exhibits a stronger ($\sim \bar{\chi}$) dependence on $\bar{\chi}$ than the weak-interaction result in equation (10.4.19a). These predictions, in fact, are corroborated by experiment, as shown in Fig. 10.10: At small $\bar{\chi}$, the observed incipient-separation pressures scale as $\bar{\chi}^{1/2}$, whereas at larger $\bar{\chi} \geq 1$, they clearly follow the linear dependence on $\bar{\chi}$ indicated, by equation (10.4.20). Thus, this switchover behavior in the $\bar{\chi}$ dependence of incipient-separation pressure (originally an empirical finding) is seen to follow from the leading asymptotic local approximation of triple-deck theory combined with the tangent-wedge approximation for the outer inviscid flow.

10.4.2.4 Interactive Heat Transfer

Knowledge of the corresponding heat-transfer disturbances in interaction zones is also of concern because of their importance in the aero-thermodynamic design of cooled hypersonic-flight vehicles and because such heat transfer is an important diagnostic in understanding the interactive flow and its separation (see Section 2.9.3). Although a significant experimental database on interactive heat transfer has accumulated, along with purely CFD-type predictions, much less has been accomplished on the triple-deck theory of the local interactive heat transfer *per se*. Rizzetta [66] treated the case of moderately supersonic compression-corner interactions using a temperature-based energy equation neglecting the viscous-dissipation effect. By reformulating the problem in terms of total enthalpy, Inger [46] subsequently extended his work to high-speed nonadiabatic flows, including a detailed examination of the various hypersonic effects throughout the triple-deck structure;

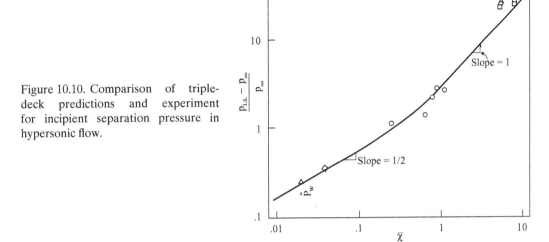

Figure 10.10. Comparison of triple-deck predictions and experiment for incipient separation pressure in hypersonic flow.

this further led to the present Howarth-Dorodnitzn formulation discussed in Section 10.3.

Typical solutions for the local interactive heat transfer in a compression-corner interaction (normalized to the upstream noninteractive value) are illustrated in Fig. 10.11. These curves (which do not allow for the large-scale hypersonic

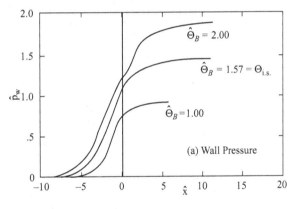

Figure 10.11. Local interactive heat-transfer predictions for laminar nonadiabatic ramp-generated interactions.

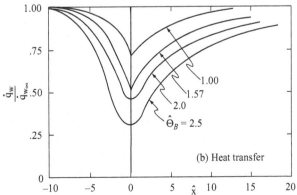

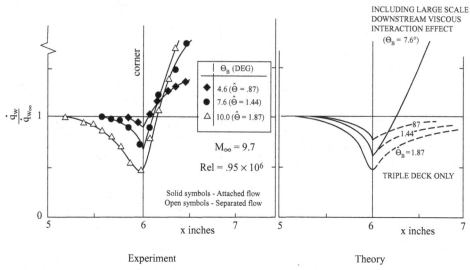

Figure 10.12. Comparison of experiment and triple-deck theory for hypersonic laminar interactions.

viscous-interaction effect that occurs downstream of the corner) clearly show how the adverse pressure-induced reduction in heating grows in direct proportion to the strength of the interaction, with a sharply peaked minimum at the shock foot (i.e., the corner) as long as the flow remains unseparated. For the stronger interactions that promote local separation, however, this peaking quickly disappears into a broader smoothed minimum.

Direct comparison of the theoretical predictions with experimental data may be made if we account for the added influence of the larger-scale viscous-inviscid effect caused by the thickening boundary layer downstream of the corner when M_∞ is large (the appropriate theory was given by Stollery [67]). This is shown in Fig. 10.12, where the local heat-transfer distributions measured by Needham [68] in a Mach 9.7 interacting corner flow of varying angles are compared with the theoretical predictions. Regarding the small-scale, triple-deck aspect of the interaction upstream and immediately at the corner, it is seen that the predicted deepening of the heat-transfer reduction with increasing interactive strength and the associated sharply peaked minimum at the shock foot are both clearly confirmed by the data. Moreover, the abrupt disappearance of the sharp minimum peak after separation indicated by the theory is also experimentally corroborated; indeed, Needham suggested that this feature, in fact, could serve as an incipient-separation criterion: Triple-deck theory provides firm support for this empirical observation.

Concerning the interaction zone downstream of the corner, it is shown in Fig. 10.12 that the locally developing large-scale viscous-interaction effect at this truly hypersonic Mach number of 9.7 completely overwhelms the wake of the small-scale, triple-deck structure, producing a very rapid pressure and heat-transfer rise in rough agreement with experimental data. This behavior is consistent with basic

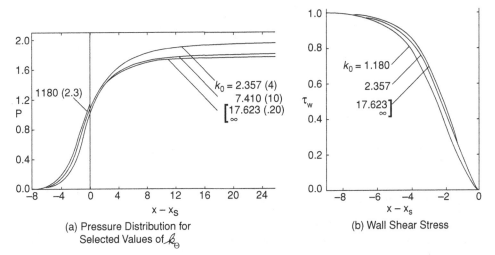

(a) Pressure Distribution for
Selected Values of k_Θ

(b) Wall Shear Stress

Figure 10.13. Effect on the transonic interactions parameter on laminar nonlinear free interactions. (a) Pressure distributions. (b) Wall shear-stress distributions.

analysis showing that cooled hypersonic boundary layers in adverse-pressure gradients actually thin out with a corresponding increase in the heat transfer [67].

This illustrates again how analytical studies prove valuable in illuminating both
experimental and CFD-based results, especially where very fine computational
meshes otherwise are required to resolve the local transfer aspects.

10.4.3 Transonic Regime

Because a laminar boundary layer subjected to a near-normal ($M_2 < 1$) shock wave
[15] separates already at $M_1 > 1.05$, the small range of lower supersonic Mach numbers is of little practical interest compared to the $M_1 < 1.3$ normal shock regime of
unseparated flow observed in the turbulent case (see Section 2.5.3). Therefore, theoretical studies of transonic-laminar interactions mostly addressed the realm of weak
waves involving purely supersonic flow ($1 < M_2 \le M_1$). Messiter et al. [69] showed
that the nonlinear effects in this realm are governed by the laminar-transonic-
interaction parameter:

$$k_0 = \left(M_\infty^2 - 1\right) \varepsilon^{-8/5} = \lambda^{-2/5} C_R^{-1/5} \left(M_\infty^2 - 1\right) Re_L^{-1/5} \qquad (10.4.21)$$

which is essentially the same as the parameter χ_T of equation (10.3.83) and is
taken to be of order unity. Following Brilliant and Adamson's study [70] of the
nonseparating case, Bodonyi and Kluwick [47] investigated free-interaction solutions of the purely supersonic-transonic triple-deck equations, including separation.
Figure 10.13 illustrates their results for the interactive pressure field showing the
influence of k_0; it is seen that upstream, where the disturbance becomes weak,
the results pass over to those predicted by linearized supersonic interaction theory
$k_0 \to \infty$, as noted previously.

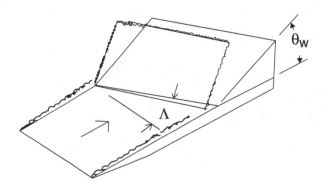

Figure 10.14. Flow configuration of an infinite span swept-ramp interaction.

10.4.4 Three-Dimensional Interactions

Compared to the many analytical and numerical studies of two-dimensional flow, relatively few addressed three-dimensional interactions: Even in the absence of separation, these prove difficult to analyze due to the need for a more complicated pressure flow–deflection relationship and the presence of a lateral-disturbance flow. However, there are a few special cases in which the problem can be simplified to render it amenable to analysis.

The simplest case involves the quasi–two-dimensional flow past a swept compression ramp comparable to an infinite-span swept wing (Fig. 10.14), wherein the streamwise flow component is constant and all interactive perturbations occur in a plane normal to the leading edge. Following an earlier numerical study of the problem by Werle et al. [71] based on interacting boundary layer theory, Gitler and Kluwick [72, 73] conducted a detailed analytical and numerical investigation based on the leading asymptotic approximation of triple-deck theory. Both studies showed that sweepback significantly increases the upstream-influence distance and decreases the incipient-separation pressure according to the following relationships:

$$\frac{\ell_u}{(\ell_u)_{\Lambda=0}} = \left(\frac{M_e^2 - 1}{M_e^2 \cos^2 \Lambda - 1} \right)^{3/8} \cos^{-3/8} \Lambda \tag{10.4.22}$$

$$\frac{p_{is}}{(p_{is})_{\Lambda=0}} = \left(\frac{M_e^2 - 1}{M_e^2 \cos^2 \Lambda - 1} \right)^{1/4} \cos^{7/4} \Lambda \tag{10.4.23}$$

These results are supported by experiment and confirm the applicability of the sweepback principle – namely, that the disturbance field in a plane normal to the ramp leading edge is two-dimensional when based on the incoming normal Mach-number component.

A more complex three-dimensional interaction problem, involving supersonic flow along a flared axisymmetric body at zero angle of attack (Fig. 10.15a), also was studied: Exploiting the absence of azimuthal (i.e., cross) flow, Gitler and Kluwick [73] provided a detailed triple-deck theory of the problem for the case in which the cylindrical-body radius is much larger than the boundary-layer thickness. Upstream of the flare, their results display the expected three-dimensional relief effects of reduced upstream influence and increased flare angle for incipient separation compared to two-dimensional ramp interactions. Downstream of the flare, the axisymmetric-spreading effect around the flare causes the local disturbance pressure

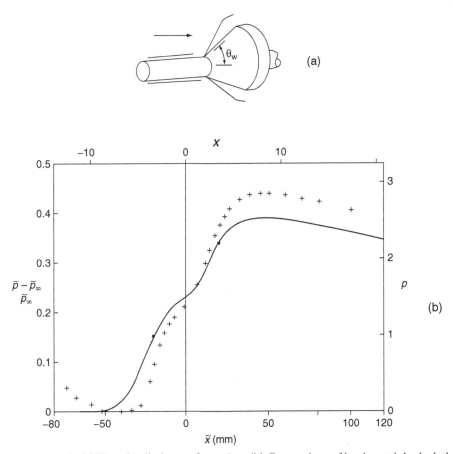

Figure 10.15. (a) Flared-cylinder configuration. (b) Comparison of laminar triple-deck theory and experiment (+) for the flared-cylinder problem in supersonic flow.

to ultimately vanish in contrast to the finite inviscid ramp shock value observed in two-dimensional flow. Allowing for finite Reynolds-number effects and the use of a linearized axisymmetric flow model of the outer-inviscid deck, this theory predicts a pressure distribution in reasonably good agreement with experimental data of LeBlanc and Ginoux [74] (Fig. 10.15b). A subsidiary issue regarding the effect of the body-radius to boundary-layer-thickness ratio (i.e., transverse-curvature effect) on the upstream free-interaction behavior was also studied in detail [75].

10.5 Application to Turbulent Interactions

10.5.1 Supersonic/Hypersonic Interactions in Asymptotic Theory

10.5.1.1 Upstream Region

The central question is the upstream influence, which involves a free-interaction–type of behavior over a streamwise distance on the order of δ_0. The leading asymptotic approximation to the local interactive pressure field of this flow is given by equation (10.3.134), which has the exponential solution in terms of $\hat{x} = x/\delta_0$ that:

$$\hat{p} = c \exp[\hat{x}/\hat{\ell}u] \qquad (10.5.1)$$

where the constant c is determined by downstream events and proportional to the forcing function magnitude, whereas $\hat{\ell u} \equiv \ell u/\delta_0$ is the upstream-influence distance ratio scaled on δ_0 as given by:

$$\hat{\ell u} = \left(M_e^2/\beta\right) I_m = \left(M_e^2/\beta\right) \int_{z_i/\delta_0}^{1} \left\{\left[1 - M_0^2(z)\right]/M_0^2(z)\right\} dz \tag{10.5.2}$$

Equation (10.5.2) is the turbulent counterpart of the nondimensional, laminar, free-interaction eigenvalue equation (10.4.12); unlike the latter, however, it depends on both the Mach number and profile shape via the integral I_m. As long as a nonvanishing lower limit is accounted for, the resulting integral is otherwise rather insensitive to the exact value of this limit when the Reynolds number is large (see Appendix B); thus, the free interaction is essentially rotationally inviscid and governed primarily by the incoming boundary-layer Mach-number shape and only slightly influenced by the underlying turbulent inner deck. Equation (10.5.2) thus suggests that the upstream influence ratio $\ell u/\delta_0$ for very-high-Reynolds-number nonseparating turbulent interactions is only weakly dependent on Reynolds number (i.e., that ℓu scales approximately as the incoming boundary-layer thickness). Because the boundary layer at $Re_L > 10^5$ is so thin, this prediction is difficult to assess experimentally for weak interactions where ℓu is very small: All of the available data pertain to strong, separated interactions with larger values of $\ell u/\delta_0$ that gradually decrease with increasing Re_L. Regarding the Mach-number effect, equation (10.5.2) predicts that $\ell u/\delta_0$ decreases with increasing M_e in agreement with experiments.

Thus, the upstream inner deck is virtually a noninteractive classical turbulent boundary-layer problem, governed by the exponential pressure gradient in equation (10.5.1), as well as the split-boundary conditions in equations (10.3.51) and (10.3.133) imposed on the leading asymptotic-momentum equation (10.3.131). Because the shear stress is already known by virtue of equations (10.3.52) and (10.5.1) as follows:

$$\tau_w'/\tau_{w_0} \simeq -2\hat{p}B_0^{-1} \simeq -2cB_0^{-1}\exp[\hat{x}/\hat{\ell u}] \tag{10.5.3}$$

the deck problem reduces to that of determining (if desired) the velocity distribution across the deck between the given upper- and lower-boundary conditions.

This upstream analysis is supported by earlier theoretical and experimental studies. For example, Inger [76] showed that along any turbulent interaction pressure-disturbance field $p(\hat{x})$, the corresponding skin-friction behavior is governed by the following general relationship:

$$\frac{\tau_w'}{\tau_{w_0}} = C_\tau\sqrt{C_{f_0}\beta}\,\frac{\hat{p}}{\tau_{w_0}}\,(\hat{x})\left[\frac{\frac{2}{3}(p')^{3/2}\,\ell u}{\int_\infty^x (p')^{3/2}\,dx}\right] \tag{10.5.4a}$$

where the constant C_τ depends only weakly on the Reynolds number. However, in the upstream-influence region governed by equation (10.5.1), the square-bracketed term in equation (10.5.4a) is exactly unity; after introducing the present rescaled variables, we thus obtain:

$$\tau_w'/\tau_{w_0} = C_\tau\sqrt{\beta}\,\hat{p}\,(\hat{x}) \tag{10.5.4b}$$

in qualitative agreement with equation (10.3.52). Equation (10.5.4b) further predicts for incipient separation ($\tau_w' = -\tau_{w_0}$) that $p_{is}/p_{0e} \sim \left(\sqrt{C_{f_0}/C_\tau}\right)\left(M_{0e}^2/\beta^{1/2}\right)$. Because typically $C_{f_0}^{1/12} \sim Re_{\delta*}^{-1/12}$ [12], p_{is}/p_{0e} thus is nearly constant over a wide range of practical Reynolds numbers $10^5 \leq Re_{\delta*} \leq 10^8$ while increasing moderately with increasing M_∞; these trends are in agreement with both experiments [76] and similitude analyses [59–61], as discussed in Section 2.7. When alternatively expressed in terms of the pressure coefficient, the foregoing gives:

$$(C_p)_{is} \sim C_{f_0}^{1/2} / \left(M_e^2 - 1\right)^{1/4} \qquad (10.5.5a)$$

which on use of the classical turbulent boundary-layer value $C_{f_0} \sim Re_L^{-1/5}$ yields

$$(C_p)_{is} \sim \left(M_e^2 - 1\right)^{-0.25} Re_L^{-1/10} \qquad (10.5.5b)$$

This result is in fairly good agreement with Erdos and Pallone's proposed semi-empirical correlation [59]:

$$(C_p)_{is} \sim \left(M_e^2 - 1\right)^{-0.31} Re_L^{-1/10} \qquad (10.5.5c)$$

The available experimental data on very-high-Reynolds-number turbulent compression-corner interactions [27] suggest that, in fact, this weak Reynolds number dependence disappears when even small amounts of separation occur.

This discussion strongly suggests that turbulent interaction pressure-distribution data may be correlated according to free-interaction concepts in a manner analogous to the laminar case discussed previously. In fact, this was successfully accomplished by several researchers [59,77], resulting in a universal nondimensional pressure-distribution curve similar in form to the laminar distribution (see Fig. 2.53).

The companion inner-deck energy equation (10.3.132) for the upstream region can be rewritten in a more revealing form by subtracting equation (10.3.131) and using the fact that $\partial \hat{p}/\partial \hat{z} \cong 0$; this yields:

$$\frac{\partial \left[\hat{H}' - (\hat{u}' + \hat{p})\right]}{\partial \hat{x}} = P_{r_T}^{-1} \frac{\partial}{\partial \hat{z}} \left\{ \hat{z} \frac{\partial}{\partial \hat{z}} \left[\hat{H}' - (\hat{u}' + \hat{p})\right] + 2(1 - P_{r_t}) \hat{z} \frac{\partial \hat{u}'}{\partial \hat{z}} \right\} \qquad (10.5.6)$$

which is to be solved subject to the outer-boundary condition that $[\hat{H}' - (\hat{u}' + \hat{p})]$ vanish as $z \to \infty$ and the inner slip-boundary conditions equations (10.3.51) and (10.3.53). As in the wall-shear stress, the local heat transfer is known from equation (10.3.54), so it remains to solve equation (10.5.6) for the enthalpy profile in a manner similar to that applied to the momentum equation. An interesting special solution can be realized when $P_{r_t} = 1$, in which case equation (10.5.6) yields the Crocco-type integral:

$$\hat{H}'(\hat{x}, \hat{y}) = C_1 + [\hat{u}'(\hat{x}, \hat{y})] + \hat{p} \qquad (10.5.7)$$

where the outer-boundary condition requires the constant $C_1 = 0$. It is shown from the inner-boundary conditions that this solution, in fact, pertains to an adiabatic wall.

10.5.1.2 Downstream Region

Here, the imposed shock-induced pressure jump comes explicitly into play in the boundary conditions. Inner-deck equations (10.3.131) and (10.3.132) are still appropriate but now are interactive with the overlying middle deck because of the pressure-integral term in equation (10.3.136). Thus, with the $\varepsilon_T M_e^2$ term neglected, we have the outer-boundary condition on the inner-deck solution that:

$$\lim_{\hat{z}\to\infty} [\hat{u}'_{\text{INNER}}(\hat{x}, \hat{z})] = -\hat{p} - \beta k^{-2}\hat{z}^{-1} \int_{0^+}^{\hat{x}} (\hat{p} - J\Delta\hat{p})\, d\hat{x} \qquad (10.5.8)$$

which implies that the corresponding disturbance vorticity $\hat{\zeta} \equiv \partial\hat{u}'/\partial\hat{z}$ decays algebraically as follows:

$$\lim_{\hat{z}\to\infty} [\hat{\zeta}(\hat{x}, \hat{z})] = -\beta k^{-2}\hat{z}^{-2} \int_{0^+}^{\hat{x}} (\hat{p} - J\Delta\hat{p})\, d\hat{x} \qquad (10.5.9)$$

Far downstream, where $\hat{p} \to \Delta\tilde{p}_s$ these relationships correctly imply that $\hat{u}'(\hat{x} \to \infty) = -\hat{p}(\hat{x} \to \infty) = -\Delta\hat{p}$, which is consistent with the small-disturbance nature of the outer flow, including the imposed shock.

The inner-deck problem may be handled in terms of the vorticity; thus, after taking $\partial/\partial z$ of Equation (10.3.136) and noting that $\partial (d\hat{p}/dx)/\partial z = 0$, we have the homogeneous partial differential equitation:

$$\frac{\partial\hat{\zeta}}{\partial\hat{x}} = \frac{\partial^2}{\partial\hat{z}^2} (2\hat{z}\hat{\zeta}) \qquad (10.5.10)$$

The corresponding inner conditions are again the slip-velocity relationships in equations (10.3.51) and (10.3.52) imposed along $\hat{z} = \hat{z}_c$, as well as the so-called compatibility relationship of satisfying the original momentum equation (10.3.131) along $\hat{z}_c$:

$$\frac{\partial\hat{u}}{\partial\hat{x}}(\hat{x}, \hat{z}_c) + \frac{d\hat{p}}{d\hat{x}} = 2\kappa \left[\hat{\zeta}(\hat{x}, \hat{z}_c) + \hat{z}_c \frac{\partial\hat{\zeta}}{\partial\hat{y}}(\hat{x}, \hat{z}_c)\right] \qquad (10.5.11)$$

As in the laminar case, this later relationship provides an additional interactive link between the external pressure and the bottom of the inner deck.

Agrawal and Messiter [31] provide an elegant asymptotic analysis of the downstream problem for the case of a slender compression ramp that includes a detailed examination of the shock diffraction–reflection process in the very small scale region $\hat{x} \leq 1$ just behind the corner. Melnick et al. [34] also studied this region in finer detail, particularly the singularities that arise at $x \to 0^+$. Although the resulting theoretical predictions of the local post-shock pressure and skin-friction disturbance are in good to fair agreement, respectively, with experiments (Fig. 10.16), the complicated nature of this theory does not lend readily to practical application, except perhaps to suggest [78] that in region $\hat{x} > 0(1)$ of interest here, the pressure perturbation relative to the final asymptotic value (i.e., the overall inviscid wedge pressure jump) decays as $(\hat{x})^{-1}$. Following this idea, we are led to explore the present formulation at $\hat{x} \gg 1$ under the assumption that:

$$\hat{p} \cong \Delta\hat{p}_s (1 + C_A \hat{x}^{-1}) \qquad (10.5.12)$$

where C_A is a nondimensional constant determined by matching with the downstream limit (e.g., at $\hat{x} = 1$) of an appropriate small-scale analysis [31] in the region $0^+ < \hat{x} \leq 0(1)$. This assumption then yields from equation (10.3.136) the outer-boundary condition that:

$$\lim_{\hat{z}\to\infty} \hat{u}'(\hat{x}, \hat{z}) = -\hat{p} - \beta\Delta\hat{p}_s k^{-2}[D_A + C_A \ell n(\hat{x})] \lim_{\hat{z}\to\infty} (1/\hat{z}) \qquad (10.5.13a)$$

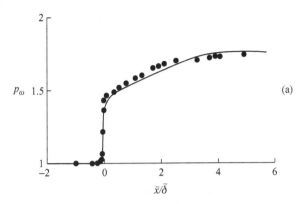

Figure 10.16. Comparison of asymptotic turbulent interaction theory predictions with experiment (shown as discrete points) downstream of a ramp.

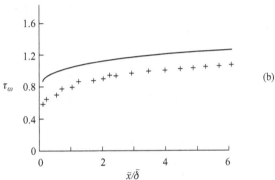

where the constant $D_A \equiv \int_{0+}^{1} \left[(\hat{p} - \Delta \hat{p}_s)/\Delta \tilde{p}_s\right] d\hat{x}$ is a known integral property of the small-scale analysis. Because the last limit term vanishes for any finite value of $\hat{x} > 1$, this result reduces to:

$$\lim_{\hat{z} \to \infty} [\hat{u}'(\hat{x} > 1, \hat{z})] \equiv \hat{u}'_e(\hat{x}) = -\Delta \hat{p}_s \left[1 + C_A \hat{x}^{-1}\right] \qquad (10.5.13b)$$

which describes the downstream decay of the inviscid-disturbance velocity $\hat{u}'_e$ toward its ultimate post-shock value associated with $\Delta \hat{p}_s$. Beneath such external behavior, we then postulate the corresponding inner-deck solution for $\hat{x} > 1$ that

$$\hat{u}'(\hat{x}, \hat{z}) \simeq -\Delta \hat{p}_s \left[1 + C_A \hat{x}^{-1}\right] + \hat{u}_i(\hat{x}, \hat{z}) \Delta \hat{p}_s \qquad (10.5.14)$$

where according to the momentum equation (10.3.131), the lateral distributions function u'_i must satisfy the isobaric turbulent diffusion-type of equation:

$$\frac{\partial \hat{u}_i}{\partial \hat{x}} = 2 \frac{\partial}{\partial \hat{z}} \left(\hat{z} \frac{\partial \hat{u}_i}{\partial \hat{z}}\right) \qquad (10.5.15)$$

subject to the outer condition that $\hat{u}_i(\hat{x}, \hat{z} \to \infty) = 0$. If desired, the solution of equation (10.5.15) provides the detailed velocity profile across the inner-deck subject to the x-dependent inner-boundary condition on $\hat{z} = \hat{z}_c$ given by equations (10.3.51) and (10.5.14). The corresponding wall-shear-stress disturbance is given independently by equations (10.3.52) and (10.5.12) as:

$$\tau'_w / \tau_{w_0} \simeq -2 \Delta \hat{p}_s B_0^{-1} \left(1 + C_A \hat{x}^{-1}\right) \qquad (\hat{x} \gg 1) \qquad (10.5.16)$$

which describe the downstream approach of τ'_w/τ_{w0} to its final negative value associated with the full shock-pressure jump. Although strictly valid only for $\hat{x} \gg 1$, the far-field results were found [78] to give good results down to $\hat{x} \geq 2$; they are also in qualitative agreement with a similar downstream analysis by Kluwick and Stross [32]. An analogous treatment of the energy relations can be conducted to obtain the $\hat{x} > 1$ behavior of the local heat-transfer disturbance along a nonadiabatic wall.

10.5.2 Transonic Flows in Asymptotic Theory

10.5.2.1 Small-Scale Features

The earliest analyses of very-high-Reynolds-number turbulent transonic interactions revealed that they involve a double-limit process of $\varepsilon_T \to 0$ and $M_\infty \to 1$ characterized by the interaction parameter $\chi_T \sim (M_\infty^2 - 1)/\varepsilon_T \sim \beta^2/\varepsilon_T \sim \beta^2/\sqrt{C_{f_0}}$. These studies focused on the detailed nonlinear transonic-disturbance–diffracted shock-wave structure within the small streamwise scales $\Delta x, \Delta y \sim 0(\varepsilon_T \delta_0)$. As reviewed by Melnick and Grossman [79] and by Délery and Marvin [14], three distinct-limit cases of χ_T corresponding to different locations on the sonic line within the boundary layer were identified (Fig. 10.17): (1) the very weak shock case $\chi_T \ll 1$, where the sonic line is near the outer edge of the boundary layer; (2) the weak shock wave case $\chi_T \sim 0(1)$, where the sonic line is farther down and well into the defect region; and (3) the moderately strong shock case with $\chi_T \gg 1$, where the sonic line is deep down in the boundary layer and into the inner deck. As pointed out in [14], there is also a fourth situation pertaining to $\chi_T \gg 1\ \varepsilon_T \to 0$.

Case (1), which entails a shock structure so spread out and weakened by the interactive boundary-layer thinning that it does not significantly penetrate the boundary layer, was investigated by Adamson and Feo [28]. Case (2) with $\chi_T \sim 0(1)$, in which the velocity changes across the incident shock are of the same order as those across the boundary layer and thus involve the numerical solution of a rotational transonic-flow equation, was studied by Melnick and Grossman [80]. Case (3), in which the velocity defect is small compared to $(M_\infty - 1)$ and the nearly straight shock deeply penetrates the boundary layer, was analyzed by Adamson et al. [81]. The fourth limit case, also pertaining to $\chi_T \gg 1$, involves a stronger normal or oblique shock with M_∞ not close to unity and extending into the supersonic regime and a near-wall sonic line. This situation was analyzed by an asymptotic approach by several authors [82, 83] and also by a small-perturbation theory [76]. Further details of these analyses are in the cited references.

10.5.2.2 Purely Supersonic Flows

When it suffices to know the flowfield on a larger streamwise scale $\Delta x \sim \delta_0$ (which is still physically quite small), application of the asymptotic equations yields valuable engineering insight. Thus, in the upstream region with a transonic inviscid flow, the free-interaction pressure distribution is governed by equation (10.3.92a), which involves the turbulent-interaction parameter:

$$\chi_{T_i} \equiv \beta^2/(\gamma + 1)\, M_e^4 \varepsilon_T \sim \left(M_{0e}^2 - 1\right)/\sqrt{C_{f_0}} \qquad (10.5.17)$$

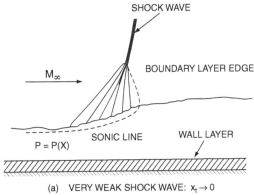

Figure 10.17. Regimes of nonseparating transonic normal-shock–turbulent boundary-layer interaction.

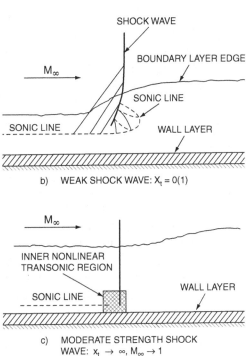

The limiting case $\chi_{T_i} \gg 1$ corresponds to linearized supersonic flow with the exponential solution (10.5.2). Otherwise, (10.3.92b) is a nonlinear equation for $\hat{p}\left(\hat{x}/\hat{\ell}_u\right)$ involving the single parameter χ_T that possesses solutions for $\hat{p} > \chi_{T_i}$.

Regarding the downstream region $x > 0$, we have a more interactive inner deck subject to boundary condition equation (10.3.92c) that:

$$\lim_{\hat{z}\to\infty} \left[\hat{u}'\left(\hat{x}, \hat{z}\right)\right] = -\hat{p} - \frac{2}{3}\beta\chi_T\left\{\frac{1 - \left[1 - \hat{p}/\chi_T\right]^{3/2}}{\kappa^2\hat{z}}\right\} \tag{10.5.18}$$

The solution of the inner-deck momentum equation (10.3.49) subject to equation (10.5.18) evidently has not been given to date.

10.5.2.3 Mixed Supersonic/Subsonic Flows

In the supersonic flow ahead of a shock, the formulation remains as given for $x > 0$ in the proceeding section. Downstream of the shock, where the inviscid flow is subsonic (see Fig. 2.34) and the pressure–flow-deflection relationship thus involves a Cauchy integral, we have instead the outer-boundary condition from equation (10.3.92d) that:

$$\hat{u}'(x > 0, \hat{y} \to \infty) = -\hat{p} - \frac{\sqrt{1 - M_2^2}}{\beta\pi} \frac{\hat{z}}{k^2} \int_0^{\hat{x}} \left\{ \int_0^\infty \left[\frac{p(\xi) - \Delta\hat{p}_s}{\hat{x} - \hat{\xi}} \right] d\hat{\xi} \right\} d\hat{x} \qquad (10.5.19)$$

where the $\sqrt{1 - M_2^2}$ term here, as with the supersonic upstream counterpart, cannot be scaled out of the analysis. Here, we require that downstream-disturbance pressure approach the full nonlinear normal-shock jump value Δp_s.

The split boundary–value problem posed by the inner-deck momentum equation with the discontinuous outer-boundary conditions of Equations (10.5.8) and (10.5.19) is a formidable one. It was solved by Inger and Mason using a rather complicated streamwise Fourier-transform method [84].

10.5.3 Three-Dimensional Effects

Due to their practical importance in high-speed propulsion and external aerodynamics problems, three-dimensional shock wave–turbulent interactions were studied extensively both experimentally and computationally for various basic flow configurations (Fig. 10.18); several detailed reviews of this work are in Settles and Dolling [85]. Because of the far greater complexity, however, only a few analytical studies of three-dimensional flows were conducted, mostly dealing with sweepback effects (see the following discussion). Of the remaining studies, we note an early theoretical study by Migotski and Morkovin [86] of the inviscid aspects of three-dimensional shock-wave reflections – including possible Mach-type reflections – associated with plane shocks intersecting a circular cylinder and conical shocks impinging on a flat surface. More recently, Gai et al. [87] conducted experiments on the latter problem for a Mach-2 conical shock impinging on a wall turbulent boundary layer. As part of this study, they include a crude theoretical model of the resulting curved upstream-influence line. It is interesting that a more complete analysis of this problem is not available, even in the simpler laminar case, which is probably due to the highly curved surface projection of the inviscid conical shock shape involved.

Regarding sweepback effects, the pioneering work is that of Stalker [88], who as part of an experimental study of swept ramps in a Mach 2.36 turbulent flow, applied the elementary sweepback principle to Lighthill's "quasi-laminar" interaction theory [2]. The results correlated the data in showing that the upstream influence of the interaction normal to the leading edge increases dramatically with sweep angle. More recent work [89, 90] on various swept-flow configurations has largely addressed shock strengths provoking separation, therefore requiring a combined experimental/CFD approach; however, some analysis has been possible for unseparated flows. In the case of swept-shock configurations involving a distinct origin or corner, for example, a theoretical study was conducted of whether the interaction process becomes locally cylindrical (i.e., properties constant along lines parallel to

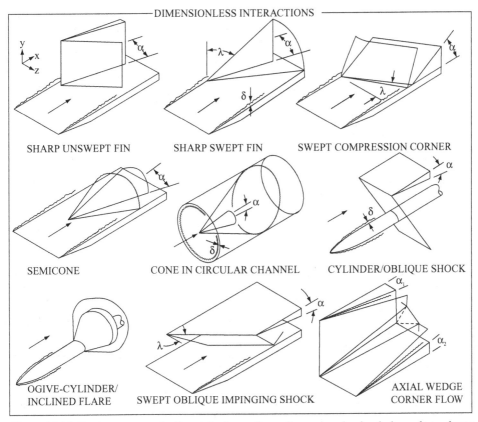

Figure 10.18. Basic types of dimensionless three-dimensional shock–boundary layer interactions.

the shock) or conical (i.e., properties constant along rays from the origin) as one moves radially outward from the corner [91] (Fig. 10.19). This issue of spanwise-disturbance propagation is important in the design, interpretation, and correlation of experimental data; in particular, it is important to understand the far-field behavior of such swept flows and establish the so-called inception distance L_i required to reach an asymptotic conical state (Fig. 10.20a). Inger [90] performed a detailed analysis of the three-dimensional interactive flow equations and determined that $L_i/\delta_0 \sim \tan \Lambda$ in agreement with Settles's experiments [91] over a wide range of attached inviscid shock strengths and boundary-layer states (Fig. 10.20b).

10.6 Limitations of the Triple-Deck Approach

10.6.1 Laminar Flows

The primary shortcoming of the triple-deck model for laminar interactions is the requirement that the Reynolds number be large enough that $\varepsilon_L = Re_L^{-1/8}$ be small compared to unity; in practice, this implies that $Re_L \geq 10^6$ (giving $\varepsilon_L < 0.18$) when, in fact, the boundary layer is usually fully turbulent. Conversely, when applied to lower Reynolds numbers in which the flow is likely to be laminar, the leading asymptotic approximation of triple-deck theory suffers from significant quantitative

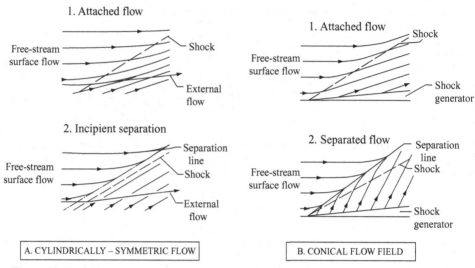

Figure 10.19. Cylindrically symmetrical versus conical interactive three-dimensional flows.

inaccuracies in the predicted pressure and skin-friction distributions, mainly due to neglect of the streamline-divergence effect that is, in fact, appreciable even when $Re_L \geq 10^6$. For example, it is shown in Fig. 10.6 that the finite-Reynolds-number effect captured by numerical solutions of both the interacting boundary layer and with the full Navier-Stokes equations, reduce the upstream influence well below that predicted by asymptotic analysis. A suggested improvement [3] in the accuracy of the latter consists of evaluating the Reynolds number and incoming boundary-layer properties at the upstream-influence station $L' = L - \ell u$ instead of L. Further improvement logically was sought by including second-order effects $\left(\sim \varepsilon_L^2\right)$ in the asymptotic analysis [54]; other than the increased complexity of the formulation,

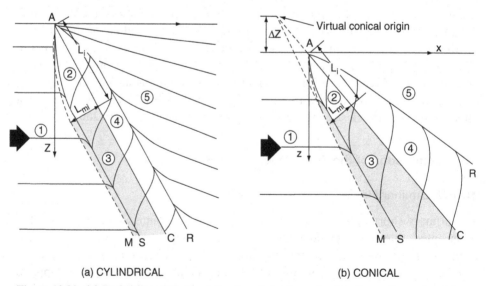

Figure 10.20. (a) Definition of the inception length in dimensionless three-dimensional inter-actions. (b) Experiment versus theoretical prediction for the inception length.

the results in fact proved less accurate than the first-order results, and it is better in practice dealing directly with a numerical solution based on interacting boundary-layer theory. Nevertheless, we reemphasize that the asymptotic approach remains of very real value in revealing the correct local velocity and coordinate scales in the interaction zone to guide purely numerical work and in providing the correct limiting behavior of the flow-structure in the $Re_L \to \infty$ limit.

Viewed in light of contemporary CFD capabilities, another practical difficulty with even the first-order triple-deck theory is the required numerical solution of the generally nonlinear partial-differential equations involved: Due to their implicit elliptic nature and the consequent need to satisfy proper downstream conditions, such solutions are no less challenging (and perhaps more expensive) than the direct use of a state-of-the-art Navier-Stokes code while still having the Reynolds-number limitations. Indeed, once separation occurs, the asymptotic approach requires the addition of more decks and even smaller local streamwise rescaled regions near separation and reattachment; without these, instabilities appear in the triple-deck solutions when significantly reversed flow occurs.

Another practical limitation on the triple-deck solutions is their extended downstream-wake behavior at $\hat{x} \gg 1$ as governed by slow algebraic-fractional power-law decays [55] in the interactive pressure and skin-friction disturbances. Although this feature is useful in efficiently implementing the downstream conditions in triple-deck numerical solutions, it proves awkward to accommodate when seeking to insert a triple-deck solution as a local interactive module in a larger scale global flowfield prediction code. Moreover, at $\hat{x} \geq 2$, we may well question the parallel shear-flow assumption [$U_0(y)$ independent of x] underlying the triple-deck theory because, in reality, the streamwise development of the baseline flow then becomes influential. Indeed, at hypersonic speeds, the viscous-interaction effect associated with the displacement-thickness growth of the baseline flow completely overwhelms the asymptotic "tail" of the triple-deck solution (see Fig. 10.12).

Finally, a limitation on the far-field–behavior aspect of the triple-deck theory is noted. As pointed out by Kluwick [33], the upper-deck inviscid-flow equations are linear if the flow is not transonic or strongly hypersonic; consequently, triple-deck solutions are not uniformly valid at long distances from the interaction region if the flow outside the boundary layer is supersonic. Linear theory predicts that the disturbances generated by the interaction process propagate along the unperturbed Mach lines; however, it is well known that even small disturbances may lead to significant distortions of Mach lines at long distances from their origin. Thus, to obtain uniformly valid results, it is necessary to account for nonlinear far-field effects when seeking to compare the calculated and observed wave patterns in this field.

10.6.2 Turbulent Flows

If the flow is unseparated and the interaction zone fairly short, the question of turbulence modeling – although far more complicated – is not a prohibitive issue in developing a turbulent-interaction triple-deck theory because the Law of the Wall/Law of the Wake and Eddy viscosity concepts provide a good engineering account of the interactive physics. All of the limitations cited in Section 10.6.1 apply here as well, except that the numerical solutions of the resulting turbulent triple-deck equations are obviously even more demanding, especially if we seek the flow details in the ultra-small-scale (Δx, $\Delta y \ll \delta_0$) shock vicinity within the boundary layer.

The issue of the accuracy of the $\varepsilon_T \to 0$ assumption underlying the asymptotic-theory approach arises again in the turbulent case when applying the results to finite Reynolds numbers $10^6 \le Re_L < 10^8$. Notwithstanding the value of such rigorous asymptotic analysis in establishing the correct limiting behavior as $Re_L \to \infty$ and the proper local scalings of the interaction field (including those of the very finely scaled events of the shock penetration deep within the boundary layer), the fact remains that the use of such asymptotic results at practical Reynolds numbers is an approximation that yields engineering results often less accurate than those based on a carefully constructed (albeit formally "nonrational") nonasymptotic small-perturbation theory [76]. Excluding hypersonic flows where the near-field inviscid-disturbance flow cannot be linearized, a nonasymptotic approach offers a valuable practical alternative in addressing viscous-inviscid interaction problems at realistic Reynolds numbers.

Appendix A   The Wall-Slip Boundary Conditions

As discussed in Section 3.2.3, the inner-boundary conditions on the fully turbulent equations (10.3.49) and (10.3.50) are obtained by locally applying the logarithmic Law of the Wall along the interaction at the value $z_c \cong 60 v_\omega / U_{\tau_0}$ pertaining to the innermost height of the Law of the Wall/Defect overlap region above the laminar sublayer (see Fig. 2.13). In carrying out this analysis, compressibility effects due to Mach number and surface cooling are treated by the reference-temperature method. Because we are dealing with events very close to the wall, this is an acceptable engineering approximation and it avoids the complications of working with the full Van Driest compressibility-transformation version of the Law of the Wall [12]. Thus, we write for the total perturbed velocity field just outside the laminar sublayer at $z = z_c$ that:

$$u_0(x, z_c) + u'(x, z_c) = (k^{-1} \ln 60 + B)[u_{\tau_0} + u'_\tau(x)] \tag{10.A.1}$$

where the constant $B \simeq 5$ includes the integrated effect of the underlying laminar sublayer and the no-slip condition. Subtracting the undisturbed flow component then yields the perturbation relationship between the slip velocity and skin-friction disturbance that:

$$u'(x, z_c) \cong B_i u'_\tau(x) \tag{10.A.2}$$

where $B_i = B + k^{-1} \ln 60$. However, from the definition $\tau_w \equiv \rho_w u_\tau^2$, we also have the small-perturbation relationship:

$$\tau'_w(x) \cong 2[u'_\tau(x)/u_{\tau_0}]\tau_{w_0} \tag{10.A.3}$$

and, hence, by equation (10.A.2):

$$u'(x, z_c) \cong B_i [\tau'_w(x)/\tau_{w_0}]u_{\tau_0}/2 \tag{10.A.4}$$

To provide a second relationship, we apply the Law of the Wall but this time along the edge of the local boundary layer $z_e(x) \cong \delta_0$, where $dU_0/dy = 0$. Neglecting the perturbation of the wake component [16] (which proves to be a second-order effect) and using the fact that $\delta \sim \varepsilon_T$, we obtain:

$$\delta'/\delta_0 \simeq (u'_\tau/U_{\tau_0}) - u'_e/U_{0e} \tag{10.A.5}$$

and, hence, from the Log Law expression along $y = \delta$, the first-order perturbation relationship:

$$u'_e(x, z_e) \simeq B_0 u'_\tau(x) \tag{10.A.6}$$

where $B_0 \equiv B + k^{-1}\ln(u_{\tau_0}\delta_0/v_w)$ and where a second-order term $\sim (\varepsilon_r/\kappa)\, u'_e/U_{0e}$ was neglected. Then, taking $u'_e(x, \delta_0) \cong -p'(x)/\rho_{0e}u_{0e}$ from equation (10.3.8) evaluated at $y = \delta$ with $dU_0/dy = 0$, equations (10.A.3) and (10.A.6) yield the following important interactive skin-friction relationship:

$$\frac{\tau'_w(x)}{\tau_{w_0}} \cong -\frac{2p'/\rho_{0e}U_{0e}^2}{B_0\varepsilon_T} \tag{10.A.7}$$

Hence, by using equation (10.A.4), the desired effective inner-slip velocity:

$$u'(x, z_c) \simeq -U_{0e}(B_i/B_0)\left(p'/\rho_{0e}U_{0e}^2\right) \tag{10.A.8}$$

An analogous procedure may be used to obtain the corresponding total-enthalpy slip needed as an inner-boundary condition on the solution of the energy equation. We thus conduct a perturbation analysis of the Temperature Law of the Wall given by:

$$T_w - \bar{T} = \left(\frac{\dot{q}_w}{P_r C_p u_\tau}\right)\left[\frac{1}{k_t}\ln\left(\frac{zu_\tau}{v_R}\right) + B_t\right] \tag{10.A.9}$$

where k_t and B_t are the thermal counterparts [92] to the parameters in equation (10.A.1). In relating the enthalpy to the temperature, we note that along the inner cutoff height z_c, the velocities are so low that kinetic-energy contribution $u^2/2$ can be neglected such that $H' \simeq C_p T'$ (this approximation, in fact, is consistent with the present treatment of the inner-deck compressibility effect for turbulent flow). Conversely, along the outer edge of the boundary layer, where $dH_0/dy = 0$ and the disturbance flow is adiabatic, we have by equation (10.3.11) that $H' = 0$ and so $C_p T' \simeq -U_0 u'$. Examining first the temperature-disturbance field along the inner boundary $z = z_c$ including the attendant heat-transfer disturbance $\dot{q}'_w$, equation (10.A.9) yields the following first-order temperature-perturbation relationship:

$$\frac{T'(x, z_c)}{T_R} \simeq \left(B_{t_i} - \kappa_t^{-1}\right)\frac{u'_\tau}{u_{\tau_0}} - B_{t_i}\frac{\dot{q}'_w}{\dot{q}'_{w_0}} \tag{10.A.10}$$

where $T_R \equiv \dot{q}_{w0}/P_r C_p u_{t0}$ is a Law of the Wall reference temperature, $B_{t_i} \equiv B_t + \kappa_t^{-1}\ln 60$, and u'_τ/u_{τ_0} is known from equation (10.A.3). Reapplying equation (10.A.9) to the temperature disturbance T'_e along the outer edge $z = \delta_0 + \delta'$ where dU_0/dz_0 and dH_0/dz_0 both vanish, we obtain after using equation (10.A.5) the second temperature-perturbation equation that:

$$\frac{T'_e(x, \delta_0)}{T_R} = \left(B_{t_0} - 2\kappa_t^{-1}\right)\frac{u'_t}{u_{t0}} + \kappa_t^{-1}\frac{u'_e}{u_{0e}} - B_{t_0}\frac{\dot{q}'_w}{\dot{q}'_{w0}} \tag{10.A.11}$$

where $B_{t_0} \equiv B_t + \kappa_t^{-1} \ln(u_{\gamma 0} \delta_0 / v_w)$ and u_e'/u_{0e} is given by equation (10.A.6). Accompanying this is the auxiliary relationship for adiabatic inviscid disturbance flow that:

$$\frac{T_e'(x, \delta_0)}{T_R} = -\left(\frac{u_{0e}^2}{C_p T_R}\right) \frac{u_{\tau 0}}{u_{0e}} \left[B_0 - \frac{2}{\kappa}\right] \tag{10.A.12}$$

Inspection shows that equations (10.A.10) and (10.A.11) as well as (10.A.12) provide a pair of relationships in the two desired unknowns $T'(x, z)$ and $\dot{q}_w'$; solving for them yields the following heat-transfer perturbation:

$$\frac{\dot{q}'}{q_{w_0}'} \simeq \left(\frac{B_{t_0} - 2\kappa_t^{-1}}{B_{t_0}}\right) \frac{u_\tau'}{u_{t_0}} + \left(\kappa_t^{-1}/B_0\right) \frac{u_e'}{u_{e0}} + \frac{u_{0e}^2}{C_p T_R} \left(\frac{u_{\tau 0}}{u_{0e}}\right) \left(\frac{B_0 - 2\kappa^{-1}}{B_{t_0}}\right) \tag{10.A.13a}$$

$$= -\lambda q_1 \left(p'/\rho_{0e} u_{0e}^2 \varepsilon_T B_0\right) + \lambda q_2 \varepsilon_T \left(u_{0e}^2/C_p T_R\right) \tag{10.A.13b}$$

where

$$\lambda q_1 \equiv \left(\frac{\kappa_\tau B_{\tau 0} - 2}{\kappa_\tau B_{\tau 0}}\right) + \varepsilon_T \left(\frac{B_0}{\kappa_\tau B_{\tau 0}}\right) \tag{10.A.13c}$$

$$\lambda q_2 \equiv \left(\frac{\kappa B_0 - 2}{\kappa B_{\tau 0}}\right) \tag{10.A.13d}$$

The corresponding enthalpy slip near the wall is found to be

$$H'(x, z_c)/c_p T_R = -\left(B_t - \kappa_t^{-1}\right)\left(p'/\rho_{0e} u_{0e}^2 \varepsilon_T B_0\right) - B_{ti} \left(\dot{q}_w'/q_{w0}\right) \tag{10.A.14}$$

Appendix B Evaluation of Boundary-Layer Profile Integrals and Related Matters

B.1 Limit Expression in the Laminar Interaction Theory

The limit expression in the last term on the right side of equation (10.3.57) is handled by noting that $z \to 0$ occurs within the linear region of the Mach-number profile near the surface. Thus, we may write this expression, assuming the inner-deck height z_i also lies in this region, as follows:

$$\lim_{z \to 0} \left[I_m(z) - \frac{2}{M_0^2(z)}\right] = \lim_{z \to 0} \left\{ \int_z^{\delta_0} \left(\frac{1 - M_0^2}{M_0^2}\right) dz - \frac{1}{\left[M_0'(0)\right]^2 z} \right\}$$

$$= \int_{z_i}^{\delta_0} \left(\frac{1 - M_0^2}{M_0^2}\right) dz + \lim_{z \to 0} \left\{ \int_z^{z_i} \frac{dz}{\left[d M_0/dz\right]_w^2 z^2} - (z_i - z) - \frac{1}{\left[d M_0/dy\right]_w^2 z^2} \right\}$$

$$= \int_{z_i}^{\delta_0} \left(\frac{1 - M_0^2}{M_0^2}\right) dz - z_i + \left[(d M_0/dz)_w\right]^{-2} \cdot \lim_{z/z_i \to 0} \left[\frac{1}{z}\left(1 - \frac{z}{z_i}\right) - \frac{1}{z}\right]$$

$$= I_m(z_i) - z_i \tag{10.B.1}$$

Because $I_m(z_i)$ is of the order δ_0 while $z_i \sim \varepsilon_L \delta_0$, the last term in equation (10.B.1) may be neglected in the first asymptotic approximation.

The comparable last term on the right-hand side of the energy-conditions equation (10.3.58) is treated by evaluating the lower limit of the integral I_m at the effective wall value of z_i given by Lighthill [2] to avoid the $M_0 \to 0$ singularity (see equation (10.B.4a)).

B.2 Evaluation of I_m for Laminar Flow

This important nondimensional Mach-number profile integral I_m defined in equation (10.3.17) may be evaluated conveniently in terms of the nondimensional laminar boundary-layer velocity profile u_0/u_e (y/δ_0) and the corresponding temperature profile given by the modified Crocco energy equation solution:

$$\frac{T_0}{T_e} = \frac{T_w}{T_e} + \left(1 - \frac{T_w}{T_e}\right)\frac{u_0}{u_e} + r\left(\frac{\gamma - 1}{2}\right)M_e^2\left(\frac{u_0}{u_e} - \frac{u_0^2}{u_e^2}\right) \tag{10.B.2}$$

where $r \cong P_r^{1/3}$ is the recovery factor. Writing $M_0^2 = U_0^2/\gamma RT_0^2 = M_e^2(u_0/u_e)^2$ (T_0/T_e) and using equation (10.B.2) in the integrand, we can thus express $I_m(z_i)$ in terms of three positive nondimensional velocity-profile integrals, as follows:

$$I_m(z_i) = \frac{T_w/T_e}{M_e^2}\int_{z_i/\delta_0}^1 \frac{d\eta}{[u_0/u_e(\eta)]^2} + \left(\frac{T_{w,\text{AD}} - T_w}{T_e M_e^2}\right)\int_{z_i/\delta_0}^1 \frac{d\eta}{[u_0/u_e(\eta)]}$$

$$- \left[1 + r\left(\frac{\gamma - 1}{2}\right)\right]\left(1 + \frac{z_i}{\delta_0}\right) \tag{10.B.3}$$

where $\eta \equiv z/\delta_0$ and $T_{w,\text{AD}}/T_e = 1 + (\gamma - 1)rM_e^2/2$ is the adiabatic wall temperature. The lower limit of the integrals is very small and understood to lie within the linear inner part of the velocity profile. In particular, this limit is accurately approximated by Lighthill's analysis [2] of the viscous-displacement effect in laminar interactions, which gives z_i as follows:

$$z_i = \frac{y_{w,\text{eff}}}{\delta_0} \cong \left[\frac{(0.78)^3\, v_w M_e^2}{\beta U_0'(0)\left[M_0'(0)\right]^2 \delta_0^4}\right]^{1/4} = \varepsilon C_L \tag{10.B.4a}$$

where $M_0'(0) = (dM_0/dz)_w$ and

$$C_L = \frac{0.830}{5.2}\frac{\mu_R}{\mu_e}\left(\frac{\dfrac{\rho_e}{\rho_w}\dfrac{T_w}{T_e}\dfrac{\mu_w}{\mu_R}}{\beta C_R^{3/2}\delta_0^3}\right)^4 C_R^{-1/2}\frac{T_e}{T_R} \tag{10.B.4b}$$

B.3 Evaluation of I_m for Turbulent Flow

Using the turbulent form of Crocco's energy-equation integral with $r = r_{\text{turb}}$, the form of equation (10.B.3) applies here as well; of course, the velocity-profile integrals involved and their lower limit have to be appropriate to a turbulent-boundary layer. Using a Law of the Wall/Wake model for this purpose, we express the integrals in terms of $u_0^* = u_0/u_\tau$ and $y^* = u_\tau y/v_w$ and therefore rewrite equation (10.B.3) as follows:

$$I_m(z_i) = \frac{T_w/T_e v_w}{u_e\delta_0}\left(\frac{u_e}{u_\tau}\right)^3\int_{z_i^*}^{z_e^*}\frac{dz^*}{u_0^{*2}(z^*)} + \left(\frac{T_{w,\text{AD}} - T_w}{T_e M_e^2}\right)\left(\frac{u_e}{u_\tau}\right)^2\left(\frac{v_w}{u_e\delta_0}\right)\int_{z_i^*}^{z_e^*}\frac{dz^*}{u_0^*(z^*)}$$

$$- \left[1 + r_t\left(\frac{\gamma - 1}{2}\right)\right]\left[1 - \frac{v_w}{u_e\delta_0}\left(\frac{u_e}{u_\tau}\right)z_i^*\right] \tag{10.B.5}$$

where we choose z_i to coincide with the effective slip height defined in Section 10.3. The two velocity-profile integrals appearing in equation (10.B.5) may be evaluated by any convenient Law of the Wall expression for $u_0^*(z)$ outside the laminar sub-layer, such as a crude but simple power law $u^* = (u_e/u_\tau)(v_w z/u_\tau \delta_0)^{1/N}$ with $N \simeq 5$ to 7, or the more sophisticated Spalding–Kleinstein expression given in White [12].

In the case of very-high-Reynolds-number turbulent interactions in which the boundary-layer profile is much fuller than the laminar case, the exact value of the inner limit has little influence on the integral I_m; for example, use of the fractional power-law velocity profile in equation (10.B.5) shows that the contribution of the inner limit to the resulting integrals is rather small.

Appendix C Summary of Constants in the Scaling Relationships for Laminar Flow

C.1 Supersonic–Hypersonic Flow

In this regime, the six scaling factors $C_x, C_y, C_\omega, C_u, C_v, C_H$ referred to in equations (10.3.102) through (10.3.107) take the following values:

$$C_x = \frac{C_R (T_R/T_{0e})^{3/2} (T_w/T_R)^{1+\frac{\omega}{2}}}{\beta^{3/4}\lambda^{5/4}} \tag{10.C.1}$$

$$C_y = \frac{C_R^{5/8} (T_R/T_{0e})^{3/2} (T_w/T_R)^{1+\frac{\omega}{2}}}{\beta^{1/4}\lambda^{3/4}} \tag{10.C.2}$$

$$C_\pi = \lambda^{1/2} C_R^{1/4}/\beta^{1/2} \tag{10.C.3}$$

$$C_u = \lambda^{1/4} C_R^{1/8} (T_w/T_{0e})^{1/2}/\beta^{1/4} \tag{10.C.4}$$

$$C_v = \lambda^{3/4}\beta^{1/4} C_R^{3/8} (T_w/T_{0e})^{1/2} \tag{10.C.5}$$

$$C_H = P_R^{1/3} (H_{ADw} - C_R T_w) C_R^{1/8} \lambda^{1/4} \left(\frac{T_w}{T_e}\right)^{1/2} \beta^{-1/4} \tag{10.C.6}$$

C.2 Adiabatic Shockless Transonic Flow

This is a distinctly different set of five scaling factors $\kappa_x, \kappa_y, \kappa_\pi, \kappa_u$ and κ_v pertaining to equations (10.3.111) through (10.3.115), as follows:

$$\kappa_x = C_R^{3/10}\lambda^{-7/5} (T_w/T_\infty)^{3/2} k_0^{-3/8} (T_w/T_R)^{1+\frac{\omega}{2}} \tag{10.C.7}$$

$$\kappa_z = C_R^{3/5}\lambda^{-4/5} (T_w/T_\infty)^{3/2} k_0^{-1/8} (T_w/T_R)^{1+\frac{\omega}{2}} \tag{10.C.8}$$

$$\kappa_\pi = C_R^{1/5}\lambda^{-2/5}\lambda^{-1/4} \tag{10.C.9}$$

$$\kappa_u = C_R^{3/10}\lambda^{-1/5} (T_w/T_\infty)^{1/2} k_0^{-1/8} \tag{10.C.10}$$

$$\kappa_v = C_R^{2/5}\lambda^{4/5} (T_w/T_\infty)^{3/2} k_0^{1/8} \tag{10.C.11}$$

where

$$k_0 = \lambda^{-2/5} C_R^{-1/5} (M_\infty^2 - 1) \varepsilon_L^{-8/5}/(\gamma - 1) M_{0e}^4 \tag{10.C.12}$$

Appendix D Nomenclature

a	amplitude factor in triple-deck solution; (equations (10.4.8), (10.4.9))
A_t	Airy function; (equations (10.4.8), (10.4.9))
B_i, B_0, B_t	Law of the Wall parameters
C_f	$\equiv 2\tau_w/\rho_{0e}U_{0e}^2$, skin-friction coefficient
Cp	$\equiv 2p'/\rho_{0e}U_{0e}^2$, pressure coefficient
C	integration constant in upstream laminar-interaction solution
c	constant in upstream turbulent-interaction solution; (equation (10.5.1))
C_p	$= \gamma R/(\gamma - 1)$, constant-pressure specific heat
C_i, C_4	constants in equations (10.4.17) and (10.4.18), respectively
$C_x, C_z, C_\pi, C_u, C_H, C_v$	coefficients in laminar supersonic triple-deck scalings; (equations (10.3.102) through (10.3.107))
C_M	$\equiv P_R^{1/3}C_zC_R^{1/2}\lambda\,(\mu_e/\mu_R)$; (equation (10.3.110))
C_R	$\equiv \mu_R T_{0e}/\mu_{0e}T_R$, Chapman–Rubesin parameter
C_δ	$\equiv 5.2C_R^{1/2}T_R/T_e$; (equation (10.3.109))
D	Van Driest damping function; (equation (10.3.31b))
f_D	variable-density functions in eddy viscosity; (equation (10.3.41))
$f(\hat{z})$	Mach-number integral distribution function; (equation 10.3.69) or (10.3.71))
$\hat{h}$	scaled nondimensional forcing-function amplitude; (see Section 10.4.1.4)
H	$\equiv C_P T + (u^2/2)$, total enthalpy
H_i	incompressible shape factor
H_R	reference enthalpy
I_m	boundary-layer Mach-number–profile integral; (equation (10.3.17))
J	jump function $[J = 0$ for $x < 0$ and $= 1$ for $x \geq 0;]$; equation (10.3.72))
k	Von Karman turbulence constant; equation (10.3.31b))
k_0	asymptotic laminar transonic interaction-parameter; (equation (10.4.21))
k_H	$= (\gamma + 1)\beta/4$, hypersonic inviscid-flow parameter; (equation (10.3.73))
$k_x, k_z, k_\pi, k_u, k_v$	coefficients in laminar transonic triple-deck scalings; (equations (10.3.111) through (10.3.115))
ℓu	upstream-influence distance reckoned from $x = L$
$\hat{\ell}u$	$= \kappa^{-1}$
$\tilde{\ell}u$	$\equiv \ell u/\delta_0$
L	reference length (inviscid shock location on body)
M	$= u/\sqrt{\gamma RT}$, Mach number
$M_{1,2}$	supersonic and subsonic Mach numbers ahead and behind normal shock, respectively

N	turbulent boundary-layer profile power-law exponent ($u \sim y^N$)
p	static pressure
Δp_s	pressure jump across inviscid shock
$\Delta \hat{p}_s$	$= \Delta p_s / \rho_e u_{0e}^2 \pi_R$, scaled pressure jump
P_{INTER}	interactive component of pressure-perturbation field; equation (10.3.72))
Pr, Pr_T	laminar and turbulent Prandtl numbers, respectively
$-\dot{q}_w$	surface-heat-transfer rate per unit area
R	molecular constant ($p = \rho R T$)
Re_L	$\equiv \rho_{0e} u_{0e} L / \mu_{0e}$, Reynolds number based on L
$Re_{\delta*}$	$\equiv \rho_{0e} u_{0e} \delta^* / \mu_{0e}$, Reynolds number based on displacement thickness
S	$= C_\delta M_e^2 / C_x \beta$, streamline-divergence parameter; (equation (10.3.109))
$S(\hat{z})$	laminar-interaction vorticity; (equation (10.4.5))
T	static temperature
T_R	$\equiv (-\dot{q}_{w_0}) / \text{Pr}_t C_p U_{\tau_0}$
u, v	streamwise and normal velocity components, respectively
U_τ	$\equiv \sqrt{(\rho_e / \rho_w) C_f / 2}\; U_{0e}$, turbulent friction velocity
U, V	Howarth-Dorodnitzen transformed velocities; (equation (10.3.25))
w	Coles's wake function
W	shifted normal-velocity variable; (equation (10.3.25))
x, y	streamwise and normal distances with respect to body, respectively
y_i	inner-deck height
$y_{w,\text{eff}}$	effective inviscid wall height of a turbulent interaction
X, Z	Howarth-Dorodnitzn transformed coordinates; (equation (10.3.25))
z	$\equiv y - y_w$, shifted normal distance; (equation (10.3.23))
$Z_{w,\text{eff}}$	value of Z corresponding to $Y_{w,\text{eff}}$
β	$\equiv \sqrt{M_{0e}^2 - 1}$
χ_{H_t}	$\equiv 2\beta K_M \varepsilon_T$, turbulent hypersonic-interaction parameter; (equation (10.3.88))
χ_t	laminar transonic-interaction parameter; (equation (10.3.83) or (10.3.84))
χ_{T_t}	$\equiv \beta^2 / (\gamma + 1) M_{0e}^4 \varepsilon_T$, turbulent transonic-interaction parameter; (equation (10.3.92))
$\bar{\chi}$	$= M_{0e}^3 (C_R / Re_L)^{1/2}$, classical hypersonic viscous-interaction parameter
δ	boundary-layer thickness
δ^*	boundary-layer displacement thickness
Δ	Z value at outer edge of turbulent inner deck
ε	general small-perturbation parameter

ε_L	$= Re_L^{-1/8}$, basic laminar-interaction small asymptotic parameter
ε_τ	$= U_{\tau_0}/U_{0e}$, basic turbulent-interaction asymptotic parameter
ε_T'	perturbation eddy viscosity; equation (10.3.31a))
γ	ideal-gas specific-heat ratio
Λ	sweepback angle
κ	universal nondimensional influence factor; (equations (10.4.5) and (10.4.6))
λ	Blasius flat-plate boundary-layer factor 0.332; (equation (10.3.100))
μ	laminar coefficient of viscosity
μ_T	turbulent eddy-viscosity; (equations (10.3.27) through (10.3.31b))
v	kinematic viscosity μ/ρ
ω	laminar viscosity-temperature dependence exponent $(\mu \sim T^w)$
π_R	interactive pressure-scaling factor $\left(\hat{p} = p'/\rho_{0e}U_{0e}^2\pi_R\right)$
ρ	density
θ	body-deflection angle
τ_w	wall-shear stress

D.1 Subscripts

AD	adiabatic wall conditions
AV	averaged value across boundary layer
e	boundary-layer edge $(y = \delta)$
0	undisturbed incoming flow
is	incipient separation
R	reference-temperature–based value; (equation (10.3.93))
S	impinging-shock value
w	conditions at body surface ("wall")

D.2 Special Symbols

$()'$	perturbation quantity
$(^)$	nondimensional scaled triple-deck variable

REFERENCES

[1] D. Malmuth. "Some Applications of Combined Asymptotics and Numerics in Fluid Mechanics and Aerodynamics." In *Asymptotics and Numerics in Transonic Aerodynamics* (ed. L. Cook) (Philadelphia, PA: SIAM, 1993), pp. 65–88.

[2] J. Lighthill. On boundary-layers upstream influence: II Supersonic flows without separation. *Proceedings of the Royal Society, London, A,* 217 (1953), 478–507.

[3] K. Stewartson and P. G. Williams. Self-induced separation. *Proceedings of the Royal Society, London, A,* 312 (1969), 181–206.

[4] Y. Neiland. Towards a theory of separation of a laminar boundary layer in supersonic stream. *Izvestia Akadmii Nauk SSSR, Mekhanika Zhidkostii Gaza*; 4 (1969), 4; see also *Fluid Dynamics*, 4, 4, 33–5.

[5] H. Babinsky and G. R. Inger. Effect of surface roughness on unseparated shock wave–turbulent boundary layer interactions. *AIAA Journal*, 40 (2002), 8, 1567–73.

[6] C. Wilcox. *Perturbation Methods in the Computer Age*. (La Canada, CA: D.C.W. Industries, 1995).

[7] J. Nietubicz, G. R. Inger, and J. E. Danberg. A theoretical and experimental investigation of a transonic projectile flowfield. *AIAA Journal*, 22 (1984), 1, 35–71.

[8] N. Cox and L. F. Crabtree. *Elements of Hypersonic Gas Dynamics*. (New York: Academic Press, 1965).

[9] D. Hayes and R. F. Probstein. *Hypersonic Flow Theory*. (New York: Academic, 1959), p. 277.

[10] Ya. Neiland. Propagation of perturbations upstream with interaction between a hypersonic flow and a boundary layer. *IZV. Akad. Nauk SSSR March. Zhid. Gaza* 4, (1970), pp. 40–9.

[11] M. Kuethe and C. Y. Chow. *Foundations of Aerodynamics*. 5th edition. (J. Wiley & Sons, 1998).

[12] F. M. White. *Viscous Fluid Flow*. 2nd edition. (New York: McGraw-Hill, 1991), pp. 511–13.

[13] B. Bush and F. E. Fendell. Asymptotic analysis of turbulent channel and boundary layer flow. *Journal of Fluid Mechanics*, 158 (1971), 657–81.

[14] J. Délery and J. G. Marvin. *Shock-Wave–Boundary-Layer Interactions*. AGARDograph #280 (1986).

[15] H. Shapiro. In *The Dynamics and Thermodynamics of Compressible Fluid Flow*, Vol. II (Ronald Press, 1954), pp. 1138–48.

[16] E. Green. Reflexion of an oblique shock wave by a turbulent boundary layer; Pt. I. *Journal of Fluid Mechanics*, 140 (1970), pp. 81–95.

[17] K. S. Yajnik. Asymptotic theory of turbulent wall boundary layer flows. *Journal of Fluid Mechanics*, 42 (1970), pp. 411–27.

[18] L. Mellor. The large reynolds number, asymptotic theory of turbulent boundary-layers. *International Journal of Engineering Science*, 10 (1972).

[19] E. Gadd, D. W. Holder, and J. D. Regan. An experimental investigation of the interaction between shock waves and boundary layers. *Proc. Royal Society of London, Series A, Mathematics and Physical Sciences*, 226 (1954), 1165, 227–53.

[20] M. Honda. A theoretical investigation of the interaction between shock waves and boundary layers. *J. AeroSpace Science*, 25 (1958), 11, 667–77.

[21] J. E. Carter. Numerical solutions of the Navier-Stokes equations for supersonic laminar flow over a two-dimensional compression corner. *NASA TR R-385* (July 1972).

[22] J. S. Shang, W. L. Hankey, and H. C. Law. Numerical simulation of shock wave- turbulent boundary layer interaction. *AIAA Journal*, 14 (1976), 10, 1451–60.

[23] I. Tani. Review of some experimental results on the response of a turbulent boundary layer to sudden perturbations. *Proc. AFOSR-IFP Stanford Conference on Computation of Turbulent Boundary Layers*, 1 (1968), 483–94.

[24] J. E. Green. Interaction between shock-waves and turbulent boundary layers. *Progress in Aerospace Science*, II (1970), 235–340.

[25] S. M. Bogdonoff and C. E. Kepler. Separation of a supersonic turbulent boundary layer. *Journal of Aeronautical Science*, 22 (1955), 414–24.

[26] D. M. Kuehn. Experimental investigation of the pressure rise for incipient separation of turbulent boundary layers in two-dimensional supersonic flow. *NASA Memo 1–21–59A* (February 1959).

[27] G. S. Settles, T. J. Fitzpatrick, and S. M. Bogdonoff. Detailed study of attached and separated compression corner flowfields in high Reynolds number supersonic flow. *AIAA Journal*, 17 (1979), 579–85.

[28] T. C. Adamson and A. Feo. Interaction between a shock wave and a turbulent layer in transonic flow. *SIAM Journal of Applied Mathematics*, 29 (1975), 7, 121–44.

[29] R. I. Sykes. An asymptotic theory of incompressible turbulent boundary-layer flow over a small hump. *Journal of Fluid Mechanics*, 101 (1980), 647–70.

[30] T. C. Adamson Jr. and A. F. Messiter. Analysis of two-dimensional interactions between shock waves and boundary layers. *Annual Review of Fluid Mechanics*, 12 (1980), 103–38.

[31] S. Agrawal and A. F. Messiter. Turbulent boundary layer interaction with a shock wave at a compression corner. *Journal of Fluid Mechanics*, 143 (1984), 23–46.

[32] A. Kluwick and N. Stross. Interaction between a weak oblique shock wave and a turbulent boundary layer in purely supersonic flow. *Acta Mechanica*, 53 (1984), 37–56.

[33] A. Kluwick. "Interacting Laminar and Turbulent Boundary Layers." In *Recent Advances in Boundary Layer Theory* (ed. A. Kluwick) (Springer XXIII, 1998), pp. 232–330.

[34] R. B. Melnick, R. L. Cusic, and M. J. Siclari. "An Asymptotic Theory of Supersonic Turbulent Interactions in a Compression Corner." In *Proceedings of IUTAM Symposium on Turbulent Shear Layer–Shock Wave Interaction* (New York: Springer, 1986), pp. 150–62.

[35] W. C. Rose and D. A. Johnson. Turbulence in shock wave-boundary layer interactions. *AIAA Journal*, 13 (1975), 7, 884–9.

[36] R. E. Davis. Perturbed turbulent flow, eddy viscosity and the generation of turbulent stresses. *Journal of Fluid Mechanics*, 63 (1974), 4, 674–93.

[37] R. G. Deissler. Evolution of a moderately-short turbulent boundary layer in a severe pressure gradient. *Journal of Fluid Mechanics*, 64 (1974), pp. 763–74.

[38] K. Stewartson. "Multistructured Boundary Layers on Flat Plates and Related Bodies." In *Advances in Applied Mechanics*, 14. (New York: Academic Press, Inc., 1974), pp. 145–239.

[39] L. Rosenhead. *Laminar Boundary Layers*. (Oxford: Clarendon Press, 1963).

[40] F. K. Moore (ed.). *Theory of Laminar Flows*. (Princeton, NJ: Princeton University. Press, 1964), pp. 214–22.

[41] M. D. Van Dyke. *Perturbation Methods in Fluid Mechanics*. (New York: Academic Press, 1975).

[42] T. Davis and M. J. Werle. "Numerical Methods for Interacting Boundary Layers." In *Proceedings of the 1976 Heat Transfer and Fluid Mechanics Institute* (Stanford, CA: Stanford University Press, 1976), pp. 317–39.

[43] O. R. Burggraf and P. W. Duck. "Spectral Computation of Triple-Deck Flows." In *Numerical and Physical Aspects of Aerodynamic Flows* (ed. T. Cebeci), (Springer-Verlag, 1982), pp. 145–58.

[44] M. Van Dyke. The combined supersonic–hypersonic similarity rule. *Journal of the Aeronautical Sciences*, 18 (1957), pp. 499–500.

[45] L. Sirovich and C. Huo. Simple waves and the transonic similarity parameter. *AIAA Journal*, 14 (1976), 8, 1125–7.

[46] G. R. Inger. Theory of local heat transfer in shock–laminar boundary layer interactions. *Journal of Thermophysics and Heat Transfer*, 12 (1998), 3, 336–42.

[47] R. J. Bodonyi and A. Kluwick. Freely interacting transonic boundary-layer. *Physics of Fluids*, 20 (1979), 9, 1432–7.

[48] W. H. Dorrance. In *Viscous Hypersonic Flow*. (New York: McGraw-Hill, 1968), pp. 134–9.

[49] K. Stewartson. *Theory of Laminar Boundary Layers in Compressible Fluids*. (Oxford: Oxford University Press, 1964).

[50] D. P. Rizzetta, O. R. Burggraf, and R. Jenson. Triple-deck solutions for viscous supersonic and hypersonic flow past corners. *Journal of Fluid Mechanics*, 89 (1978), 535–52.

[51] R. M. Kerimberkov, A. I. Ruban, and I. D. A. Walker. Hypersonic boundary layer separation on a cold wall. *Journal of Fluid Mechanics*, 274 (1994), 163–95.

[52] O. R. Burggraf, D. Rizzetta, M. J. Werle, and V. N. Vatsa. Effect of Reynolds number on laminar separation of a supersonic stream. *AIAA Journal*, 17 (1979), 4, 336–45.

[53] M. V. Hassaini, B. S. Baldwin, and R. W. MacCormack. Asymptotic features of shock wave boundary layer interaction. *AIAA Journal*, 18 (1980), 8, 1014–16.

[54] S. A. Ragab and A. H. Nayfeh. Second order asymptotic solution for laminar separation. *Physic of Fluids*, 23 (1980), 6, 1091–100.

[55] F. T. Smith. Laminar flow over a small bump on a flat plate. *Journal of Fluid Mechanics*, 57 (1973), 4, 803–24.

[56] D. R. Chapman, D. M. Kuehn, and H. K. Larson. Investigation of separated flows in supersonic and subsonic streams with emphasis on the effects of transition. *NACA Report* 1356 (1958).

[57] K. Oswatitsch and K. Wiegardt. *Theoretishe Untersuchengen Uber Stationäre Postentialsfromangen und Grenzschichten. Bericht der Lilienthal-Gusellschaft für Luftfahrtftirschung.* (1941); see also *NACA* TM-1189 (1948).

[58] A. P. Rothmayer and F. T. Smith. "Free Interactions and Breakaway Separation." In *Handbook of Fluid Dynamics* (CRC Press, 1998), pp. 24-1–24-22.

[59] J. Erdos and A. Pallone. "Shock Boundary-Layer Interaction and Flow Separation." In *Proceedings of the Heat Transfer and Fluid Mechanics Institute* (Stanford, CA: Stanford University Press, 1962).

[60] R. J. Hakkinen, G. L. Trilling, and S. S. Abarbanel. The interaction of an oblique shock wave with a laminar boundary layer. *NASA Memo* 2–18–59W (1959).

[61] G. R. Inger. Similitude properties of high speed laminar and turbulent boundary layer incipient separation. *AIAA Journal*, 15 (1977), 5, 619–23.

[62] A. Nayfeh, H. L. Reed, and S. A. Ragab. Flow over plates with suction through porous strips. *AIAA Journal*, 20 (1982), 5, 587–8.

[63] G. Inger and P. A. Gnoffo. Analytical and computational study of wall temperature jumps in supersonic flow. *AIAA Journal*, 39 (2001), 1, 79–87.

[64] M. J. Werle and V. N. Vatsa. Numerical solution of interacting supersonic boundary layer flows including separation effects. *U.S. Air Force Report* ARC-73–01 62 (1973).

[65] J. E. Lewis, T. Kubota, and L. Lees. Experimental investigation of supersonic laminar two-dimensional boundary-layer separation in a compression corner with and without cooling. *AIAA Journal*, 6 (1968), 1, 7–14.

[66] D. P. Rizzetta. Asymptotic solutions of the energy equation for viscous supersonic flow past corners. *Physics of Fluids*, 22 (1979), 1, 218–23.

[67] J. L. Stollery. Hypersonic viscous interaction on curved surfaces. *Journal of Fluid Mechanics*, 43 (1970), 497–511.

[68] D. A. Needham. A heat-transfer criterion for the detection of incipient separation in hypersonic flow. *AIAA Journal*, 3 (1965), 4, 781–3.

[69] A. F. Messiter, A. Feo, and R. E. Melnik. Shock-wave strength for separation of a laminar boundary-layer at transonic speeds. *AIAA Journal*, 9 (1971), 6, 1197–8.

[70] H. M. Brilliant and T. C. Adamson. Shock-wave-boundary-layer interactions in laminar transonic flow. *AIAA Journal*, 12 (1974), 3, 323–9.

[71] M. J. Werle, V. N. Vatsa, and S. D. Bertke. Sweep effects on supersonic separated flows: A numerical study. *AIAA Journal*, 11 (1973), 12, 1763–5.

[72] Ph. Gittler and A. Kluwick. Interacting laminar boundary layers in quasi-two dimensional flow. *Fluid Dynamics Research* 5 (1989), 29–47.

[73] Ph. Gittler and A. Kluwick. Triple-deck solutions for supersonic flows past flared cylinders. *Journal of Fluid Mechanics*, 179 (1987), 469–87.

[74] R. Leblanc and J. Ginoux. Influence of cross flow on two-dimensional separation. *Von Karman Institute for Fluid Dynamics Technical Note* 62, Belgium (1970).

[75] A. Kluwick, Ph. Gittler, and R. J. Bodonyi. Freely interacting axisymmetric boundary layers on bodies of revolution. *Quarterly Journal of Applied Mathematics*, 38 (1985), 4, 575–90.

[76] G. R. Inger. "Nonasymptotic Theory of Unseparated Turbulent Boundary Layer–Shock Wave Interaction." In *Numerical and Physical Aspects of Aerodynamic Flows* (ed. T. Cebeci) (Springer-Verlag, 1981), pp. 159–69.

[77] P. Carriere, M. Sirieix, and J.-L. Solignac. *Propriétés de similitude des phénomènes de décollement laminaires ou turbulents en écoulement supersonique non uniforme*. In *Proceedings of 12th International Congress of Applied Mechanics*, (Stanford University, August 1968) and *ONER TP* No. 659F (1968).

[78] A. F. Messiter and T. C. Adamson Jr. "A Study of the Interaction of a Normal Shock-Wave with a Turbulent Boundary-Layer at Transonic Speeds." In *NASA Langley Research Center Advanced Technology Airfoil Research*, 1 (1978), 1, 271–9.

[79] R. E. Melnik. Turbulent interactions on airfoils at transonic speeds: Recent developments. *AGARD CP* 291, Paper 10 (1981).

[80] R. E. Melnick and B. Grossman. "Further Developments in an Analysis of the Interaction of a Weak Normal Shock Wave with a Turbulent Boundary Layer." In *Proceedings of Symposium Transonicum II* (Springer-Verlag, 1975), pp. 262–72.

[81] T. C. Adamson Jr., M. S. Liou, and A. F. Messiter. Interaction between a normal shock-wave and a turbulent boundary-layer at high transonic speeds. *NASA-CR*-3194 (1980).

[82] A. F. Messiter. Interaction between a normal shock wave and a turbulent boundary layer at high transonic speeds. Part I: Pressure distribution. *Journal of Applied Mathematics and Physics (ZAMP)*, 31 (1980), 2, 204–26; see also (with appendices) *NASA CR* 3194 (1980).

[83] T. C. Adamson and M. S. Liou. Interaction between a normal shock wave and a turbulent boundary layer at high transonic speeds. Part II: Wall shear stress. *Journal of Applied Mathematics and Physics (ZAMP)*, 31 (1980), 2, 227–46; see also *NASA CR* 3194 (1980).

[84] G. R. Inger and W. H. Mason. Analytical theory of transonic normal shock–turbulent boundary-layer interaction. *AIAA Journal*, 14 (1976), 9, 1266–72.

[85] G. S. Settles and D. S. Dolling. "Swept Shock Wave Boundary Layer Interactions." In *Tactical Missile Aerodynamics* (New York: AIAA Progress in Astronautics and Astronautics, 104, 1986), ch. 8, pp. 297–379.

[86] E. Migotsky and M. V. Morkovin. Three-dimensional shock-wave reflection. *Journal Aeronautical Sciences*, 18 (1951), 7, 484–9.

[87] S. I. Gai and S. L. Teh. Interaction between a conical shock wave and a plane turbulent boundary layer. *AIAA Journal*, 28 (2000), 7, 804–11.

[88] R. J. Stalker. Sweepback effects in turbulent boundary layer shock wave interaction. *Journal Aeronautical Sciences*, 8 (1960), 5, 348–56.

[89] H. Kubota and J. L. Stollery. An experimental study of the interaction between a glancing shock wave and a turbulent boundary layer. *Journal of Fluid Mechanics*, 116 (1982), 431–58.

[90] G. R. Inger. Spanwise propagation of upstream influence in conical swept shock–boundary layer interactions. *AIAA Journal*, 25 (1987), 2, 287–93.

[91] G. S. Settles. "On the Inception Lengths of Swept Shock Wave–Turbulent Boundary Layer Interactions." In *Proceedings of IUTAM Symposium on Turbulent Shear Layer–Shock Wave Interactions* (New York: Springer, 1986), pp. 203–15.

[92] J. A. Schetz. *Boundary Layer Analysis* (Englewood Cliffs, NJ: Prentice Hall, 1993), p. 433.

Index

Printed in the United States
By Bookmasters